Silviculture: Forest Ecology

Silviculture: Forest Ecology

Editor: Rachel Ledger

States Academic Press,
109 South 5th Street,
Brooklyn, NY 11249, USA

Visit us on the World Wide Web at:
www.statesacademicpress.com

© States Academic Press, 2022

This book contains information obtained from authentic and highly regarded sources. Copyright for all individual chapters remain with the respective authors as indicated. All chapters are published with permission under the Creative Commons Attribution License or equivalent. A wide variety of references are listed. Permission and sources are indicated; for detailed attributions, please refer to the permissions page and list of contributors. Reasonable efforts have been made to publish reliable data and information, but the authors, editors and publisher cannot assume any responsibility for the validity of all materials or the consequences of their use.

ISBN: 978-1-63989-482-6 (Hardback)

Trademark Notice: Registered trademark of products or corporate names are used only for explanation and identification without intent to infringe.

Cataloging-in-Publication Data

Silviculture : forest ecology / edited by Rachel Ledger.
   p. cm.
Includes bibliographical references and index.
ISBN 978-1-63989-482-6
1. Forest ecology. 2. Forests and forestry. I. Ledger, Rachel.
QH541.5.F6 S55 2022
577.3--dc23

Table of Contents

	Preface	VII
Chapter 1	**Empirical modelling of chestnut coppice yield for Cimini and Vicani mountains (Central Italy)**	1
	Alice Angelini, Walter Mattioli, Paolo Merlini, Piermaria Corona and Luigi Portoghesi	
Chapter 2	**Forest genetic resources to support global bioeconomy**	7
	Saša Orlović, Mladen Ivanković, Vlatko Andonoski, Srdjan Stojnić and Vasilije Isajev	
Chapter 3	**Structural attributes of stand overstory and light under the canopy**	18
	Alice Angelini, Piermaria Corona, Francesco Chianucci and Luigi Portoghesi	
Chapter 4	**Forest biotechnology advances to support global bioeconomy**	27
	Antoine Harfouche, Sacha Khoury, Francesco Fabbrini and Giuseppe Scarascia Mugnozza	
Chapter 5	**Long-term development of experimental mixtures of Scots pine (*Pinus sylvestris* L.) and silver birch (*Betula pendula* Roth.)**	36
	Bill Mason and Thomas Connolly	
Chapter 6	**Cultivation techniques in a 34 years old farming oak (*Quercus robur* L.) plantation in the Arno valley (Tuscany)**	44
	Serena Ravagni, Claudio Bidini, Elisa Bianchetto, Angelo Vitone and Francesco Pelleri	
Chapter 7	**Climate change impact on a mixed lowland oak stand**	51
	Dejan Stojanović, Tom Levanič, Bratislav Matović and Andres Bravo-Oviedo	
Chapter 8	**Tree-oriented silviculture for valuable timber production in mixed Turkey oak (*Quercus cerris* L.) coppices**	57
	Diego Giuliarelli, Elena Mingarelli, Piermaria Corona, Francesco Pelleri, Alessandro Alivernini and Francesco Chianucci	
Chapter 9	**Coppice forests, or the changeable aspect of things**	64
	Gianfranco Fabbio	
Chapter 10	**Explore inhabitants' perceptions of wildfire and mitigation behaviours in the *Cerrado* biome, a fire-prone area of Brazil**	89
	Giovanni Santopuoli, Jader Nunes Cachoeira, Marco Marchetti, Marcelo Ribeiro Viola and Marcos Giongo	
Chapter 11	**Figuring the features of the Roman Campagna: recent landscape structural**	101
	Luca Salvati, Lorenza Gasparella, Michele Munafò, Raoul Romano and Anna Barbati	
Chapter 12	**An approach to public involvement in forest landscape planning**	110
	Isabella De Meo, Fabrizio Ferretti, Alessandro Paletto and Maria Giulia Cantiani	

Chapter 13   **Geospatial analysis of woodland fire occurrence and recurrence in Italy** 123
Leone Davide Mancini, Anna Barbati and Piermaria Corona

Chapter 14   **Single-entry volume table for *Pinus brutia* in a planted peri-urban forest** 130
Kyriaki Kitikidou, Elias Milios and Kalliopi Radoglou

Chapter 15   **Use of Sentinel-2 for forest classification in Mediterranean environments** 136
Nicola Puletti, Francesco Chianucci and Cristiano Castaldi

Chapter 16   **The role of dominant tree cover and silvicultural practices on the post-fire recovery of Mediterranean afforestations** 143
Ilaria Cutino, Salvatore Pasta, Concetta Valeria Maggiore, Emilio Badalamenti and Tommaso La Mantia

Chapter 17   **Genetic resources and forestry in the Mediterranean region in relation to global change** 155
Fulvio Ducci

Chapter 18   **Multifunctionality assessment in forest planning at landscape level: the study case of Matese Mountain Community (Italy)** 179
Umberto Di Salvatore, Fabrizio Ferretti, Paolo Cantiani, Alessandro Paletto, Isabella De Meo and Ugo Chiavetta

Chapter 19   **Ozone dynamics in a Mediterranean Holm oak forest: comparison among transition periods characterized by different amounts of precipitation** 189
Flavia Savi and Silvano Fares

Chapter 20   **Comparing growth rate in a mixed plantation (walnut, poplar and nurse trees) with different planting designs: results from an experimental plantation in northern Italy** 195
Francesco Pelleri, Serena Ravagni, Elisa Bianchetto and Claudio Bidini

Chapter 21   **Comparison between people perceptions and preferences towards forest stand characteristics in Italy and Ukraine** 204
Oksana Pelyukh, Alessandro Paletto and Lyudmyla Zahvoyska

Chapter 22   **Natural capital and bioeconomy: challenges and opportunities for forestry** 215
Marco Marchetti, Matteo Vizzarri, Bruno Lasserre, Lorenzo Sallustio and Angela Tavone

Chapter 23   **Marginal/peripheral populations of forest tree species and their conservation status** 227
Fulvio Ducci, Ilaria Cutino, Maria Cristina Monteverdi, Nicolas Picard and Roberta Proietti

Permissions

List of Contributors

Index

Preface

In my initial years as a student, I used to run to the library at every possible instance to grab a book and learn something new. Books were my primary source of knowledge and I would not have come such a long way without all that I learnt from them. Thus, when I was approached to edit this book; I became understandably nostalgic. It was an absolute honor to be considered worthy of guiding the current generation as well as those to come. I put all my knowledge and hard work into making this book most beneficial for its readers.

Silviculture refers to the methods and techniques used to control the structure, growth and quality of forests. These techniques are used to meet the timber production requirements, along with other values and needs of the society. Silviculture systems can be broadly classified into high forest, compound coppice, coppice with standards and short rotation coppice. It is divided into the three primary phases, namely, harvesting, regeneration and tending. Regeneration can be natural through self-sown seeds or through artificially sown seeds or planted seedlings. Regeneration is affected by the growth potential of the seeds and the surrounding environment. The pre-harvest treatment of forest crop trees is known as tending. It can be carried out at any stage post the initial seeding. This may involve the treatment of the crop itself or the treatment or surrounding competitive vegetation. This book is compiled in such a manner, that it will provide in-depth knowledge about the theory and practice of silviculture. It aims to present researches that have transformed this discipline and aided its advancement. This book includes contributions of experts and scientists which will provide innovative insights into this field.

I wish to thank my publisher for supporting me at every step. I would also like to thank all the authors who have contributed their researches in this book. I hope this book will be a valuable contribution to the progress of the field.

Editor

Empirical modelling of chestnut coppice yield for Cimini and Vicani mountains (Central Italy)

Alice Angelini[1], Walter Mattioli[1*], Paolo Merlini[2], Piermaria Corona[2], Luigi Portoghesi[1]

Abstract - The prescription of stand rotation according to site conditions and economic targets requires yield information that expresses stand productivity under different site-classes. This is particularly relevant for the optimal management of chestnut coppices allowing the production of timber assortments sized differently. This paper reports a new yield model built for chestnut coppices of the Cimini and Vicani mountains (Central Italy), according to three site index classes. The model focuses on the development of stands after an intense thinning carried out at the ages of 13-14 years. The model is compared with two previous yield tables built for chestnut coppice forests living in the same area and including one site index class only. The study stresses the high productivity of chestnut coppices growing on the volcanic soils of Cimini and Vicani mountains and shows how the growth course following intense thinning allows to get stems with large mean volume at the end of rotation. In the light of the most recent studies on the causes of ring shake, i.e. the most relevant defect of chestnut wood, the negative consequences on timber quality originating from the current thinning regime are also outlined.

Keywords - forest management, yield table, thinning, rotation length

Introduction

Coppice is a silvicultural system extensively applied in Italy and in other Mediterranean countries to get quite exclusively firewood under relatively short rotations. Coppice management is based on cultivation techniques that generally result in the simplification of forest composition and structure to the benefit of wood production (Ciancio and Nocentini 2004).

According to the National Forest Inventory (Gasparini and Tabacchi 2011), coppices cover about 35% (i.e. 3.7 million hectares) of the forest area. Within the coppice area, chestnut (*Castanea sativa* Mill.) coppices are a case apart because this fast-growing species produces a wide range of easily marketable and quality assortments for agriculture, constructions, furniture and leather tanning industry.

In Italy, chestnut forests cover an area of 0.8 million hectares (Gasparini and Tabacchi 2011): 70% are managed as coppices with standards and the remaining 30% as orchards for fruit production. In the early fifties of the last century, the relationship between the two forms of cultivation was reversed compared to nowadays and it has been turning over time following the socio-economical changes affecting the rural areas and the spread of two destructive fungi causing, respectively, the chestnut blight (*Cryphonectria parasitica*) and the ink disease (*Phytophthora cambivora*). The domestic production of chestnut wood barely exceeds 0.9 Mm^3 per year, 63% being represented by timber assortments (ISTAT 2002). The current annual increment of chestnut stands is 5 Mm^3 (Gasparini and Tabacchi 2011): that means that the harvested volume could be significantly increased.

The rising demand of renewable resources, the versatility of the raw material also supported by technological innovation by the wood industry, and the need to diversify the productive activities in rural areas, are all together favourable factors to increase domestic chestnut wood yield. This perspective might be however hindered by the fragmentation of private forest land ownership and the inadequate organization of wood supply chain that makes difficult to guarantee the continuity in the supply and the quality of available products, as well (Pettenella 2001).

In the past, timber assortments from chestnut coppices were frequently characterized by relatively low economic value, mainly because of the small average size of the shoots at the end of cutting cycle, since relatively short rotations (16 to 20 years) were adopted, usually without any thinning. By contrast, in recent years, new management schemes based on longer rotations and selective thinnings, accord-

---

[1] University of Tuscia - Department for Innovation in Biological, Agro-food and Forest systems (DIBAF), Viterbo, Italy
[2] Consiglio per la Ricerca e la sperimentazione in Agricoltura, Forestry Research Centre (CRA-SEL), Arezzo, Italy
* Corresponding author: walter.mattioli@unitus.it

ing to site condition and socio-economic context, have been experimentally developed to increase the timber value (Amorini and Manetti 2002, Lemaire 2009a). Research trials showed that gradual thinnings are being able to promote regular and moderate growth of tree rings, thus reducing the occurrence of ring shake, a chestnut wood defect that can significantly lower its commercial value (Fonti et al. 2002, Cosseau and Lemaire 2009).

The improvement of the quality of chestnut timber from coppices should take into duly account rotation lengths suited to growth potential under various site conditions. This asks for effective tools to assess the growth potential of chestnut coppice stands with respect to a given site.

In Italy, the available yield tables show the high productivity as well as the variability of chestnut coppices in terms of mean volume increment, it varying between 9 and 21 m³ per hectare per year (Corona et al. 2002). However, most of the available yield tables are not framed by site-classes, thus limiting their use as operative support to silvicultural and management decision processes. This is the case of chestnut coppices on the Cimini and Vicani mountains in Central Italy, growing under a suitable environment for chestnut but variable as to ground slope, aspect and soil conditions. Two yield tables are available for these coppices: the first one is referred to the entire area of the Cimini and Vicani mountains (Cantiani 1965), whilst the second one is referred to coppices growing on the North-West slopes of these mountains (Lamani 1993); both tables are established with reference to an unique site-class and take into account rotations and thinning regimes no longer applied.

In the light of these considerations, the objectives of this study are to: a) building up a new empirical yield model for chestnut coppices of the Cimini and Vicani mountains, framed by site-classes and based on the most common thinning regime currently applied; b) comparing the new model with the preexisting ones to integrate knowledge about chestnut coppices in the studied woodland.

Study area

The Cimini and Vicani mountains are located in the north of Latium region. Pyroclastic rocks and lavas form the geological substrate of the relieves (Chiocchini 2006); soils, classified as andisols and identified as "black soils", are fertile, deep and loose with acid pH (Bernetti 1959). Mean annual temperature is 12.8°C. The hottest month is July with an average temperature of 22°C, and January is the coldest one (4.2°C). The minimum temperature is just below zero (-2.1°C). Annual rainfall ranges between 1,250 mm and 1,550 mm; summer drought is not actually recorded thanks to the humid air coming from the Thyrrenian Sea and the nearby Vico Lake, although a sub-arid period between July and August may occur (Blasi 1994).

The forest area is about 220 km², mostly covered by oak and chestnut coppices. Turkey oak (*Quercus cerris* L.) and beech (*Fagus sylvatica* L.) high forests characterize the flat top of Mount Cimino (1,053 m a.s.l.) and the west side of the Vico Lake valley. Public forest estates (mostly municipal) extend for several hundred hectares whilst the private ownership is very fragmented. Chestnut coppices grow in monospecific stands on about 8,000 hectares, between 550 m and 950 m a.s.l.. Their monospecificity is due either to the applied silvicultural practices and to the environmental conditions (distinctively, the volcanic soils where chestnut finds its optimal growth conditions).

An average of about 200 hectares of chestnut coppices are harvested or thinned per year. Rotation length ranges between 16 and 20 years (19 years, on average) under private properties and between 18 and 25 years (23 years, on average) under public properties. Chestnut coppices are usually thinned once at ages of 13-14 years. In a few public estates a second thinning is occasionally carried out 3-5 years before the end of the rotation. This study is focused on modelling coppice growth after the unique thinning usually applied.

Materials and methods

Three forest estates where chestnut coppices are prevailing were considered: the municipal property of Soriano nel Cimino along the slopes of Mount Cimino; the regional property near the village of San Martino al Cimino; the Mount Palanzana, where many small privately-owned stands are located. In each forest estate, five thinned stands of various ages (from 14 to 31 years) were selected. The boundaries of the selected stands were identified on colour ADS40 2008 aerial orthophotos (nominal geometric resolution: 1 m).

Three circular sample plots with 10-m radius were established in each stand, for a total of 45 plots. The location of plot center was randomly selected and reached in the field by GPS with sub-metric accuracy. In each sample plot the following attributes were measured: diameter at breast height (dbh) of live and dead shoots and standards, height of a sample of 15 shoots and the age of shoots, assessed by a tree corer on the stems of average dbh.

The following stand parameters were calculated: number of shoots (live and dead) per hectare, number of standards per hectare, mean and dominant dbh of shoot and standard layer, mean and top height of shoots, wood volume of shoots and of standards. Top height was calculated as the regression height of the shoots with quadratic mean diameter of the 100 thickest shoots per hectare (Van Laar and Akca 2008). The wood volume was calculated through the volume tables by Castellani et al. (1984), the same used by Lamani (1993).

The collected data were used to establish a model predicting the top height of shoots (TH, in meters) as a function of stand age (A, in years). The anamorphic model proposed by Schumacher (1939) was adopted, as suggested by von Gadow and Hui (1999):

$$TH = b_0 \cdot \exp(-b_1 \cdot \frac{1}{A}) \quad [1]$$

From this relationship, the site-index curve was derived (Sharma et al. 2002, Skovsgaard and Vanclay 2008), that is the predictor of site index (SI) from TH measured at a given age. SI is a standardized indicator of stand productivity, i.e. the stand top height at a selected index age. In this case, the age of 20 was chosen as index, since this is the average age of the sampled stands:

$$SI = TH * (\exp(9.777/A)/1.63044) \quad [2]$$

Then, regressions were established to model the stand growing stock (V, in m³ per hectare) and the number of shoots per hectare (N) as a function of A and SI under a bio-mathematically sound perspective (Corona 1995):

$$\hat{V} = SI^{c_1} * \exp(-\frac{c_2}{SI * A^2}) \quad [3]$$

$$\hat{N} = m_0 + m_1/\sqrt{A} + m_2 * SI \quad [4]$$

where c_1, c_2, m_0, m_1 and m_2 are the model coefficients to be estimated.

From V and N the mean volume of a single shoot ($\hat{v}$, in m³) can be calculated:

$$\hat{v} = \hat{V}/\hat{N} \quad [5]$$

Since number of standards proved to be no-correlated with A (r = -0.08) and with SI (r = -0.1), it was deemed suitable to keep constant this stand parameter, i.e. equal to the average number of standards found in sample plots.

The total volume of standards is equal to their number times the mean volume of a single standard ($\hat{v}_{st}$, in m³), which, in turn, was predicted as a function of SI and A under a bio-mathematically sound perspective:

$$\hat{v}_{st} = SI^{f_1} * \exp(-\frac{f_2}{SI * \sqrt{A}}) \quad [6]$$

where f_1 and f_2 are the model coefficients to be estimated.

Results

Tab. 1 shows the statistical parameters of [1]-[3]-[4] and [6] models; all the model coefficients are characterized by $p<0.01$. Tab. 2 reports the resulting yield for chestnut coppices of the Cimini and Vicani mountains according to SI values equal to 14 m, 17 m and 20 m, corresponding to the site index range in the studied area (Fig. 1).

Site-index stresses the positive effect of thinning on stand volume and, especially, on mean volume of a single shoot, which doubles at age of 30 years moving from a site class to the upper one (Fig. 2). Mean increment culminates between 18 (SI=20) and 22 (SI=14) years, ranging from 7.2 to 13 m³ per hectare per year. In the yield table of Lamani (1993) mean increment culminates at the age of 21 with 13.8 m³ per hectare per year and the curve of top height as a function of age overlaps quite well the same relationship in our new yield model with SI=17.

The comparison with the yield table of Cantiani (1965) is difficult because height values are lacking here; then, the Eichhorn's rule (Eichhorn 1902, in Skovsgaard and Vanclay 2008) was applied. According to this rule, total production of wood volume at a given stand height is independent of age and site for a given site and species. It was used to attribute a SI value: in the Cantiani yield table the total volume at the age of 18 is 367 m³ per hectare; in the yield table of Lamani this volume corresponds to an age of 21-22 and to a top height of about 17 m. In our new model top height of 17 m and age of 18 are equivalent to a SI between 17 and 20 m. This result is in agreement with data sampling in the most productive chestnut forests only, as for the yield table by Cantiani (1965).

Table 1 - Statistical parameters of yield models established for chestnut coppices of the Cimini and Vicani mountains.

Model	Coefficients	Value	Stand. Err.	R^2
[1]	b_0	27.333	2.280	0.458
	b_1	-9.777	1.589	
[3]	c_1	2.103	0.021	0.743
	c_2	3269.445	388.500	
[4]	m_0	2564.157	978.050	0.412
	m_1	11599.748	3277.014	
	m_2	-197.046	41.359	
[6]	f_1	0.663	0.169	0.375
	f_2	186.192	40.640	

What most distinguishes the three yield models is the curve describing the development of shoot number as a function of age, which highlights the difference in the rotation length and thinning occurrence (Fig. 3). The yield table by Cantiani stipulates

Table 2 - Yield predictions for chestnut coppices of the Cimini and Vicani mountains with reference to site-indexes of 14, 17 and 20 m (TH = top height; V = standing volume; N = number of shoots; v = mean volume; MAI = mean annual increment of standing volume).

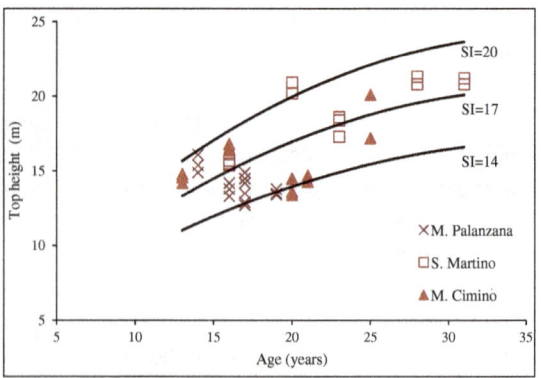

Figure 1 - Top height as a function of age and site-index in chestnut coppices of the Cimini and Vicani mountains. Values measured in the sample plots are also showed.

| | | | Site Index = 14 | | | | |
| | | | SHOOTS | | | STANDARDS | |
age (years)	TH (m)	V (m³ha⁻¹)	N (nha⁻¹)	v (m³)	MAI (m³ha⁻¹year⁻¹)	v (m³ha⁻¹)	MAI (m³ha⁻¹year⁻¹)
15	11.9	91	2801	0.033	6.1	6.3	0.4
16	12.4	103	2705	0.038	6.5	7.0	0.4
17	12.8	115	2619	0.044	6.7	7.8	0.5
18	13.3	125	2540	0.050	7.0	8.5	0.5
19	13.6	135	2467	0.055	7.1	9.3	0.5
20	14.0	143	2399	0.060	7.2	10.0	0.5
21	14.3	151	2337	0.065	7.2	10.7	0.5
22	14.6	159	2279	0.070	7.2	11.5	0.5
23	14.9	165	2224	0.074	7.2	12.2	0.5
24	15.2	171	2173	0.079	7.2	13.0	0.5
25	15.4	177	2125	0.083	7.1	13.7	0.6
26	15.7	182	2080	0.088	7.0	14.4	0.6
27	15.9	187	2038	0.092	6.9	15.1	0.6
28	16.1	191	1998	0.096	6.8	15.8	0.6
29	16.3	195	1960	0.100	6.7	16.6	0.6
30	16.5	198	1923	0.103	6.6	17.3	0.6

| | | | Site Index = 17 | | | | |
| | | | SHOOTS | | | STANDARDS | |
age (years)	TH (m)	V (m³ha⁻¹)	N (nha⁻¹)	v (m³)	MAI (m³ha⁻¹year⁻¹)	v (m³ha⁻¹)	MAI (m³ha⁻¹year⁻¹)
15	14.4	165	2209	0.075	11.0	13.2	0.9
16	15.0	183	2114	0.087	11.4	14.4	0.9
17	15.6	199	2028	0.098	11.7	15.6	0.9
18	16.1	214	1948	0.110	11.9	16.8	0.9
19	16.6	227	1876	0.121	12.0	18.0	1.0
20	17.0	239	1808	0.132	12.0	19.2	1.0
21	17.4	250	1746	0.143	11.9	20.4	1.0
22	17.8	260	1687	0.154	11.8	21.5	1.0
23	18.1	269	1633	0.165	11.7	22.7	1.0
24	18.4	277	1582	0.175	11.6	23.8	1.0
25	18.7	284	1534	0.185	11.4	24.9	1.0
26	19.0	291	1489	0.196	11.2	25.9	1.0
27	19.3	297	1447	0.205	11.0	27.0	1.0
28	19.5	303	1407	0.215	10.8	28.1	1.0
29	19.8	308	1368	0.225	10.6	29.1	1.0
30	20.0	312	1332	0.234	10.4	30.1	1.0

| | | | Site Index = 20 | | | | |
| | | | SHOOTS | | | STANDARDS | |
age (years)	TH (m)	V (m³ha⁻¹)	N (nha⁻¹)	v (m³)	MAI (m³ha⁻¹year⁻¹)	v (m³ha⁻¹)	MAI (m³ha⁻¹year⁻¹)
15	17.0	187	1618	0.116	12.5	22.4	1.5
16	17.7	204	1523	0.134	12.8	24.2	1.5
17	18.3	220	1437	0.153	12.9	25.9	1.5
18	18.9	234	1357	0.172	13.0	27.6	1.5
19	19.5	246	1284	0.192	13.0	29.3	1.5
20	20.0	257	1217	0.211	12.9	30.9	1.6
21	20.5	267	1155	0.231	12.7	32.5	1.6
22	20.9	276	1096	0.252	12.6	34.0	1.6
23	21.3	284	1042	0.273	12.4	35.6	1.6
24	21.7	291	991	0.293	12.1	37.0	1.5
25	22.1	298	943	0.316	12.0	38.5	1.5
26	22.4	304	898	0.339	11.7	39.9	1.5
27	22.7	309	856	0.361	11.5	41.3	1.5
28	23.0	314	815	0.386	11.2	42.7	1.5
29	23.3	319	777	0.411	11.0	44.0	1.5
30	23.5	323	741	0.436	10.8	45.3	1.5

two thinnings: at 6 years 25% of shoots is harvested, whereas a second thinning at the age of 12 years reduces further stem number of 30%. The model by Lamani stipulates only one thinning at 13 years, with a removal of 37%. In the management practice of the last period, the intensity of the only thinning at the age of 13-14 years has further increased to get a higher stumpage value, and this condition is present in the new model here established. It is noteworthy that, even following the heavy thinning currently applied, tree mortality in coppice stands remains high: from 31% (SI=14) to 51% (SI=20) of the shoots become suppressed from the age of 15 years to the age of 30 years.

The number of standards is relatively low and it does not significantly affect stand volume: forest owners attribute to such trees, grown relatively isolated, a low economic value because of the epicormic branches and of the high probability of ring shake occurrence.

Discussion and conclusions

The interest of the new yield model presented here for chestnut coppices of the Cimini and Vicani mountains relies upon the high potential productivity of these stands in terms of woody mass and average size of the shoots at the end of rotation, as well.

Stand growth is largely influenced by site-class. The top height-age curves corresponding to SI=14, SI=17 and SI=20 overlap the first three site-classes of chestnut yield tables established by Lemaire (2009b) in France characterizing stands managed to produce high value timber assortments. This is an interesting evidence since many small woodworking companies located in the nearby of the Cimini and Vicani mountains import a lot of chestnut timber from France, neglecting local production because of an alleged low quality.

The economic value of chestnut stands does not only depend on wood productivity. The technologi-

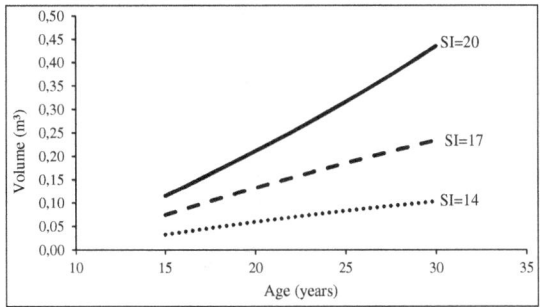

Figure 2 - Mean wood volume of shoots as a function of stand age and site index in chestnut coppices of the Cimini and Vicani mountains.

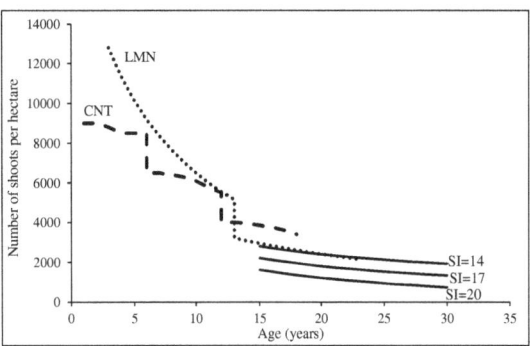

Figure 3 - Chestnut coppices of the Cimini and Vicani mountains: number of shoots per hectare as a function of stand age and site-index predicted by the new yield model (solid line) compared with the yield tables by Cantiani (CNT; dashed line; two thinnings) and by Lamani (LMN; dotted line; one thinning).

cal quality of timber influence greatly the income of forest owner, as well. Spina and Romagnoli (2010) observed that chestnut wood from the coppices of the Cimini and Vicani mountains is more affected by ring shake than wood coming from other chestnut coppice areas in Italy. This defect limits the use of chestnut wood for the most valuable assortments, especially those used as structural elements of buildings and furniture. Several interacting factors affect the phenomenon of ring shake under a single basic principle: an imbalance between the tensions being activated in the wood and the opposing cohesion forces (Fonti et al. 2002). Silvicultural practices can be very effective in preventing the defect or in reducing its incidence. Various studies conducted in different countries showed that abrupt changes in radial growth favour the ring shake occurrence (Cousseau and Lemaire 2009). Thus, it is essential to keep the diameter annual increment sustained and regular, e.g. to apply thinning before the growth rate declines because of competition (Cousseau and Lemaire 2009). For shoots growing in unthinned chestnut coppice stands in Central Italy, a critical age for ring shake establishment proves generally to be between the age of 12 and 14 years (Spina and Romagnoli 2010). Becagli et al. (2006) highlight the importance of anticipating the age of thinning to counteract ring shake occurrence. On the other hand, too intense thinnings may affect negatively chestnut wood quality: if ring width exceeds 6 mm, wood density and its own properties, tend to decrease (Fioravanti 1999).

The new yield model presented here points out that a unique intense thinning allows to get on average large-sized stems, but this might result in a lower quality of the final product. The intense resprouting after clearcutting, the soil fertility that speeds up self-thinning process and the incidence of diseases emphasize an early shoots' mortality. Thus, the regular reduction of interindividual competition would allow both the increase of stem quality and the decrease of deadwood accumulation and fire risk (Agee and Skinner 2005).

The new yield model integrates well the previous models developed by Cantiani (1965) and by Lamani (1993), while taking into account site-index. It improves the knowledge on chestnut coppices and provides to the forest owners of the Cimini and Vicani mountains an useful tool for making decisions about optimal rotation. It may be helpful in introducing innovation in chestnut coppice silviculture and supporting the enhancement of timber supply to gain more attention even from the most demanding woodworking companies.

Acknowledgements

Thanks are due to the anonymous reviewers for their constructive criticisms.

References

Agee J.K., Skinner C.N. 2005 - *Basic principles of forest fuel reduction treatments*. Forestry Ecology and Management 211: 83-96. http://dx.doi.org/10.1016/j.foreco.2005.01.034

Amorini E., Manetti M.C. 2002 - *Selvicoltura nei cedui di castagno. Sostenibilità della gestione e produzione legnosa di qualità*: In: Ciancio O., Nocentini S.: "Il bosco ceduo in Italia". A.I.S.F., Firenze: 219-248.

Becagli C., Amorini E., Manetti M.C. 2006 - *Incidenza della cipollatura in popolamenti cedui di castagno da legno del Monte Amiata*. Annali CRA - Istituto Sperimentale Selvicoltura 33: 245-256.

Bernetti G. 1959 - *Note sul clima e sui terreni forestali dei Monti Cimini (Viterbo)*. In: "Scritti geo-pedologici in onore di Paolo Principi". Tip. Coppini, Firenze: 6-40.

Blasi C. 1994 - *Fitoclimatologia del Lazio*. Fitosociologia 27: 151-175.

Cantiani M. 1965 - *Tavola alsometrica dei cedui di castagno dei Monti Cimini*. Ricerche sperimentali di Dendrometria e Auxometria, Fasc. IV: 27-31.

Castellani C., Scrinzi G., Tabacchi G., Tosi V. 1984 - *Inventario Forestale Nazionale Italiano (IFNI). Tavole di cubatura a doppia entrata*. Ministero dell'Agricoltura e delle Foreste - Istituto Sperimentale per l'Assestamento Forestale e per l'Alpicoltura, Trento.

Chiocchini U. 2006 - *La geologia della città di Viterbo*. Gangemi Editore S.P.A., Roma, 192 p.

Ciancio O., Nocentini S. 2004 - *Il bosco ceduo. Selvicoltura, Assestamento, Gestione*. A.I.S.F., Firenze, 721 p.

Corona P. 1995 - *Criteri di corroborazione dei modelli di regressione*. L'Italia Forestale e Montana 4: 390-403.

Corona P., Chirici G., Vannuccini M. 2002 - *Contributo conoscitivo sugli aspetti dendrometrici, auxometrici e gestionali dei cedui italiani*. In: Ciancio O., Nocentini S.: "Il bosco ceduo in Italia". A.I.S.F., Firenze: 73-124.

Cosseau G., Lemaire J. 2009 - *Diradamenti tardivi e rischio di cipollatura*. In: "Dossier Castagno: Selvicoltura e Cipollatura". Sherwood 151: 25-28.

Eichhorn F. 1902 - *Ertragstafeln für die Weißtanne*. Verlag von Julius Springer, Berlin, 81 p.

Fioravanti M. 1999 - *Valutazione tecnologica della influenza delle pratiche selvicolturali sulla qualità del legno*. In: "Il legno di Castagno e di Douglasia della Toscana. Qualità del legno e selvicoltura. Classificazione e valori caratteristici del legname strutturale". Quaderno Arsia 9/99. Agenzia Regionale per lo Sviluppo e l'Innovazione nel Settore Agricolo-forestale, Firenze, 23-40.

Fonti P., Macchioni N., Thibaut B. 2002 - *Ring shake in chestnut* (Castanea sativa *Mill.*): *state of the art*. Annals of Forest Science 59: 129-140.

Gasparini P., Tabacchi G. (Eds.) 2011- *L'Inventario Nazionale delle Foreste e dei serbatoi forestali di Carbonio INFC 2005. Secondo inventario forestale nazionale italiano. Metodi e risultati*. Ministero delle Politiche Agricole, Alimentari e Forestali; Corpo Forestale dello Stato; Consiglio per la Ricerca e Sperimentazione in Agricoltura, Unità di ricerca per il Monitoraggio e la Pianificazione Forestale. Edagricole – Il Sole 24 Ore, Bologna, 653 p.

ISTAT 2002 - *Coltivazioni agricole e foreste. Anno 1997*. Collana Informazioni, Roma, 160 p.

Lamani F. 1993 - *Tavola alsometrica dei cedui di castagno dei Comuni di Viterbo e Vetralla*. Master Thesis, Università degli Studi della Tuscia, Viterbo.

Lemaire J. 2009a - *Diradamenti dinamici - una selvicoltura ottimale nei cedui di castagno*. In: "Dossier Castagno: Selvicoltura e Cipollatura". Sherwood 151: 32-35.

Lemaire J. 2009b - *Produttività dei cedui e trattamenti selvicolturali*. In: "Dossier Castagno: Selvicoltura e Cipollatura". Sherwood 151: 13-16.

Pettenella D. 2001 - *Marketing perspectives and instruments for chestnut products and services*. Forest Snow and Landscape Research 76(3): 511-517.

Schumacher F.X. 1939 - *A new growth curve and its application to timber-yield studies*. Journal of Forestry 37: 819-820.

Sharma M., Amateis R.L., Burkhart H.E. 2002 - *Top height definition and its effects on site index determination in thinned and unthinned loblolly pine plantations*. Forest Ecology and Management 168: 163-175. http://dx.doi.org/10.1016/S0378-1127(01)00737-X

Skovsgaard J.P., Vanclay J.K. 2008 - *Forest site productivity: a review of the evolution of dendrometric concepts for even-aged stands*. Forestry 81(1): 13-31. http://dx.doi.org/10.1093/forestry/cpm041

Spina S., Romagnoli M. 2010 - *Characterization of ring shake defect in chestnut* (Castanea sativa *Mill.*) *wood in the Lazio Region (Italy)*. Forestry 83(3): 315-327. http://dx.doi.org/10.1093/forestry/cpq014

Van Laar A., Akca A. 2008 - *Forest Mensuration*. Springer, Dordrecht, 383 p.

Von Gadow K., Hui G. 1999 - *Modelling forest development*. Kluwer Academic Publishers, Dordrecht, 218 p.

Forest genetic resources to support global bioeconomy

Saša Orlović[1*], Mladen Ivanković[2], Vlatko Andonoski[3], Srdjan Stojnić[1], Vasilije Isajev[4]

Abstract - A biobased economy implies sustainable and effective use of the biomass. This includes new products from forestry. The sustainable production, use, consumption and waste management of biomass all contribute to a bioeconomy (The European Bioeconomy in 2030).
In the context of bioeconomy the conservation of forest genetic resources assumes a key significance in overcoming global challenges such as climate change. Forests are expected to play a key role in climate change mitigation, but they will only be able to fulfil that role if the trees themselves are able to survive and adapt to changing climate conditions. Genetic diversity provides the fundamental basis for the evolution of forest tree species and for their adaptation to change. The enormous range of goods and services provided by trees and forests is both a function of and testimony to the genetic variability contained within them. Conserving forest genetic resources is therefore vital, as they constitute a unique and irreplaceable resource for the future, including for sustainable economic growth and progress and environmental adaption (The State of the Worlds Forest Genetic Resources 2014). Previous research of population characteristics and the effects of natural and artificial selection on the genetic structure of populations contribute to the conservation and enhancement of the gene pool of the native tree species. The balance model of the population genetic structure reveals the new properties of the populations and requires further investigations, especially of the relations of subpopulations, half-sib families and organisms and the effect of variable factors of the environment, on the exchange of genetic material within natural and cultural populations.
Being of national and international significance, these resources require intensive protection and enhancement *in situ* and *ex situ*. In this paper a general introduction is given to conservation of forest genetic resources in Serbia, Croatia and Macedonia in the context of bioeconomy. Based on the current situation of conservation of forest genetic resources, some strategic suggestions concerning the future development of genetic conservation is given, taking into consideration the conservation objectives and future trends of great impact on existing forest genetic resources.

Keywords - Genetics resources, conservation, bioeconomy

Introduction

The concept of "bioeconomy" or "bio-based society" has become an important component of national, EU and global policies. The social, economic and biological challenges we face, and the scarcity of natural resources combined with climatological changes, necessitate new approaches to knowledge and innovation as well as to knowledge-based policies. The transformation to a bio-based economy means a transition from a fossil fuel-based economy to a more resource-efficient economy based on renewable materials produced through sustainable use of ecosystem services from land and water. A greater focus on research and innovation can provide us with new products developed from biomass that will replace fossil material, combat climate change, reduce waste and create new jobs.

A bio-based economy (bioeconomy) can be defined as an economy based on (The European Bio-economy in 2030):

a) The sustainable production of biomass to enhance the use of biomass products within a number of different sectors of society. The objective is to reduce climate effects and the use of fossil-based raw materials.

b) Increased added value for biomass materials, concomitant with a reduction in energy consumption and recovery of nutrients and energy as additional end-products. The objective is to optimise the value and contribution of ecosystem services to the economy.

Climate change influences both the forest as an ecosystem and also sustainable wood production. The forests need to be adapted to climate change, to continue to secure both their function for use, protection and recreation, and also the role that wood and the forest play in protecting the climate (The State of the Worlds Forest Genetic Resources 2014).

An important objective of the conservation of genetic resources is to maintain the adaptedness of organisms to changing environmental conditions. By conserving sufficient amounts of heritable variation in different species and thus their evolutionary

---

[1] University of Novi Sad, Institute of Lowland Forestry and Environment, Novi Sad, Serbia
[2] Croatian Forest Research Institute, Jastrebarsko, Croatia
[3] University Ss. Cyril and Methodius - Faculty of Forestry, Skopje, Macedonia
[4] University of Belgrade, Faculty of Forestry, Belgrade, Serbia
* corresponding author: sasao@uns.ac.rs

potentials, life can continue under changing and even new conditions. The possibilities for future generations to meet their varying demands are thus secured (Rajora and Mosseler 2001).

The genetic variation of most agricultural and horticultural crops as well as of farm animals can be collected and conserved in so-called gene banks. However, forest genetic resources are usually conserved as living trees in growing forests.

Considering the objective of preserving the broadest genetic diversity, not only the most representative trees populations or important single trees should be subject to gene conservation. Populations from marginal localities also need to be conserved, despite their lower economic importance as such populations and trees may carry genes of importance for breeding (adaptability, resistance). *In situ* and *ex situ* measures are necessary to complement each other (Andonovski and Velkovski 2011).

Conservation, testing and utilization of tree species gene pool involves several successive *in situ* and *ex situ* activities such as: a) study of the nature of phenotypic variability in large and small populations, b) improvement of mass and individual selection techniques, c) application of intervarietal and distant hybridization, d) analysis of morphometric characters, e) familiarity with the correlation of growth characteristics and the development of the analyzed genotypes and their progeny (Isajev et al. 1988). Activities contributing to gene pool conservation and utilization imply its conservation *ex situ* - through the reproduction of forest populations and superior genotypes by establishing specialized seed sources, arboretums, live archives, provenance tests, progeny tests, clonal tests and seed orchards (Gustafsson 1950, Jovanović 1972, Isajev et al. 1995, Skrøppa 2005). *Ex situ* populations of forest trees in Serbia, Croatia and Macedonia are established in the aim of protection and directed utilization of the gene pool of physically endangered populations or individuals, as a supporting activity to conservation *in situ* and for the provision of readily available constant supplies of genetically improved reproductive material.

The paper presents preliminary results of multiannual analyses which are being carried out in specialized plantations of different tree species aimed at testing of genotypes of parent individuals as well as their half-sib lines. The results of multiannual analyses lead to a better knowledge of the production and adaptation potentials of analyzed species. Seed orchards and pilot plots, being the specialized plantations, should contribute not only to the conversion of the potential genetic variability into free one, as the base of directed utilization of tree gene pool, but also they are the polygons for testing and conservation of species biodiversity.

Conservation of genetic resources *in situ* in Serbia

In order to conserve gene pool *in situ*, in Serbia there are six National Parks, ten Regional Parks, 50 Reserves and 158 Seed Stands of major economic species of broadleaves and conifers. National Parks are: Fruska Gora – 25'300 ha, Djerdap – 63'608 ha, Tara – 19'200 ha, Kopaonik – 11'810 ha, and Shara – 39'000 ha (total 158'918 ha). In the National Parks Fruska Gora and Djerdap, mostly the communities of deciduous tree and shrub species are represented, in the Parks Tara, Kopaonik and Shara, the communities of coniferous species are predominant. Seed stands, along with the production of good quality seed stock, are also intended for the conservation of the gene pool of tree and shrub species in situ, as they contain plus and normal trees and the trees which represent the average of the population. Seed stands are listed in the Proposed Register of Forest Seed Sources in Serbia, with separate lists of broadleaf and coniferous species. The Register includes 151 coniferous seed sources - total area 1'476.1 ha, of which 36.9 % are fir, 28.5 % spruce, 14.2 % Scots pine, 11.8 % Austrian pine, 4.1 % Serbian spruce. There are 65 seed stands of broadleaf species covering 1'665.2 ha, of which 30.5 % are Sessile oak stands, 31.1 % Pendunculate oak, 17.1 % Common beech, 21 % Turkish hazel, 8.4 % lime and 2 % other species. The Registers of seed stands were made twenty years ago, so on the occasion of their revision, only the best populations, i.e. the best trees should be chosen by which the greatest possible portion of the desirable intraspecies variability will be included. Also the work should be intensified on the study and zonation of seed utilization from the zones containing seed stands of major economic species of trees.

Conservation of genetic resources *ex situ* in Serbia

The opportunities for the singling out of Nature Reserves, and in this respect, the protection of forest natural resources, are proportionally limited. For these reasons, Norway spruce provenance test and specialized sources and plantations have been established in Serbia to conserve the genetic resources of woody species *ex situ*, by the reproduction of selected species in forest populations. They are: (a) seed orchards, (c) provenance tests, and (d) progeny tests. Each of the above categories has a special purpose, which is coordinated with the needs of gene pool protection and enhancement.

Seedling seed orchards were established from 50 Serbian spruce half-sib families on the area of 2.7 ha (Isajev et al. 1990), Austrian pine - 40 families on 3 ha (Tucović and Isajev, 1991), divulge wild cherry - 30 families on 1.5 ha, Pedunculate Oak – 76 ramets (Erdeši 1996, Orlović et al. 1999 and 2002), Pedunculate Oak – 122 families (Orlović et al. 2001). The yield in these plantations has not yet reached the level of commercial exploitation. However, the progeny from seedling seed orchards will be more variable than that from clonal ones, and consequently seedling seed orchards are better for the purpose of gene pool conservation.

Norway spruce provenance test in Serbia

Morphological variability and changeability of physiological properties of intraspecial taxons of Norway spruce are described, among other researches, on the bases of provenance tests in Europe and North America (Lines 1967, Kleinschmit 1970, König 2005). The results obtained in the analyses of these tests enabled better familiarity with the production and adaptive potential of the Norway spruce gene pool and the ecological factors which determine the range of its horizontal and vertical distribution were established (Isajev et al. 1990).

For the establishment of a provenance test in the vicinity of Ivanjica 3 locations with different altitude, exposition and area as well as different site characteristics were chosen (Isajev et al. 1992). In the test plantations, five Norway spruce provenances from Serbia are comparatively examined: Golija, Zlatar, Cemerno, Radocelo and Kopaonik, as well as three Slovenian provenances - Jelovica, Menina and Masun. On the basis of data of the Hydro meteorological service in Belgrade regarding the area of Ivanjica, climatic conditions of all three locations were studied. In the very areas in which the test plantations were established topographic surveys were carried out and a total of 8 soil profiles were opened.

The first test plantation is in the department 51, at an altitude of 570 to 610 m, northern exposition and the total area of 2.02 ha, on deep acid soil (dystric cambisol) and the geological base were schists. The second test plantation with the area of 0.65 ha is in the department 38 of FMU Kovilje- Rabrovica: south-eastern exposition, altitude of 1'105 to 1'125 m on dystric cambisol on schists. The third test plantation was established in the department 46, section C in FMU Golija at an altitude of 1'560 to 1'570 m. The exposition was north-eastern and the area was 0.73 ha, the soil was podzolic brown. The number of Norway spruce seedlings planted in test plantations was 2'442.

The results of prior analyses show that the impact of natural selection is the most distinguished in field experiments realized at three altitudinal levels, in which Serbian provenances demonstrate greater adaptability to very different ecological condition (Ivetić 2004). It was recorded that even in sites of submontane beech which is not within the natural range of spruce its growth and adaptability were successful which proves that besides its natural optimum in the zone of spruce belt *Picetum abietis serbicum* s.l., its technogenic optimum can also be reached in sites of other species. The results obtained from all three altitudinal belts contribute to the explanation why Norway spruce in Serbia has a specific climatogenous belt compared to other countries of the Balkan Peninsula. From the expert perspective prior researches play an important role in the economy because they facilitate the choice of a provenance, or a group of provenances, suitable for certain sites as part of planning of afforestation and reclamation works in degraded stands and sites.

Austrian pine seed orchard in Serbia

Large areas of bare land and degraded sites which require urgent afforestation as well as the capability of Austrian pine to achieve good results in extremely bad sites, set forth the need for organized seed production and seed forests as priority aims. In the aim of fulfilling these needs adequately, there is a need for intensified scientific and expert activities on the establishment and management of Austrian pine seed orchards.

Generative seed orchard of Austrian pine in Jelova Gora, with the area of 2.70 ha on the site *Fagetum montanum* Rud. was established in 1991 from 5'422 two-year-old seedlings in 40 half-sib lines of the test trees selected in seed forests Sargan-Mokra Gora and Crni Vrh-Priboj (Fig. 1) (Isajev et al. 1992). Austrian pine seed orchard with 40 half-sib lines, with three repetitions each, in each of the five subplantations and dynamic environmental factors- altitudinal difference of 20 m, two expositions and two soil types is the first generative seed orchard of metapopulation structure established in Serbia. Its structure enables the realization of genetic and development mechanisms and mechanisms of regulation on the one hand and realization of the effects of ecological mechanisms on the other. The above mentioned will benefit gene pool conservation of this species.

Multi-annual research involved detailed study of the variability of nine morphological seedling parameters - seedling height, annual height increment, root-collar diameter, diameter of horizontal crown projection and the number of branches in the

Figure 1 - Generative seed orchard of Austrian pine in Jelova Gora.

last three whorls. The obtained data revealed great variability even in case of slight site changes, as well as great adaptability of the incorporated planting stock. Analysis of correlation confirmed significant positive correlation of almost all elements of growth.

The research of qualitative characteristics also illustrates great genetic diversity. The following was observed: flowering at the age of five years, seedlings with grey needles, three-needle pines, smaller sized needles, proliferation.

The investigation of variability of the root system of Austrian pine seedlings created a base for the selection of genotypes with favourable characteristics in the sense of better adaptation to shock after transplanting or growing in extremely arid stands.

The analyses of the participation of photosynthetically active pigments - chlorophyll a and b and carotenoids are among the first analyses of that kind carried out regarding conifers in this country. These researches contribute to the improvement of familiarity with the correlation of the analyzed photosynthetic matters and basic elements of growth. Prior research of the variability of juvenile Austrian pine and the correlation of morphological and physiological parameters have their multiple importance for science as well as for practical application, being the base for the improvement of the good quality seed and planting stock with favourable characteristics. On the basis of one-way analysis of variance of all the examined morphometric properties it was concluded that significant differences appear at the inter-line and inter-provenance level which indicates that the differences among half-sib lines and provenances are not the consequence of random errors i.e. random variations. By using an LSD - test groups of half-sib lines according to the years of research were homogenized, which confirmed statistically significant differences and the superiority and inferiority of certain lines, previously determined on the basis of relative percentage of average heights and half-sib line diameters in average values of certain provenances. Connecting of clusters of half-sib lines with the highest and the lowest average values of all the examined quantitative parameters within the seed forest (1) is at a shorter total distance than in the seed forest (2) on the basis of which it is also confirmed that the sample of seed forest (1) shows higher homogeneity and lower intra-population variety. The applied statistical analysis indicates great genetic diversity within the seed orchard, but clear differences between certain half-sib lines cannot be strictly determined because that would require full-sib progeny tests.

The obtained results are important for the directed utilization of the gene pool of this species and as directions for the improvement of techniques used in the establishment of Austrian pine seed orchards of the second, third and later generations.

Serbian spruce seed orchard

On the basis of the applied analyses of multi-

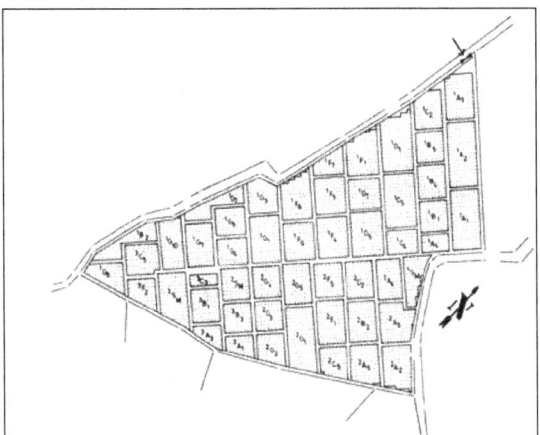

Figure 2 - Generative seed orchards of Serbian spruce (*Picea omorika*/Panć/Purkyne).

annual research of the genetic variability of Serbian spruce (Isajev 1987), generative Serbian spruce seed orchard was established in Western Serbia in 1987 (village Godovik, near Pozega), on an area of 2.70 ha. As much as 5'959 seedlings (age 2+3) were incorporated into the seed orchard. The seedlings from the same family were planted in the same block, with random distribution in the form of a square-shaped planting scheme 2 x 2 m (Fig. 2).

The applied selection and hybridization produced reliable data on general and specific values of half-sib lines and incorporated genotypes. In this seed orchard, based on the planting scheme in which genotypes of the same line are in one block, we made possible, for the first time simultaneously, the three basic types of Serbian spruce reproduction: inbreeding, outbreeding and uniparetal. The study data are a base for further work on the directed utilization of Serbian spruce genetic potential.

By multiannual analyses of individual and line variability of a great number of continuous and discontinuous vegetative and reproductive characters, parental genotypes were selected in order to direct seed crop parameters by spontaneous and controlled hybridization. The study of several flowering parameters, intra- and inter- half-sib lines, such as abundance and regularity of micro- and macro-strobile formation, differentiation of genotypes into functionally male, i.e. female ones, analysis of pollen quality and seedling analysis, resulted in valuable information necessary for successful hybridization. After controlled hybridization performed by the model of incomplete diallel cross, which included 48 different parental combinations, we studied a part of genotype structure of parent individuals and their hybrid combinations, based on the analysis in the salts of soluble proteins of seeds, as the most common polymorph markers at the level of gene products. In these analyses, we used the seed from free pollination of half-sib lines, which were functionally female i.e., male, and the seed of parental genotypes and their hybrid combinations. In parental genotypes and their hybrid combinations, the obtained electrophoregrams were used for the calculation of the coefficient of similarity, at the levels mother-hybrid, father-hybrid and mother-father. Electrophoregram analysis of hybrid combinations shows the existence of different types of protein fractions - bands: bands common to both parents, bands originating one from father, one from mother (codominance of parent gene expression in the hybrid), bands originating from mother only, bands originating from father only, and bands specific for the hybrid.

Based on the results of multiannual study, we differentiated such hybrid combinations which show the highest mean values of the analyzed morphometric characters of the cones, i.e. seed crop, as well as better parental genotypes which show good general and specific combining ability. The obtained data were the basis for the construction of the model of experimental clonal seed orchard for the production of Serbian spruce intraspecific hybrids.

Balkan maple seed orchard in Serbia

Forest Estate "Golija" from Ivanjica established a Balkan maple (*Acer heldeichii*) seed orchard in 1994 (Isajev 1994). The location of the orchard is on the site *as. Fagetum montanum Rud.s* in the forest management unit "Kovilje-Rabrovica", department 12. The altitude of the location is 950 to 1020 m and the exposition is north-eastern. The area of the seed orchard is 1.05 ha (Ćurčić 1997 and 1999). The planting stock was produced in the forest nursery in Ivanjica from the seeds of 26 seed trees, selected in Golija, which had above-average morphologic (technical) and physiological (abundance and regularity of seed yield) characteristics in the population (Isajev et al. 1994).

Planting of seedlings of the age 2+0 was carried out in the spring of 1994. Distribution of planting was planned and realized in blocks. There are 6 blocks in total, 4 of which have a regular rectangular shape and 2 are in the shape of a scalene triangle. The irregular shape of the blocks depends on the shape of the area determined for seed orchard establishment. The realized distribution of the seedlings in blocks is based on the so-called metapopulation strategy of the establishment of generative seed orchards of forest trees (Tucović and Isajev 1991).

Planting of 2'962 seedlings from 26 half-sib families created the base for further works on the testing and becoming familiar with the gene pool of the populations in which mother trees were selected, as well as for the improvement of this interesting and precious species of our valuable broadleaves.

After the application of the appropriate tending, good quality seed for further reproduction will be produced. On the basis of multiple analysis of the genetic value of the incorporated genotypes, decisions on further works on the improvement of Balkan maple and the establishment of seed orchards of future generations will be made.

Taking into account the advantages of the lower, warmer locations for the establishment of seed orchards due to their favourable effect regarding the abundance and frequency of flowering and seed yield and the applied method of metapopulation structure, it can be expected with certainty that the established generative seed orchard of Balkan maple near Ivanjica achieves the expected production of good quality seed.

Pedunculate oak seed orchards in Serbia

There are two seed orchards of Pedunculate oak (*Quercus robur* L.) in Serbia – clonal and generative. They are established at the territories of Forest Units "Morović" and "Klenak", which belong to the Public Enterprise "Vojvodinašume", Forest Estate "Sremska Mitrovica".

Clonal seed orchard was established in the period between 1979 – 1983 (Erdeši 1996). It is established of phenotypically superior genotypes from natural populations. The main criteria for selection of stems were the straightness of stem, branching, fast growth and resistance to oak powdery mildew (*Erysiphe alphitoides* (Griffon & Maubl.) U. Braun & S. Takam.). The seed orchard was established at the area of 7 ha, from 86 genotypes, which were multiplied by grafting into 2'520 remets. Depending on the scion thickness, five techniques of grafting were applied: simple copulation, English copulation, "mjesok" – little sack, cutting and "goat leg". The seed orchard is composed of four varieties indigenous in the valley of the Sava river: early pedunculate oak (*Q. robur* var. *praecox*), typical pedunculate oak (*Q. robur* var. *typica*) and two varieties of late pedunculate oak (*Q. robur* var. *tardiflora* and *Q. robur* var. *tardissima*) (Orlović et al. 1999, 2001 and 2002). Clonal seed orcharad has been an object of numerous researches focused on genetic variability of acorn and leaf morphological and anatomical characteristics (Nikolić and Orlović 2002, Nikolić et al. 2003, 2005, 2006 and 2010). Likewise to the previously mentioned researches, in order to give the first insights into the genetic structure and diversity in the clonal seed stand, the aim of recently conducted study was to characterise the genetic structure related to different phenology of sampled oak genotypes using a system of established microsatellite molecular markers. Leaves from fifteen individuals were sampled from four different varieties of pedunculate oak (*praecox*, *typica*, *tardiflora* and *tardissima*). Seven microsatellite primer sets were used designed to be specific to the sequences flanking the $(GA/CT)_n$ and $(AG/TC)_n$ dinucleotide repeat motives in oak genome. *Quercus* species have revealed high levels of polymorphism suggested that these markers are well suited for studies of genetic diversity within oak population and between different varieties. Successful amplification of all observed microsatellite loci revealed allelic polymorphism between and within all varieties established the variety specific genetic structure (Galović et al. 2014).

Generative seed orchard is founded in the period between 2000 – 2004 at the area of 10 ha. It is established from the acorn that was sown. Acorn is collected from clonal seed orchard previously mentioned and phenotypically superior genotypes from natural populations. Orchard is composed from 129 families, in different number of replication (min. six replications), so the total number of plants (genotypes) in orchard amount 2'585. The space between trees is 7 x 5 m. Similarly to clonal seed orchard, all for varieties indigenous in the valley of the Sava river is represented in the generative orchard. Researches in the orchard have been started recently, collection of acorn from various families and establishing of progeny test.

European beech provenance trials in Serbia

European beech (*Fagus sylvatica* L.) provenance trials in Serbia were established in the spring of 2007. One of the trials is situated on the territory of National park "Fruška Gora" (Northern Serbia), while the second one is located on the territory of the Scientific Centre of the University of Belgrade, Faculty of Forestry – "Majdanpečka Domena", in Debeli Lug (Eastern Serbia) (Stojnić et al. 2012a). The trials are founded within the European network of beech provenance trials. On that occasion, in order to study the genetic variation relevant for adaptation among provenances in the Balkan region, 20 provenances of Croatia, Serbia and Bosnia and additional 12 for comparison from Austria, Germany, Hungary, Italy, Switzerland, and Romania were planted, of which 15 provenances are common to all trials. A total of seven provenance trials were established in Bosnia and Herzegovina, Croatia, Serbia (2), Germany (2) and Italy (von Wuehlisch et al. 2010). The main objectives in these trials could be arranged into four groups: 1) tree improvement, 2) gene conservation, 3) evolution biology and 4) stimulation of European co-operation in forest research (von Wuehlisch 2004).

Previous studies in the provenance trials in Serbia have been focused mainly on the examination of genetic variation within and among different provenances, as well as research of phenotypic plasticity. The aforementioned studies have included numerous parameters that could be, roughly, classified into: physiological, biochemical, morphological, and the parameters of the anatomy of wood and leaves. The research results indicate the existence of significant genetic variation both within and between different provenances, as well as ecotypic pattern of genetic variation (Stojnić et al. 2010, 2012b and 2013, Štajner et al. 2013). Also, given that some authors believe that in order to improve the adaptability of the population, special attention should be paid to phenotypic plasticity, as an alternative to genetic adaptability (Šijačić-Nikolić and Milovanović 2010), attention has been devoted to the study of phenotypic plasticity of wood anatomical structure. The results showed an existence of a plastic response of provenances, as well as the ability of provenances originating from moist sites to adapt to drier habitat conditions (Stojnić et al. 2013).

In situ conservation of forest genetic resources in Macedonia

In Macedonia, *in situ* gene conservation is mainly achieved through the establishment of protected areas and so-called gene reserve forests. In addition to these there are also long-term genetic studies and breeding populations (Andonovski 1995).

National parks and nature protected areas are of great importance for maintaining or improving the forest genetic resources. These areas in Macedonia are classified as follows (Andonoski 2011):

A comprehensive resource inventory on the nature protected areas was set up and management regulations were established for the natural reserves.

According to the Law on Forests, selected natural seed stands for production of seeds belong to category of forests with special purposes and they are under special management regime. During the latest period, an increased effort to conserve and enhance the forest genetic resources has been undertaken on the basis of present knowledge about variability and heritability.

The first mass and individual selection in Macedonia was performed in 1962-1965 and the following coniferous seed stands of were selected and registered (Andonoski 1994):

Table 2 - Selected seed stands (conifers).

Species	Area (ha) Total	Area (ha) Reduced	Age (years)	Provenance
Abies borisii-regis	182.9	84.4	81	Indigenous
Pinus nigra ssp. *pallasiana*	258.9	137.5	73	Indigenous
Pinus silverstris	45.5	32.8	80	Indigenous
Pinus peuce	5.0	3.4	95	Indigenous
Pseudotsuga menziesii	2.8	2.1	35	Exotic

In 2008, new program for conservation of forest genetic resources in Macedonia started with the revision of the current seed stands of various economically important native and exotic tree species. This program includes preregistration of the current seed stands and seed orchards and registration of new, including those of broadleaved species. Following is the table of registered seed stands under the latest Law on forest reproductive material:

Table 3 - Registered seed stands under the latest Law on forest reproductive material.

Species	Number of seed stands	Area (ha)
Pinus nigra ssp. *Pallasiana* (native)	8	218.7
Pinus sylvestris (native)	7	131.8
Abies borisii regis (native)	16	375.9
Pinus peuce (native)	3	68.3
Larix decidua (exotic)	3	34.9
Pseudotsuga menziesii (exotic)	9	59.2
Sequoiadendron giganteum (exotic)	1	4.4
Robinia pseudoacacia (exotic)	1	1.2
Fagus moesiaca (native)	7	326.1
Quercus petraea (native)	2	35.2
Total	57	1'255.7

Table 1 - National parks and nature protected areas (coniferous species).

Name	Area (ha)	Description
National park "Pelister"	12'500.0	The best preserved natural stand of Macedonian pine (*Pinus peuce*)
National park "Mavrovo"	73'088.0	Natural stand of fir and spruce (*Abies borisii-regis*, *Picea abies*)
Nature reserve "Rozden"	3.5	Crimean pine (*Pinus nigra* ssp. *pallasiana*)
Nature reserve "Tumba"	5.0	Fir (*Abies borisii-regis*)
Nature reserve "Golem Kozjak"	4.0	Scots pine (*Pinus silvestris*)
Nature reserve "Popova sapka"	5.2	Norway spruce (*Picea abies*)
Nature reserve "Rupa"	7.6	Fir (*Abies borisii-regis*)
Natural reserve "Tsam Tsiflik"	490.0	Crimean pine (*Pinus nigra* ssp. *pallasiana*)
Natural reserve "Rutsica"	1'785.0	Dwarf mugo pine (*Pinus mugo* var. *mughus*)

In situ forest genetic conservation in Macedonia includes the "dynamic" approach which encourages the adaptation of forest trees to the changing environment through naturally occurring evolutionary processes. This can maximize adaptability with the sufficient number of trees in the genetic resource population (Andonoski 1974).

Ex situ conservation of forest genetic resources in Macedonia

Ex situ conservation of forest genetic resources

in Macedonia includes establishment of *ex situ* gene conservation stands, seed orchards, clone archives or individual trees. Conservation of individual coniferous tree species was carried out using "plus" trees selected for the development of tree improvement programs. The following "plus" trees were selected (Andonoski 1988):

Abies borisii-regis	42 "plus" trees
Pinus silvestris	62 "plus" trees
Pinus nigra ssp. pallasiana	82 "plus" trees
Pinus peuce	20 "plus" trees

On the basis of these selected "plus" trees the following Seed orchards were established:

Table 4 - Seed orchards (conifers).

Species	Type	Year of rising	Area (ha)	Fructification
Pinus peuce	clonal	1963	1.1	full
Abies borisii-regis	clonal	1963	0.5	full
Pinus silvestris	clonal	1978-1980	2.5	full
Pinus silvestris	generative	1978	5.0	full
Pinus nigra ssp. pallasiana	clonal	1988	1.5	started

Outlook

In the past more emphasis was placed on the conservation and study of "plus" trees, so now it is necessary to focus on study and conservation of the most valuable populations. The majority of gene reserves were selected in the early 1960, thus it needs repeated inventory with biochemical, cytological and molecular genetic methods.

It is necessary to get more information about genetic structure and differentiation of the tree species.

In situ conservation activities should be integrated part of the regular forest management. The major challenges for the conservation of forest genetic resources in Macedonia include population decline and population structure changes due to forest removal and conversion of forest land to other uses, forest fragmentation, forestry practices, climate change, disease conditions, introduced pests, atmospheric pollution, and introgressive hybridization. Developing scientifically sound conservation strategies, maintaining minimum viable population sizes, and deployment of genetically engineered organisms represent other important challenges in conservation. Both *in situ* and *ex situ* forest genetic resource conservation strategies must include the use of various biochemical and molecular genetic markers, adaptive traits, and genetic diversity measures. So, major opportunities for conservation of forest genetic resources in Macedonia include: use of molecular genetic markers and adaptive traits for developing conservation strategies; *in situ* conservation through natural reserves, protected areas, and sustainable forest management practices; *ex situ* conservation through germplasm banks, common garden archives, seed banks, DNA banks, and tissue culture and cryopreservation; incorporation of disease, pest, and stress tolerance traits through genetic transformation; plantation forestry; and ecological restoration of rare or declining tree species and populations.

Conservation of forest genetic resources in Croatia

Croatia, with its area of forest and forest land (2.49 million hectares, which is 44% of mainland Croatia) has 260 indigenous wood species. 50% belongs to forest ecosystem and 60 of them make economical richness of Croatian forests while there is more than 100 species which are added to the forest ecosystem to implement their biodiversity. Conservation of genetic diversity of our forest species represents the foundation of a sustainable forest management and preservation of natural structure of our forest stands, currently making 95% of the total woodland area. Croatia's richness in diversity of geographical regions has resulted in various ecological types and a large number of forest trees that are directly affected by habitat degradation, different types of soil, air and water pollution, excessive use of some more valuable species of forest trees, increasing impact of global climatic changes, as well as by anthropogenic effects (Kajba et al. 2006).

The need for conservation of genetic variability is related to the species pertaining to social broadleaves which are economically the most prevailing species (Pedunculate oak, Sessile oak, and Common beech). Among the conifers, Silver fir (*Abies alba*) is the most endangered species, with more than 70% of its population being permanently damaged. Other native coniferous species must be preserved from the deprivation of genetic variability as well.

Conservation of noble broadleaves should encompass a larger number of species from various genuses (*Fraxinus, Alnus, Ulmus, Prunus, Juglans, Castanea, Sorbus, Acer, Malus, Pyrus, Tilia*). They are partially endangered because of their exposure to different diseases and pests, as well as by continuous exploration caused by their technical value. Changes in hydrological conditions of our rivers have generated difficulties in restoration of riparian forests, and decreased genetic variability of European black and white poplar in their habitats. In coastal areas of our country, there is a need for conservation of genetic resources of Dalmatian black pine (*P. nigra* ssp. *dalmatica*) and our Mediterranean oaks.

Genetic diversity conservation of various species of forest trees is conducted within the programs that include in situ dynamic methods and ex situ static methods.

Conservation of native species within in situ method is based on concept of status quo of natural conditions protections on local environment, where is optimal alleles frequency to survive and reproduction in that environment reached. Starting point is that local population of certain species are resistant and adapted to environment stress, diseases and injurers impact. Conservation of genetic diversity researches contains knowledge about the smallest population size (MVP) which is required for their relative safe survival according to genetic, demographic, environmental and other factors. According to the size and type of areal of each species (continuous and discontinuous areal, genetic drift, etc.), we need to define number of subpopulation and units which will successfully present, include and conserve complete variability of each species. Conservation within *in situ* method differentiate populations in categories of protected objects of biological and landscape diversity, natural forest stand and population which already are or will be excluded from regular management (e.g. seed stands).

Protection by *ex situ* method represents forest tree species conservation out of their natural habitat. This method is used parallel to *in situ* method, especially with species where conservation of population or their parts is not possible. For that cause, setting of experimental surfaces with *ex situ* methods is required and includes researches on provenance, progeny and clone tests. Genetic diversity of each species can be saved by establishing collections (provenance trials, progeny tests, clonal archives, clonal seed orchard, seed bank, pollen and plant tissue banks) using this method.

In Croatia there are more than 350 forest seed objects: forest stands, seed stands, clonal seed orchard and group of tree. Croatian Forest Research Institute (CFRI), as official body, according Forest reproductive material legislative, supervises the production and marketing of forest reproductive material. Also Croatian forest research institute set up a register of forest seed objects constituting a gene bank of forest trees of Republic Croatia. Total area of all seed stands (category of seed: selected) in Croatia is 3'898.35 ha and for conservation of genetic diversity (gene bank) are suggest 1'103.60 ha (*in situ*). During the past 50 years, researchers Croatian Forest Research Institute were founded by dozens of provenance experiments. In this paper, we mention the provenance experiments involved in the gene bank of forest trees Croatia, where it is still carried out by scientific research like: Pedunculate oak provenance trials established in 1988, 2008 and 2010 (Gračan 1999, Perić et al. 2006, Ivanković et al. 2011), Silver fir provenance trials established in 2000 (Ivanković 2003) and International beech provenance trials established in 1998 and 2007 (Gračan and Ivanković 2001, Gračan et al. 2006, Jazbec et al. 2007, Ivanković et al. 2008).

In Table 5 is list of Clonal seed orchard nominated for registration in Genetic bank, while in Table 6 is list of provenance trials which are included in genetic bank (*ex situ*).

Table 5 - List of Clonal seed orchard nominated for registration in Genetic bank.

Species	No. of orchards	Type	Year of rising	Area (ha)
Pinus sylvestris	2	clonal	1966	3.0
Pinus nigra	2	clonal	2006	1.5
Larix europea	2	clonal	1985	2.5
Quercus robur	4	clonal	1996, 2000, 2001, 2008	47.0
Quercus petrea	1	clonal	2008	7.3
Alnus glutinosa	2	clonal	1985	1.7
Fraxinus angustifolia	2	clonal	2005, 2007	3.5
Prunus avium	1	clonal	2001	3.0
Pinus strobus	1	clonal	1965	-

According the same legislative, Croatian forest research institute take care about forming and conservation of reserve forest seed material in seed bank.

Table 6 - List of provenance trials which are included in Genetic bank.

Tree species	No. of trials	Year of establishing
Quercus robur	4	1988, 2008, 2010, 2010
Fagus sylvatica	2	1998, 2007
Abies alba	1	2000
Larix decidua	1	1959
Pinus sylvestris	1	1959
Pinus nigra	1	1959
Picea omorika	1	1959
Picea abies	1	1959

Conservation of forest trees genes represents maintenance of the evolutionary created adaptation potential of a particular species, i.e. its forest community and the entire forest ecosystem. For the purpose of conservation of forest genetic reso¬urces, we must protect the existing genetic variability, its adaptability to processes of natural evolution and forest tree breeding, as well as improve our knowledge and ways of identification of those individuals that have developed tolerance to certain diseases and pests. That way, we will be able to prevent a decrease in genetic resources of the endangered species. The research should be supplemented with data including making of species inventories, legislation, practical use, coordination on national and paneuropean level, together with promotion of public awareness on the importance of conservation of the endangered species in forest ecosystems.

Conclusions

Beside gene pool conservation and testing *in situ-* in natural populations in different sites, conditions for biodiversity testing as well as the familiarity with the range of potential variability *ex situ* are provided by establishing of separate plantations. Starting from the floristic, genetic and applicative potential of Serbian spruce, Norway spruce, Austrian pine, Balkan maple, Pedunculate Oak, European beech, the paper presents multi-annual researches aimed at becoming familiar with their gene pool as well as its conservation and directed utilization by the establishment of specialized plantations.

The obtained results enable better familiarity with the potential of production and adaptability of the analyzed species. Seed orchards and pilot seed forests, as specific plantations, should contribute to the conversion of potential genetic variability into the free one, as the base for directed utilization of tree gene pool, and serve as polygons for testing and preservation of biodiversity of these species.

The presented research objectives and methods and the results of the genetic valuation of forest tree species are up-to-date methods in gene pool conservation and testing, as well as planning and establishment of future plantation communities of these species.

The activities on conservation and use forest genetic resources lead to produce superior genotypes which are important for increasing of wood production and climate change mitigation. Those activities support global bio-economy by enhancement of use genetic potential of forest trees.

References

Andonoski A. 1988 - *The state of Macedonian forest genofond and measures for its conservation*. In: X Congress of dendrologists, Sofia, Bulgaria: 144-152.

Andonoski A. 1974 - *Application of genetical principles on the production of forest seeds*. Forest review 5,6: 43-67, Skopje, Macedonia.

Andonoski A. 1994 - *Genetics and tree improvement*. Skopje, Macedonia: 226-261.

Andonoski A. 1995 - *Adaptive capability of some introduced forest tree species*. Master thesis: 82-94, Skopje, Macedonia.

Andonoski A. 1998 - *Multy-Country Forestry Program (National report for the Republic of Macedonia)*, Skopje, Macedonia:18-34.

Andonovski V., Velkovski N. 2011 – *Conservation of Macedonian pine (Pinus peuce Griseb.) genetic resources in Pelister National Par.* Journal of Protected Mountain Areas Research and Management 3 (1): 49-53.

Ćurčić G., Isajev V., Tošić M. 1999 - *Generativna semenska plantaža planinskog javora (Acer heldreichii Orph.) kod Ivanjice – pilot objekt – za dalje oplemenjivanje vrste*. Drugi Kongres Genetičara Srbije. Knjiga Abstrakta 215.

Ćurčić, G. 1997- *Generativna semenska plantaža planinskog javora (Acer heldreichii Orph.) "Perkovići-Strane" kod Ivanjice*. Diplomski rad. str. 1-43. Beograd.

Erdeši J. 1996 - *Vegetativna semenska plantaža hrasta lužnjaka (Quercus robur L.) – izvođački projekat*. JP " Srbijašume„ Beograd, Šumsko gazdinstvo Sremska Mitrovica. Sremska Mitrovica.

European Commission's Seventh Framework Programme (FP7) 2008 – *The European Bioeconomy in 2030, BECOTEPS –Bio-Economy Technology Platforms:* 5-22.

Food and Agriculture Organization of the United Nations 2014 - *The State of the Worlds Forest Genetic Resources*. FAO Rome.

Galović V., Orlović S., Zorić M., Kovačević B., Vasić S. 2014 - *Different phenology induced genotype diversity of Q. robur L. in the seed orchard in Srem provenance, Republic of Serbia*. International Conference "Natural reseources, green technology, and sustainable development „, Zagreb, November 26-28, 2014 (in print).

Gračan J., Ivanković, M. 2001 - *Prvi rezultati uspijevanja provenijencija obične bukve (Fagus sylvatica L.) u Hrvatskoj*. Znanost o potrajnom gospodarenju Hrvatskim šumama / Matić, Slavko; Krpan P. B. Ante; Gračan Joso (ur.). Zagreb: Šumarski fakultet; Jastrebarsko, Šumarski institute: 636.

Gračan J., Ivanković M., Marjanović H., Perić S. 2006 - *Istraživanje uspijevanja provenijencija domaćih i stranih vrsta drveća s osvrtom na međunarodni pokus provenijencija obične bukve (Fagus sylvatica L.)*. Radovi. Izvanredno izdanje 9: 337-352.

Gustafsson A. 1950 - *Conifer Seed Plantations: Their Structure and Genetical Principles*. Proceedings of the III World Forestry Conres No 3, Helsinki: 128-138.

Isajev V. 1987 - *Oplemenjivanje omorike (Picea omorika /Panč/ Purkyne) na genetičko selekcionim osnovama*. Doktorska disertacije. Šumarski fakultet Beograd: 285-381.

Isajev V., Delić S., Mančić A., Šijačić M. 1995 - *Unapređenje osnovnih Šumskih kultura u funkciji razvoja Srbije*. Potencijali šuma i šumskih područja i njihov značaj za razvoj Srbije. Šumarski fakultet Beograd: 63-69.

Isajev V., Đukić M. 1990 - *Značaj provenijeničnih testova za vrednovanje genetsko-fiziološkog kvaliteta sastojina smrče na Kopaoniku*. Zbornik radova sa nučno-stručnog skupa "Priroda Kopaonika- zaštita i korišćenje: 223-235.

Isajev V., Šijačić M., Vilotić D. 1994 - *Varijabilnost makroskopskih i mikroskopskih karakteristika dvogodišnjih sadnica 26 familija polusrodnika planinskog javora (Acer heldreichii Orph.)*. Šumarstvo 3-4: 21-28.

Isajev V., Tucović A. 1992 - *Provenijenični test smrče na tri lokaliteta kod Ivanjice*. Izvođački projekat Beograd: 1-52.

Isajev V., Tucović A., Šijačić-Nikolić M. 1998 - *Očuvanje i unapređenje korišćenja genofonda endemoreliktnih vrsta četinara Srbije*. Zaštita prirode 50: 327-334.

Ivanković M. 2003 - *Varijabilnost nekih svojstava obične jele (Abies alba Mill.) u pokusu provenijencija "Brloško"*. Radovi Šumarski institut, Jastrebarsko, 38: 159-176.

Ivanković M., Bogdan S., Božič G. 2008 - *Varijabilnost visinskog rasta obične bukve (Fagus sylvatica L.) u testovima provenijencija u Hrvatskoj i Sloveniji*. Šumarski list 11-12: 529-541.

Ivanković M., Popović M., Bogdan S. 2011 - *Varijabilnost morfoloških svojstava žireva i visina sadnica hrasta lužnjaka (Quercus robur L.) iz sjemenskih sastojina u Hrvatskoj*. Šumarski list: 46-58.

Ivetić V. 2004 - *Uticaj staništa i provenijencija na razvoj juvenilnih kultura smrče (Picea abies /L./ Karst) na Goliji*. Magistarski rad. Šumarski fakultet: 1-131.

Jazbec A., Šegotić K., Ivanković M., Marjanović H., Perić S. 2007 - *Ranking of European beech provenances in Croatia using statistical analysis and analytical hierarchy process*. Forestry 80(2): 151-162.

Jovanović M. 1972 - *Proizvodnja šumskog selekcionisanog semena u semenskim plantažama*. Aktuelni problemi šumarstva, drvne industrije i hortikulture: 191-199.

Kajba, D., Gračan, J., Ivanković M., Bogdan S., Gradečki-Poštenjak M., Littvay T., Katičić I. 2006 - *Očuvanje genofonda šumskih vrsta drveća u Hrvatskoj*. Glasnik za šumske pokuse. 5; 235-249.

Kleinschmit J. 1970 - *Present knowledge in spruce provenance and species hybridiation potential*. IUFRO Norway spruce meeting, Bucharest, S 2.03.11- S 2.02.11: 187-201.

König A.O. 2005 - *Provenence research:evaluating the spatial pattern of genetic variation*. In: Conservation and Management of Forest Genetics Resources in Europe. Arbora Publishers, Zvolen: 275-290.

Lines R. 1967 - *Standardiyation of methods for provenence research and testing*. Proceedings 14th IUFRO Congr. 3: 672-718.

Mataruga M. 1997 - *Međuzavisnost osobina i razvoja sadnica crnog bora (Pinus nigra Arn.) u semenskoj plantaži na Jelovoj Gori*. Magistarski rad: 1-112.

Mataruga M., Isajev V. 1998 - *Mogućnost testiranja i očuvanja biodiverziteta crnog bora u specijalizovanim kulturama*. Zaštita prirode 50: 63-71

Nikolić N., Krstić B., Pajević S., Orlović S. 2006 - *Varijabilnost osobina lista kod različitih genotipova hrasta lužnjaka (Quercus robur L.)*. Zbornik Matice srpske za prirodne nauke 111: 95-105.

Nikolić N., Merkulov Lj., Krstić B., Orlović S. 2003 - *A comparative analysis of stomata and leaf trichome characteristics in Quercus robur L. genotypes*. Zbornik matice srpske za prirodne nauke 105: 51-59.

Nikolić N., Merkulov Lj., Pajević S., Krstić B. 2005 - *Variability of leaf anatomical characteristics in Pedunculate oak genotypes (Quercus robur L.)*. Gruev, B., Nikolova, M., Donev A. 2005 - Proceedings of the Balkan scientific conference of biology, Plovdiv, Bulgaria, 19-21, May 2005: 240–247.

Nikolić N.P., Merkulov L.S., Krstić B.Đ., Pajević S.P., Borišev M.K., Orlović S. 2010 - *Varijabilnost anatomskih osobina žira kod genotipova hrasta lužnjaka (Quercus robur L.)*. Zbornik Matice srpske za prirodne nauke 118: 47-58.

Nikolić N.P., Orlović S. 2002 - *Genotipska varijabilnost morfoloških osobina žira hrasta lužnjaka (Quercus robur L.)*. Zbornik Matice srpske za prirodne nauke 102: 53-58.

Orlović S., Radivojević S., Erdeši J., Obućina Z., Janjatović, G. 1999 - *Seed orchards of intra-species forms as a way for maintain and increasing genetic variability of Pedunculate oak in Yugoslavia*. International Conference "Recent advances on oak health in Europe", Warsaw, Book of abstracts 31.

Orlović S., Erdeši J., Radivojević S., Janjatović G. 2001 - *Semenske plantaže hrasta lužnjaka (Quercus robur L.) – osnov za dalje oplemenjivanje u ravnom Sremu*. Šumarstvo 1: 1-9.

Orlović S., Klašnja B., Galić Z., Pilipović A. 2002 - *Conservation of Pedunculate oak (Quercus robur L.) in Yugoslavia*. Proceedings DYGEN Conference «Dynamics and conservation of genetic diversity in forest ecosystems»: 210.

Perić S., Jazbec A., Medak J., Topić V., Ivanković M. 2006 - *Analysis of biomass of 16th Pedunculate Oak Provenances*. Periodicum Biologorum Vol. 108, No 6, 631-709.

Rajora, O.P., Mosseler A. 2001. - *Challenges and opportunities for conservation of forest genetic resources*. Euphytica 118 (2): 197-212.

Skrøppa T. 2005 - *Ex situ conservatin methods*. In: Conservation and Management of Forest Genetics Resources in Europe. Arbora Publishers, Zvolen: 567-583.

Stojnić S., Orlović S., Pilipović A., Kebert M., Šijačić-Nikolić M., Vilotić D. 2010 - *Variability of physiological parameters of different European beech provenances in international provenance trials in Serbia*. Acta Silvatica & Lignaria Hungarica 6: 135-142.

Stojnić S., Orlović S., Galić Z., Vasić V., Vilotić D., Knežević M., Šijačić-Nikolić M. 2012a - *Stanišne i klimatske karakteristike u provenijeničnim testovima bukve na Fruškoj gori i u Debelom Lugu*. Topola 189/190: 145-162.

Stojnić S., Orlović S., Pilipović A., Vilotić D., Šijačić-Nikolić M., Miljković D. 2012b - *Variation in leaf physiology among three provenances of European beech (Fagus sylvatica L.) in provenance trial in Serbia*. Genetika 44: 341-353.

Stojnić S. 2013a - *Varijabilnost anatomskih, fizioloških i morfoloških karakteristika različitih provenijencija bukve u Srbiji*. Doktorska disertacija, Univerzitet u Beogradu, Šumarski fakultet: 340p..

Stojnić S., Sass-Klaassen U., Orlović S., Matović B., Eilmann B. 2013b - *Plastic growth response of European beech provenances to dry site conditions*. IAWA Journal 34: 475-484.

Šijačić-Nikolić M., Milovanović J. 2010 - *Konzervacija i usmereno korišćenje šumskih genetičkih resursa*. Univerzitet u Beogradu Šumarski fakultet, Beograd, 200p..

Štajner D., Orlović S., Popović B., Kebert M., Stojnić S., Klašnja B. 2013 - *Chemical parameters of oxidative stress adaptability in beech*. Journal of Chemistry, Article ID 592695, doi:10.1155/2013/592695.

Tucović A., Isajev V. 1982 - *Uticaj različitih tipova oprašivanja na neka svojstva šišarica i semena omorike*. Glasnik Šumarskog fakulteta, Serija C Pejzažna arhitektura 59: 59-65.

Tucović A., Isajev V. 1983 - *Neka uporedna opažanja u kulturama omorike odgajenim na različitim staništima SR Srbije*. Glasnik Šumarskog fakulteta, Serija A. 60: 77-87.

Tucović A., Isajev V. 1991 - *Metapopulaciona strategija osnivanja generativnih semenskih plantaža drveća*. Zbornik radova Prošlost, sadašnjost i budućnost srpskog šumarstva kao činioca razvoja Srbije: 313-323.

Von Wuehlisch G., Ballian D., Bogdan S., Forstreuter M., Giannini R., Götz B., Ivankovic M., Orlovic S., Pilipovic A., Sijacic Nikolic M. 2010 - *Early results from provenance trials with European beech established 2007*. COST E52 "Evaluation of Beech Genetic Resources for Sustainable Forestry" Final Meeting. Book of abstracts. 4-6th May 2010, Burgos, Spain: 21.

Von Wuehlisch, G. 2004 - *Series of International Provenance Trials of European Beech*. Proceedings from the 7th International Beech Symposium IUFRO Research Group 1.10.00 "Improvement and Silviculture of Beech". 10-20 May 2004, Tehran, Iran: 135-144.

Structural attributes of stand overstory and light under the canopy

Alice Angelini[1*], Piermaria Corona[2], Francesco Chianucci[2], Luigi Portoghesi[1]

Abstract - This paper reviews the literature relating to the relationship between light availability in the understory and the main qualitative and quantitative attributes of stand overstory usually considered in forest management and planning (species composition, density, tree sizes, etc.) as well as their changes as consequences of harvesting. The paper is divided in two sections: the first one reviews studies which investigated the influence of species composition on understory light conditions; the second part examines research on the relationships among stand parameters determined from mensurational field data and the radiation on understory layer. The objective was to highlight which are the most significant stand traits and management features to build more practical models for predicting light regimes in any forest stand and, in more general terms, to support forest managers in planning and designing silvicultural treatments that retain structure in different way in order to meet different objectives.

Keywords - Understory light, structural attributes, overstory stand, forest management

Introduction

The recognition of forest as complex system among scientists and communities (Levin 1998, Kuuluvainen 2009, Ciancio and Nocentini 2011, Puettmann et al. 2013) has increasingly raised the necessity to develop new strategies for managing woodlands and make them more suited to face the challenges of global change (Franklin et al. 2002, Larsen and Nielsen 2007, Millar et al. 2007, Puettmann 2011, O'Hara and Ramage 2013, Wagner et al. 2014). Different approaches, new tools and decision criteria on analyzing forest stands and designing silvicultural systems are being developed and improved (O'Hara 1998, Koch and Skovsgaard 1999, Gamborg and Larsen 2004, Pommerening and Murphy 2004, Meitner et al. 2005, Puettmann et al. 2009, Geldenhuys 2010, Messier & Puettmann 2011, Bradford and Kastendick 2010). Most of proposals aim to get multi-aged, mixed forests with heterogeneous structure consisting of a fine-scale mosaic of cohorts of trees, with different species, size, age and development stage and temporal continuity of natural regeneration of trees. Single and group selection silvicultural systems with very variable retention of live and dead trees, emulating natural disturbance regimes, were proposed in order to modify overstory cover and create spatially differentiated microclimate conditions, particularly in terms of understory light availability. Light directly or indirectly affects other environmental parameters such as temperature, humidity, wind speed, soil condition, and can be even an effective indicator of the differences in stand structure across forests (Larcher 2003).

Foresters are aware that understory light availability plays a crucial role in driving forest dynamics, since it influences several aspects of plant regeneration and growth processes, such as seed germination, plant recruitment, early establishment of seedlings, young tree survival (Beaudet et al. 2011, Bartemucci et al. 2006). The benefits of managing light levels in the understory also include the control of shrub/herb layers growth either to suppress them as competitors or to promote their richness as source of biodiversity (Lieffers et al. 1999, McKanzie et al. 2000, Whigham et al. 2004, Royo and Carson 2006, Hart and Chen 2006, Gilliam et al. 2007, Moelder et al. 2008, Tinya et al. 2009).

The importance of light in forest ecology justifies the attention that researchers have devoted to it. A considerable work in reviewing knowledge on description and prediction of understory light was done by Lieffers et al. (1999). The paper synthesized much literature relating to light dynamics in northern and boreal forests, considering the factors affecting light transmission through the canopy, instruments and techniques for measurement and models for prediction of light in stands. Objective estimation of light transmission in different stand structures would be very useful to support the ap-

---

[1] University of Tuscia, Department for Innovation in Biological, Agro-food and Forest systems (DIBAF), Viterbo, Italy
[2] Consiglio per la ricerca in agricoltura e l'analisi dell'economia agraria, Forestry Research Centre (CRA-SEL), Arezzo, Italy
* corresponding author: aliceangelini@hotmail.it

plication of silvicultural treatments that aims to modify light conditions within stands by thinning (Chianucci and Cutini 2012, Drever and Lertzman 2001). However, the high cost of instrumentation and more time-consuming procedures required to estimate factors to use as model inputs (e.g., foliage inclination distribution, foliage clumping, canopy structure and stem mapping) often favoured the use of more readily available mensurational variables as independent variables.

Over the last two decades, several articles have been dedicated to the relationship between light availability in the understory and the main qualitative and quantitative attributes of stand overstory usually considered in forest management and planning (e.g., species composition, density, tree sizes, etc.) as well as their changes as consequences of harvesting (e.g., Canham et al. 1990, Chianucci and Cutini 2013, Lieffers et al. 1999, Thomas et al. 1999). We reviewed such literature to individuate the most significant stand traits and management features able to predict light regimes in any forest stand; such information would support forest managers in planning and designing silvicultural treatments that retain structure in different way in order to meet different objectives.

The paper is divided in two sections: the first one reviews studies which investigated the influence of species composition on understory light conditions; the second part examines the relationships among stand parameters determined from mensurational field data and the radiation on understory layer.

Forest composition and understory light

Silvicultural practices modify tree species composition, simultaneously modulating overstory canopy cover and therefore the availability of light under the canopy (Barbier et al. 2008, Chianucci and Cutini 2013).

Light transmittance also varies considerably among tree species, partly because their light demanding strategies (Montgomery and Chadzon 2001), so that the relative proportion of some categories of species (deciduous or coniferous, shade tolerant or intolerant) in mixed stands may explain, at least in part, the spatial and temporal variability of understory light (Hart and Chen 2006, Barbier et al. 2008). Light demanding species (both deciduous and coniferous) transmit more light than shade tolerant species; in terms of canopy attributes, these species generally exhibits lower canopy density, higher between-crowns clumping (canopy nonrandomness) and higher crown porosity (the fraction of gap within crown envelopes; Kucharik et al. 1999). Conversely, shade tolerant species can reach higher canopy density, less between-crowns clumping and lower crown porosity (Chianucci and Cutini 2013, Macfarlane et al. 2007), with resulting lower light transmittance (Canham et al. 1994, Messier et al. 1998, Messier et al. 1999, Beaudet et al. 2002, Coates et al. 2003, Le Francois et al. 2008).

Forest canopy structure and light transmittance in mixed-species stands are the results of complex interactions which may lead to denser canopy space filling and more complete light interception (Pretzsch and Schütze 2005). However, not all studies came to the same results: Drever and Lertzman (2003), in coastal Douglas-fir stands that varied in abundance and distribution of retained trees after partial cutting of different intensity, found that species composition was only weakly related to the amount of light in the understory. In that case the higher canopy openness than in intact forests dominated by different species (Canham et al. 1994, Hunter et al. 1999) highlighted that in managed forests other structural features affect light availability in the understory.

Among the canopy properties, spatial arrangement of branches and leaves, leaf angle distribution and leaf orientation, leaf size and other optical properties of leaves, play an important role in affecting overstory transmittance (Valladares and Pearcy 1999, Falster and Westoby 2003, Hardy et al. 2004, Gendron et al 2006, Barbier et al. 2008) as indicated by the Beer-Lambert's law (Equation 1, based on Nilson 1971):

$$P(\theta) = \exp\left(\frac{-G(\theta) \times L_t \times \Omega(\theta)}{\cos\theta}\right) \quad (1)$$

Where $P(\theta)$ is the radiation transmitted through the canopy, $G(\theta)$ is the foliage projection function, which is dependent on leaf angle distribution, L_t is the plant area index, including foliar and woody vegetation, (θ) is the foliage clumping index an $1/\cos(\theta)$ is the path length at zenith angle θ. The inversion of Beer-Lambert law is often used to extract many of these attributes (Chianucci et al. 2014b, Nilson 1999, Monsi and Saeki 2005, Pisek et al. 2013) from optical measurements of radiation.

Aussenac (2000) showed that the inclination angle of leaves with respect to canopy thickness, for *Fagus sylvatica* L. and *Quercus petraea* Liebl., follows Beer's law, and also that beech adapts better to excess and very low radiation than oak. This type of tropism can also be seen in conifers. Species exhibiting more horizontal leaf angle distribution intercept more light than species having more vertical distribution. Some studies (Oker-Blom and Kellomaki 1982, Pisek et al. 2013) have shown that broadleaf species at northern latitude exhibit a planophile leaf angle distribution (i.e., leaves have predominantly horizontal leaf angle distribution). Hikosaka and Hirose (1997) observed a greater capacity of species with planophile foliage orientation to shade out the species with vertical foliage orientation while simultaneously having a higher foliage tolerance as well. In plants of chaparral vegetation, Valladares and Pearcy (1999) highlighted the influence of leaf orientation on the heterogeneity of the light environments; upper, south-facing leaves intercepted greater daily light than leaves of any other orientation. For many coniferous species (ponderosa pine, Douglas fir and western hemlock), the distribution and arrangement of foliage on shade shoots can greatly increase light interception, and therefore photosynthesis in the lower canopy (Bond et al. 1999). Needle clustering and penumbral effects of small size leaf also affect light penetration, interception, and photosynthesis (Stenberg et al. 1999).

Variation of light resources in the understory environment might also be observed in relation to the leaf phenology due to the seasonality that differs among species (Gendron et al. 1998, Hart and Chen 2006). Before leaf expansion, and following leaf senescence, deciduous canopies have much higher light transmission than all other stand types (Ross et al. 1986). Even for that reason deciduous forests are considered to have a marked seasonal light variability than evergreen forests (Gendron et al. 2001, Yirdaw and Luukkanen 2004).

Komiyama et al. (2001a) reported that differential overstory leaf flushing patterns contributed to the formation of a patchy understory. Also Kato and Komiyama (2002) found that the heterogeneity of light conditions that occurred in a deciduous broad-leaved forest in late spring resulted from the different timing of leaf flushing by different tree species. In particular, heterogeneity is the main cause of the patchy distribution of understory plants. Effectively, direct spring sunlight penetrating should result in a positive correlation in terms of spatial distribution between late-flushing trees and understory plants (Komiyama et al. 2001a).

In general, we can sustain that species-specific attributes, such as crown structure, determine significant effect on the amount, quality and spatial variation of light transmittance (Yirdaw and Luukkanen 2004, Pretzsh et al. 2014) and consequently a simple but profound effect on forest succession (Canham et al. 1994, Canham et al. 1999). For example, crown depth (Canham et al. 1994, Beaudet et al. 2002, Beaudet et al. 2011, Ametzegui et al. 2012) and crown width (Canham et al. 1999), which were higher in shade tolerant species, influences the ratio of PAR to global radiation inside the canopy (Ross and Sulev 2000). Nevertheless, size and spacing of the crowns, or rather canopy openness, regardless of species, were of primary importance to the interspecific variation in openness of individual crowns, (Canham et al. 1999, Beaudet et al. 2002), revealing as a good predictor of the below-canopy transmitted diffuse and global solar radiation in old-growth and uneven-aged evergreen forest (Promis et al. 2009).

The crown structure of a tree is even more crucial in mixed stands where different species demonstrate their abilities to acclimate their structures in order to benefit of the resources more efficiently or obstruct the access of competitors to the same resources (Pretzsch 2009, Bayer and Pretzsch 2013). A morphological plasticity may results in crown and canopy structures in mixed stands which differ considerably from those observed in pure stands.

Effectively, in pure stands all individuals compete with similar behavior for the growing space and resources involving a homogenization of canopy structure with low canopy depth and size-asymmetric competition (Grams and Andersen 2007). Differently, in mixed stands the complementarity of species in terms of light ecology allow trees to have more canopy space to occupy without mechanical abrasion or penetration of neighboring crowns (Pretzsch 2014). However, the ability of trees to intercept light decreases with environmental stress (Waring and Schlesinger 1985). In general, light transmission is higher for species of Boreal forests other than for species in warmer and wetter temperate deciduous forests or conifer forests of the Northwestern America (Lieffers et al. 1999).

Mensurational attributes of stands and understory light

Understory light availability, frequently expressed as canopy openness (the proportion of the sky hemisphere not obscured by vegetation when viewed from a single point; Jenning et al. 1999), is a measure of great utility to foresters since it can be used to guide the level of canopy manipulation necessary for successful natural regeneration.

Understory light and its spatial distribution can be manipulated, at least in part, by designing and shaping harvesting according to the overstory structure of a forest stand (Battaglia et al. 2002, Beaudet et al. 2011). Therefore, knowing the interplay between stand structure and light is fundamental for managing forests. An accurate description of allometric functions and their relationships with radiance would provide foresters precious information for silvicultural decisions. Among stand structural attributes determined from readily available field data, those describing stand density, such as sum of DBH, basal area and number of trees, are usually the most considered in similar studies. For example. Comeau et al. (2001) observed that in white spruce-aspen dominated boreal mixed stands with high initial tree densities, the decline in understory light levels is likely to occur more rapidly, resulting in the potential for substantial reductions in growth and survival of understory spruce due to competition for light and physical damage to spruce as a consequence of aspen mortality by self-thinning. At another level, Drever and Letzerman (2003) found a significant correlation in a coastal Douglas fir forest in British Columbia between light transmittance and stem density, volume of retained trees, summed DBH and summed height. However, the predictive capacity of these variables was much better for high light levels (> 50 % of full sun) than for low levels of light (< 20 % of full sun).

Basal area is frequently used as independent variable to explain light transmittance through the canopy (Nilson et al. 1999), although the radiative transfer may differ between young and old stands and the possible difference of overstory structure and site conditions should be considered (Comeau et al. 2001). In mixed-species forests plot basal area should be not enough informative and separate coefficients should be developed for each species, at least the dominant species: this was the case of mixed aspen-conifer forests in British Columbia (Comeau et al. 2006) where basal area of deciduous species was significantly related to understory light, unlike conifer basal area. This contrasted with results from birch-conifer stands in the same areas, where the inclusion of conifer basal area improves the relationship with light. In another study, Sonohat et al. (2004) found a negative exponential relationship between light transmittance and stand basal area in even-aged stands of Douglas fir, Norway spruce, larch and Scots pine, which explained between 56% and 80% of transmittance variation according to the species, and 82% for all species pooled data.

Such relationship between basal area and canopy transmittance was often explored in relation to silvicultural practices. In the case of Sitka spruce thinned stands, studies have individuated a basal area < 30 m^2/ha to provide the minimum light requirements, i.e. 15% of incident light, for the growth of Sitka spruce seedlings (Hale 2001, Hale 2003, Page et al. 2001, Malcom et al. 2001). However, some authors (Beaudet et al. 2011, Battaglia et al. 2002, Sprugel et al., 2009), showed that harvesting in a stand does not necessarily increase light transmission proportionally to the reduction in basal area. In effect, the spatial arrangement of the residual trees (and hence the spatial pattern of harvest) also plays a very important role. Battaglia et al. (2002) demonstrated that increasing the aggregation of residual basal area, not only increases the mean stand level understory light availability but also increases the variation of light resulting in more heterogeneous understory light environments.

In old growth and second growth forests in lowland Costa Rica, Montgomery and Chadzon (2001) did not find strong relationships between measures of forest structure and light availability, although the strength of these relationships differed between forest types. In both the studied forests, understory light availability at 0.75 m decreased with increased sapling and shrub density, but was not significantly influenced by local tree density or basal area. Similar trends were found in an old-growth and uneven-aged forest of *Nothofagus betuloides* (Promis et al. 2009). However, by combining basal area, crown projection, crown volume, and stand volume, it was possible to explain a large amount of the variability of the below-canopy transmitted, diffuse and global radiation.

A study carried out by Valladares et al. (2006) in the holm oak (*Quercus ilex*) woodlands of the Western Mediterranean basin, characterized by low mean canopy height (2.4 m), high stem density (14,500 stems ha^{-1}) and intermediate basal area, showed that canopy height exhibited a more significant correlation with understory light (particularly with indirect light), than stem density and basal area although only in the tree-dominated zone of the plot. However, since the potential of canopy height as a predictor of understory light was low due to the large fraction of unexplained variance, the incorporation of other canopy features (e.g. leaf angle distribution, leaf and branch clustering) would likely increase significantly the accuracy of the estimation of understory light based on canopy structure.

Results from a study by Heithecker and Halpern (2007) suggested that levels of light at the forest floor within aggregate retained trees can be surprisingly similar to those inside the forest; the aggregates significantly reduced Photosynthetic Photon Flux Density PPFD) in the adjacent harvested area to distances of 10-30 m.

Therefore, it is evident that spatial aggregation or rather the spatial distribution of stem density for retained trees strongly regulates the abundance and spatial variation of light in the understory (Coates et al. 2003). Changes concern the quantity and quality of light, as well as its directionality, so that more of the forest floor receives direct solar radiation and sunflecks become longer and more intense (Lieffers et al. 1999, Gendron et al. 2001).

Conclusions

The many studies concerning the relationships between transmittance and structural attributes in forest stands carried out over the last two decades confirmed the great interest in predicting understory light conditions by using attributes readily available from field data. Different bioclimatic zones (boreal vs tropical), stands structure (plantation vs natural, even-aged vs uneven-aged, young vs old-growth), species composition (pure vs mixed), and silvicultural treatments (clearcutting vs partial cutting) were taken into account in these studies. However, most of the research was carried out in boreal forests, likely because light was considered one of the most critical factors for successional dynamics in this environment.

On the whole, the results of the examined studies highlighted that different traits of forest overstory affect light intensity in the understory, even more in heterogeneous stands with continuous canopy cover. Composition, density and structure of overstory are the characteristics mainly correlated to light transmittance (Fig. 1).

The weight of each of them seems to depend on the degree of complexity of the stand. In evenaged, unthinned and monospecific stands with homogeneous canopy covers and regular spatial distribution of trees, understory light conditions much depend on species specific traits such as shade-tolerance, which in turn is strictly linked with crown properties (depth, width, leaf angle distribution, etc.). In regular mixed forests, tree composition controls the amount of radiance under the canopy and the spatial and temporal distribution of light especially if evergreen and deciduous species or deciduous species with different phenology are a significant part of the mixture. In managed forests, canopy openness can be manipulated by silvicultural practices changing stand density attributes; basal area is amongst the most important in predicting light as long as understory radiation fall above 20% and especially

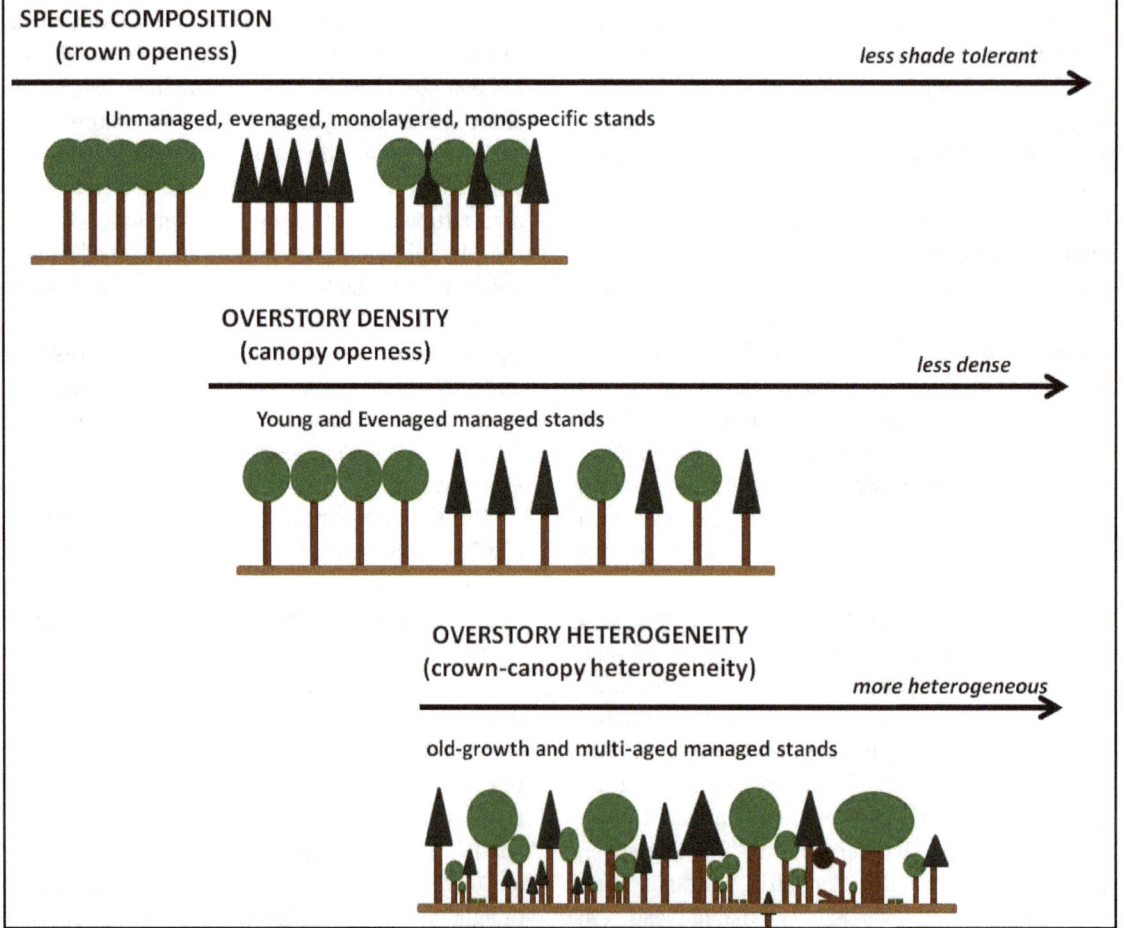

Figure 1: Diagram showing the three stand overstory characteristics affecting understory light.
When the stand profile is closed, simple, homogeneous the passage of light through the canopy depends on the characteristics of the crowns, then on tree composition. When silvicultural practices open the crown cover, composition being the same, stand density comes into play too. In forests with heterogeneous structure the light in the understorey is the result of the variability in the arrangement of stems in space and their size, which is added to the composition and density.
The weight of tree composition is higher in populations of tolerant species, that of density in the more open stands, that of structure in the very heterogeneous forest.
The three profiles are staggered to highlight that the three factors add up and interact with increasing diversification of the profile.

in young stands. Harvesting does not necessarily increase light transmission proportionally to the reduction in basal area and spatial arrangement of the residual trees plays a very important role (Beaudet et al. 2011, Battaglia et al. 2002, Sprugel et al. 2009).

For low average light levels typical of uneven-aged and old-growth forests, horizontal and vertical stand structure attributes need to be considered for increasing the accuracy of prediction. In such conditions, light transmission through canopy and the occurrence of patchy or homogeneous understory is controlled by the complex interplay of overstory composition, density and structure.

In conclusion, basal area can be viewed as the preferable light predictor for managing young stands with homogeneous structure. The experiences suggested considering additional parameters descriptive of tree size, such as DBH, height and volume, in order to increase the accuracy of light predictions in case of older stand or in presence of a layer of suppressed trees.

A different and more complex task is providing a significant estimate of radiation below canopy in stands characterized by heterogeneous vertical and horizontal structures. The few studies carried out until now didn't provide a clear overview of appropriate attributes for an accurate prediction of understory light in these types of forest structure. Therefore, considering the increasing importance of creating and maintaining stand structure heterogeneity through silvicultural treatments in order to enhance the resilience of forest ecosystems facing the global change, more research would be necessary to deepen this topic.

The continuous development of technologies is increasingly allowing their access to researchers and forest managers with relative low costs. The use of proximal and remote sensing technologies (Aschoff et al. 2004, Chianucci et al. 2014a, Danson et al. 2007, Maas et al. 2008) could represent in the future a valid solution for improving this type of studies and integrating field data with a more detailed information of canopy structure.

Acknowledgments

The authors want to thank the anonymous reviewers for their helpful suggestions.

References

Ameztegui A., Coll L., Benavides R., Valladares F., Paquette A. 2012 - *Understory light predictions in mixed conifer mountain forests: Role of aspect-induced variation in crown geometry and openness.* Forest Ecology and Management 276: 52-61. doi:10.1016/j.foreco.2012.03.021

Aschoff T., Thies M., Spiecker H. 2004 - *Describing forest stands using terrestrial laser-scanning.* In: Conference Proceedings ISPRS Conference. ISPRS International Archives of Photogrammetry, Remote Sensing and Spatial Information Sciences, Istanbul, Turkey, 12–23 July, vol. XXXV, Part B: 237–241.

Astrup R., Larson B. C. 2006 - *Regional variability of species-specific crown openness for aspen and spruce in western boreal Canada.* Forest Ecology and Management 228: 241-250. doi:10.1016/j.foreco.2006.02.048

Aussenac G. 2000 - *Interactions between forest stands and microclimate: Ecophysiological aspects and consequences for silviculture.* Annals of Forest Science 57: 287–301. doi: 10.1051/forest:2000119

Barbier S., Gosselin F., Balandier P. 2008 - *Influence of tree species on understory vegetation diversity and mechanisms involved-A critical review for temperate and boreal forests.* Forest Ecology and Management 254: 1–15. doi:10.1016/j.foreco.2007.09.038

Battaglia M.A., Mou P., Palik B., Mitchell R.J. 2002 - *The effect of spatially variable overstory on the understory light environment of an open-canopied longleaf pine forest.* Canadian Journal of Forest Research 32 (11): 1984-1991. doi:10.1139/x02-087

Bayer D., Seifert S., Pretzsch H. 2013 - *Structural crown properties of Norway spruce (Picea abies [L.] Karst.) and European beech (Fagus sylvatica [L.]) in mixed versus pure stands revealed by terrestrial laser scanning.* Trees 27: 1035-1047. doi: 10.1007/s00468-013-0854-4

Beaudet M., Messier C., Canham C.D. 2002 - *Predictions of understorey light conditions in northern hardwood forests following parameterization, sensitivity analysis, and tests of the SORTIE light model.* Forest Ecology and Management 165: 235-248.

Beaudet M., Harvey B. D., Messier C., Coates K. D., Poulin J., Kneeshaw D. D., Brais S.,

Bergeron Y. 2011 - *Managing understory light conditions in boreal mixedwoods through variation in the intensity and spatial pattern of harvest: A modelling approach.* Forest ecology and Management 261: 84-94. doi: 10.1016/j.foreco.2010.09.033

Bond B. J., Farnsworth B. T., Coulombe R. A., Winner W. E. 1999 - *Foliage physiology and biochemistry in response to light gradients in conifers with varying shade tolerance.* Oecologia 120: 183-192. doi: 10.1007/s004420050847

Bradford J. B., Kastendick D. N. 2010 - *Age-related patterns of forest complexity and carbon storage in pine and aspen-birch ecosystems of northern Minnesota, USA.* Canadian Journal of Forest Research 40 (3): 401-409. doi: 1 0.1 1 39/X 10-002

Canham C. D., Denslow J. S., Platt W. J., Runkle J. R., Spies T. A., White P. S. 1990 - *Light regimes beneath closed canopies and tree-fall gaps in temperate and tropical forests.* Canadian Journal of Forest Research 205: 620-631.

Canham C. D., Finzi A. C., Pacala S. W., Burbank D. H. 1994 - *Causes and consequences of resource heterogeneity in forest: interspecific variation in light trasmission by canopy trees*. Canadian Journal of Forest Research 24 (2): 337-349. doi: 10.1139/x94-046

Canhaman C., Coates K. D., Bartemucci P., Quaglia S. 1999 - *Measurement and modeling of spatially explicit variation in light transmission through interior cedar-hemlock forests of British Columbia*. Canadian Journal of Forest Research 29 (11): 1775-1783. doi: 10.1139/x99-151

Chianucci F., Cutini A. 2012 - *Digital hemispherical photography for estimating forest canopy properties: current controversies and opportunities*. iForest – Biogeosciences and Forestry 5: 290-295.

Chianucci F., Cutini A. 2013 - *Estimation of canopy properties in deciduous forests with digital hemispherical and cover photography*. Agricultural and Forest Meteorology 168: 130-1393.

Chianucci A., Cutini A., Corona P., Puletti N. 2014a - *Estimation of leaf area index in understory deciduous trees using digital photography*. Agricultural and Forest Meteorology 198: 259-264

Chianucci F., Macfarlane C., Pisek J., Cutini A., Casa R. 2014b - *Estimation of foliage clumping from the LAI-2000 Plant Canopy Analyzer: effect of view caps*. Trees (early view) doi. 10.1007/s00468-014-1115-x

Ciancio O. 2011 - *Systemic silviculture: philosophical, epistemological and methodological aspects*. L'Italia Forestale e Montana 66 (3): 181-190. doi: 10.4129/ifm.2011.3.01

Coates K. D., Canhaman C. D., Beaudet M., Sachs, D. L. Messier C. 2003 - *Use of a spatially explicit individual-tree model (SORTIE/BC) to explore the implications of patchiness in structurally complex forests*. Forest Ecology and Management 186: 297-310. doi:10.1016/S0378-1127(03)00301-3

Comeau P. G. 2001- *Relationships between stand parameters and understorey light in boreal aspen stands*. BC Journal of Ecosystems and Management 2: 1-8.

Comeau P. G., Heineman J. L. 2003 - *Predicting understory light microclimate from stand parameters in young paper birch (Betula papyrifera Marsh.) stands*. Forest Ecology and Management 180: 303-315. doi: 10.1016/S0378-1127(02)00581-9

Comeau P., Heineman J., Newsome T. 2006 - *Evaluation of relationships between understory light and aspen basal area in the British Columbia central interior*. Forest Ecology and Management 226: 80-87. doi:10.1016/j.foreco.2005.12.060

Danson F. M., Hetherington D., Morsdorf F., Koetz B., Allgöwer B. 2007 - *Forest Canopy Gap Fraction From Terrestrial Laser Scanning*. Geoscience and Remote Sensing Letters, IEEE 4: 157-160. doi: 10.1109/LGRS.2006.887064

Drever C. R., Lertzman K. P. 2001 - *Light-growth responses of coastal Douglas-fir and western redcedar saplings under different regimes of soil moisture and nutrients*. Canadian Journal of Forest Research 31 (12): 2124-2133. doi: 10.1139/cjfr-31-12-2124

Drever C. R., Lertzman K. P. 2003 - *Effects of a wide gradient of retained tree structure on understory light in coastal Douglas-fir forests*. Canadian Journal of Forest Research 33 (1): 137–146. doi: 10.1139/X02-167

Falster D. S., Westoby M. 2003 - *Leaf size and angle vary widely across species: what consequences for light interception?* New Phytologist 158: 509-525. doi: 10.1046/j.1469-8137.2003.00765.x

Franklin J. F., Spies T. A., Van Pelt R., Carey A. B., Thornburgh D. A., Berg D. R., Lindenmayer D. B., Harmon M. E., Keeton W. S., Shaw D. C., Ken B., Jiquan C. 2002 - *Disturbances and structural development of natural forest ecosystems with silvicultural implications, using Douglas-fir forests as an example*. Forest Ecology and Management 155: 399-423. doi:10.1016/S0378-1127(01)00575-8

Gamborg C., Larsen J. B. 2003 - *'Back to nature' - a sustainable future for forestry?* Forest Ecology and Management 179: 559-571. doi: 10.1016/S0378-1127(02)00553-4

Geldenhuys C. J. 2010 - *Managing forest complexity through application of disturbance- recovery knowledge in development of silvicultural systems and ecological rehabilitation in natural forest systems in Africa*. Journal of Forest Research 15: 3-13. doi: 10.1007/s10310-009-0159-z

Gendron F., Messier C., Comeau P. G. 1998 - *Comparison of various methods for estimating the mean growing season percent photosynthetic photon flux density in forests*. Agricultural and Forest Meteorology 92: 55-70. doi:10.1016/S0168-1923(98)00082-3

Gendron F., Messier C., Comeau P.G. 2001 - *Temporal variations in the understorey photosynthetic photon flux density of a deciduous stand: the effects of canopy development, solar elevation, and sky conditions*. Agricultural and Forest Meteorology 106: 23-40. doi:10.1016/S0168-1923(00)00174-X

Gendron F., Messier C., Lo E., Comeau P. G. 2006 - *The angular distribution of diffuse photosynthetically active radiation under different sky conditions in the open and within deciduous and conifer forest stands of Quebec and British Columbia, Canada*. Annals of Forest Science 63: 43-53. doi: 10.1051/forest:2005096

Gilliam F. S. 2007 - *The ecological significance of the herbaceous layer in temperate forest ecosystems*. BioScience 57 (10): 845-858. doi: 10.1641/B571007

Grams T.E., Andersen C.P. 2007 - *Competition for resources in trees: physiological versus morphological plasticity*. Progress in Botany 69: 356-381. doi: 10.1007/978-3-540-36832-8_16

Hale S.E. 2001 - *Light regime beneath Sitka spruce plantations in northern Britain: preliminary results*. Forest Ecology and Management 151: 61–66. doi:10.1016/S0378-1127(00)00696-4

Hale S. E. 2003 - *The effect of thinning intensity on the below-canopy light environment in a Sitka spruce plantation*. Forest Ecology and Management 179: 341-349. doi:10.1016/S0378-1127(02)00540-6

Hardy J.P., Melloh R., Koenig G., Marks D., Winstral A., Pomeroy J.W., Link T. 2004 - *Solar radiation transmission through conifer canopies* - Agricultural and Forest Meteorology 126: 257-270. doi:10.1016/j.agrformet.2004.06.012

Hart S. A., Chen H. Y. H. 2006 - *Understory vegetation dynamics of North American boreal forests*. Critical Reviews in Plant Sciences 25: 381-397. doi: 10.1080/07352680600819286

Heithecker T. D., Halpern C. B. 2007 - *Edge-related gradients in microclimate in forest aggregates following structural retention harvests in western Washington*. Forest Ecology and Management 248: 163-173. doi:10.1016/j.foreco.2007.05.003

Hikosaka K., Hirose T. 1997 - *Leaf angle as a strategy for light competition: optimal and evolutionarily stable light-extinction coefficient within a leaf canopy*. Ecoscience 4: 501-507.

Hunter J.C., Parker V.T., Barbour M.G. 1999 - *Understory light and gap dynamics in an old-growth forested watershed in coastal California*. Madrono 46: 1-6.

Jennings S.B., Brown N.D., Sheil D. 1999 - *Assessing forest canopies and understorey illumination: canopy closure, canopy cover and other measures*. Forestry 72 (1): 59–74. doi: 10.1093/forestry/72.1.59

Kato S., Komiyama A. 2002 - *Spatial and seasonal heterogeneity in understory light conditions caused by differential leaf flushing of deciduous overstory trees*. Ecological Research 17: 687-693. doi: 10.1046/j.1440-1703.2002.00529.x

Koch N. E., Skovsgaard J.P 1999 - *Sustainable management of planted forests: some comparisons between Central Europe and the United States*. New Forests 17: 11–22. doi: 10.1023/A:1006520809425

Komiyama A., Kato S., Teranishi M. 2001a - *Differential overstory leaf flushing contributes to the formation of patchy understory*. Journal of Forest Research 6: 163–171. doi: 10.1007/BF02767088

Kuuluvainen T. 2009 - *Forest Management and Biodiversity Conservation Based on Natural Ecosystem Dynamics in Northern Europe: The Complexity Challenge*. A Journal of the Human Environment 38 (6): 309-315. doi: http://dx.doi.org/10.1579/08-A-490.1

Larcher W. 2003 - *Physiological plant ecology – Ecophisiology and stress physiology of functional groups*. 4th Edition. Springer, 513 p.

Larsen J.B., Nielsen A.B. 2007 - *Nature-based forest management - Where are we going? Elaborating forest development types in and with practice*. Forest Ecology and Management 238: 107–117. doi:10.1016/j.foreco.2006.09.087

Lefrançois M.L., Beaudet M., Messier C. 2008 - *Crown openness as influenced by tree and site characteristics for yellow birch, sugar maple, and eastern hemlock*. Canadian Journal of Forest Research 38 (3): 488-497. doi :10.1139/X07-177

Levin S. A. 1998 - *Ecosystems and the biosphere as complex adaptive systems*. Ecosystems 1: 431-436. doi: 10.1007/s100219900037

Lieffers V.J., Messier C., Stadt K.J., Gendron F., Comeau P.G. 1999 - *Predicting and managing light in the understory of boreal forests*. Canadian Journal of Forest Research 29 (6): 796-811. doi: 10.1139/x98-165

Maas H.G., Bienert A., Scheller S. Keane E. 2008 - *Automatic forest inventory parameter determination from terrestrial laser scanner data*. International Journal of Remote Sensing 29: 1579-1593. doi: 10.1080/01431160701736406

Macfarlane C., Hoffman M., Eamus D., Kerp N., Higginson S., McMurtrie R., Adams M.A. 2007 - *Estimation of leaf area index in eucalypt forest using digital photography*. Agricultural and Forest Meteorology 143: 176–188.

Malcom D. C., Mason W. L., Clarke G. C. 2001 - *The transformation of conifer forests in Britain-regeneration, gap size and silvicultural systems* - Forest Ecology and Management 151: 7-23. doi:10.1016/S0378-1127(00)00692-7

McKenzie D., Halpern C. B., Nelson C. R. 2000 - *Overstory influences on herb and shrub communities in mature forests of western Washington, U.S.A.* Canadian Journal of Forest Research 30 (10): 1655-1666. doi: 10.1139/x00-091

Meitner M. J., Sheppard S. R.J., Cavens D., Gandy R., Picard P., Harshaw H., Harrison D. 2005 - *The multiple roles of environmental data visualization in evaluating alternative forest management strategies*. Computers and Electronics in Agriculture 49: 192-205. doi:10.1016/j.compag.2005.03.002

Messier C., Parent S., Bergeron Y. 1998 - *Effects of overstory and understory vegetation on the understory light environment in mixed boreal forests*. Journal of Vegetation Science 9: 511-520. doi: 10.2307/3237266

Messier C., Doucet R., Ruel J.C., Claveau Y., Kelly C., Lechowicz, M.J. 1999 - *Functional ecology of advance regeneration in relation to light in boreal forests*. Canadian Journal of Forest Research 29 (6): 812-823. doi: 10.1139/x99-070

Messier C., Puettmann K. 2011 - *Forest as complex adaptive systems: implications for forest management and modelling*. Italian Journal of Forest and Mountain Environments 66 (3): 249-258. doi: 10.4129/ifm.2011.3.11

Millar C. I., Stephenson N. L., Stephens S. L. 2007 - *Climate change and forests of the future: managing in the face of uncertainty*. Ecological Applications 17: 2145-2151. http://dx.doi.org/10.1890/06-1715.1

Molder A., Bernhardt-Romermann M., Schmidt W. 2008 - *Herb-layer diversity in deciduous forests: raised by tree richness or beaten by beech?* Forest Ecology and Management 256: 272-281. doi:10.1016/j.foreco.2008.04.012

Monsi M., Saeki T. 2005 - *On the factor light in plant communities and its importance for matter production*. Annals of Botany 95: 549–567. doi:10.1093/aob/mci052

Montgomery R. A., Chadzon R. L. 2001 - *Forest structure, canopy architecture, and light transmittance in tropical wet forests*. Ecology 82 (10): 2707-2718.

Nilson T. 1999 - *Inversion of gap frequency data in forest stands*. Agricoltural and Forest Meteorology 98-99: 437-448. doi:10.1016/S0168-1923(99)00114-8

Oker-Blom, P., Kellomäki S. 1982 - *Theoretical computations on the role of crown shape in the absorption of light by forest trees*. Mathematical Biosciences 59: 291-311.

O'Hara Kevin L. 1998 - *Silviculture for structure diversity – a new look at multiaged systems*. Journal of Forestry 96: 4-10.

O'Hara Kevin L., Ramage Benjamin S. 2013 - *Silviculture in an uncertain world: utilizing multi-aged management systems to integrate disturbance*. Forestry 86 (4): 401-410. doi: 10.1093/forestry/cpt012

Page L.M., Cameron A.D., Clarke G.C. 2001- *Influence of overstorey basal area on density and growth of advance regeneration of Sitka spruce in variably thinned stands*. Forest Ecology and Management 151: 25–35. doi:10.1016/S0378-1127(00)00693-9

Pisek J., Sonnentag O., Richardson A. D., Mõttus M. 2013 - *Is the spherical leaf inclination angle distribution a valid assumption for temperate and boreal broadleaf tree species?* Agricultural and Forest Meteorology 169: 186–194. doi:10.1016/j.agrformet.2012.10.011

Pommerening A., Murphy S.T. 2004 - *A review of the history, definitions and methods of continuous cover forestry with special attention to afforestation and restocking*. Forestry 77 (1): 27-44. doi: 10.1093/forestry/77.1.27

Pretzsch H., Schütze G. 2009 - *Transgressive overyielding in mixed compared with pure stands of Norway spruce and European beech in Central Europe: Evidence on stand level and explanation on individual tree level*. European Journal of Forest Research 128: 183-204. doi: 10.1007/s10342-008-0215-9

Pretzsch H. 2014 - *Canopy space filling and tree crown morphology in mixed-species stands compared with monocultures*. Forest Ecology and Management 327: 251-264. doi:10.1016/j.foreco.2014.04.027

Pretzsch H., Schütze G. 2005 - *Crown allometry and growing space efficiency of Norway spruce (Picea abies [L.] Karst.) and European beech (Fagus sylvatica L.) in pure and mixed stands*. Plant Biology 7 (6): 628-639.

Promis A., Schindler D., Reif A., Cruz G. 2009 - *Solar radiation transmission in and around canopy gaps in an uneven-aged Nothofagus betuloides forest*. International Journal of Biometeorology 53: 355-367. doi:10.1007/s00484-009-0222-7

Puettmann K. J., Coates K. D., Messier C. 2009 - *A critique of silviculture: managing for complexity*. Island Press, Washington, D.C., USA, 189 p.

Puettmann K. 2011 - *Silvicultural challenges and options in the context of global change: "simple" fixes and opportunities for new management approaches*. Journal of Forestry 109 (6): 321-331.

Puettmann K. J., Messier C., Coates K. D. 2013 - *Managing forests as complex adaptive systems: introductory concepts and applications*. In: Messier C., Puettmann K. J., Coates K. D.: "Managing forests as complex adaptive systems. Building resilience to the challenge of global change". Routledge, Abingdon: 3-16

Ross M. S., Flanagan L. B., La Roi G. H. 1986 - *Seasonal and successional changes in light quality and quantity in the understory of boreal forest ecosystems*. Canadian Journal of Botany 64 (11): 2792-2799. doi:10.1139/b86-373

Ross J., Sulev M. 2000 - *Sources of errors in measurements of PAR*. Agricultural and Forest Meteorology 100: 103-125. doi: 10.1016/S0168-1923(99)00144-6

Royo A. A., Carson W. P. 2006 - *On the formation of dense understory layers in forests worldwide: consequences and implications for forest dynamics, biodiversity, and succession*. Canadian Journal of Forest Research 36 (6): 1345-1362. doi: 10.1139/x06-025

Sonohat G., Balandier P., Ruchaud F. 2004 - *Predicting solar radiation transmittance in the understory of even-aged coniferous stands in temperate forests*. Annals of Forest Science 61 (7): 629-641. doi: 10.1051/forest:2004061

Sprugel D.G., Rascher K.G., Gersonde R., Dovciak M., Lutz J.A., Halpern C.B. 2009 - *Spatially explicit modeling of overstory manipulations in young forests: effects on stand structure and light*. Ecological Modelling 220: 3565-3575. doi:10.1016/j.ecolmodel.2009.07.029

Stenberg P., Kangas T., Smolander H., Linder S. 1999 - *Shoot structure, canopy openness, and light interception in Norway spruce*. Plant, Cell & Environment 22: 1133-1142. doi: 10.1046/j.1365-3040.1999.00484.x.

Thomas S. C., Halpern C. B., Falk D. A., Liguori D. A., Austin K. A. 1999 - *Plant diversity in managed forests: understory responses to thinning and fertilization*. Ecological Applications 9: 864-879.

Tinya F., Márialigeti S. , Király I. , Németh B. , Ódor P. 2009 - *The effect of light conditions on herbs, bryophytes and seedlings of temperate mixed forests in Orség, Western Hungary*. Plant Ecology 204: 69-81. doi: 10.1007/s11258-008-9566-z

Valladares F., Pearcy R.W. 1999 - *The geometry of light interception by shoots of Heteromeles arbutifolia: morphological and physiological consequences for individual leaves*. Oecologia 121:171-182. doi: 10.1007/s004420050919

Valladares F, Guzman B. 2006 - *Canopy structure and spatial heterogeneity of understory light in an abandoned Holm oak woodland*. Annals of Forest Science 63 (7): 749–761. doi:10.1051/forest:2006056

Wagner S., Nocentini S., Huth F., Hoogstra-Klein M. 2014 - *Forest management approaches for coping with the Uncertainty of Climate Change: trade-offs in service provisioning and adaptability*. Ecology and Society 19 (1): 32. http://dx.doi.org/10.5751/ES-06213-190132

Waring R. H., Schlesinger W. H. 1985 - *Forest ecosystems. Concepts and management*. Academic Press London, 340 p.

Whigham D. F. 2004 - *Ecology of woodland herbs in temperate deciduous forests*. Annual Review of Ecology, Evolution and Systematics 35: 583-621. doi: 10.1139/x00-091

Yirdaw E., Luukkanen O. 2004 - *Photosynthetically active radiation transmittance of forest plantation canopies in the Ethiopian highlands*. Forest Ecology and Management 188: 17-24. doi:10.1016/j.foreco.2003.07.024

Forest biotechnology advances to support global bioeconomy

Antoine Harfouche[1*], Sacha Khoury[1], Francesco Fabbrini[1], Giuseppe Scarascia Mugnozza[1]

Abstract - The world is shifting to an innovation economy and forest biotechnology can play a major role in the bio-economy by providing farmers, producers, and consumers with tools that can better advance this transition. First-generation or conventional biofuels are primarily produced from food crops and are therefore limited in their ability to meet challenges for petroleum-product substitution and climate change mitigation, and to overcome the food-versus-fuel dilemma. In the longer term, forest lignocellulosic biomass will provide a unique renewable resource for large-scale production of bioenergy, biofuels and bio-products. These second-generation or advanced biofuels and bio-products have also the potential to avoid many of the issues facing the first-generation biofuels, particularly the competition concerning land and water used for food production. To expand the range of natural biological resources the rapidly evolving tools of biotechnology can ameliorate the conversion process, lower the conversion costs and also enhance target yield of forest biomass feedstock and the product of interest. Therefore, linking forest biotechnology with industrial biotechnology presents a promising approach to convert woody lignocellulosic biomass into biofuels and bio-products. Major advances and applications of forest biotechnology that are being achieved to competitively position forest biomass feedstocks with corn and other food crops are outlined. Finally, recommendations for future work are discussed.

Keywords - Forest biotechnology, bio-economy, biofuels, bio-materials

Introduction

Bioeconomy encompasses all economic activities that are fueled by research and innovation in the biological sciences. It is a large and rapidly growing segment of the global economy that provides significant public benefit (OECD 2009). Also, it has emerged as a worldwide priority because of its tremendous potential for growth and the numerous societal and environmental benefits it offers. Bioeconomy can help to reduce the world dependence on fossil fuels, address key environmental challenges, transform manufacturing processes, and increase the productivity and scope of the agricultural and silvicultural sectors while creating new jobs and innovative industries.

Food, feed, energy, and industrial products demand will increase concomitantly with human population and climate change. This production therefore needs to be high enough and, at the same time, minimize or prevent harm to the environment. Yet, this balance cannot be achieved with current strategies. Moreover, food and energy crises have often occurred in the past. In recent years, the rapidly increasing demand for food (i.e., for human populations and livestock) along with biofuels has led to food price volatility (Battisti et al. 2009).

Based on recent findings, new avenues for forest breeding which take into account the integration of modern genetic and genomic techniques with conventional breeding will expedite forest tree improvement (Harfouche et al. 2012a and 2014). In this context, we argue that forest breeders and land owners have the opportunity to make use of modern biotechnologies in an innovative ecologically sound silviculture.

In addition, as we are advancing towards a global economy that is attempting to find ways to break its addiction to fossil fuels and develop an economy based on renewable biological materials and energy sources, forest biomass is being actively explored as a possible substitute in many parts of the world.

Forest trees constitute about 82% of the continental biomass (Roy et al. 2001) and harbor more than 50% of the terrestrial biodiversity (Neale and Kremer 2011). They also help to mitigate climate change, and provide a wide range of products that meet human needs, including wood, biomass, paper, fuel, and bio-materials (Harfouche et al. 2012a). However, forest tress grown in the field are usually exposed to environmental stress. Current and predicted climatic conditions, such as prolonged drought, increased soil and water salinity, and low-temperature episodes pose a serious danger to forest productivity, affecting tree growth and survival (Harfouche et al. 2014).

Conventional breeding has been successful at ameliorating several phenotypic characteristics that

---

[1] Department for Innovation in Biological, Agro-food and Forest systems (DIBAF), University of Tuscia, Italy
* corresponding author: aharfouche@unitus.it

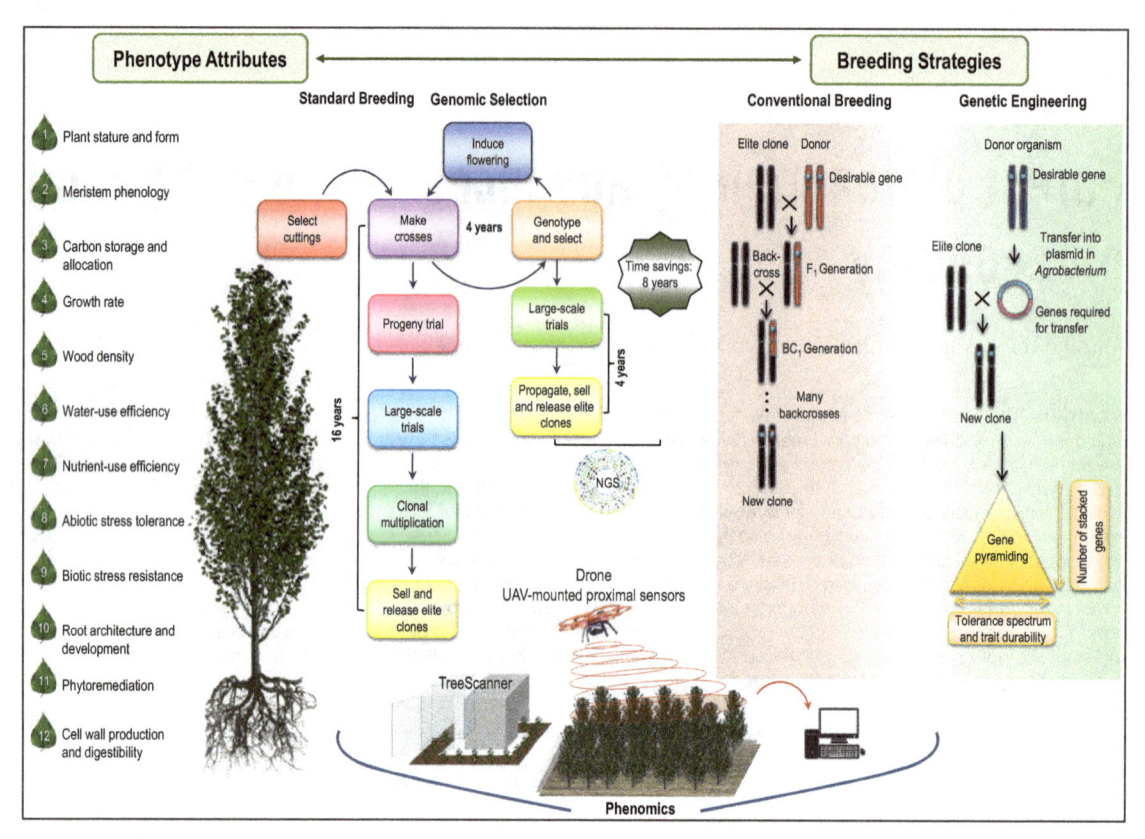

Figure 1 - A schematic representation of integrated forest breeding strategies in next generation genomics and phenomics era focusing on poplar as the tree model species for biofuel research. Incorporation of Next-Generation Sequencing (NGS) based multi-omics approaches with genetic engineering, Genomic Selection and phenomics can be used to speed up tree breeding and save time as compared with conventional breeding. These new varieties will maximize yield per acre with minimum inputs and exhibit value-added traits that enhance their use as biofuel and bio-based feedstocks.

impact tree growth, such as crown architecture and partial resistance to biotic stress. But, continued improvement of forest trees by traditional means is slow due to their long generation times and large genomes (Neale and Kremer 2011). Nevertheless, over the past two decades, genetic improvement opportunities have been broadened by genetic engineering targeting important traits in model forest-tree species (Harfouche et al. 2011). This stems from the urgent need to better understand how forest trees adapt to severe environmental conditions in order to develop new and improved varieties that are able to sustain productivity and meet future demands for commercial products. For example, the completion of a draft genome sequence of the poplar (*Populus trichocarpa* Torr. & Gray ex Brayshaw; Tuskan et al. 2006) has advanced forest tree genetics to unprecedented levels. Recent releases of the genome sequences of white spruce (*Picea glauca* (Moench) Voss; Birol et al. 2013), Norway spruce (*P. abies* L.; Nystedt et al. 2013) and Eucalyptus (*Eucalyptus grandis*; Myburg et al. 2014) and the anticipated release of the Pinus genome will undoubtedly lead to more rapid advances.

Despite the enormous promise of advanced forest biotechnology, it has yet to gain traction as a viable breeding alternative, mainly because of its exhaustive biosafety regulations and the divided public opinion over it. Besides, understanding the relationship between genotype and phenotype and dissecting complex polygenic traits in forest trees becomes a central goal in tree genetic improvement.

In addition, innovative technologies, such as genomic selection (GS) and next-generation Ecotilling (Harfouche et al. 2012a), are now proving to be important strategic approaches to improving our understanding of various processes in forest trees and the role of key genes associated with their regulation (Resende et al. 2012, Vanholme et al. 2013). Although there have been considerable advances in our understanding of lignin biosynthesis and monomer composition, most of the genetic engineering efforts have been restricted to poplars, and overcoming biomass recalcitrance due to novel engineering of lignin pathways in other forest trees is still in its infancy.

This article provides a summary of recent achievements in forest biotechnology with an overview on the integrative approach to accelerate the development of improved forest feedstocks for a sustainable bioeconomy (Fig. 1). Furthermore, it discusses new approaches and concepts (Fig. 2, 3 and 4) to advance the forest-based bio-economy and issues that should be considered when devel-

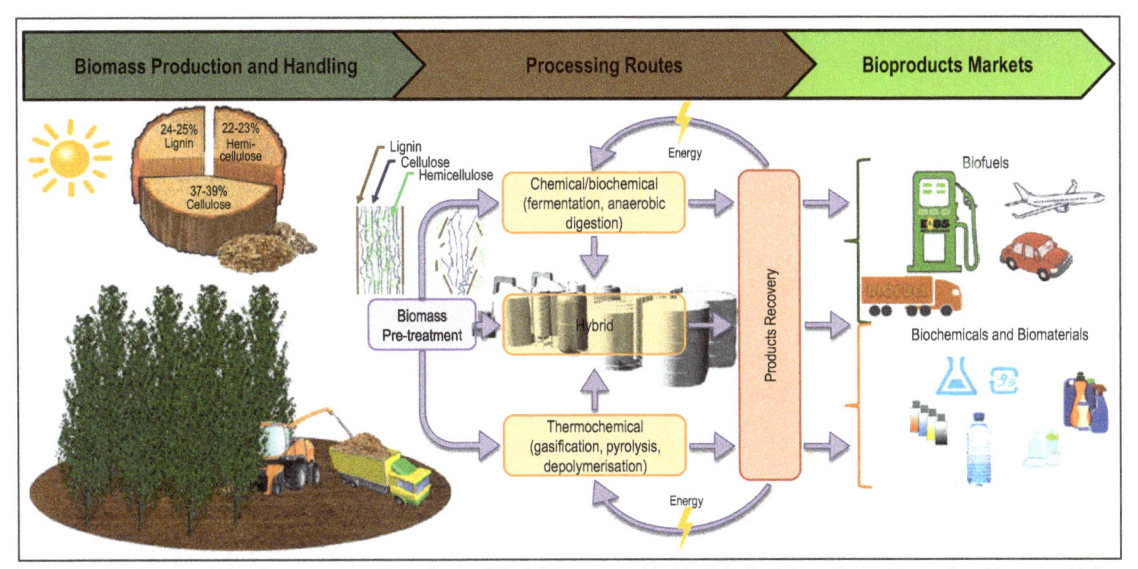

Figure 2 - The integrated biorefinery concept. Forest lignocellulosic biomass are ideal feedstocks that can be used to produce bioenergy, biofuels and bio-products, using thermochemical or biochemical conversion routes, or a combination of both.

oping and deploying biotech trees with improved attributes.

The emerging global bioeconomy

Bioeconomy involves the sustainable production of renewable biological resources and their conversion into food, feed, biofuels, bioenergy, and bio-products by harnessing the power of biotechnology, but at the same time preserving the environment and ecosystem services. In this article, bio-based production encompasses items produced from forest biomass feedstocks including biofuels, bioenergy and bio-products or bio-materials (largely bio-chemicals and bio-plastics) (Fig. 2).

Until recently, in most countries the focus has been to a great extent on first-generation biofuels production. However, second-generation biofuels and bio-materials offer exciting opportunities for future manufacturing to replace existing fossil-based materials with bio-based, therefore, contributing to greenhouse gas (GHG) emissions reductions. Yet, green credentials are not enough to justify their place in the market. Therefore, the technical, economic and social performances of these materials have to be considered (Philp and Pavanan 2013). In addition, bio-based production promises high-value jobs. Carus et al. (2011) have estimated that materials use of biomass can directly support 5-10 times more employment and 4-9 times the value-added compared with energy uses, principally due to longer, more complex supply chains for material use. Moreover, a report commissioned for The Blue Green Alliance estimated that shifting 20% of current plastics production into bio-plastics would create a net 104,000 jobs in the US economy (Heintz and Pollin 2011).

However, if a bioeconomy is to succeed in any country, it should rely on international cooperation and trade (Philp and Pavanan 2013). On the one hand, the drivers behind the development of bio-based production are global: climate change mitigation, energy security and independence, the attraction and creation of new jobs associated to rural regeneration. On the other hand, global food security is a grand challenge facing society, and there are ways in which energy and food production come into direct competition (Seidenberger et al. 2008).

Bioeconomy is becoming a reality in many parts of the world as it offers great opportunities and solutions to tackle major societal, environmental and economic challenges. A global bio-economy that is also based on agricultural and forest biomass is emerging in Europe, the United States and Canada that offers an avenue toward a more low-carbon green economy. Exploitation of non-food feedstocks such as forest biomass is gaining importance for this sustainable production.

The bioeconomy concept is currently flexible and it is interpreted differently in different countries and regions. While, many countries have already published national strategies and visions on the bio-economy (i.e., The Bio-economy to 2030: Designing a Policy Agenda by OECD, the National Bio-economy Blueprint in the United States, Innovating for Sustainable Growth: A Bio-economy for Europe, The Canadian Blueprint: Beyond Moose and Mountains) some have established organizations and networks to stimulate and develop it. Though, sustainability is recognized as important, the driving force behind the bioeconomy is the opportunity for economic growth and innovation.

The role of forest biotechnology in a sustainable bioeconomy future

The tools of biotechnology have the potential to produce a new generation of genetically improved bioenergy crops that are engineered to either produce high biomass yield and digestibility, or offer protection to bioenergy crops against environmental stresses, or a combination of all attributes (Fig. 1).

Lignocellulosic biofuels promise to resolve the most significant problems associated with existing first-generation biofuels. For example, Littlewood et al. (2014) have recently shown that bioethanol from poplar biomass feedstock is a commercially viable alternative to fossil fuel in the European Union. A techno-economic modeling to compare the price of bioethanol produced from short rotation coppice (SRC) poplar feedstocks under two leading processing technologies (dilute acid and liquid hot water), in five European countries (Sweden, France, Italy, Slovakia, and Spain) has been used. In a forward-looking scenario, genetically engineering poplar with a reduced lignin content showed potential to enhance the competitiveness of bioethanol with conventional fuel by reducing overall costs by approximately 41% in four out of the five countries modeled (Littlewood et al. 2014). Current research and development (R&D) also focus on evaluating poplar biomass production potential in a SRC. Such research is critical for investigating the performance of novel poplar genotypes deriving from standard breeding programs with potential for commercial biomass production over multiple coppice rotations. Results by Sabatti et al. (2014) showed that poplar biomass production differed significantly among rotations starting from 16 tons/ha/year in the first, peaking at 20 tons/ha/year in the second, and decreasing to 17 tons/ha/year in the third rotation. This will ultimately lead to a more efficient economic feasibility of utilizing tree woody biomass for biofuels and bio-materials.

Miscanthus can greatly reduce the land intensity of biofuel production. While only 4.5 dry tons of harvestable corn grain are extracted from each acre of corn grain, 13 dry tons of harvestable biomass of Miscanthus is produced per acre. Thirteen-hundred gallons of cellulosic ethanol can be produced per acre of Miscanthus biomass plantation. Only 450 gallons of corn-ethanol are yielded per acre of corn. In the United States, a hypothetical scenario to produce 35 billion gallons of bioethanol, using corn as a feedstock would demand one-quarter of all harvested cropland. However, using Miscanthus bioenergy crops would need less than one-tenth (Heaton et al. 2008).

Biotechnologically-improved bioenergy crops such as poplar can also be grown on marginal and drought-prone lands where major crops are less productive. This would permit to ease the competition on land and water resources to be used for food and feed. Recently, it has been shown that the constitutive overexpression of the wintersweet (*Chimonanthus praecox*) fatty acyl–acyl carrier protein thioesterase (CpFATB) in poplar activates an oxidative signal cascade and leads to drought tolerance in the transgenic plants. The genetically engineered poplar maintained significantly higher photosynthetic rates, suggesting that changes in fatty-acid composition and saturation levels may be involved in leaf tolerance to dehydration during drought stress (Zhang et al. 2013). Another important study reported that gene stacking by overexpressing multiple resistance genes enhanced tolerance to environmental stresses in transgenic poplar. The transgenic lines harboring effector genes: *vgb*, encoding aerobe *Vitreoscilla hemoglobin*; *SacB*, encoding a levansucrase that is involved in fructan biosynthesis in *Bacillus subtilis*; and *JERF36*, a tomato gene encoding jasmonate/ethylene-responsive factor protein exhibited higher growth than the controls, as demonstrated by greater height, basal diameter, and biomass than the corresponding non-transgenics. This improved growth could be primarily due to higher water-use efficiency and fructan levels, and better root architecture under drought and salinity stress (Su et al. 2011).

These selected recent advances in maximizing tree tolerance to drought stress will allow an important bioenergy crop to be bred so it will grow in less than ideal soils and climate. Together, these results demonstrate the potential of forest biotechnology for improving environmental stress tolerance and biomass processability in forest trees.

Equally important, GS is extremely appealing in forest trees due to the prospects of improving accuracy when selecting for traits with low heritability (e.g., biomass productivity and abiotic stress tolerance) and where long generation times and late-expressing, complex traits are involved (Grattapaglia and Resende 2010). However, successful application of GS in tree breeding programs aimed at developing trees that are tolerant to environmental stresses or to dissect quantitative trait variation will require comprehensive physiological information that rely on rigorous phenotyping. Therefore, high-throughput phenotyping of morpho-physiological traits will require the utilization of sophisticated, non-destructive imaging techniques in a multi-

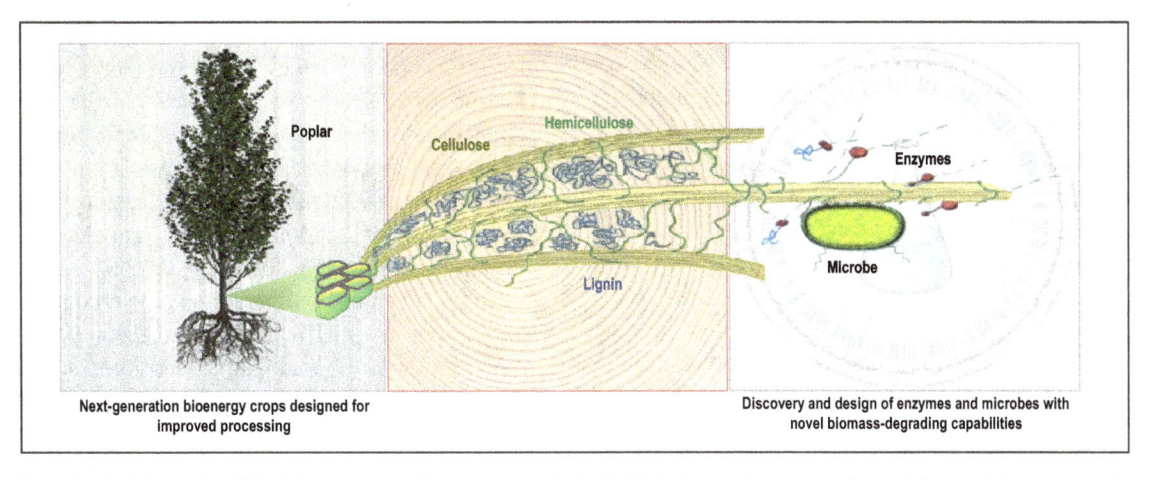

Figure 3 - An integrated multidisciplinary approach linking forest and industrial biotechnology to overcome the recalcitrance of biomass toward processing.

spectral approach. For example, near-infrared spectroscopy, canopy spectral reflectance, and infrared thermography can be used to assess biomass productivity and plant water status, and to detect environmental stresses at the individual-tree level. These sensors can be mounted on drones, which can be directed with a global positioning system to enhance the precision and accuracy of phenotyping under field conditions (Harfouche et al. 2014).

This phenomics approach further creates opportunities to overcome the field-based phenotyping bottleneck and generate phenotyping sets, such as environmental stress tolerance and biomass estimation. Ultimately, it will help to uncover phenotype-to-genotype relationships and their relevance for improving tolerance to environmental stresses in forest trees (Fig. 1). GS coupled with phenomics offers great promise for forest breeders in accelerating the genetic improvement of forest trees as bioenergy feedstocks.

Biotechnology has the potential to not only produce more productive bioenergy crops and minimize inputs, but also to develop more efficient biofuels and bio-materials conversion processes. This offers a great cause for optimism that the global bioeconomy challenges of the new century can be met.

Linking forest and industrial biotechnology to accelerate drive towards sustainable bioeconomy

Undoubtedly, one of the greatest impediments to commercializing second-generation biofuels along with bio-materials that has yet to be solved is that most of the conversion processes are not yet ripe, costly and time-consuming. An integrated second-generation biorefinery can use either a biochemical or thermochemical process, or a hybrid of both, in order to process efficiently biomass feedstocks into multi-purpose products with great added-value (Fig. 2). The biomass-tailored thermochemical conversion system relies on heating the feedstock to high temperatures with little or no oxygen. Whereas, the biochemical system relies on the use of microorganisms and enzymes to process the biomass and on the genetic engineering of these microorganisms for more efficient biological conversion. The concept of coupling green (plant) and white (industrial) biotechnology is proposed here (Fig. 3). R&D focusing on the synergistic interaction between these two biotechnologies are therefore of paramount importance. Plant biotechnology is, on one hand, playing an important role in the development of advanced biomass feedstocks for a bioenergy and bio-products industry. On another hand, industrial biotechnology is involved in the conversion of these renewable and sustainable resources into a wide range of biofuels and bio-materials (Fig. 2 and 3).

Lignocellulosic biomass are mostly composed of cellulose, hemicellulose, and lignin, which serve to maintain the structural integrity of plant cells (Fig. 2 and 3). Lignocellulosic woody materials have great potential for biofuel and bio-materials production. The plant cell wall polysaccharides can be used as a feedstock for biofuel production after being broken down into simple sugars (saccharification) (Ragauskas et al. 2006, Solomon et al. 2007), but lignin strongly impedes this process (Boerjan et al. 2003). A highly degradation-resistant phenolic polymer, lignin is part of a complex matrix in which cellulose microfibrils are embedded. The inhibition of saccharification enzymes by lignin may result from the reduced accessibility of cellulose microfibrils, as well as the adsorbtion of hydrolytic enzymes to the lignin polymer (Weng et al. 2008). Furthermore, current chemical and physical strategies to remove lignin from biomass, such as pretreatment with steam or acid, result in the formation of compounds which can inhibit downstream processes of saccharification and fermentation (Hamelinck et al.

2005). All together, these lignin properties are hard to deal with and make its biosynthesis a key control point in biomass degradation and in determining the efficiency of biofuels production (Weng et al. 2008).

In addition to work being conducted *in planta* using genetic engineering, with the goal of manipulating lignin content and monomer composition, another line of experimentation is to discover novel strategies for lignin degradation. Therefore, R&D aimed at increasing the efficiency and decreasing cost of lignocellulosic biomass pretreatment is currently a high priority along with the metagenomic discovery of biomass-degrading genes and genomes. For example, recent deep sequencing data sets in the cow rumen microbes provided a substantially expanded catalog of genes and genomes participating in the deconstruction of cellulosic biomass (Hess et al. 2011). In addition, recent release of the genome sequence of white rot fungus (*Phanerochaete chrysosporium*) has shown that the genome of this fungus encodes hundreds of enzymes potentially dedicated to lignin degradation (Martinez et al. 2004, Vanden Wymelenberg et al. 2006). Thus, harnessing this rich biological diversity that has recently started with the metagenome sequencing of the gut flora of *Nasutitermes* is an important step forward for lignin degradation (Warnecke et al. 2007). Likewise, a deeper understanding of termite lignocellulose digestion by metagenomics could shed light on the enzymatic mechanisms useful for biomass delignification.

Building skills for Europe's bioeconomy: the role of biotechnology and entrepreneurship

Bioeconomy is one of the world's most educated industries. These creative scientific minds drive the global bioeconomy where Small and Medium Size Enterprises (SMEs) constitute ~80% of the companies in most developed countries' bio-economy. Besides, the bio-economy in Europe is currently worth more than € 2 trillion a year and employs over 22 million people, predominantly in rural areas and often in SMEs (Ernst & Young 2012).

Entrepreneurship education is now recognized as an important part of fostering entrepreneurial activity in the European Union. Entrepreneurship in biotechnology has a great potential to maximize the impact and commercial potential of the bio-economy (Fig. 4). The significant growth of the biotechnology sector over the last decade means that biotechnology enterprises are seen as playing a vital role in creating solutions and jobs in the future. For example, Harfouche et al. (2010) have recently shed light on how to protect biotech-based innovation for the development of feedstock for second-generation biofuels and strategies for technology transfer to show the important role biotechnology Intellectual Property plays in the global bioeconomy (Harfouche et al. 2012b).

Important prerequisites for a competitive bio-economy is the availability of well-trained workforce with the necessary entrepreneurial mindset and business skillsets. To move this forward, European efforts have recently sought to develop new learning and teaching model for entrepreneurship education in biotechnology to train the next-generation of talents to turn vision into reality. A two-year Knowledge Alliance project funded by the European Commission which brings together a knowledge and innovation community through partnership across Europe (www.bioinno.eu) for biotechnology

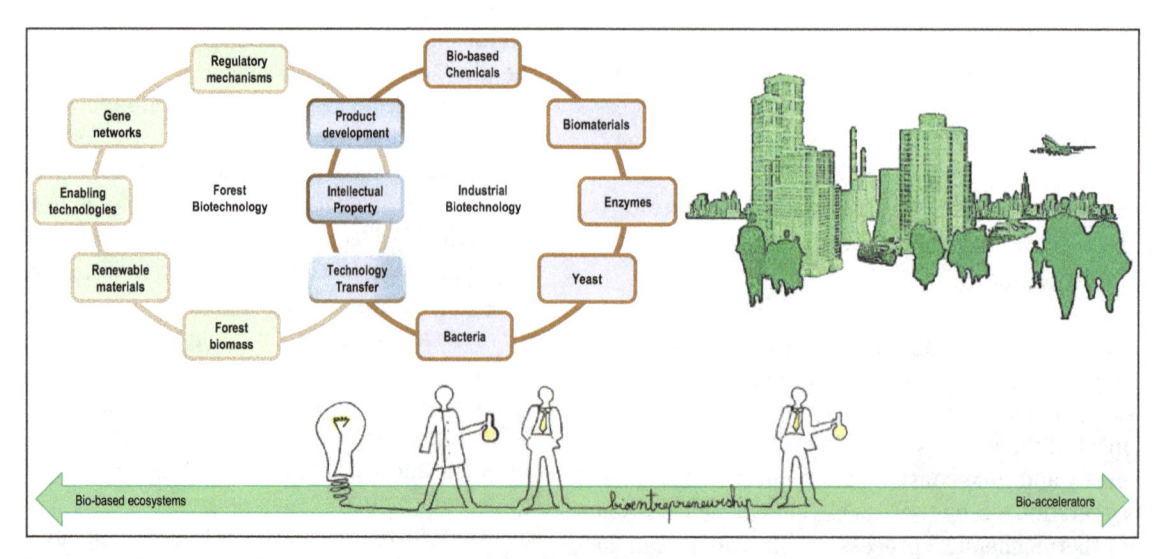

Figure 4 - Building skills for Europe's bio-economy. An increased focus on entrepreneurship, translational sciences, regulatory science, product development, and technology transfer in biotechnology and forestry can help accelerate movement of bioinventions out of laboratories and into markets.

entrepreneurship education has recently initiated. This program will focus on teaching biotechnology entrepreneurship with an emphasis on innovative biotechnology applications in agriculture, forestry, and bio-based economy.

With innovations in biotechnology at the core of the success of global bio-economy, and the world's need for a more accessible and translational science, this challenge will have to be tackled by governments and industry associations. Greater investment in R&D and integration of bio-entrepreneurship education and traineeship are also necessary to react to global bioeconomy challenges.

What next?

Future R&D directions and key actions that need to be taken to tackle the managerial, economic and political challenges facing the forest-based bioeconomy are highlighted.

(i) We need to adopt an all-inclusive approach among forest biomass developers, landowners, biofuels producers, end users and policymakers. This will enhance our ability to develop bioenergy crops for the growing bio-economy agenda. Government interventions with subsidies for production, consumption and R&D are instrumental in the promotion of second-generation biofuels.

(ii) With unprecedented recent technological advances in the areas of genomics and phenomics, we are now well poised to capitalize on these strategies to speed up the development of ideal forest biomass feedstocks for bio-economy. Ultimately, this holistic approach will deepen our understanding of forest tree breeding and enhance our ability to develop desired tree phenotypes.

(iii) Water supply is obviously another crucial factor in sustainability of forest biomass production. The scale of its importance is worth highlighting. As it is often necessary to grow forest trees on marginal land, where water and nutrient resources are limiting, it will become increasingly important to improve water and nutrient use efficiency in forest trees using biotechnology to ensure sufficiency and sustainability. To reap these research-proven benefits, biosafety regulations must be improved and public acceptance must be properly addressed.

(iv) Another major obstacle to industrial-scale biofuels production from lignocellulosic biomass lies in the inefficient deconstruction of plant cell wall material. Metagenomics aimed at retrieving novel enzymes from naturally evolved biomass-degrading microbial communities coupled with in planta engineering should aid in the optimization of biofuel and bio-materials production and the development of advanced bioenergy crops.

(v) High octane fuel blends have a great potential to expand the market for advanced second-generation biofuels, increase engine efficiency, and reduce GHG emissions from transportation sector. Before these benefits can be realized, key market and regulatory challenges must be overcome.

With growing global commitments to energy security and climate change mitigation, the world have a great opportunity to reap the benefits of economic growth, jobs creation, and environmental improvements that bio-economy plans promise. We hope these endeavors will encourage greater international coordination and cooperation from both public and private sectors for R&D.

Concluding remarks

The world's ambitious plan to reduce its carbon footprint has led to the emergence of a new bio-economy, one in which non-food forest-based biofuels and bio-products have a significant advantage over fuels and products that are non-renewable and require large amounts of fossil fuel energy to manufacture. Yet, this will require more forest biomass resources and a rigorous forest management, in order to be well positioned to ensure sufficiency and sustainability in the production of new bio-products while creating new jobs and preserving the ecosystem services (e.g., water-quality protection, as well as wildlife conservation).

By producing biofuels and bio-based materials from wood, forest products companies can capture new markets, and support rural communities and government services.

There are numerous viable strategies to convert forest biomass into biofuels and bio-materials. Yet, to maximize jobs and economic returns, these strategies have to be integrated with the existing forest products industry.

In this article, we propose to integrate novel genetics, genomics and phenomics with conventional breeding to expedite forest tree improvement. This integrative approach could prove a useful tool for speeding up future forest breeding programs with the aim of sustainable woody biomass production. For example, the use of genome-wide selection is an emerging approach that will revolutionize the applications of tree breeding. Phenotypic selection or marker-assisted breeding protocols can be replaced by selection, based on whole-genome predictions in which phenotyping updates the model to build up the prediction accuracy. Ultimately, GS could substantially shorten generation time through rapid cycles of breeding, selection and propagation.

Second-generation biofuels produced from forest lignocellulosic biomass represent a renewable, more carbon-balanced alternative to both fossil fuels and corn-derived or sugarcane-derived biofuels. However, forest biomass recalcitrance to saccharification is one of the major impediments to high-yield and cost-effective production of biofuels and value-added bio-chemicals and bio-materials from lignocellulosic feedstocks. Due to the natural recalcitrance of lignocellulose, coupling forest and industrial biotechnology will further improve the conversion process from biomass to second-generation biofuels and bio-products (Fig. 2 and 3). Decreasing lignin content and/or modifying lignin monomer composition of forest biomass by genetic engineering is believed to mitigate biomass recalcitrance and improve conversion efficiency of tree biomass. Likewise, industrial biotechnology involving sequencing the genomes of natural microbes will lead to important insights relevant to biofuels production. Lignin degradation in nature may provide novel resources for the delignification of forest lignocellulosic biomass feedstocks. It is strongly believed that the genome of these microorganisms encode hundreds of enzymes potentially dedicated to lignin degradation. Finally, a better biosafety regulation over the momentous tree genetic engineering and novel breeding technologies and their long-term economic impact would bring valuable contributions towards developing an economically sustainable biofuels and biomaterials markets worldwide.

Acknowledgements

We thank Professor Piermaria Corona for the invitation to write this article. We assert that we have no business interests or relationships that could be viewed as a conflict of interest. Our research on forest biotechnology, and bio-innovation and entrepreneurship is supported by the Brain Gain Program (*Rita Levi Montalcini Rientro dei cervelli*) of the Italian Ministry of Education, University and Research (A.H.), and grants from the European Commission's Seventh Framework Program (WATBIO FP7 - 311929), and Erasmus Multilateral European Knowledge Alliances (BIOINNO 539427) projects to A.H. S.K. is supported by a German Federal Enterprise for International Cooperation GIZ Master fellowship.

References

Battisti D.S., Naylor R.L. 2009 - *Historical warnings of future food insecurity with unprecedented seasonal heat*. Science 323: 240-244.

Birol I., Raymond A., Jackman S.D., Pleasance S., Coope R., Taylor G.A., Yuen M.M.S., Keeling C.I., Brand D. et al. 2013 - *Assembling the 20 Gb white spruce (Picea glauca) genome from whole-genome shotgun sequencing data*. Bioinformatics 29: 1492-1497.

Boerjan W., Ralph J., Baucher M. 2003 - *Lignin biosynthesis*. Annual Review of Plant Biology 54: 519-546.

Carus M., Carrez D., Kaeb H., Venus J. 2011 - *Level Playing Field for Biobased Chemistry and Materials*. Nova Institute 2011-04-18 Policy paper, 8 p.

Ernst & Young 2012 - *What Europe has to offer biotechnology companies: bioeconomy industry overview*. Report of the Ernst & Young Global Company [Online]. Available: http://www.europabio.org/sites/default/files/europabio_-_ernst_young_report___what_europe_has_to_offer_biotechnology_companies.pdf [2014]

Grattapaglia D., Resende M. 2010 - *Genomic selection in forest tree breeding*. Tree Genetics & Genomes 7: 241-255.

Hamelinck C.N., van Hooijdonk G., Faaij A.P.C. 2005 - *Ethanol from lignocellulosic biomass: techno-economic performance in short-, middle- and long-term*. Biomass & Bioenergy 28: 384-410.

Harfouche A., Grant K., Selig M., Tsai D., Meilan R. 2010 - *Protecting innovation: genomics-based intellectual property for the development of feedstock for second-generation biofuels*. Recent Patents on DNA & Gene Sequences 4: 94-105.

Harfouche A., Meilan R., Altman A. 2011 - *Tree genetic engineering and applications to sustainable forestry and biomass production*. Trends in Biotechnology 29: 9-17.

Harfouche A., Meilan R., Kirst M., Morgante M., Boerjan W., Sabatti M., Scarascia Mugnozza G. 2012a - *Accelerating the domestication of forest trees in a changing world*. Trends in Plant Science 17: 64-72.

Harfouche A., Meilan R., Grant K., Shier V. 2012b - *Intellectual property rights of biotechnologically improved plants*. In: "Plant Biotechnology and Agriculture: Prospects for the 21st Century". Altman A., Hasegawa P. Ed., Elsevier and Academic Press, San Diego: 525-539.

Harfouche A., Meilan R., Altman A. 2014 - *Molecular and physiological responses to abiotic stress in forest trees and their relevance to tree improvement*. Tree Physiology 00: 1-18. doi:10.1093/treephys/tpu012.

Heaton E.A., Dohleman F.G., Long S.P. 2008 - *Meeting US biofuel goals with less land: The potential of Miscanthus*. Global Change Biology 14: 2000-2014.

Heintz J., Pollin R. 2011 - *The Economic Benefits of a Green Chemical Industry in the United States: Renewing Manufacturing Jobs While Protecting Health and the Environment*. Report of the Political Economy Research Institute, University of Massachusetts, Amherst, MA.

Hess M., Sczyrba A., Egan R., Kim T-W., Chokhawala H., Schroth G., Luo S., Clark D.S., Chen F. et al. 2011 - *Metagenomic discovery of biomass-degrading genes and genomes from cow rumen*. Science 331: 463-467.

Littlewood J., Guo M., Boerjan W., Murphy R.J. 2014 - *Bioethanol from poplar: a commercially viable alternative to fossil fuel in the European Union.* Biotechnology for Biofuels 7: 113.

Martinez D., Larrondo L.F., Putnam N., Gelpke M.D., Huang K., Chapman J., Helfenbein K.G., Ramaiya P., Detter J.C. et al. 2004 - *Genome sequence of the lignocellulose degrading fungus Phanerochaete chrysosporium strain RP78.* Nature Biotechnology 22: 695-700.

Myburg A.A., Grattapaglia D., Tuskan G.A., Hellsten U., Hayes R.D., Grimwood J., Jenkins J., Lindquist E., Tice H. et al. 2014 - *The genome of Eucalyptus grandis.* Nature 509: 356-362.

Neale D.B., Kremer A. 2011 - *Forest tree genomics: growing resources and applications.* Nature Reviews Genetics 12: 111-122.

Nystedt B., Street N., Wetterbom A., Zuccolo A., Lin Y.C., Scofield D.G., Vezzi F., Delhomme N., Giacomello S. et al. 2013 - *The Norway spruce genome sequence and conifer genome evolution.* Nature 497: 579-584.

OECD 2009 - *The Bioeconomy to 2030: designing a policy agenda.* International Futures Project. OECD Publishing; doi: 10.1787/9789264056886-en. [Online]. Available: http://www.oecd.org/document/48/0,3746,en_2649_36831301_42864368_1_1_1_1,00.html [2014]

Philp J.C., Pavanan K.C. 2013 - *Bio-based Production in a Bioeconomy.* Asian Biotechnology and Development Review 15: 81-88.

Ragauskas A.J., Williams C.K., Davison B.H., Britovsek G., Cairney J., Eckert C.A., Frederick W.J. Jr, Hallett J.P., Leak D.J. et al. 2006 - *The path forward for biofuels and biomaterials.* Science 311: 484-489.

Resende M.F.R. Jr, Muñoz P., Acosta J.J., Peter G.F., Davis J.M., Grattapaglia D., Resende M.D.V., Kirst M. 2012 - *Accelerating the domestication of trees using genomic selection: Accuracy of prediction models across ages and environments.* New Phytologist 193: 617-624.

Roy J., Saugier B., Mooney H.A. 2001 - *Terrestrial Global Productivity.* Academic Press London.

Sabatti M., Fabbrini F., Harfouche A., Beritognolo I., Mareschi L., Carlini M., Paris P., Scarascia-Mugnozza G. 2014 - *Evaluation of biomass production potential and heating value of hybrid poplar genotypes in a short-rotation culture in Italy.* Industrial Crops and Products 61: 62-73.

Seidenberger T., Thrän D., Offermann R., Seyfert U., Buchhorn M., Zeddes J. 2008 - Global biomass potential - *Investigation and assessment of data, Remote sensing in biomass potential research, Country specific energy crop potentials.* Report of the German Biomass Research Centre.

Solomon B.D., Barnes J.R., Halvorsen K.E. 2007 - *Grain and cellulosic ethanol: history, economics, and energy policy.* Biomass & Bioenergy 31: 416-425.

Su X., Chu Y., Li H., Hou Y., Zhang B., Huang Q., Hu Z., Huang R., Tian Y. 2011 - *Expression of multiple resistance genes enhances tolerance to environmental stressors in transgenic poplar (Populus × euramericana 'Guariento').* PLoS ONE 6: e24614.

Tuskan G.A., Difazio S., Jansson S., Bohlmann J., Grigoriev I., Hellsten U., Putnam N., Ralph S., Rombauts S. et al. 2006 - *The genome of black cottonwood, Populus trichocarpa (Torr. & Gray).* Science 313: 1596-1604.

Vanden Wymelenberg A., Sabat G., Mozuch M., Kersten P.J., Cullen D., Blanchette R.A. 2006 - *Structure, organization, and transcriptional regulation of a family of copper radical oxidase genes in the lignin-degrading basidiomycete Phanerochaete chrysosporium.* Applied and Environmental Microbiology 72: 4871-4877.

Vanholme B., Cesarino I., Goeminne G., Kim H., Marroni F., Van Acker R., Vanholme R., Morreel K., Ivens B. et al. 2013 - *Breeding with rare defective alleles (BRDA): a natural Populus nigra HCT mutant with modified lignin as a case study.* New Phytologist 198: 765-776.

Warnecke F., Luginbuhl P., Ivanova N., Ghassemian M., Richardson T.H., Stege J.T., Cayouette M., McHardy A.C., Djordjevic G. et al. 2007 - *Metagenomic and functional analysis of hindgut microbiota of a wood-feeding higher termite.* Nature 450: 560-517.

Weng J.K., Li X., Bonawitz N.D., Chapple C. 2008 - *Emerging strategies of lignin engineering and degradation for cellulosic biofuel production.* Current Opinion in Biotechnology 19: 166-172.

Zhang L., Liu M., Qiao G., Jiang J., Jiang Y., Zhuo R. 2013 - *Transgenic poplar 'NL895' expressing CpFATB gene shows enhanced tolerance to drought stress.* Acta Physiologiae Plantarum 35: 603-613.

Long-term development of experimental mixtures of Scots pine (*Pinus sylvestris* L.) and silver birch (*Betula pendula* Roth.) in northern Britain

Bill Mason[1*], Thomas Connolly[1]

Abstract - The Caledonian pinewoods of northern Scotland are a priority conservation habitat in Europe which are dominated by Scots pine (*Pinus sylvestris*), but varying proportions of a number of broadleaved species such as silver birch (*Betula pendula*) can occur in these forests. Better understanding of the dynamics of mixed Scots pine-birch stands would be helpful in informing current initiatives to restore and increase the area of the pinewood ecosystem. Some evidence is provided by two experiments established in the 1960s which compared plots of pure Scots pine and pure birch with two treatments where the two species were mixed in 3:1 and 1:1 ratios. Some fifty years later, Scots pine was the more vigorous of the two species in these experiments, being both taller and significantly larger in diameter. The highest basal area was generally found in the pure Scots pine plots and the values in the mixed plots tended to be intermediate between those of the two component species. Examination of the growth in the mixed plots showed a slight, but non-significant, tendency towards overyielding. This appeared to be due to Scots pine growth being better than predicted, while that of birch was slightly less than predicted. These results suggest that in these mixtures, which are composed of two light demanding species, the main mechanism driving long-term performance is inter species competition and there is little evidence of any complementary interaction. These results suggest that any strategy seeking to increase the long-term representation of broadleaves such as birch in the Caledonian pinewoods will need to create discrete blocks that are large enough to withstand the competitive pressures exerted by the pine.

Keywords - Mixtures, Pinus sylvestris, Betula pendula, competition

Introduction

There has been increasing interest in growing tree species in mixed stands for reasons such as adapting forests to climate change, providing greater biodiversity, and enhancing the visual attractiveness of forests (Quine et al. 2013). However, successful establishment and management of mixed species forests depends on understanding the characteristics of the component species (e.g. growth habit, shade tolerance) and the way in which their mutual interactions change over time (Pretzsch 2009, chapter 9). Paquette and Messier (2011) suggested that beneficial interactions between tree species may be more important in stressful environments such as the boreal forests while reviews of facilitation in wider plant communities have also highlighted the need for taking environmental gradients into account (Brooker et al. 2008). The complexity of these interactions suggests that, despite recent reports of the benefits of mixed stands for the provision of a range of ecosystem services including productivity (Felton et al. 2010, Zhang et al. 2012, Gamfeldt et al. 2013), it may be problematic to extrapolate potential performance of mixtures from one climatic region or site type to another.

Forests of the British Isles and adjoining regions of Atlantic Europe are mostly characterised by single species plantations of fast growing non-native conifers grown on relatively short rotations (Mason 2007, Mason and Perks 2011). Recent data (Forestry Commission 2003) suggest that the total area of mixed-species stands (defined as where no single species occupies more than 80 per cent of the stand) was only around 200,000 ha or about 8 per cent of the forest area of Great Britain. In the last decade there has been increasing recognition of the potential role of growing tree species in mixture as part of a strategy of adapting British forests to projected climate change (Read et al. 2009, p. 174-175). The UK Forestry Standard, which sets out the national basis for sustainable forest management, encourages forestry practices which promote greater species diversity such as fostering of mixed stands (Anon. 2011, p. 96). In addition, recent guidance in Wales and Scotland supports the wider use of a range of

---

[1] Forest Research UK, United Kingdom
[*] bill.mason@forestry.gsi.gov.uk

species mixtures (Anon. 2010, Grant et al. 2012). Nevertheless, the limited experience of the creation and management of mixtures in British forestry makes it difficult to be certain about the regions of the British Isles where mixtures may be most effective, the particular species combinations that should be deployed, and the interactions between management practice and mixture development over time.

One forest type where the role of mixtures has been discussed for several decades is the Caledonian pinewoods of northern Scotland, which are recognised by the European Union Habitats Directive as being of special conservation value (Mason et al. 2004). These forests are dominated by Scots pine (*Pinus sylvestris* L.) at all stages of stand development, but some stands can contain variable amounts of several broadleaved species including birches (*Betula pendula* and *B. pubescens*), aspen (*Populus tremula*) and rowan (*Sorbus aucuparia*) (Edwards and Mason 2006). In broad terms, these pinewoods can be divided into two categories: the remnants of genuinely native pinewoods amounting to about 17,800 ha and a more extensive area of Scots pine dominated plantations amounting to about 101,000 ha (Mason et al. 2004). The remnant pinewoods are managed primarily for biodiversity and landscape while the plantations are managed for a range of ecosystem services including timber production (of sawlogs and small roundwood) on a rotation of 70-100 years. However, the considerable age of many trees in the remnant pinewoods (Edwards and Mason, 2006) plus concerns over regeneration failure in these stands means that sensitive management of the plantations will be important for the long-term continuity of the pinewood ecosystem in northern Scotland. Given earlier studies showing beneficial effects of birch species on soil properties of acid heathland soils (Dimbleby 1952, Gardiner 1968, Miles 1981), it has been proposed that incorporating a proportion of birch into Scots pine plantation stands would improve soil and tree nutrition with consequent benefits for stand productivity, and possibly other ecosystem services. However, there is little published evidence that can be used to examine this proposition.

A study in south-eastern Norway compared productivity of nine paired plots of pure Scots pine with that found in mixtures of Scots pine and birch (Frivold and Frank 2002). Volume growth in the mixtures was less than that in pure stands, although the differences were not significant. Hynynen et al. (2011) investigated performance of mixed Scots pine and silver birch stands of mid rotation age on 14 sites in eastern Finland. Over a 19 year period, they found that volume increment decreased with increasing amounts of birch. However, an earlier report from Finland had suggested 10-14% increases in productivity from Scots pine/birch mixtures over the respective pure stands (Mielikäinen 1980). Models suggested that this increase appeared to diminish between 30 and 70 years of age with an optimum proportion of birch of no more than 20 per cent (Mielikäinen 1996). In an overview of the theory and performance of two species mixtures in Europe, Pretzsch (2005) also suggested that the performance of mixtures of light demanding species such as Scots pine and birch could be strongly affected by site conditions, noting an apparent loss of increment in birch in more oceanic conditions.

The only relevant British study described two experiments with Scots pine-birch mixtures where basal area declined with increasing proportion of birch (Malcolm and Mason 1999). The authors suggested that Scots pine appeared to be benefitting in mixture at the expense of birch. These results were obtained in 30-years-old stands that had only recently closed canopy while the studies by Frivold and Frank (2002) and Hynynen et al. (2011) also involved stands that were mostly under 50 years of age. Given that relative productivity of species can change with age (Pretzsch 2005), it would be dangerous to extrapolate long-term performance of two species mixtures from growth in the early stem exclusion phase (*sensu* Oliver and Larson 1996). In this paper, we report on the further development of Scots pine-silver birch mixtures in the two experiments previously examined by Malcolm and Mason (1999) when the trees were about 50 years of age, which is about two-thirds of normal rotation age for Scots pine in Britain (Mason et al. 2004).

Materials and Methods

The two experiments described in this paper were located at Ceannacroc in north-west Scotland (57° 7' N, 4° 45' W) and at Hambleton in north-east England (54° 15' N, 0° 30' W). The Ceannacroc experiment was planted on a peaty podsol on undulating terrain at 150 m elevation with annual rainfall of 1500 mm while the Hambleton experiment was sited on a podsolic ironpan soil on level ground at an elevation of 210 m with an annual rainfall of 810 mm. Both sites were used for sheep grazing before planting in 1960 (Ceannacroc) and 1961 (Hambleton). At time of planting, both sites were characterised by heathland vegetation with heather (*Calluna vulgaris* L.) being the dominant species. Soil fertility of both experiments would be classed as 'very poor' using the Ecological Site Classification (ESC) (Pyatt et al. 2001), but soil moisture would be classed as 'very moist' at Ceannacroc and 'slightly dry' at Hambleton.

Both sites were cultivated before planting to reduce vegetation competition using a shallow (c. 20 cm deep) single mouldboard plough. The same experimental design was used at both locations with 4 treatments being compared, namely pure Scots pine, pure silver birch, a 1:1 mixture of both species, and a 3:1 mixture of Scots pine and silver birch. These treatments were laid out in a randomised block design with three replications of each treatment using a plot size of 0.2 ha with 900 plants per plot at a spacing of 1.5 m between and within rows. The mixture treatments were achieved by planting alternative 25 plant plots (5 by 5 trees) of each species in a chequer-board pattern. This design would be considered as a 'replacement series' (Sackville Hamilton 1994) since the focus of investigation is on the effect of contrasting species proportions at a constant spacing.

All replicates were located in close proximity to one another at the Hambleton site, but at Ceannacroc one block was located 900 m to the east on a similar site type. At Ceannacroc, all birch trees were fertilised in 1962 at a rate equivalent to 8 kg P ha^{-1}, but no other remedial treatments were undertaken at either site during the establishment phase. At Ceannacroc there was an unauthorised thinning in 2002 which removed a number of Scots pine trees from all plots where this species was present. There has been windblow of isolated trees within this experiment since 2002. The Hambleton experiment was thinned in 1998 and in 2003. These were thinnings from below which removed suppressed and sub-dominant trees, amounting to between 5 and 15 per cent of the basal area in each treatment.

The early assessment history was described by Malcolm and Mason (1999), but essentially involved measurements of height growth at 3, 6 and 10 years after planting, and estimates of basal area and standing volume at around 31-32 years of age. Subsequent assessments at 40, 45 and 55 years (Ceannacroc) and 38, 43 and 48 years (Hambleton) measured dbh of all trees in an internal 0.09 ha assessment plot to calculate basal area plus also providing estimates of top height. The only exception was at Ceannacroc where inspection of the 32 year data revealed very poor growth in one plot of the 3:1 mixture in block two which had been planted on a wet peaty soil: this plot was excluded in the later measurements. The variable thinning history described above with no precise measure of material removed has meant that we have had to use current basal area as a measure of productivity in these experiments.

Analysis of the data followed procedures used recently in examination of results from the long-term mixtures experiment at Gisburn (Mason and Connolly 2014). In brief, this involved comparing species performance pure and in mixture assuming that performance of an individual species in a mixed plot could be treated as an independent value. We then compared the overall performance of the pure species and the two mixed treatments using standard analysis of variance procedures. This was extended to compare the performance of the mixture treatments with what would have been expected from the growth of the species in the pure plots. For this purpose we calculated a *delta* statistic which is derived as (*actual basal area – predicted basal area*) where the *actual* value is the observed performance in mixture while the *predicted* value is based on the species performance in the pure plots. A delta statistic of zero implies that mixed stand performance conforms to the predictions derived from that of the component species in pure stands, a negative value indicates that performance in mixture is less than would be predicted, while a positive value is a sign of enhanced productivity in the mixed stand. We calculated the delta statistic for each mixture combination in each replicate and analysed the results with ANOVA. We also examined the results of the various mixture combinations using methods for presenting results from a replacement series (Kelty 1992).

Positive mixing effects can be shown when the productivity of a mixture is greater than that of pure stands of the individual component species. Such effects are classed either as 'overyielding' where the productivity of the mixture is more than the average of the pure stands or 'transgressive overyielding' where the mixture outyields the most productive of the pure species (Pretzsch 2009).

Results

At both sites, and when averaged over all treatments, there were major difference between the growth of Scots pine and silver birch, with trees of the first species generally being significantly taller and larger at most ages of assessment (Table 1). The only exception was in the first decade after planting when birch trees tended to be taller than the pines. Based upon top height measurements at the last assessment, productivity was similar at both sites being 10 m^3 ha^{-1} yr^{-1} for Scots pine and 4 m^3 ha^{-1} yr^{-1} for birch (Edwards and Christie 1981).

In contrast to the major difference found between the species at most ages, there were relatively few interactions between species and mixture treatments (Table 2). Those that occurred were due to birch trees growing in one or both of the mixture treatments being appreciably taller than those growing in the pure birch plots (e.g. at Hambleton in year 32). At the time of the last assessment, the density of

Table 1 - Comparative growth of Scots pine (SP) and silver birch (Bl) planted pure and in varying mixture proportions at different ages of stand development in two separate experiments in Scotland and northern England. Results are averaged over all treatments.

Parameter		Height (m)							Diameter (cm)			
Age (years)		3	6	10	32	40/38	45/43	55/48	32	40/38	45/43	55/48
Experiment	Treatment											
Ceannacroc	SP	0.2	1.1	2.6	13.5	-	-	18.9	14.5	17.8	21.4	22.3
	Bl	0.3	0.9	1.9	13.5	-	-	17.6	12.9	11.5	15.1	15.4
	Significance	***	**	***	ns	-	-	ns	**	***	***	*
	SED	0.01	0.04	0.1	0.4	-	-	0.8	0.5	1.1	1.0	3.1
	5%LSD	0.02	0.09	0.2	0.9	-	-	1.8	1.2	2.5	2.2	6.9
Hambleton	SP	0.5	1.3	3.1	13.4	15.3	16.1	17.2	13.9	16.7	17.0	19.2
	Bl	0.9	1.7	3.0	11.8	14.6	15.6	16.1	8.3	10.5	10.9	12.0
	Significance	***	***	ns	***	*	ns	**	***	***	***	***
	SED	0.02	0.05	0.1	0.2	0.3	0.2	0.2	0.3	0.5	0.7	0.7
	5%LSD	0.04	0.1	0.2	0.4	0.6	0.5	0.5	0.6	1.0	1.6	1.4

Notes:
1. Where two ages are given for an assessment, the first refers to the Ceannacroc experiment and the second to the Hambleton one.
2. Significance is defined as: ***= $p<0.001$, **= $p<0.01$, *= $p<0.05$, ns=non-significant.

Table 2 - Height and diameter growth of Scots pine (SP) and silver birch (Bl) planted pure and in varying mixture proportions at different ages of stand development in two separate experiments in Scotland and northern England. Also stand density at the last assessment.

Parameter		Height (m)							Diameter (cm)				Density (stems ha⁻¹)
Age (years)		3	6	10	32	40/38	45/43	55/48	32	40/38	45/43	55/48	55/48
Experiment	Treatment												
Ceannacroc	SP pure	0.2	1.0	2.5	13.0	-	-	18.6	13.6	16.4	20.7	24.0	807
	SP3:Bl1	0.2	1.1	2.6	13.5	-	-	18.3	14.8	18.1	21.1	25.2	467
	SP1:Bl1	0.3	1.1	2.7	14.1	-	-	19.7	15.1	18.5	22.4	26.1	459
	Bl pure	0.4	0.9	1.7	12.4	-	-	18.6	12.4	13.0	14.6	16.8	1374
	Bl1:SP3	0.3	0.8	1.8	14.2	-	-	16.1	14.4	9.4	16.4	19.1	194
	Bl1:SP1	0.3	1.0	2.1	14.0	-	-	18.2	11.9	12.2	14.3	16.8	364
	Significance	**	ns	ns	ns	-	-	ns	ns	ns	ns	ns	-
	SED	0.01	0.1	0.2	0.8	-	-	1.5	1.0	2.2	1.9	5.4	-
	5%LSD	0.03	0.2	0.3	1.7	-	-	3.5	2,3	5.0	4.3	11.9	-
Hambleton	SP pure	0.5	1.3	3.0	13.3	15.4	16.5	17.6	13.3	15.9	16.7	18.6	1585
	SP3:Bl1	0.5	1.3	3.0	13.4	15.4	16.0	17.4	13.6	16.4	17.3	19.0	1261
	SP1:Bl1	0.4	1.3	3.1	13.4	15.2	15.7	16.7	14.8	17.7	16.8	20.1	922
	Bl pure	0.9	1.7	3.0	10.4	13.8	15.0	15.9	7.9	10.4	10.8	11.9	1931
	Bl1:SP3	0.9	1.7	3.0	12.1	15.0	16.0	15.9	8.2	10.7	10.7	12.1	305
	Bl1:SP1	0.9	1.7	3.1	12.7	15.0	15.8	16.5	8.9	10.5	11.1	12.1	663
	Significance	ns	ns	ns	**	ns	*	*	ns	ns	ns	ns	-
	SED	0.03	0.1	0.1	0.3	0.5	0.4	0.4	0.5	0.8	1.3	1.1	-
	5%LSD	0.07	0.2	0.3	0.6	1.1	0.9	0.9	1.1	1.7	2.8	2.5	-

Notes:
1. In mixed plots, the value shown in a given row is for the first species listed, i.e. in SP3:Bl1 the value refers to the Scots pine component of the mixture.
2. Where two ages are given for an assessment, the first refers to the Ceannacroc experiment and the second to the Hambleton one.
3. Height measure is a mean height for years 3-10 and a top height measure thereafter.
4. Significance is defined as: **= $p<0.01$, *= $p<0.05$, ns=non-significant.

the pure birch treatment at both sites was appreciably higher than that of the pure pine plots (Table 2). The overall density of the mixture plots was similar to that found in the pure Scots pine, but the pine was the dominant component of the mixture. Thus at the last assessment date, the percentage of Scots pine stems per mixture was 71 per cent (3:1 mixture) and 56 per cent (1:1 mixture) at Ceannacroc: equivalent figures for Hambleton were 81 per cent and 58 per cent (Table 2).

In the Hambleton experiment, basal area production was significantly higher in pure Scots pine than in pure birch at all ages of assessment (Table 3). The values for the two mixtures were intermediate between the two pure plots, but were never significantly different from each other. The 1:1 mixture always had a significantly lower production than the pure Scots pine treatment, but the differences between the latter and the 3:1 treatment were smaller. However, production in the two mixture treatments was always higher than in the pure birch treatment. Results at Ceannacroc were much more variable, reflecting the impact of the unauthorised thinning when the trees were 42-years-old. Until that time, results reflected those from Hambleton with highest values in the pure Scots pine, lowest in the pure birch, and the two mixtures intermediate between the two pure plots. However, at the two later dates, there was little difference between any of the treatments, reflecting the preferential removal of the

Table 3 - Basal area production of Scots pine (SP) and birch (BI) grown pure and in varying proportions in mixture in two different experiments in Scotland and northern England.

Parameter		Basal area (m² ha⁻¹)			
Age (years)		32	40/38	45/43	55/48
Experiment	Treatment				
Ceannacroc	SP pure	41.0	49.7	29.0	34.9
	SP3:BI1	41.9	33.6	32.4	28.8
	SP1:BI1	34.2	42.3	32.1	32.8
	BI pure	19.8	27.5	30.7	33.3
	Significance	**	ns	ns	ns
	SED	2.6	9.5	3.6	7.1
	5%LSD	6.7	22.4	9.1	16.2
Hambleton	SP pure	42.3	42.0	44.9	44.8
	SP3:BI1	36.2	38.3	41.0	41.3
	SP1:BI1	34.2	35.6	30.0	38.5
	BI pure	24.4	23.0	24.7	24.6
	Significance	**	***	*	***
	SED	2.9	1.4	5.6	1.3
	5%LSD	7.1	3.2	13.8	3.1

Notes:
1. Where two ages are given for an assessment, the first refers to the Ceannacroc experiment and the second to the Hambleton one.
2. Significance is defined as: ***= $p<0.001$, **= $p<0.01$, *= $p<0.05$, ns=non-significant.

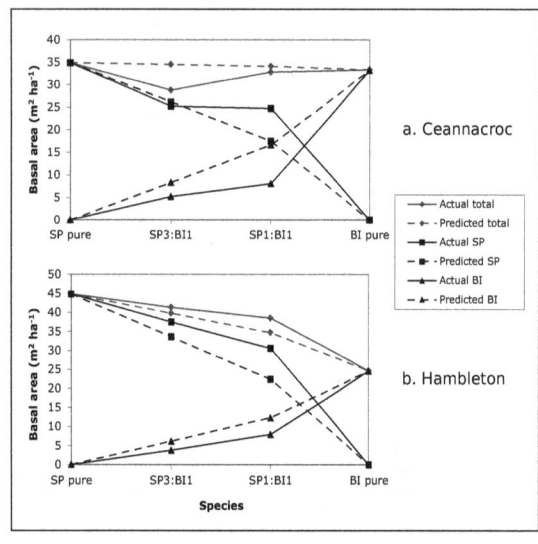

Figure 2 - Graphs of the basal area (m² ha⁻¹) at 55 years at Ceannacroc (Fig. 2a) and 48 years (Hambleton) (Fig. 2b) in two Scots pine-birch mixtures compared with the performance in pure plots of these species. Solid lines show the actual productivity in each treatment while the broken lines indicate the expected outturn if intra- and inter-specific interactions were equivalent.

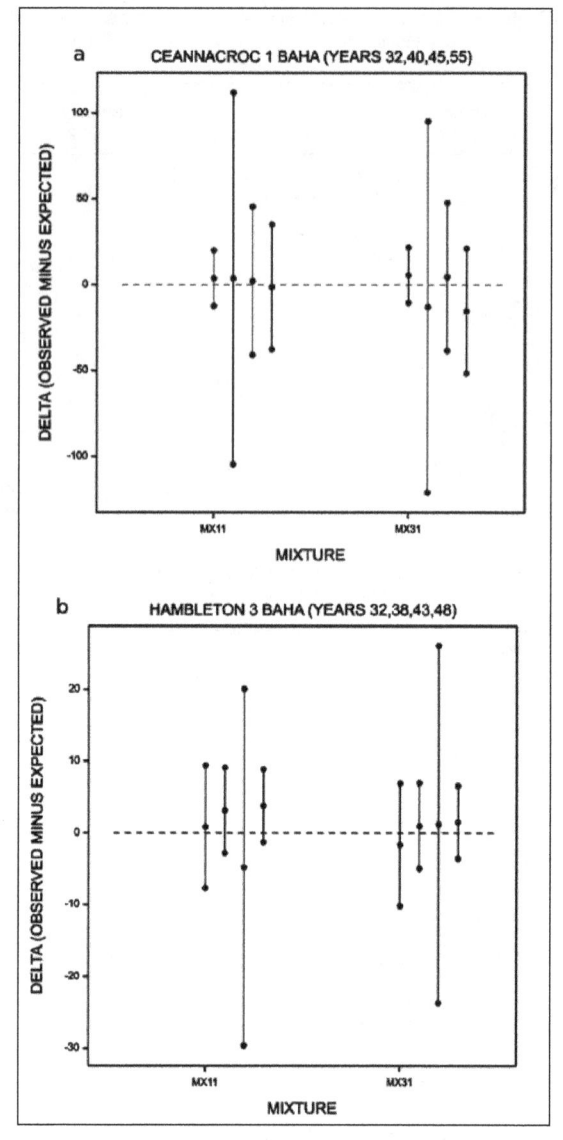

Figure 1 - Graphs showing the values of the delta statistic for differences in basal area (m² ha⁻¹) between the two Scots pine-birch mixture treatments and expected values based on the performance of the pure species plots in the experiments at Ceannacroc (Fig 1a) and Hambleton (Fig. 1b). Values are shown for the 1:1 and 3:1 Scots pine: birch mixture combinations in four different years covering tree ages 32-55 (Ceannacroc) and 32-48 (Hambleton). At each age of assessment, the mean delta value and the 95 per cent confidence interval is presented.

pine in the thinning. At both sites, Scots pine had the highest proportion of basal area in the mixtures, comprising 83 per cent (3:1 mixture) and 75 per cent (1:1 mixture) at Ceannacroc, compared to 91 per cent and 80 per cent respectively at Hambleton.

Examination of the growth of the mixed plots compared to predictions based on performance of Scots pine and silver birch in the pure plots (Fig. 1a and 1b) revealed a general tendency for performance of the mixed plots to be slightly better than predicted (i.e. delta statistic >0), but these differences were never significant. There was also little evidence of any difference between the two mixture proportions. There were a couple of assessments where there was substantial variation around the predictions, namely year 40 at Ceannacroc and year 43 at Hambleton. The latter almost certainly reflects the thinning carried out in 2003, but the cause of the former is unclear. In both experiments the Scots pine component was more productive in mixture than predicted whereas the reverse applied to the birch (Fig. 2). This differential performance between the species was most apparent in the 1:1 mixture.

Discussion

Although both Scots pine and birches are widely distributed in northern Europe and are often found growing in mixture (Hynynen et al. 2010), there is a surprising lack of long-term experimental evidence to indicate how stands composed of these two light-demanding, pioneer species might interact. These two experiments were planted in parts of Britain which experience different climates, with annual rainfall at Ceannacroc being at least twice that recorded for Hambleton, yet there was relatively little difference in tree growth and productivity between the two sites. This suggests that, despite the variation caused by the unauthorised removal of Scots pine in the Ceannacroc experiment, the pattern of growth in the mixtures would have been quite similar at both locations. For the rest of this discussion, we mainly focus upon the Hambleton experiment to try to understand the processes influencing the patterns of growth and development in these mixtures.

After the initial establishment phase, Scots pine was taller and larger than birch throughout the life of these stands, and so came to dominate the mixed plots (Table 2). There was a period around years 30-40 at Hambleton where birch appeared to grow taller in mixture as also reported from Scandinavian studies (Mielikäinen 1980, Kaitaniemi and Lintunen 2010) but this trend did not persist in the later years. As a result of this differential growth between the species, there was a slight suggestion of overyielding in mixture (Fig. 2 - Hambleton) due to the greater productivity in the Scots pine more than offsetting the lower production in the birch. The poorer performance of birch in mixture is also evident at Ceannacroc (Fig. 2) despite the likelihood that the removal of the pine in thinning would have reduced the amount of inter-specific competition. However, as yet the overall improved performance in mixture has been small and not significantly different from what would have been expected based on the performance of the pure plots (Fig. 1). At Hambleton, there was also evidence that overall basal area production in mixture declined with increasing proportion of birch (Table 3), in line with results recorded in Scandinavia (Frivold and Frank 2002, Hynynen et al. 2011). The slower rate of self-thinning in the pure birch plots (Table 2) will also have influenced the smaller diameters and lower heights recorded for this species compared to Scots pine (Table 1).

Examination of tree species' interactions in mixed stands typically distinguishes three types of response, namely 'competition', 'competitive reduction' and 'facilitation' (Forrester 2014). The first of these responses occurs when one species has a negative impact on the growth or survival of another. The second arises where competition between species is less intense than competition within species, normally because of differential resource use by the component species of the mixture. Facilitation arises when the species interact in such a way that the growth of at least one of the species is positively affected. The second and third response can be difficult to distinguish and therefore the combined response is sometimes referred to as 'complementarity' (Forrester 2014). Although previous reports had shown slight changes in soil properties (e.g. a small increase in pH) with increasing proportions of birch (Malcolm and Mason 1999), the lack of any significant overyielding effect in the mixtures suggests that facilitation is unlikely to have occurred in these experiments.

Therefore, it seems likely that the response observed in the mixtures in these experiments represents a balance between competition between the two species, and competitive reduction in that the Scots pine appears to benefit from reduced intra-species competition due to the presence of birch in the mixtures. A further indication of the latter process is provided by the densities observed in the mixed stands at Hambleton, where there was negligible difference between the combined species density in the pure Scots pine and in the two mixed plots (Table 2) whereas stocking of the pure birch was some 20 per cent higher. The slower rate of self-thinning and lower vigour recorded in the pure silver birch plots would accord with the view that this species performs less well in oceanic Europe (Pretzsch 2005) and reflects the recommendation that dense birch stands should be heavily thinned to maintain vigour and improve timber quality (Cameron 1996). Site quality could also have influenced the outcome if the sites were too nutrient poor for good birch growth, since Scandinavian experience is that typical pine sites are too poor for silver birch (Hynynen et al. 2010). However, evaluation of species potential on these sites using the British ESC system (Pyatt et al. 2001) suggests that growth of both Scots pine and silver birch would be less than optimal (grading of 'suitable' in ESC), with limitations imposed either by lack of soil nutrients or excessive soil moisture.

These mixture experiments are now of an age that is close to two-thirds of that found in a standard rotation for Scots pine in Britain, yet there is no evidence that the magnitude of the limited positive interaction in the mixed plots has altered over time (Fig. 1). This may reflect the fact that the two species are both light demanding and have other similar functional traits which mean that they have limited ecological combining ability (Kelty 2006). The pattern of mixing used in the design may also have influenced the development of the mixtures and the extent of any overyielding since the 'chequerboard' layout will have resulted in small pockets of intense within-species competition alternating with less intense areas of between species competition along the edges of the species groups. Thus, analysis of a similar chequerboard mixture experiment with Norway spruce (*Picea abies* L. (Karst.)) and Scots pine, showed that diameter growth of individual Norway spruce trees was negatively affected by increasing numbers of Scots pine, but there was no effect of increasing numbers of Norway spruce (Yanai, 1992).

Conclusion

As noted earlier, one practical benefit from these experiments is to help improve understanding of the dynamics of the Scots pine-birch mixtures that can develop within the Caledonian pinewoods of northern Scotland, especially in the more oceanic western part of the pinewood zone (Edwards and Mason 2006). The results presented here do not suggest that there is much likelihood of a long-term coexistence of Scots pine and birch in intimate single storeyed mixtures, but rather that the more vigorous growth of the pine will tend to progressively eliminate the admixed broadleaved species. Any plan to enhance the proportion of birch in the Caledonian pinewoods would seemingly need to develop small blocks of birch within a pine matrix that were large enough to withstand the competitive pressure exerted by the pine and which could act as a future seed source.

Acknowledgements

This paper was prepared within the frame of the EU Cost Action FP1206 - European Mixed Forests: integrating scientific knowledge in sustainable forest management. We are grateful to two anonymous referees and the journal's editor for helpful comments on the draft manuscript. We acknowledge the major contribution of the staff of Forest Research's Technical Support Unit in maintaining and assessing these experiments.

References

Anonymous 2010 - *A Guide for Increasing Tree Species Diversity in Wales*. Forestry Commission Wales, Aberystwyth. 41p.

Anonymous 2011 - *The UK Forestry Standard*, 116 p. http://www.forestry.gov.uk/pdf/FCFC001.pdf/$FILE/FCFC001.pdf . Accessed on May 10 2015.

Brooker R.W., Maestre F.T., Callaway R.M., Lortie C.L., Cavieres L.A., Kunstler G., Liancourt, P., Tielborger, K., Travis, J.M.J., Anthelme, F., Armas, C., Coll, L., Corcket, E., Delzon, S., Forey, E., Kikvidze, Z., Olofsson, J., Pugnaire, F., Quiroz, C.L., Saccone, P., Schiffers, K., Seifan, M., Touzard, B., Michalet, R. 2008 - *Facilitation in plant communities: the past, the present, and the future*. Journal of Ecology 96: 18-34.

Cameron A.D. 1996 – *Managing birch woodlands for the production of quality timber*. Forestry 69: 357-371.

Dimbleby G.W. 1952 - *Soil regeneration on the north-east Yorkshire moors*. Journal of Ecology 40: 331-341.

Edwards C.E., Mason W.L. 2006 - *Stand structure and dynamics of four native Scots pine (Pinus sylvestris L.) woodlands in northern Scotland*. Forestry 79: 261-277.

Edwards P., Christie J.M. 1981 - *Yield models for forest management*. Forestry Commission Booklet 48, HMSO, London, UK.

Felton A., Lindbladh M., Brunet J., Fritz O., 2010 - *Replacing coniferous monocultures with mixed-species production stands: an assessment of the potential benefits for forest biodiversity in northern Europe*. Forest Ecology and Management 260: 939-947.

Forestry Commission 2003 - *National Inventory of Woodland and Trees: Great Britain*, 68 p. http://www.forestry.gov.uk/pdf/nigreatbritain.pdf/$FILE/nigreatbritain.pdf Accessed on May 15, 2015.

Forrester D.I. 2014 - *The spatial and temporal dynamics of species interactions in mixed-species forests: from pattern to process*. Forest Ecology and Management 312: 282-292.

Frivold L.H., Frank J. 2002 - *Growth of mixed birch-coniferous stands in relation to pure coniferous stands at similar sites in south-eastern Norway*. Scandinavian Journal of Forest Research 17 (2): 139-149.

Gardiner A.S. 1968 - *The reputation of birch for soil improvement*. Research and Development Paper 67, Forestry Commission, Edinburgh.

Gamfeldt L., Snall T., Bagchi R., Jonsson M., Gustaffson L., Kjellander P., Ruiz-Jaen, M.C., Froberg, M., Stendahl, J., Philipson, C.D., Mikusinski, G., Andersson, E., Westerlund, B., Andren, H., Moberg, F., Moen, J., Bengtsson, J. 2013 - *Higher levels of multiple ecosystem services are found in forests with more tree species*. Nature Communications 4: 1340.

Grant A., Worrell R., Wilson S., McG., Ray D., Mason W.L. 2012 - *Achieving diversity in Scotland's forest landscapes*. Forestry Commission Scotland Practice Guide, Forestry Commission, Edinburgh, 30 p.

Hynynen J., Niemisto P., Vihera-Aarnio A., Brunner A., Hein S., Velling P. 2010 - *Silviculture of birch (Betula pendula Roth and Betula pubescens Ehrh.) in northern Europe*. Forestry 83: 103-119.

Hynynen J., Repola J., Mielikainen K. 2011 - *The effects of species mixture on the growth and yield of mid-rotation mixed stands of Scots pine and silver birch*. Forest Ecology and Management 262: 1174-1183.

Kaitaniemi P., Lintunen A. 2010 - *Neighbour identity and competition influence tree growth in Scots pine, Siberian larch, and silver birch*. Annals of Forest Science 67: 604. DOI: 10.1051/forest/2010017

Kelty M.J. 1992 - *Comparative productivity of monocultures and mixed species stands*. In: "The Ecology and Silviculture of mixed-species forests". Kelty M.J., Larson B.C., Oliver C.D. Eds., Kluwer Academic Publishers: 125-141.

Kelty M.J. 2006 - *The role of species mixtures in plantation forestry*. Forest Ecology and Management 233: 195-204.

Malcolm D.C., Mason, W.L. 1999 - *Experimental mixtures of Scots pine and birch: 30 year effects on production, vegetation and soils*. In: 'Management of mixed-species forest: silviculture and economics'. Olsthoorn A.F.M. et al., Eds. IBN Scientific Contributions 15, IBN-DLO, Wageningen; 79-87.

Mason W.L. 2007 - *Changes in the management of British forests between 1945 and 2000 and possible future trends.* Ibis 149: 41–52.

Mason W.L., Hampson A., Edwards C. 2004 - *Managing the Pinewoods of Scotland.* Forestry Commission, Edinburgh, 234 p.

Mason W.L., Perks M.P. 2011 - *Sitka spruce (Picea sitchensis) forests in Atlantic Europe: changes in forest management and possible consequences for carbon sequestration.* Scandinavian Journal of Forest Research, supplement 11: 72-81.

Mason W.L., Connolly, T. 2014 - *Mixtures with spruce species can be more productive than monocultures: evidence from the Gisburn experiment in Britain.* Forestry 87 (2): 209-217.

Mielikäinen K. 1980 - *Structure and development of mixed pine and birch stands.* Commun. Inst. For. Fenn. 99:1-82 (Finnish with English summary).

Mielikäinen K. 1996 – *Approaches to managing birch-dominated mixed stands in Finland.* In: 'Silviculture of temperate and boreal broadleaf-conifer mixtures'.Comeau, P.G., Thomas, K.D. Eds. Land Management Handbook 36, BC Ministry of Forests, Victoria, British Columbia, Canada, 8-14.

Miles J. 1981 - *Effect of birch on moorlands.* Institute of Terrestrial Ecology, Cambridge. 18 p.

Oliver C.D., Larson B.C. 1996 - *Forest stand dynamics.* McGraw-Hill, New York.

Paquette A., Messier C., 2011 - *The effect of biodiversity on tree productivity: from temperate to boreal forests.* Global Ecology and Biogeography 20: 170-180.

Pretzsch H. 2005 - *Diversity and Productivity in Forests: evidence from long-term experimental plots.* In: 'Forest diversity and function: temperate and boreal forests'. Scherer-Lorentzen M., Korner Ch., Schulze E.D. Eds. Springer-Verlag, Berlin, 41-64.

Pretzsch, H. 2009 - *Forest dynamics, growth and yield.* Springer-Verlag, Berlin, p664.

Pyatt D.G., Ray D., Fletcher J. 2001 - *An ecological site classification for forestry in Great Britain.* Forestry Commission Bulletin 124. Forestry Commission, Edinburgh, UK. 74. p

Quine C.P., Bailey S.A., Watts K. 2013 - *Sustainable forest management in a time of ecosystem services frameworks: common ground and consequences.* Journal of Applied Ecology 50: 863-867.

Read D.J., Freer-Smith P.H., Morison J.I.L., Hanley N., West C.C., Snowdon P. (eds) 2009 - *Combating climate change - A role for UK Forests. An assessment of the potential of the UK's trees and woodlands to mitigate and adapt to climate change.* The Stationery Office, Edinburgh (UK).

Sackville Hamilton, N.R. 1994 - *Replacement and additive designs for plant competition studies.* Journal of Applied Ecology 31: 599-603.

Yanai R.D. 1992 - *Competitive interactions between Norway spruce and Scots pine at Gisburn Forest, NW England.* Forestry 65: 435-451.

Zhang Y., Chen H.Y.H., Reich P.B. 2012 - *Forest productivity increases with evenness, species richness and trait variation: a global meta-analysis.* Journal of Ecology 100: 742-749.

Cultivation techniques in a 34 years old farming oak (*Quercus robur* L.) plantation in the Arno valley (Tuscany)

Serena Ravagni[1*], Claudio Bidini[1], Elisa Bianchetto[2], Angelo Vitone[3], Francesco Pelleri[1]

Abstract - This report aims to provide a description of the cultivation techniques adopted at the oldest oak pure plantation (age 34) established within the environmental restoration plan of soil dumps at the Santa Barbara ENEL Company opencast mine in Cavriglia (AR). The goals of the initial plan, following which the plantation was carried out, were to (i) verify the possibility of restoring the soil dump by using tree farming plantations and produce a range of valuable timber assortments; (ii) test the growth potential of the oak species in a pure plantation. The plantation, carried out with a stem density of 1,111 trees per hectare, was managed by targeted practices (pruning up to the stem height of 4-5 m and then undertaking four thinnings). A geometric-selective thinning was applied first and, later, periodical thinning from above was implemented, releasing, as a result, about 70 crop trees per hectare. In the meanwhile, it was possible to monitor the growth parameters concerning dbh, tree height and crown diameter. The stem quality and the presence of epicormic branches were evaluated in 2013 and the relationship between the presence of epicormic branches and tree characteristics were also analyzed. The wood production was compared with other European plantations, especially from France. Today, 34 years after the plantation got started, the site is an interesting case-study of tree farming as it plays a consistent role within the environmental restoration of the area. The applied management system allowed to reach a noticeable wood production level and also valuable timber assortments for industrial use. Crop trees (70 per hectare) reached a mean dbh of 38.1 cm, the height of 22.3 m at the age of 34.

Keywords - English oak, thinning, valuable timber, tree farming, plantation.

Introduction

English oak (*Quercus robur* L.) is a quite widespread species in Europe, with geographical distribution from the Atlantic coast of France and northern Portugal up to the Urals and the Caucasus, and from Britain and southern Scandinavia up to northern Greece, the Italian peninsula and the Pyrenees. In Italy, it is present in almost all the regions, especially in the North and mainly in the plains. English oak needs a constant and continuous level of water availability in the soil, proving to be susceptible to drought. In dry years, in fact, the more sized trees, may easily show typical phenomena of desiccation, in the upper canopy.

English oak prefers soils with shallow and stable groundwater, tolerating periodic flooding of the root system up to 2-3 months. Such behavior is made easier by the rather shallow roots, this allowing bearing the lack of oxygen in the soil, but making them less suited to preventing drought occurrence. The species is therefore naturally located at the base of slopes and at valley bottoms (Frattegiani 1996, Lemaire 2010, Mori et al. 2007, Sevrin 1997).

English oak is a very light-demanding species and it can be considered a pioneer tree, because of its ability to colonize open spaces and abandoned fields. Being so light-demanding, it requires a dynamic silvicultural approach with frequent thinnings, ensuring an adequate crown development and a regular diameter growth, in order to get the best growth performance (Lemaire 2010).

In Europe, English oak is cultivated for valuable timber production. At this purpose, many experimental trials have been set up to define suited management criteria and get valuable timber production in a shorter time-span than according to customary management system (Nebout 2006). All these trials are characterized by an initial phase (qualifying period) where the tree stand is maintained at high stem density to favor the natural pruning. This phase is necessary to get straight stems without branches. The best crop trees are then selected and frequent thinnings from above are carried out all around them

---

[1] Consiglio per la ricerca in agricoltura e l'analisi dell'economia agraria, Forestry Research Centre (CREA-SEL), Arezzo, Italy
[2] Consiglio per la Ricerca e la sperimentazione in Agricoltura e l'analisi dell'economia agraria, Centro per l' Agrobiologia e la Pedologia (CREA-ABP), Firenze, Italy
[3] Dottorando, Dipartimento di Bio-scienze e Territorio (DiBT), Università degli Studi del Molise
* corresponding author: serena.ravagni@gmail.com

(thinning phase) to get the free growth of crown and, at the same time, maintain the diameter growth of selected trees high and costant.

Experiences about this issue can be found in Britain since the 1950s with dense oak plantations aged 20 (Jobling and Pearce 1977, Kerr 1996). At the moment, similar management criteria are being applied in Central Europe (Lemaire 2010, Perin and Claessens 2009, Nebout 2006) and we also have, in Italy, a few cases aimed at getting valuable, large-sized stems in a shorter time, as compared with traditional management (Corazzesi et al. 2010).

Furthermore, while in central and northern Europe we have a good general ecological, genetic, eco-physiological, technological, silvicultural and operational knowledge concerning valuable oaks, in the Mediterranean area this background is limited and, moreover, the biological response of these oak species is not well-known. In spite of this, the residual oak stands in southern Europe are extremely important as a relict source of genetic variability (Aa.Vv. 1999, Ducci 2007).

In Italy, English oak is particularly widespread in the lowlands and also in the alluvial plain of the Arno valley. Its presence has been greatly reduced since Middle Ages, following the diffusion of human settlements and the population increase, which was due to the progressive deforestation in favor of the agricultural practice. In addition, the widespread coppicing system did not support the species because of its lower sprouting ability compared with the other tree species associated in mixed forests.

Since the 1950s, the progressive re-diffusion of secondary forests, due to the abandonment of agriculture and pasture activity in the mountains and hilly marginal areas, has been noticeable, whilst more recently a growing attention towards the protection of lowland forests and the establishment of new forests has been developed. Since the 1980s, new plantations have been carried out in the Po valley (Pividori et al. 2015) under the financial support of both Regional governments and the European Union (set-aside, EEC Regulation 2080/92 and Rural Development Plans).

The widespread cultivation of English oak is linked to its timber value, workability and aesthetical features. Moreover, since a few decades, the market of valuable broadleaved trees acknowledges increasing prices to high quality trunks for veneers and furniture. In this latter case the market requires straight and healthy, cylindrical trunks, free of knots, with a larger than 50 cm diameter and a regular growth course, i.e. the awaited goals to be pursued when cultivating valuable tree species.

Since the late 1970s, in the Valdarno Aretino, over 240 hectares of plantations for timber production were carried out in a close cooperation between the Forestry Research Institute (now CREA-SEL) and ENEL Co., according to the Plan of environmental restoration of the wide landfill mining, stocking the thermal power station of Santa Barbara (AR).

English oak was the most used tree species in these reforestation activity. The reasons for this are twofold: on the one hand, English oak was chosen because of its own feature of pioneer species suited to the barren soil of the mining area, on the other hand, to test the opportunity to use the species, poorly known in Italy, in tree farming plantation for valuable timber production (Buresti 1984).

The plantation here analyzed and others carried out in the Santa Barbara district, resulted to be an important training for the experimental activity in tree farming. The significance of this site lies, therefore, in testing new pruning techniques and different thinning trials in addition to verifying the oak potential in the concerned area. The outputs of this experience were later applied at other Italian sites. Given the lack of specific models and information about planting and management for similar growth environments, the applied criteria described in this paper should be considered and evaluated as a fully experimental trial. This paper does not intend therefore to propose a model, but to report methods and results achieved so far.

The early goals may be summarized as follows: (i) restoration of an environment heavily modified, ensuring, first of all, the ground cover and then the recovery of biological activity in the soil, which was greatly reduced, if not absent, at the time of planting; (ii) checking the cultivation of English oak at this site and defining management techniques suited to produce valuable timber in a life-span of 40-50 years with the traits required by industrial processing.

Materials and methods

Study site

The area is characterized by an average annual rainfall of 927 mm and an average temperature of 13° C, with a dry period in July. The soil, at the time of planting, had very special characteristics, resulting from the accumulation of inert layers; soil texture was silty-clay, with a sub-acid pH and a balanced amount of the main nutrients (N, P, K), but also with a reduced presence of calcium (Buresti 1984).

Plantation design

The pure plantation was established in November 1979 in an area of 2,700 m^2, according to a square design with a spacing of 3 m and with a density of 1,111 trees per hectare. 1 year old oak seedlings were used, choosing a provenance which was close to the site (the forest of Renacci). Extensive deep ploughing and hoeing all around the seedlings were implemented for two years after planting, to reduce weeds and shrubs competition. Pruning was carried out up to the height of 5-6 meters over the following years. Annual dbh inventory was carried out since 2000. Total height, tree crown insertion and dbh of dominant trees were periodically measured.

Thinning

The plantation underwent four thinning operations. The first thinning was carried out at the age of 13 (winter 1992) according to a mixed geometric-selective design. Tree crowns began to touch one another and it was necessary to intervene to prevent the occurrence of competition for light, which would negatively affect their diametrical growth.

Given the young age of the plantation and its still evident homogeneity, a geometric thinning was applied with rare exceptions. 50% of the trees were felled, following alternate diagonal rows; the selective criterion was applied only in few cases to preserve good-shaped trees (Buresti et al. 1993).

A second thinning was performed in 1996 with a selective criterion, still removing about the 50% of the trees. At this time, the worst and less vital phenotypes were felled, and also a relatively even distribution of the trees on the ground was maintained. Four years after the first thinning, tree crowns were already in contact but no significant reductions in diametrical growth were recorded. The surveys

View of the plot at the age of 24 years following the third thinning.

showed the progress of individual differentiation regarding tree vitality and stem shape and this is the reason why a selective design was applied for the second thinning (Buresti et al. 2000).

Following the French and Belgian experiences (Sevrin 1997, Baar at al. 2005, Baar 2010, De Potter et al. 2012), the third thinning did not concern the whole plantation, but a few crop trees: 70 superior phenotypes per hectare were chosen and a thinning from above was performed around them.

According to Lemaire (2010), a regular and sustained diametric growth of the crop trees is provided by wide and well-lighted crowns, with green branches for at least half of total stem height.

Three main criteria were followed for selecting the crop trees:

View of the plot at the age of 34 years following the fourth thinning.

Final crop tree. Age 34 years.

1. Average distance between the trees of about 12 meters;
2. Preference for the most vital and dominant trees;
3. Preference for the trees with the following stem qualities: straightness, knots absence, lack of injuries or pathogens.

Following the selection of crop trees, further thinnings were carried out in 2003 and in 2013, surrounding competitors were progressively felled to increase the space available for the crown development. About 40% of the trees were removed in 2003, and about 33% in 2013.

At each thinning operation, the stem volume of the felled trees was measured analyzing the stem sections, up to the top diameter of 5 cm, in order to build up a local volume function of the English oak plantations in the Arno valley (Marchi et al., forthcoming).

Epicormic branches

One year after the last thinning, a survey of the epicormic branches distribution was conducted to record their presence, their age and distribution along the stems. The data were analyzed according to the Pearson's χ^2 test. The aim of the analysis was to identify the possible relationship between the presence of epicormic branches, the tree position within the plantation (edge tree, inner tree) and the Dch/H (crown diameter to total tree height ratio).

Results and discussion

Mensurational parameters

The analysis of dbh and the total tree height datasets was focused on the crop trees, i.e. on trees concerned since the third thinning. Fig. 1 and 2 show the regular and sustained increase both of dbh and tree height, their values being on average 38.1 cm and 22.3 m, respectively, at the age of 34.

The analysis of the dbh current level (c.a.i.) of the crop trees (Fig. 3) highlights its decrease since 2001 (age 22), which was triggered because of the starting competition for the light. To avoid this effect and the loss of crown reaction, a further thinning was recommended. The thinning, carried out in 2003, at the age of 24, had beneficial effects, i.e. it gave rise to the positive trees reaction to the major light availability and we had the recovery of c.a.i. to value higher than 1 cm.

Dbh c.a.i. kept rather stable and sustained up to 2010, with a heavy reduction in the very dry year 2007 (age 28). Since 2010, the reduction of dbh c.a.i. would have required a timely thinning, which was carried out only in 2013, when a collapse of average dbh c.a.i. up to 0.5 cm yr^{-1} took place.

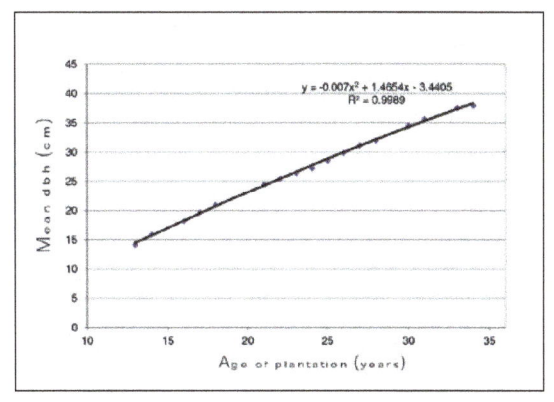

Figure 1 - Crop trees: trend of mean dbh.

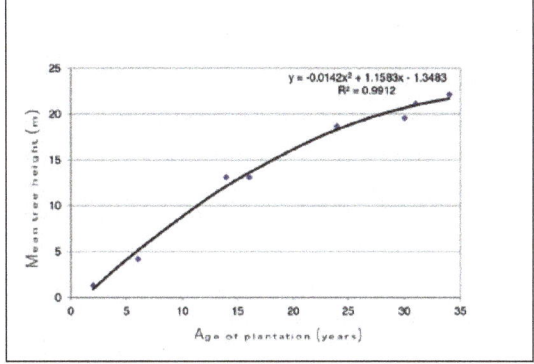

Figure 2 - Crop trees: trend of mean height.

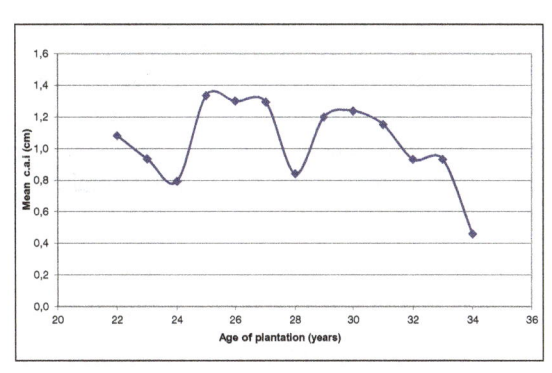

Figure 3 - Crop trees: dbh c.a.i. trend.

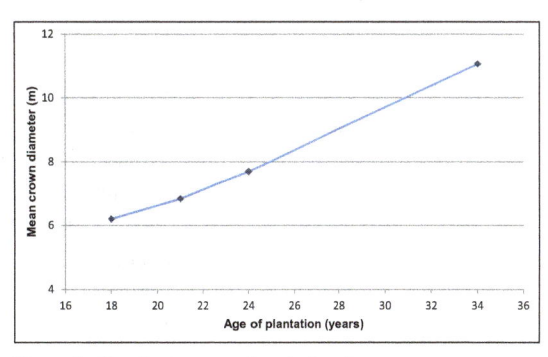

Figure 4 - Crop trees: crown diameter trend.

If we take into account the pattern of crop trees' crown diameter increase (Fig. 4), there is evidence of the regular growth, without any remarkable change over time.

In a few European papers some mensurational parameters related to common oak stands (in free-

Table 1 - Variation of mensurational parameters with plantation age.

Year	plant. age	n. trees (n ha⁻¹)	mean dbh (cm)	tree height (m)	tree volume (m³ ha⁻¹)	removal (m³ ha⁻¹)	canopy cover (%)
1993 pre-thin.	14	1,111	12.5	11.0	73.06		/
1993 post-thin.	14	588	14.2	10.9	40.17	32.89	/
1996 pre-thin.	18	588	17.2	12.4	84.63		/
1996 post-thin.	18	281	20.0	/	43.74	40.89	63%
2003 pre-thin.	24	277	25.6	17.8	113.88		79%
2003 post-thin.	24	164	27.9	/	73.90	39.98	57.6%
2013 pre-thin.	34	157	36.9	22.2	146.04		84.7%
2013 post-thin.	34	105	37.7	22.3	101.82	44.22	72.5%

growing and unthinned) are being compared. These parameters are defined as "shape parameters" and provide an indication of English oak growth pattern (Lemaire 2010, Perin et al. 2009, Jobling et al. 1977).

A synthetic competition index, applied to determine the competition level as a function of the crown width, is given by crown diameter to total height ratio (Dch/H). The right balance is achieved with values varying between 50% and 60%.

At the Santa Barbara plantation, Dch/H values between 40% and 60% were measured, i.e. very close to the optimal readings highlighted by the French trials. Even the index height of crown insertion to total height ratio (Hi/H) provides optimal figures, lower than 40% and typical of trees with deep crowns.

Accounting for stem volume felled at the different thinning times reported in Tab. 1 and Fig. 5, its value was fairly constant at each intervention, with a total removal of 158 m³ per hectare. Total mass reaches 260 m³ ha⁻¹.

Epicormic branches

Data analysis say that, at the present time, the 33% of trees do not host any sort of epicormic branches. The 57% of the stem sprouts are one year old and produced after the last thinning (2013). The 53% of one year sprouts were present on two trees. These are not located at the plantation edge and do not show a low Dch/H relationship. In the case, the genetic component is probably prevailing and plays an basic role in the phenomenon (Servin 1997, Attocchi 2013a). The χ^2 analysis highlighted that trees with Dch/H values lower than 0.5 showed an increased frequency of sprouts. This index is considered reliable to evaluate the tendency to the emission of epicormic branches. Dch/H value in the range 50-60% highlights crowns wide and deep enough to be less prone to new epicormic branches production in English oak (Lemaire 2010).

Pruning, and especially thinning play an important role in the management of oak plantations as a stimulus to the production of new epicormic shoots (Attocchi 2013a). From medium to low tree density plantations, pruning is necessary to get valuable timber productions whilst, in the most dense plantings, natural pruning is usually prevailing and it is integrated only when necessary (Spiecker 1991, Weaver and Spiecker 1993, Attocchi 2013b). The risk of newly-established epicormic shoots may be reduced by early and progressive pruning, carried out only around final crop trees. At each pruning time, the lower branches have to be removed only to get a clean bole free of branches up to 50% of total stem height. At this type of plantation a clean bole of 5-6 m (25-30% of final height) may be considered an awaited goal. Where possible, a basic role is to preserve the dominated layer and favor, in this way, both natural or artificial establishment of an understory (which is characterized by shade-tolerant species, able to reduce the direct enlightenment of the stems). Recently in a few European countries, mixed plantation are being preferred, intercropping oaks or other valuable broadleaved species with nurse trees (alders, hornbeams, limes, etc.) and shrubs like (hazel, elder, etc) (Hochbichler 1993, Buresti et al. 2006).

Stem quality

Following the last thinning, stems of selected trees were individually classified on the basis of the stem quality at the end of the crop cycle (Nosenzo et al. 2008). The 41% of the crop trees had a first stem, 2.5 m long, attributable to Class A, i.e. suitable for the more profitable uses, such as veneer. The 47% had a first stem attributable to Class B, suitable

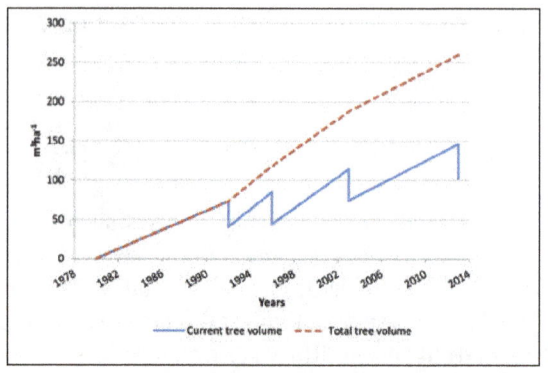

Figure 5 - Current and total tree volume trends.

for fine saw timber, while the 12% had a first part attributable to class C, i.e. to standard saw timber.

Conclusions

The soil condition at the time of plantation was extremely poor: no vegetation cover in any form or organic layer were present. The current well-established tree farming plantation fully satisfies the manifold goals of environmental recovery expected by the owner, ENEL Co., on the landfill grounds, i.e. stabilize the soil; limit the erosive action of running water on bare soil; form again an organic layer; restore the biological growth medium and the landscape. At the time of the implementation, and according to the already achieved experience, the chances of a successful establishment of the plantation were really uncertain, in terms of both tree species choice (design, pure or mixture) and physical environment of introduction.

Under this perspective, the trial may be considered as a fully experimental implementation. 34 years later, not only the goal of mining area restoration, but also the results in terms of trees' growth performance have been reached. The limited practices implemented in the soil after the plantation, confirm the ability of English oak to act as a pioneer species and to colonize difficult and poor soils, provided that a good water availability is being ensured.

Dbh c.a.i. varying from 0.8 to 1.2 cm is a quite good performance both in terms of growth and of growth-steadiness along the full time-span up to 2011. The sharp reduction occurred over the last two years is basically linked to the delay of the fourth thinning operation and to the particularly unfavorable rainfall pattern in 2011 and 2012. At the age of 34, crop trees have achieved an average diameter of 38.1 cm and an average height value of 22.3 m; these performances are fully comparable with those of the French forests - first site-class (Lemaire 2010). The total wood production at the site is about 260 m^3ha^{-1}, where the standing volume is about two-thirds of the intermediate yield at the age of 34. This ratio being an own attribute of this type of tree farming.

Due attention has to be paid over the cultivation-span to crowns cover fulfilment, given their prolonged compression causes the death of long-shaded lower branches, the consequence being the loss of trees' ability to react further to thinning occurrence and the reduction in diameter growth.

The rule to keep deep enough crowns in the plantation management has to be underlined, as well as to get and maintain Dch/H values above or close to 0.50. This rule has probably also limited the sprouting of epicormic branches along the stem which often occurs as a result of delayed thinning, producing a reduction of wood quality.

Even if today a few design features of the original plantation could be improved further to current experience, the case-study may be considered a relevant trial for two reasons. The first relates to the potential of English oak which, if properly managed, can ensure interesting results in terms of wood production on relatively short rotations. The second highlights the importance, for the plantation design, of giving each crop tree a suitable space for crown development (Buresti et al. 2006, Buresti Lattes and Mori 2009, Buresti Lattes and Mori 2012). It is fundamental to increase this individual available space over time, as a function of the progressive expansion of the crowns. This is, actually, the only way to ensure the active photosynthetic area, capable of maintaining constant and high radial stem increment.

Acknowledgements

A special thank to the ENEL Co. for the continuative cooperation in the conduct of the trial, to Enrico Buresti, our friend, teacher and former colleague who planned and started the trials, to the technical staff of CREA-SEL, especially to Eligio Bucchioni and Walter Cresti.

We acknowledge also the anonymous referees for their helpful comments contributing the improvement of the paper.

References

Aa.Vv. 1999 - *Conservation and valorisation of valuable oaks in Mediterranean European union countries* - COST RTD project.

Attocchi G. 2013a - *Pruning effects on the production of new epicormic branches: a case study in young stands of pedunculate oak (Quercus robur L.)*. In: Proceeding of 4th International Scientific Conference on Hardwood Processing (ISCHP 13), 7th-9th October 2013, Florence, Italy: 44-49.

Attocchi G. 2013b - *Effects of pruning and stand density on the production of new epicormic shoots in young stands of pedunculate oak (Quercus robur L.)*. Annals of Forest Science 70: 663-673.

Barr F. 2010 - *Synthèse de réflexions sur la sylviculture d'arbres-objectif en peuplement irrégulier ou équienne, mélangé ou non*. Service public de Wallonie, Direction générale opérationnelle - Agriculture, Ressources naturelles et Environnement, Liège, 56 p.

Baar F., Balleux P., Claessens H., Ponette Q., Snoeck B. 2005 - *Sylviculture d'arbres-objectif en hêtre et chêne: mise en place d'un dispositif de parcelles de demonstration et d'expérimentation*. Forét Wallonne 78: 34-46.

Buresti E. 1984 - *Il restauro forestale delle discariche minerarie dell'ENEL. Miniera di S. Barbara nel Valdarno*. Annali dell' Istituto Sperimentale per la Selvicoltura XV: 157-171.

Buresti. E., Frattegiani M., Sestini L. 1993 - *Prove di diradamento in un impianto di farnia (Quercus robur L.) in Valdarno. Moduli colturali tra arboricoltura e selvicoltura.* Note di informazione sulla ricerca forestale III (2): 7-8.

Buresti E., De Meo I., Pelleri F. 2000- *Criteri e risultati di un diradamento in un impianto di arboricoltura da legno farnia (Quercus robur L.)* Annali dell'Istituto Sperimentale per la Selvicoltura XXIX : 29-40.

Buresti E., Mori P., Pelleri F., Ravagni S. 2006 - *Enseignements de 30 années de rechercher sur les plantations mélangées en Italie.* Forêt-enterprise 170: 51-55.

Buresti Lattes E., Mori P. 2009 - *Impianti policiclici permanenti. L'arboricoltura da legno si avvicina al bosco.* Sherwood - Foreste ed alberi oggi 150: 5-8.

Buresti Lattes E., Mori P. 2012- *Piantagioni policicliche. Elementi di progettazione e collaudo.* Sherwood - Foreste ed alberi oggi 189: 12-16.

Corazzesi A., Tani A., Pelleri F. 2010 - *Effetto della consociazione e del diradamento in un impianto di arboricoltura da legno con latifoglie di pregio dopo oltre 20 anni dall'impianto.* Annali CRA- Centro di Ricerca per la Selvicoltura 36: 37-48.

De Potter B., Perin J., Ponette Q., Claessens H. 2012- *Détourage d'arbres-objectif: enseignements des dispositifs installés en Wallonie après six années.* Forêt Wallonne 119: 43-54.

Ducci F. (a cura di) 2007- *Le risorse genetiche della farnia della Val Padana.* Cra Issel, Regione Lombardia, Ersaf, casa editrice Le Balze, Montepulciano (SI), 143 p.

Frattegiani M. 1996 - *La farnia.* Sherwood - Foreste ed alberi oggi 16: 19-22.

Hochbichler E. 1993 - *Methods of oak silviculture in Austria.* Annals of Forest Science 50: 583-591.

Kerr G. 1996 - *The effect of heavy or 'free growth' thinning on oak (Quercus petraea and Q. robur).* Forestry 69 (4): 303-316.

Jobling J., Pearce M.L. 1977 - *Free growth of oak.* Forestry Commission Forest Record 113. HMSO, London, 16 p.

Lemaire J. 2010 - *Le chene autrement. Produire du chêne de qualitè en moins de 100 ans en futaie reguiliere.* Guide tecniche IDF, 176 p.

Marchi M., Ravagni S., Pelleri F. 2015 - *Volume function for the tree farming English Oak plantations of the Valdarno (Tuscany-Italy).* Annals of Silvicultural Research, (forthcoming).

Mori P., Bruschini S., Buresti E., Giulietti V., Grifoni F., Pelleri F., Ravagni S., Berti S., Crivellaro A. 2007 - *La selvicoltura delle specie sporadiche in Toscana. Supporti tecnici alla Legge Regionale Forestale della Toscana.* ARSIA Firenze, 355 p.

Nebout J.P. 2006 - *Des chênes en croissance libre: bilan et perspectives.* Bollettin 3 (LII) Société Forestière de Franche-Comtê: 103-135.

Nosenzo A., Berretti R., Boetto G. 2008 - *Piantagioni da legno: valutazione degli assortimenti ritraibili.* Sherwood - Foreste ed alberi oggi 145: 15-20.

Perin J., Claessens H. 2009 - *Considerations sur la designation et le dètourage en chênes et hetre.* Fôret Wallonne 98: 39-52.

Pividori M., Marcolin E., Marcon A., Piccinin N. 2015 - *Prove di diradamento in impianti di bosco planiziale della Pianura veneta orientale.* Annals of Silvicultural Research 39 (1): 46-54.

Sevrin E. 1997 - *Chênes sessile et péduncolé.* Institut pour le développement forestier, 97 p.

Spiecker H. 1991 - *Controlling the diameter growth and the natural pruning of sessile and pedunculater oaks (Quercus petraea (Matt.) Liebel. and Quercus robur L.).* Selbestverlag der Landesforstverwaltung Baden-Württemberg. Diss. University of Freiburg, Stuttgart, 135p.

Weaver G.T., Spiecker H. 1993 - *Silviculture of high-quality oaks: questions and future research needs.* Annals of Forest Science 50: 531-534.

Climate change impact on a mixed lowland oak stand in Serbia

Dejan Stojanović[1*]; Tom Levanič[2]; Bratislav Matović[1]; Andres Bravo-Oviedo[3]

Abstract - Climatic changes and bad environmental conditions may lead to forests vitality loss and even mortality. This is the reason why increased sanitary felling operations were performed in mixed oak forests in northern Serbia in 2013 in order to solve the severe dieback which affected some Pedunculate oak (*Quercus robur* L.) and Turkey oak (*Quercus cerris* L.) stands, after the very dry years 2011 and 2012. Dendrochronological techniques were applied to both these oak species collected in a stand, to examine the impact of temperature, precipitation and ground water level on forest growth and investigate the potential causes of the dieback. Differences in tree-ring patterns between surviving and dead trees were not significant according to t-value (from 5.68 to 14.20) and *Gleichläufigkeit* coefficient (from 76% to 82%), this meaning no distinctive responses of the two ecologically different oak species. As for radial increment, pedunculate and Turkey oak trees showed a similar response to environmental variables in this mixed stand. The Simple Pearson's correlation analysis, which was conducted, showed that among three basic environmental variables (the mean monthly air temperature, the monthly sum of precipitation and the mean monthly water level, proxy of ground water level), the water level of Danube river in May and the temperature in April were statistically related to the growth of the four tree groups: (i) pedunculate oak vital, (ii) pedunculate oak dead, (iii) Turkey oak vital and (iv) Turkey oak dead trees, for the period 1961-2010 ($p<0.05$, $n=60$). Similar phenomena had already been observed in the Sava River basin for the growth of pure pedunculate oak forests. The long-term decline of the Danube River water level may be related to climate variations and to the changes of water management, river bed, as well as land-use. Together with the increase of temperature, this decline of the water level, and its potential unavailability in the soil, represents a serious challenge for the mixed oak forests silviculture in the Danube basin.

Keywords - *Quercus cerris*, *Quercus robur*, dendroecology, Danube, dieback.

Introduction

The causes of forests decline are complex and uncertain, as the issue involves different abiotic and biotic factors which are predisposing, inciting and contributing to the decline itself (Manion 1991). Some projections indicate that the global land area that is experiencing heat waves may double by 2020, and quadruple by 2040. This may impact on a wide variety of tree processes such as photosynthesis, leaf area development, stomatal conductance, and transpiration among others (Teskey et al. 2014). Drought-induced forest decline may affect carbon, energy and water balance, with an adverse effect on ecosystem services (Martinez-Vilalta et al. 2012). In fact, many cases of drought-induced forest decline in the European continental climate zone have been already reported (Spathelf et al. 2014).

During the last decades, oak decline and mortality were frequently recorded in lowlands of Serbia. There were several hypotheses about the causes of oak dieback (Stojanović et al. 2013): (i) construction of protective embankment along rivers, which prevented occasional flooding in forests which had been flooded in the past; (ii) inappropriate forest management measures; (iii) change of climatic factors; (iv) attacks of pests and diseases.

Medarević et al. (2009) first reviewed the occurrence of oak mortality in this area stating that economic benefits were reduced from 64% to 95% of what expected due to this problem.

Furthermore, Bauer et al. (2013) provided some useful details regarding sanitary felling on permanent plots.

Stojanović et al. (2013), using dendro-ecological methods, rejected the hypothesis about the negative impact of protective embankments along the Sava River as the main cause of pedunculate oak decline and dieback in the Srem region. Effectively other scientists like Matić (1989) had already argued that intensive thinning operations may lead to oak mortality and Vajda (1948) had recognized the change of climatic trends as the potential cause of occasional oak diebacks in the region.

Application of dendrochronology in the research

---

[1] Institute of Lowland Forestry and Environment, University of Novi Sad, Antona Cehova 13d, Novi Sad, Serbia
[2] Slovenian Forestry Institute, Vecna pot 2, Ljubljana, Slovenia
[3] Forest Research Center - National Institute of Agricultural and Food Research and Technology & Sustainable Forest Management Research Institute University of Valladolid-INIA, Madrid, Spain
[*] corresponding author: dejan.stojanovic@uns.ac.rs

on forest decline is not actually a new approach (Cook et al. 1987), also for oak species outside Europe (Dwyer et al. 1995). For instance, Levanič et al. (2011) investigated pedunculate oak mortality in Slovenia using advanced dendrochronological methods.

Moreover, Stojanović et al. (2014a) recognized the significance of Danube water level for the growth of oak forests, with preliminary results, later extended in this study. Stojanović et al. (2014b) found the same linear trend in the Sava River water level and in the tree-ring widths and provided projections of the Sava future dynamics.

Finally, Stojanović et al. (2015) investigated the statistical relationships between the growth of pedunculate oak stands, the Sava water level, the temperature and the precipitation. Water level in the nearby river, as a proxy of ground water table depth, appeared to be positively correlated with tree growth, while air temperature was negatively correlated.

The aim of this study is to consider more in depth both the drivers of growth (temperature, precipitation and water level) and decline of the observed mixed pedunculate-Turkey oak stand, as well as to analyse the patterns among different tree groups: (i) pedunculate oak vital, (ii) pedunculate oak dead, (iii) Turkey oak vital and (iv) Turkey oak dead trees.

The novelty of this research is in the evaluation of mixed Turkey-pedunculate oak stand, in the parallel analysis of living and dead trees, as well as in the evaluation of Danube River impact to the forest. Besides the evaluated parameters, it may be that air pollution, paired with climate change, played a role in the decline of these forests (Bytnerowicz et al. 2007), but long-term time-series of air pollutants are not available in the area. Our hypotheses were to find out close relationships between environmental factors and growth; distinctive response between Turkey and Pedunculate oak, as well as the difference among living and dead trees.

Materials and Methods

The samples were taken in the late 2013 and early 2014 from a mature stand experiencing severe dieback (Branjevina 08i, Forest Management Unit "Sombor", 45° 28' N, 19° 10' E). Its distance from the Danube River is about 5 km. Such stands in Serbia are managed according to the shelterwood system, with several thinnings and the establishment of a new generation at the end of the cycle. 10 dead and 10 vital trees of pedunculate oak (*Quercus robur* L.) and Turkey oak (*Quercus cerris* L.) were selected. The latter species was dominant because of the larger number of trees.

The stand was arranged according to a single tree mixture. Ecological condition within the stand can be considered uniform. The age of the trees was 120 years, according to forestry plans. Cross-sections were taken at 1/5 height (between 5 and 7 m), because it is assumed that this part of the trunk has the most balanced growth due to water and sugars flow. The samples were dried, cut and polished with sandpaper. They were first scanned in high resolution using the ATRICS system (Levanič 2007) and then the tree-ring width was measured using WinDENDRO.

The chronologies were cross-dated and synchronized with the PAST-5™ dendrochronological software, using both visual on-screen comparisons and statistical parameters (t-value after Baillie and Pilcher (tBP), Baillie and Pilcher 1973, and Gleichläufigkeit coefficient (GLK%), Eckstein and Bauch 1969).

Individual tree-ring widths (TRW) were standardized using ARSTAN for Windows (Cook and Holmes 1999) to remove age-related trends (Cook 1985) and averaged into four tree-ring chronologies, representing each one of the studied groups. ARSTAN was also used for the calculation of all the basic statistical parameters of the tree-ring widths.

BootRes package (Zang 2010) allowed the calculation of the bootstrapped Simple Pearson's correlation between environmental factors (water level of the Danube River, air temperature and precipitation) and TRW residuals for the period 1961-2010. We analysed the months of the year prior to ring formation up to the end of the growing season in the year of ring formation (displayed on the x-axis marked as small letters and capital letters, respectively). Significant correlations ($p<0.05$, n=60 years) were displayed with darker colour.

R package berryFunction was used to construct climate diagrams, according to Walter and Lieth (Boessenkool 2015). The mean monthly climate data and Danube water level data were obtained from the Hydro-meteorological Service of the Republic of Serbia for the station Sombor (45° 46' N, 19° 09' E) and the measuring point Bezdan (45° 50' N, 18° 51' E), respectively.

Results

The differences between the chronologies of the surviving and the dead trees and the tree species were not significant in dendrochronological terms (Tab. 1). All the four analysed groups showed a declining trend over the past decades (Fig. 1).

With respect to the tree-ring widths, pedunculate and Turkey oak trees, in this mixed stand, showed a similar growth pattern. According to tree-rings, 7

Table 1 - Statistical parameters among the four groups of oaks. Values of GLK coefficient >75.00% and t_{BP} >4.00 mean that the pairs of chronologies are similar.

	Pedunculate oak - live	Pedunculate oak - dead	Turkey oak - live	Turkey oak - dead
Pedunculate oak - live	x	81.60	78.20	76.00
Pedunculate oak - dead	10.50	x	79.10	78.60
Turkey oak - live	9.27	10.90	x	82.00
Turkey oak - dead	5.68	12.80	14.20	x

out of 10 Turkey oaks, that experienced mortality, died in 2012, while just 2 out of 10 pedunculate oaks died the same year and all the others in 2013.

Time-series and climate diagrams, according to Walter and Lieth, show the increasing air temperature trend, the decreasing water level trend and the change of precipitation regime (more intensive extreme events) in the period 1991-2010, as compared with the reference period 1961-1990 (Fig. 2, 3 and 4).

The mean annual air temperature in the period 1991-2010 increased by 0.8 °C, while the annual amount of precipitation dropped by 55 mm in 20 years. Besides the change in climate statistics, we also observed seasonal changes in the temporal distribution and in the amount of precipitation, with a heavy decrease in winter and early spring. This might lead to a decreased water accumulation in the soil and an increased drought stress in the trees, especially at the peak of the growing period.

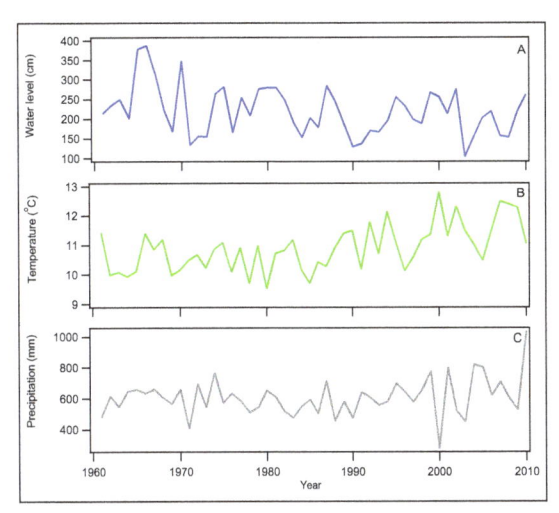

Figure 2 - Time-series of (A) mean annual Danube water level (Bezdan station), (B) mean annual temperature (Sombor station) and (C) annual sum of precipitation (Sombor station) for the period 1961-2010.

An in-depth analysis of the water level data of Danube River shows that, beside a decrease of more than 50 cm in the height of water level in the summer months (June, July and August) there is also a shift in the peak of the highest water level from summer months (1961-1990) to spring months (1991-2010).

A correlation analysis between the tree-ring indices and the three environmental variables, i.e. air temperature, precipitation and water level of Dan-

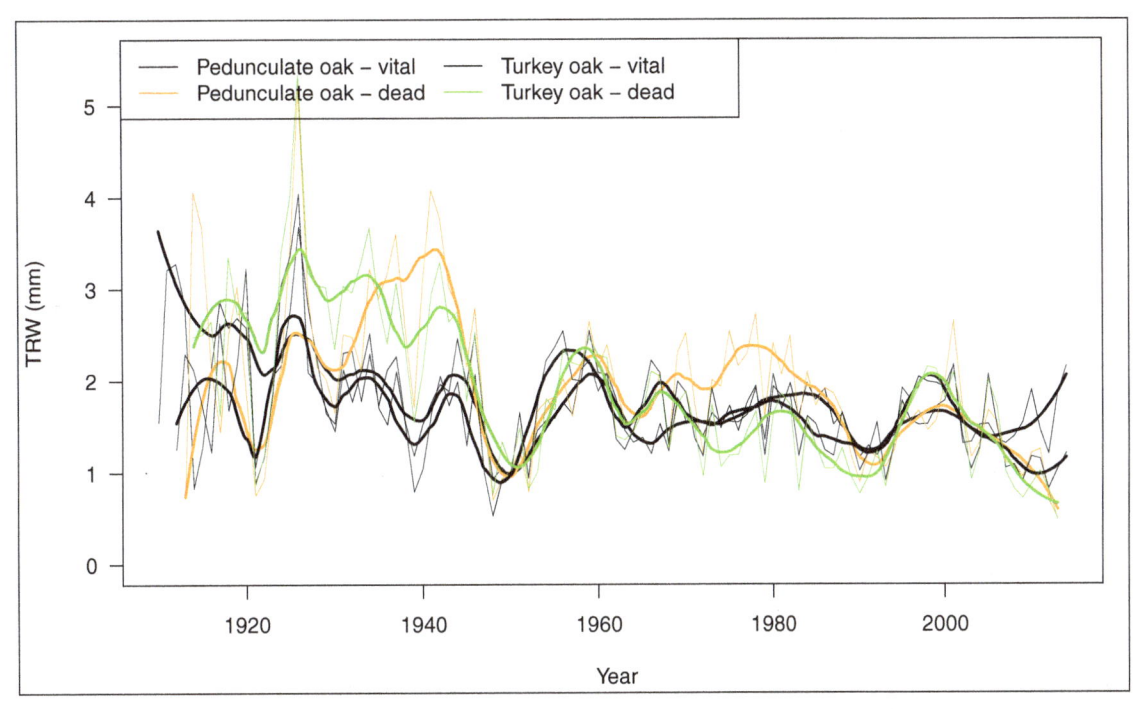

Figure 1 - Mean tree-ring width chronologies (thin lines) of the four tree groups (pedunculate and Turkey oak, vital and dead trees) at the Branjevina stand. Spline curves (thick lines) describe the low frequency growth trend. Each tree-ring width chronology is based on 10 trees.

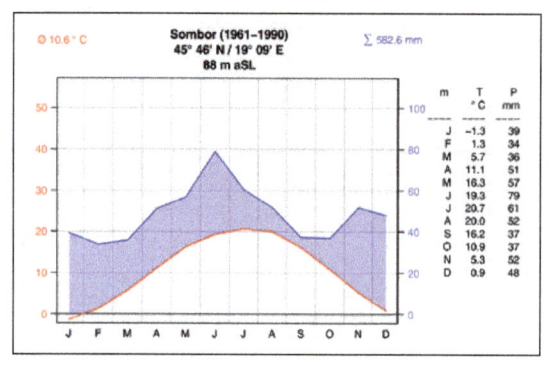

Figure 3 - Climate diagram at the Sombor station for the period 1961-1990.

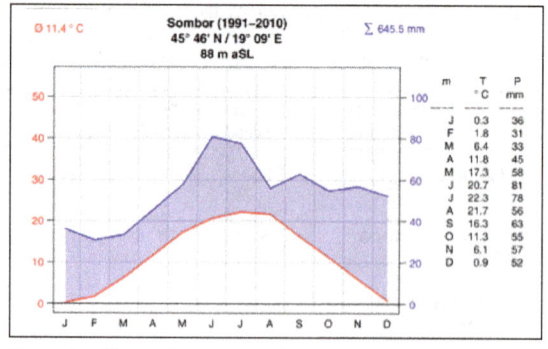

Figure 4 - Climate diagram at the Sombor station for the period 1991-2010.

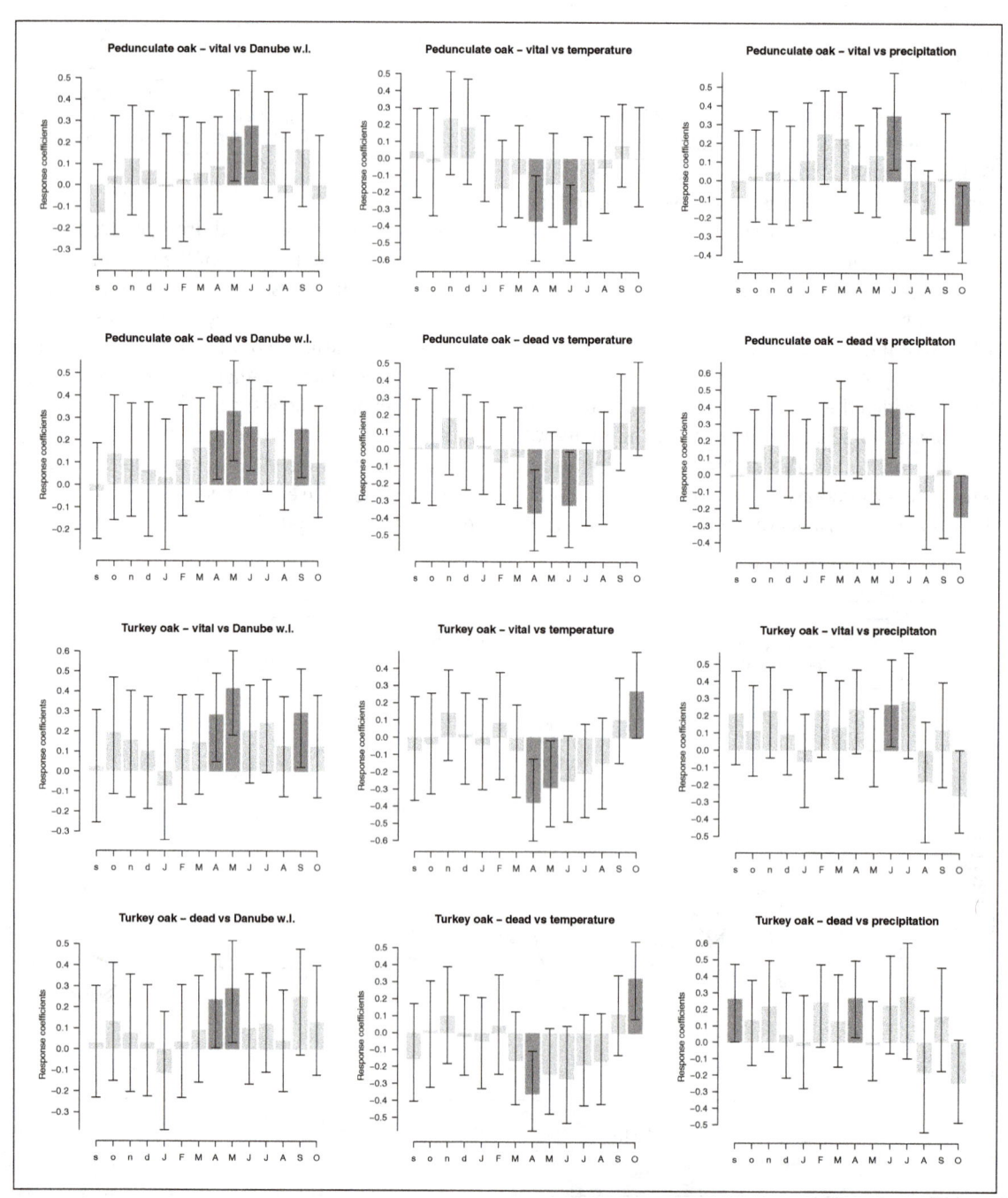

Figure 5 - Bootstrapped Pearson's correlation between tree-ring width residuals and Danube water level (left), air temperature (middle) and precipitation (right) at the Sombor station in the period 1961-2010 for the four groups of trees (top-bottom): (i) pedunculate oak vital, (ii) pedunculate oak dead, (iii) Turkey oak vital and (iv) Turkey oak dead. Months marked by small letters are from the year prior to the growth and capital letters represents the year of the growth. Dark colour represents significant correlation at $p<0.05$ (n=60).

ube River for the period 1961-2010, showed that the water level in May (as a proxy of the ground water level) and the air temperature in April were significantly correlated for all the four tree groups (Fig. 5). The precipitation in June correlated significantly with the growth of pedunculate oak tree groups for the observed period.

Discussion

The results confirm the hypothesis of a relationship between the growth of mixed oak forests and specific environmental variables (the water level in May and the temperature in April for pedunculate and Turkey oak). The observed change (decrease of water level, increase of air temperature and change of precipitation regime, Fig. 2) took place along with the increased mortality in oak forests.

A few other studies support the findings of this research. For instance, Stojanović et al. (2014c) focused on pedunculate oak as one of the most potentially endangered tree species in Serbia, according to different climate change scenarios and the use of ecological niche modelling. Since the extremely dry years 2011 and 2012, about 7% of the growing stock in a compartment of 400 ha was cut in sanitary felling operations (more than 10,000 m^3 of wood) in 2013, in the wider area concerned with this research (Public Enterprise *Vojvodinašume* - Serbia).

A similar phenomenon regarding the impact of environmental variables on tree growth was recently observed in the Sava River basin. Stojanović et al. (2015) found that the Sava River water level and the air temperature in April, May, June, July and August played a key role in the growth of pedunculate oak in the lowlands. A new finding was that the relationship between tree-ring growth and water level has smoothly weakened during the last decades, according to a running correlation analysis, whilst the one between growth and precipitation became more evident, which led to the conclusion that the water level of Sava river, and consequently the ground water level, became so low that roots could not reach the ground water anymore and the trees needed to rely only on precipitation.

The similar growing patterns of pedunculate and Turkey oak, within the investigated mixed stand (Table 1), draw attention to the non-distinctive response of two ecologically different species, this leading to reject the former hypothesis. The phenomenon may be explained by the functional redundancy (Rosenfeld 2002), i.e. the observational evidence that a few species perform similar roles in communities and ecosystems, with the implication that they may be substituted without compromising the ecosystem processes (Lawton and Brown 1993). The full mixture of the two oak species may also explain their common growth response. The analysis of C and O isotopes will, in case, further contribute to the understanding of physiological traits within the investigated tree species and tree groupings.

The long-term decline of Danube River water level may be related to multiple factors: the climate change, the water management, the river bed and/or land-use (Stojanović et al. 2014a). Teskey et al. (2014) pointed out that the drought stress followed by heat waves can lead to an increased tree mortality. With regards to this statement, the increase of air temperature and the change of precipitation regime should be accounted in future studies, where, in particular, the change in precipitation regime means namely less rain during the winter months and a decrease of the overall amount of precipitation (Fig. 2 and 3).

Here, the Danube water level was used as the proxy of ground water table. According to Stojanović et al. (2015), the Sava River water level was highly correlated with the ground water table throughout a sixty-year period. The correlation was above 0.7 with a two-month lag. Since the Danube and the Sava River are both flowing through lowlands and the oak forests are similar with respect to their distance from the rivers and management practice, we assumed that similar ecological impacts exist.

These are the variables to be accounted in the context of future forest management: the observed trend in the Danube River water level drawdown, the air temperature increase, the precipitation regime change, as well as the growth decline of mixed oak forests.

Conclusions

- The recorded differences in tree-growth patterns between pedunculate oak and Turkey oak, both for living and dying trees, were not statistically significant in dendrochronological terms in the analysed context, in opposition to own ecological requirements, according to general forestry knowledge.
- The correlation analysis showed that, according to three basic environmental variables (i.e. the mean monthly temperature, the mean amount of precipitation and the mean monthly water level), the Danube river water level in May (as a proxy of ground water level) and the air temperature in April were statistically related to the growth of all the four studied tree groups over the period 1961-2010.

Acknowledgements

This study was supported by the project "Improvement of lowland forest management" financed by the *Vojvodinašume* public forest enterprise, short-term scientific mission in Ljubljana for Dr. Dejan Stojanović by the COST Action FP 1206 - EUMIXFOR, project "Studying climate change and its influence on the environment: impacts, adaptation and mitigation" (III 43007) financed by the Ministry of Education and Science of the Republic of Serbia and by the Bilateral cooperation Serbia-Slovenia (451-03-3095/2014-09/50).

The authors wish to thank the anonymous reviewers for the helpful revision of the paper.

References

Baillie M.G.L., Pilcher J.R. 1973 - *A simple cross-dating program for tree-ring research.* Tree-Ring Bulletin 33: 7-14.

Bauer A., Bobinac M., Andrašev S., Rončević S. 2013 - *Devitalization and sanitation fellings on permanent sample plots in the stands of pedunculate oak in Morović in the period 1994-2011.* Bulletin of the Faculty of Forestry 107: 7-26 (in Serbian).

Boessenkool B. 2015 – *berryFunction: Function Collection Related to Plotting and Hydrology.* R package version 1.8.1

Bytnerowicz A., Omasa K., Paoletti E. 2007 - *Integrated effects of air pollution and climate change on forests: a northern hemisphere perspective.* Environmental Pollution 147 (3): 438-445.

Cook E.R. 1985 - *Time series analysis approach to tree ring standardization.* Dissertation, University of Arizona, Tucson.

Cook E. R., Johnson A. H., Blasing T. J. 1987 - *Forest decline: modeling the effect of climate in tree rings.* Tree Physiology 3 (1): 27-40.

Cook E.R., Holmes R.L. 1999 - *Program ARSTAN—chronology development with statistical analysis (user's manual for program ARSTAN).* Report, Laboratory of Tree-Ring Research, University of Arizona, Tucson.

Dwyer J.P., Cutter B.E., Wetteroff J.J. 1995 - *A dendrochronological study of black and scarlet oak decline in the Missouri Ozarks.* Forest Ecology and Management 75 (1): 69-75.

Eckstein D., Bauch J. 1969 - *Beitrag zur Rationalisierung eines dendrochronologischen Verfahrens und zur Analyse seiner Aussagesicherheit.* Forstwissenschaftliches Centralblatt 88 (1): 230-250.

Lawton J.H., Brown V.K. 1993 - *Redundancy in ecosystems.* In: Schulze E.-D. and Mooney H. A. (eds), Biodiversity and ecosystem function. Springer : 255–270.

Levanič T. 2007 - *ATRICS-A new system for image acquisition in dendrochronology.* Tree-Ring Research 63 (2): 117-122 http://dx.doi.org/10.3959/1536-1098-63.2.117

Levanič T., Čater M., McDowell N.G. 2011 - *Associations between growth, wood anatomy, carbon isotope discrimination and mortality in a Quercus robur forest.* Tree physiology 31: 298–308 http://dx.doi.org/10.1093/treephys/tpq111

Manion P.D. 1991 - *Tree disease concepts*, 2nd ed. Prentice Hall Inc, Englewood Cliffs

Martínez-Vilalta J., Lloret F., Breshears D.D. 2012 - *Drought-induced forest decline: causes, scope and implications.* Biology letters 8 (5): 689-691.

Matić S. 1989 - *The intensity of thinning and its impact on the stability, productivity and rejuvenation of oak stands.* Glasnik za šumske pokuse 25: 261-278 (in Serbo-Croatian).

Medarević M., Banković S., Cvetković Đ., Abjanović Z. 2009- *Problem of forest dying in Gornji Srem.* Šumarstvo 61 (3-4): 61-73 (In Serbian).

Rosenfeld J.S. 2002 - *Functional redundancy in ecology and conservation.* Oikos 98 (1): 156-162.

Spathelf P., Van Der Maaten E., Van Der Maaten-Theunissen M., Campioli M., Dobrowolska D. 2014 - *Climate change impacts in European forests: the expert views of local observers.* Annals of forest science 71 (2): 131-137.

Stojanović D., Levanič T., Orlović S., Matović B. 2013 - *On the use of the state-of-the-art dendroecological methods with the aim of better understanding of impact of Sava river protective embankment establishment to pedunculate oak dieback in Srem.* Topola (191-192): 83-90 (in Serbian).

Stojanović D., Levanič T., Matović B., Galić Z., Bačkalić T. 2014a - *The Danube water level as a driver of poor growth and vitality of trees in the mixed pedunculate oak-Turkey oak stand.* Šumarstvo 3-4: 155-162 (In Serbian).

Stojanović D., Levanič T., Matović B., Plavšić J. 2014b – *Trends in growth and vitality of pedunculate oak forests in Srem from the aspect future Sava river water level change.* Topola (193-194): 107-115 (in Serbian).

Stojanović D. B., Matović B., Orlović S., Kržič A., Trudić B., Galić Z., Stojnić S., Pekeč S. 2014c - *Future of the Main Important Forest Tree Species in Serbia from the Climate Change Perspective.* SEEFOR (Southeast European forestry) 5 (2): 117-124.

Stojanović D., Levanič T., Matović B., Orlović S. 2015 - *Growth decrease and mortality of oak floodplain forests as a response to change of water regime and climate.* European Journal of Forest Research 134 (3): 555-567.

Teskey R., Wertin T., Bauweraerts I., Ameye M., McGuire M.A., Steppe K. 2014 - *Responses of tree species to heat waves and extreme heat events.* Plant, cell & environment doi: 10.1111/pce.12417

Vajda Z. 1948 - *What are the causes of drying oak in Sava and Drava river basins.* Šumarski list 4: 105-113 (in Serbo-Croatian).

Zang C. 2010 - *BootRes: Bootstrapped response and correlation functions.* R package version 0.3.

Tree-oriented silviculture for valuable timber production in mixed Turkey oak (*Quercus cerris* L.) coppices in Italy

Diego Giuliarelli[1], Elena Mingarelli[2], Piermaria Corona[2], Francesco Pelleri[2], Alessandro Alivernini[3], Francesco Chianucci [2*]

Abstract - Coppice management in Italy has traditionally focused on a single or few dominating tree species. Tree-oriented silviculture can represent an alternative management system to get high value timber production in mixed coppice forests. This study illustrates an application of the tree-oriented silvicultural approach in Turkey oak (*Quercus cerris* L.) coppice forests. The rationale behind the proposed silvicultural approach is to combine traditional coppicing and localized, single-tree practices to favor sporadic trees with valuable timber production. At this purpose, a limited number of target trees are selected and favored by localized thinning. In this study, the effectiveness of the proposed tree-oriented approach was compared with the customary coppice management by a financial evaluation. Results showed that the tree-oriented approach is a reliable silvicultural alternative for supporting valuable timber production in mixed oak coppice forests.

Keywords - single-tree selection, localized thinning, valuable tree, mixed forests, sporadic tree species

Introduction

Coppice is a widespread silvicultural system in Mediterranean European countries where it covers about 23 million hectares (FOREST EUROPE, UNECE and FAO 2011). In Italy, coppice is the most frequently adopted silvicultural system in private forests, and it amounts to about 56% of the total forest area (http://www.inventarioforestale.org). The success of coppice system can be explained considering the advantages to forest owners, like simple management, easy and rapid natural regeneration, faster initial growth rate than the high forest system (Ciancio et al. 2006). Deciduous Turkey oak (*Quercus cerris* L.) occupies the intermediate vegetation belt between sclerophyllous and mountain broadleaved forest over one million hectares (Barbati et al. 2014). Turkey oak represents an economically relevant species with regards to coppice management. On the whole, tree species composition in the dominated Turkey oak forests usually reflects the natural vegetation, even though the diffusion of a few species (e.g. maples, ashes, service trees, wild service trees, hornbeams) has been frequently reduced by management in the past.

With the exception of chestnut woods, coppice management has been traditionally focused on the production of fuelwood and charcoal, which in the past have represented fundamental resources for people living in rural areas (Chianucci et al. 2016b). This approach has traditionally favored the wood production by dominant tree species, at the expenses of often neglected sporadic ones (Chianucci et al. 2016a). More recent changes in the management perspective aimed at integrating economic, social and environmental aspects have led to consider more sustainable silvicultural approaches for coppice woods (Corona 2014). In this line, sporadic tree species may have a potentially interesting role from the ecological and productive point of view (Spiecker 2006, Mori et al. 2007, Mori and Pelleri 2014, Manetti et al. 2016).

Sporadic tree species are less competitive than dominant tree species, and thus their conservation and valorization require specific, tree-oriented silvicultural approaches (Mori et al. 2007). The tree-oriented silvicultural concept has been developed in Central Europe for managing oaks, beech and spruce high forests (e.g. Abetz 1993, Sevrin 1994, Bastien and Wilhelm 2000, Abetz and Kladtke 2002, Wilhelm 2003, Oosterbaan et al. 2008, Spiecker et al. 2009) and specifically for protecting and valorizing of sporadic tree species (e.g. Spiecker 2003, Sansone et al. 2012, Pelleri et al. 2013, Mori and Pelleri 2014).

The objective of tree-oriented silviculture is obtaining high-quality timber assortments in a relatively short rotation period (Oosterbaan et al. 2008) simultaneously minimizing the operational costs. A

---
[1] Università degli Studi della Tuscia, Viterbo
[2] Consiglio per la ricerca in agricoltura e l'analisi dell'economia agraria, Centro di ricerca per le foreste e il legno, Arezzo
[3] Consiglio per la ricerca in agricoltura e l'analisi dell'economia agraria, Centro di ricerca per lo studio delle relazioni tra pianta e suolo, Roma
* francesco.chianucci@crea.gov.it

limited number of target trees are selected, and their growth is supported by reducing surrounding competitors by frequent, repeated thinning from above (Kerr 1996, Perin and Claessens 2009, Lemaire 2010). Such interventions allow high crown enlargement, and therefore high and uniform diameter growth of released trees. Recent studies have demonstrated that this approach can be successfully applied to coppice woods (Mori and Pelleri 2014; Manetti et al. 2016).

In this technical note we illustrated the application of the tree-oriented silviculture approach in mixed (deciduous) Turkey oak coppice stands. Specific aims are:
i) to provide an overview of the proposed tree-oriented silviculture approach;
ii) to evaluate the financial feasibility of the investments required for the implementation of proposed approach at the forest district level;
iii) to illustrate a case-study of practical implementation of the silvicultural approach.

Description of the adopted silvicultural scheme

The silvicultural scheme here proposed is based on a two-fold system of intervention, designed to integrate the tree-oriented concept with customary coppicing. In Central Italy, customary coppice rotation for Turkey oak stands is 20 years, which approximately corresponds to the culmination of mean annual volume increment under average site conditions (Bianchi and La Marca 1984). Customary coppicing practices are combined with other interventions specifically focused on fostering tree growth of a few selected sporadic tree species (hereafter target trees). These tree-oriented interventions differ according to the development of target trees, as indicated by Sansone et al. (2012) and Mori and Pelleri (2014). Basically, there are three different stages of growth, here indicated as T1, T2 and T3 (Fig. 1), which may occur at different times within the rotation.

(T1) The first tree-oriented interventions occur at mid coppice rotation (at 10, 30, 50, 70, 90, 110 years, …). In this stage the young target trees are selected, and their growth is favored by localized thinning from above of the main competitors to support uniform crown enlargement (Table 1). Pruning of target trees may also be carried out in this stage to accelerate the qualification of a branch-free bole and reach more than 2.5 meters length.

(T2) The second stage of growth of target trees occurs at the end of coppice rotation (at 20, 40, 60, 80, 100, 120 years, …). During this stage, a protective ring is being created around the target trees to favor individual crown enlargement with localized thinning every 6-10 years (Fig. 1). The minimum diameter of protective ring corresponds to the mean height of coppice standards. If the target trees are close to each other, a single protective ring is created around them. Pruning of target trees may also be carried out in this phase to complete the qualification of the branch-free bole.

(T3) The third stage of growth occurs when the target trees reached a size adequate to compete with the dominant trees. During this stage, the protective ring is removed and the target trees are left to grow further until they reach the awaited harvesting size. The felling of the target tree coincides with the coppicing period (at the ages of 40, 60, 80, 100, 120 years).

The proposed approach is similar to the one proposed t by Mori et al. (2007) and Mori and Pelleri

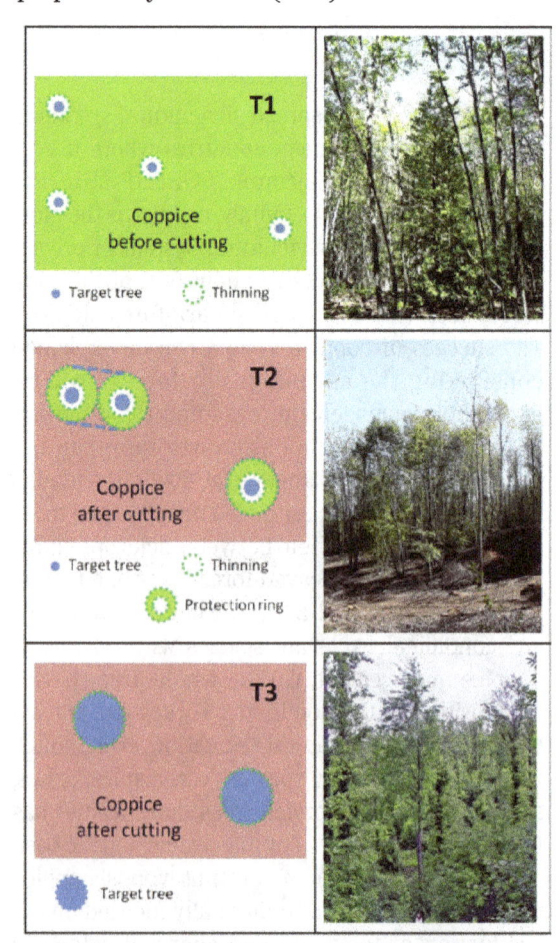

Figure 1 – Growth stage T1: localized thinning undertaken to promote target valuable sporadic tree species in a young coppice. Growth stage T2: coppice stand cut at the rotation end where a number of trees is being left as a protective ring around the target trees. Growth stage T3: coppice stand cut at the rotation end where protective trees are being removed to allow free growth of target trees.

(2014) for Turkey oak coppices, but differs as for the transition period herein of 80 years (previous studies indicated 72 years). In the following stage, the number of target individuals (selected, thinned and felled) of sporadic tree species is expected to remain constant over time and homogenously distributed in the three stages of growth. A synthetic description of the model is reported in Table 1 considering an observation time of 120 years. The main features of the proposed scheme are:
- about 6 target trees per hectare selected at the each mid coppice rotation for timber production (at 10, 30, 50, 70, 90, 110 years);
- coppicing every 20 years (coppice rotation);
- harvesting of about 6 mature target trees per hectare at the end of the transition period (80 years and every 20 years after 80 years);

The above features need to be considered as general guidelines, and they need to be calibrated and adapted to the actual site and stand bio-ecological conditions.

1. Financial assessment

Methodological considerations

The feasibility of the investments required for the implementation of the proposed silvicultural scheme was carried out according to the financial evaluation methods proposed by Andrighetto and Pettenella (2013) and Marone et al. (2014). The financial evaluation is aimed at verifying the profitability of the approach compared with customary coppicing, with regards to the following indicators:
- Net present value (NPV);
- Internal rate of return (IRR), i.e. the discount rate corresponding to a zero NPV;
- Payback period;
- Cash flows;
- Revenues to costs ratio (R/C).

The analysis has considered an evaluation period of 120 years. This was motivated because the profitability of the single-tree approach needs to be evaluated after the end of the transition period, when the single-tree approach is meant to provide a constant production through time (6 target trees every 20 years after 80 years).

Thinning cost was estimated based on real data collected in demonstrative areas of the LIFE project PProSpoT (Marone and Fratini 2013, see also http://www.pprospot.it). Thinning cost of sporadic tree species was included within coppicing costs. To avoid the influence of fluctuations in market prices

Table 1 - The proposed silvicultural (tree-oriented) scheme for Turkey oak coppice stands in Central Italy.

Phase		T1	T2	T1	T2-T3	T1	T2-T3	T1	T2-T3	T1	T2-T3	T1	T2-T3
					Transition period						Ordinary regime		
Operations	0	10	20	30	40	50	60	70	80	90	100	110	120
Selection and marking of the target trees T1 for timber production (n/ha^{-1})	6			6		6		6		6		6	
Selection target trees T1 for biodiversity (n/ha^{-1})	2												
Marking main competitors (n/ha^{-1})		8	8	14	6	12	6	12	6	12	6	12	6
Pruning target trees T1 (n)	6			6		6		6		6		6	
Localized thinning/girdling of the competitors of target trees T1 (n/ha^{-1}))	8			6		6		6		6		6	
Individuation of protection rings (n/ha^{-1})			8		6		6		6		6		6
Pruning target trees T2 (n)			6		6		6		6		6		6
Localized thinning inside protection rings T2 (n/ha^{-1})		8	8	6	6	6	6	6	6	6	6	6	6
Logging woods resulting from thinning (n)			8	14	6	12	6	12	6	12	6	12	6
Release of target trees T3 (n/ha^{-1})					8		8		8		8		8
Felling and logging of mature target trees (n)									6		6		6
Coppice harvesting (ha)	The whole surface		Net surface without the protection rings		Net surface without the protection rings		Net surface without the protection rings		Net surface without the protection rings		Net surface without the protection rings		Net surface without the protection rings
Total number of target trees (n/ha^{-1})	0	8	8	14	14	20	20	26	20	26	20	26	20

Table 2 - Market prices (€ m⁻³) of roundwood from sporadic tree species (from Marone et al. 2014 *modified*).

Tree species	First timber quality class	Second timber quality class	Third timber quality class
Cherry	340	226	113
Service tree	665	443	221
Wild apple	300	200	100
Field maple	300	200	100
Wild pear	665	443	221
Field elm	340	226	113

on stumpage, we considered a null stumpage value by assuming the coppicing costs and the stumpage price as 5,000 € ha⁻¹ and 50 € ton⁻¹, respectively, and a coppice yield of 100 tons ha⁻¹ at the end of coppice rotation. Indirect costs have been identified in the reduction of coppice yield due to the release of target trees, which was assumed reducing the exploitable coppice area by 180 m² per target tree.

Timber market prices considered were selected from comparable local stand and site conditions (Table 2). To perform the calculation of the financial indicators, we adopted a discount rate decreasing over time (3.5% after the first 30 years; 3.0% between 31 and 75 years; 2.5% between 76 and 120 years), consistently with the approach proposed by HM Treasury (2011) for a proper evaluation of public resources in investments lasting several decades.

The revenue of single-tree silviculture heavily depends on the target tree species. Table 3 lists result from a field survey in Viterbo (Central Italy) of the main production and composition from the most diffuse sporadic tree species in a coppice stand representative of mean conditions in Turkey oak coppices in Central Italy; these values were considered in the financial evaluation. The post-intervention increments were verified through appropriate monitoring activities carried out in Tuscany, three years after the interventions in various forest stands within the LIFE project PProSpoT. The wood material obtained by coppice thinning to favor T1 and T2 trees and the wood obtained from pruning was considered negligible and thus not considered in the revenue.

Financial simulations

Financial simulations were carried out by analyzing the influence of the following aspects: coppice productivity under canopy cover of target trees and their protective rings; transition period and target timber size; age of selection of target trees and type of selected individuals.

Coppice productivity under canopy cover of target trees and their protective rings

A target tree must grow isolated for the period required to reach a commercially profitable stem diameter. This may trigger shading phenomena in the surrounding coppice shoots depending by the crown size of the target tree, its crown porosity, and the light requirements of coppice shoots (Chianucci 2016). In financial terms, the occurrence of shading phenomena represents an opportunity cost, consisting in the loss of part of coppice productivity. This production loss depends on the number of target trees per surface unit, and it then stabilizes after the transition period (i.e. after 80 years Table 4). Even in case that no productivity loss occurs due to shading effects, there is still a surface loss due to the coverage of the protective rings. Therefore, a reduction in coppice productivity during the transition period (80 years) was considered as part of the financial evaluation (Table 5). A 50% productivity reduction in

Table 3 - Stem diameter at breast height and volume (m³ tree⁻¹) foreseen by the application of the proposed tree-oriented silvicultural approach during two transition periods (minimum: 60 years; optimal: 80 years) in average site conditions and for different sporadic tree species (expected assortment = 6 m). For cherry (*Prunus avium* L.) it is assumed a growing cycle of 60 years, considering the relatively high growth rate and to prevent plant diseases often occurring at older ages in this species.

Tree species	Transition period 60 years		80 years	
	Diameter (cm)	Volume (m³)	Diameter (cm)	Volume (m³)
Prunus avium L.	72	2.2	-	-
Sorbus domestica L.	36	0.4	48	1.0
Malus sylvestris Mill.	36	0.4	48	1.0
Acer campestre L.	42	0.5	56	1.3
Pyrus pyraster Burgsd.	36	0.4	48	1.0
Ulmus minor Mill.	36	0.4	48	1.0

Table 4 - Coppice income loss (expressed in percent) in relation to the reduction of coppice wood production in protective rings due to shading effects by the target trees in Turkey oak coppices in Central Italy.

Transition period (years)	Wood production in the protective rings compared to customary coppicing				
	0%	25%	50%	75%	100%
20	14	14	14	14	14
40	25	22	18	14	11
60	36	30	23	17	11
80	47	38	29	20	11
100	47	38	29	20	11
120	47	38	29	20	11

Table 5 - Financial indicators of the proposed tree-oriented silvicultural approach as a function of wood productivity reduction in protective rings of the coppice stand due to the shading effect by the target trees.

Wood production in the protective rings compared with customary coppicing	R/C	IRR (%)	NPV (euro ha⁻¹)	NPV (euro ha⁻¹ year⁻¹)	Year when cash flow becomes positive	Payback period (years)
0%	1.01	2.46	100.16	0.83	80	100
25%	1.01	2.63	145.60	1.21	80	100
50%	1.02	3.00	249.20	2.08	80	100
75%	1.02	3.02	259.57	2.16	80	100
100%	1.03	3.35	354.08	2.95	80	80

the protective rings around the target trees is used for all the financial assessments below reported.

Transition period and target timber size

Based on previous studies (Table 3), we considered 60 years as a minimum transition period to reach profitable timber from sporadic species in the proposed silvicultural scheme, while the optimum transition period was foreseen after 80 years, because it offers higher financial performances, leading to permanent property improvement and an increase in income capability. A sixty year transition period may be recommended on fertile sites and in situations where target trees are characterized by relatively fast growth species (e.g. *Prunus avium* L.) (Table 6).

Target trees selection period and type of selected individuals

Financial convenience is compared with respect to 80 year transition period with target trees selected at different times (Table 1). Three situations were compared:

- Early selection: target trees consist of coppice shoots selected at mid coppice rotation (T1);
- Medium-late selection: target trees consist of coppice shoots selected at the end of each coppice rotation (T2):
- Late selection: target trees consist of coppice shoots and standards selected at the end of coppice rotation (i.e. a few target trees from T2 and a few from T3).

Late target trees selection is theoretically more convenient (Table 7), since the starting time of production guarantees both a positive cash flow and a shorter payback period. However, late selection can only be carried out when there are enough target trees able to produce a sufficient amount of valuable timber assortments. This condition is not often met under most Turkey oak coppices in Central Italy.

Financial assessment of implementing the tree-oriented silvicultural scheme at a case-study level

Financial feasibility of the investment required for the implementation of the adopted silvicultural scheme was verified over an area of 12.5 hectares in

Table 6 - Financial indicators of the proposed tree-oriented silvicultural approach as a function of transition period length.

Duration of transition period (years)	R/C	IRR (%)	NPV (€ ha^{-1})	NPV (€ ha^{-1} year^{-1})	Year when cash flow becomes positive	Payback period (years)
60	1.00	2.65	26.24	0.22	60	100
80	1.02	3.00	249.20	2.08	80	100

Table 7 - Financial indicators of the proposed tree-oriented silvicultural approach as a function of time of target tree selection.

Time of selection	R/C	IRR (%)	NPV (€ ha^{-1})	NPV (€ ha^{-1} year^{-1})	Year when cash flow becomes positive	Payback period (years)
Early (T1 trees)	1.02	3.00	249.20	2.08	80	100
Medium Late (T2 trees)	1.07	4.30	730.68	6.09	60	60
Late (T2+T3 trees)	1.12	6.52	1.298.50	10.82	40	40

a Turkey oak coppice in Grotte S. Stefano (Viterbo, Central Italy). Target trees selection was carried out at the end of the first coppice rotation (late selection). Early selection is then applied during subsequent coppice rotations, every 10 years. A transition period of 80 years was considered and wood production within the protective ring was estimated equal to 50% of customary coppice regime. The harvestable roundwood as a function of the transition time (years) is reported in Table 8. The first selection of target tree species has not taken advantage of the early silvicultural operations during the selection phase to get valuable timber assortments. Nevertheless, convenience indicators gave positive outcomes: R/C =1.01; IRR = 2.54%; NPV = 126 € ha^{-1}. Payback time is 100 years and the cash flow turns positive after 60 years (Table 8).

Discussion

This study showed that enhancing sporadic tree species able to produce valuable timber, coupled

Table 8 - Harvestable roundwood (expressed in m³ha^{-1}) from the proposed tree-oriented silvicultural approach as a function of years from the first intervention (transition period) in the case-study (I = first timber quality class; II = second timber quality class; III = third timber quality class).

Tree species	Transition period (years)									
	40			60			80			100
	I	II	III	I	II	III	I	II	III	I
Service tree	0.02	0.06	0.02	0.31	0.21	0.42	3.50	1.70	0.67	2.98
Wild service tree					0.03	0.09		0.09	0.08	0.37
Wild apple		0.02					0.06			0.06
Field maple						0.14	0.22		0.08	0.22
Wild pear		0.02					0.15			0.10
Field helm				0.56	0.39	0.22	2.67	1.18	0.22	2.20

with coppice management, is more convenient than customary coppice management. Main advantages of the proposed tree-oriented silvicultural approach rely on the improvement of the property, on its profitability and low introduction costs, mainly due to a reduction of coppice revenues rather than of actual cash outflow.

The highest financial performances are guaranteed by silvicultural practices which largely maximize the revenue to cost balance, especially in the late selection of target trees (Fig. 2), rarely occurring anyway in the practice. On the other hand, the early selection of target trees represents the easiest silvicultural choice, and has the highest potential to yield valuable timber production at the end of the transition period, especially on fertile sites. In the early selection of target trees, however, the revenue to cost ratios are relatively low, which may limit the attractiveness of the investment in this silvicultural approach. Other critical issues linked to the introduction of the proposed tree-oriented approach in Turkey oak coppice silviculture are mainly referred to the: (i) operational costs for localized (target trees) silvicultural practices during coppice rotation; (ii) relatively long investment return time; (iii) market uncertainty, with possible substantial changes of the stumpage and market price of sporadic tree species timber. For these reasons, the financial support by European Union Common Agricultural Policy, and namely Rural Development Plans, is deemed desirable for allowing the implementation of tree- oriented silvicultural approach: particularly, the incentives could enable the intensification of silvicultural tending, mandatory to increase the technological quality of wood material.

Acknowledgements

This study was carried out under the project SELVALB "The tree silviculture to increase the value of forests in some areas of the territory of Tuscia", funded by Measure 124 "Cooperation for development of new products, processes and technologies in the agricultural sector, food and forestry" of the Rural Development Programme (RDP) 2007/2013 of the Lazio Region, Italy (concession Act no. 59/124/10 of 12.12.2014 - Question no. 8475921080 - recipient Department for Innovation in Biological systems, food and agriculture and Forestry (DIBAF), University of Tuscia in Viterbo). Francesco Chianucci was funded by the MiPAAF - Italian National Rural Network. Language translation from Italian to English was made by Nicolò Camarretta.

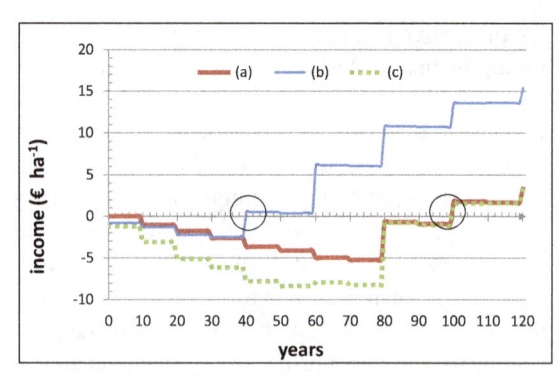

Figure 2 - Cash flow and payback period (circles) according to the proposed tree-oriented silvicultural approach with respect to the cases described Table 7 (a = early selection of the target trees; b = late selection of target trees) and at Table 8 (c).

References

Abetz P. 1993 - *L'arbre d'avenir et son traitement sylvicole en Allemagne*. Revue Forestiere Francaise 45 (5): 551-560.

Abetz P., Klädtke J. 2002 - *The target tree management system*. Forstw. Cbl. 121: 73-82.

Andrighetto N., Pettenella D. 2013 - *Financial evaluation of the tree-oriented silviculture. The software for the evaluation of the investments proposed by PProSpot*. Sherwood 195: 1-4. [online English version] http://www.pprospot.it/english-products.htm.

Barbati A., Marchetti M., Chirici G., Corona P. 2014 - *European Forest Types and Forest Europe SFM indicators: tools for monitoring progress on forest biodiversity conservation*. Forest Ecology and Management 321: 145-157.

Bastien Y., Wilhelm G.J. 2000 - *Une sylviculture d'arbres pour produire des gros bois de qualité*. Revue Forestière Française 5: 407-424.

Bianchi M., La Marca O. 1984 - *I cedui di cerro nella provincia di Viterbo. Ricerche dendrometriche ed allometriche in relazione ad una ipotesi di matricinatura intensiva*. Ricerche sperimentali di dendrometria ed auxometria 10: 41-70.

Chianucci F. 2016. *A note on estimating canopy cover from digital cover and hemispherical photography*. Silva Fennica 50, doi: 10.14214/sf.1518.

Chianucci F., Minari E., Fardusi M.J., Merlini P., Cutini A., Corona P., Mason F. 2016a - *Relationships between overstory and understory structure and diversity in semi-natural mixed floodplain forests at Bosco Fontana (Italy)*. iForest-Biogeosciences and Forestry (early view). doi: 0.3832/ifor1789-009 [online 2016-08-21]

Chianucci F., Salvati L., Giannini T., Chiavetta U., Corona P., Cutini A. 2016b - *Long-term (1992-2014) response to thinning in a beech (Fagus sylvatica L.) coppice stand under conversion to high forests in Central Italy*. Silva Fennica 50. doi: 10.14214/sf.1549

Ciancio O., Corona P., Lamonaca A., Portoghesi L., Travaglini D. 2006 - *Conversion of clearcut beech coppices into high forests with continuous cover: a case study in central Italy*. Forest Ecology and Management 3: 235-240.

Corona P. 2014 - *Forestry research to support the transition towards a bio-based economy*. Annals of Silvicultural Research 38: 37-38.

FOREST EUROPE, UNECE and FAO 2011 - *State of the Europe's Forests 2011*. Status and trends in sustainable forest management in Europe, 341 p.

HM Treasury 2011 - *The Green Book: Appraisal and Evaluation in Central Government*. www.gov.uk/government/uploads/system/uploads/attachment_data/file/220541/green_book_complete.pdf

Kerr G. 1996 - *The effect of heavy or 'free growth' thinning on oak (Quercus petraea and Quercus robur)*. Forestry 69, 4: 303-316.

Lemaire J. 2010 - *Le chêne autrement. Produire du chêne de qualité en moins de 100 ans en futaie régulière*. Guide technique, Institut pour le Développement Forestier, Paris 176 p.

Manetti M.C., Becagli C., Sansone D., Pelleri F. 2016 - *Tree-oriented silviculture: a new approach for coppice stands*. iForest-Biogeosciences and Forestry (early view). doi: 10.3832/ifor1827-009 [online 2016-08-04]

Marone E., Fratini R. Andrighetto N, Pettenella D, Bruschini S. 2014 - *The economy of sporadic tree species. Financial evaluation of the tree-oriented silviculture: the results of the PproSpoT Project*.

Marone E., Fratini R. 2013 - *Tree-oriented silviculture in an oak coppice. Estimation of financial profitability and possible public funding*. Sherwood 198: 25-30. [online English version] http://www.pprospot.it/english-products.htm

Mori P., Bruschini S., Buresti E., Giulietti V., Grifoni F., Pelleri F., Ravagni S., Berti S., Crivellaro A. 2007 - *La selvicoltura delle specie sporadiche in Toscana*. Supporti tecnici alla Legge Regionale Forestale della Toscana, 3. ARSIA Firenze: 355 p. http://www.regione.toscana.it/documents/10180/13328713/4_Manuale-specie-sporadiche-Toscana.pdf/aa46b002-3609-4681-89e9-69b0043f01bb

Mori P., Pelleri F. (eds.) 2014 - *Silviculture for sporadic tree species. Extended summary of the technical manual for tree-oriented silviculture proposed by the LIFE+ project and PProSpoT*. Compagnia delle Foreste, Arezzo [online English version] http://www.pprospot.it/english-products.htm

Oosterbaan A., Hochbichler E., Nicolescu V.N., Spiecker H. 2008 *Silvicultural principles, phases and measures in growing valuable broadleaved tree species*. 11 p. [online] http://www.valbro.uni-freiburg.de/

Pelleri F., Sansone D., Bianchetto E., Bidini C., Sichi A. 2013 - *Selvicoltura d'albero in fustaie di faggio: valorizzazione delle specie sporadiche e coltivazione della specie dominante*. Sherwood 190: 43-47. [online English version] http://www.pprospot.it/english-products.htm

Perin J, Claessens H., 2009. *Considerations sur la designation et le dètourage en chênes et hetre*. Foret wallonne, 98: 39-52.

Sansone D., Bianchetto E., Bidini C., Ravagni S., Nitti D., Samola A., Pelleri F. 2012 - *Tree-oriented silviculture in young coppices Silvicultural practices to enhance sporadic species: the LIFE+ PPRoSpoT project experience*. Sherwood 185: 1-6. [online English version] http://www.pprospot.it/english-products.htm

Sevrin E. 1994 - *Chênes sessile et péduncolé*. Institut pour le Développement Forestier, Paris 96 p.

Spiecker H. 2003 - *Silvicultural management in maintaining biodiversity and resistance of forests in Europe-temperate zone*. Journal of Environmental Management, 67: 55-65.

Spiecker H. 2006 - *Minority tree species: a challenge for a multi-purpose forestry*. In: Nature based forestry in central Europe. Alternative to industrial forestry and strict preservation. Studia Forestalia Slovenica, 126: 47-59.

Spiecker H., Hein S., Makkonen-Spiecker K., Thies M. 2009 - *Valuable broadleaved forests in Europe*. EFI Research Report 22, 256 p.

Wilhelm G.J. 2003 - *Qualification-grossissement: la stratégie sylvicole de Rhénanie-Palatinat*. Rend Des-Vous techniques, Office National des Forêt 1: 4-9.

Coppice forests, or the changeable aspect of things

Gianfranco Fabbio[1*]

Abstract - Coppiced forests were the main source of firewood, brushwood, and charcoal for rural and urban settlements' basic needs such as cooking food and domestic heating for thousands of years and up to the mid-20th century in many European countries and, specifically, in Mediterranean countries. The global diffusion of fossil fuels reduced this leadership and the coppice system turned, to some extent, to a reminder of the past. Nowadays, the ongoing global changes and the related green-economy issues call for resilient systems and effective bio-energy producers. These issues have caused a second turning point and the coppice has returned fifty years later to play a role. A review of the silvicultural system has been carried out with a special focus on the changes which have occurred in between, taking Italy as a consistent case-study. The analysis is mainly framed upon the long-term research trials established by the CREA-Forestry Research Centre in the late sixties, to find out adaptive management strategies and overcome the system's crisis. The findings and further knowledge achieved so far on the dynamics and functioning of coppice forests in the outgrown phase, both as natural evolutive patterns and silviculture-driven processes, are highlighted in this paper. They provide useful tools to handle the management shift regarding forthcoming issues, i.e. the current role attributable to the coppice system within the changing environment and the renewable energy demand. The basic features of each management area and their complementarities within the current framework are outlined.

Keywords - silvicultural system, natural dynamics, pro-active silviculture, sustainability, past management, future forestry, Italy

Introduction

Coppices have imprinted the broad-leaved forest landscape across Europe since the establishment of early human settlements. Coppice is an anthropogenic system created and optimised for small-sized wood production over several million hectares. The main products; firewood and charcoal, have had a global use because they assisted people's common, daily needs such as cooking food and domestic heating, whilst manufacturing produced a further, huge demand for energy over the last centuries. The peak of coppice exploitation took place during the first industrial revolution whilst its role reduced following the diffusion of fossil fuels since the mid-1900's.

Coppice forests are a significant part of Europe's semi-natural forests (about 70%) (Forest Europe 2011), characterise the forest landscape of the five EU Mediterranean countries over about 8.5 million hectares (Morandini 1994), and cover more than 3.6 million hectares in Italy (Gasparini and Tabacchi 2011). Italian coppices account for almost 19% of the coppices in the EU28, which in turn represent 83% and 52% in the whole of Europe and at global level, respectively (UN/ECE-FAO 2000, Mairota et al. 2016a).

Coppice forests are therefore a significant element of forest landscapes throughout Europe. The landscape is an appropriate management unit because it considers the interrelatedness of component segments. The spatial heterogeneity made of a mosaic of structurally different forest patches, the presence of different age classes, and the implementation of contact or transition zones among contrasting ecosystems are all conditions that favour environmental variability and, therefore, biological diversity (Scarascia-Mugnozza et al. 2000).

Italy may be taken as a consistent case-study between the Mediterranean region and neighbouring continental countries because of its large coppice forest coverage, the diversity of growth environments, the number of tree species that exist, and the evidence of large changes which have occurred over the last two centuries. The fragmented forest ownership structure with many private (73%) small-sized forest holdings is also a common trait in Europe (Forest Europe 2011).

Cultivation techniques have been well-documented since the Middle Ages (Piussi 1980 and 1982, Szabo et al. 2015), but there is also evidence of the late conversion of wide, high forest areas into coppices between the 1800's and 1900's following the

---

[1] Consiglio per la ricerca in agricoltura e l'analisi dell'economia agraria, Centro di ricerca per le foreste e il legno, Arezzo (Italy)
* gianfranco.fabbio@crea.gov.it
§ Partly funded by LIFE14 ENV/IT/000514 www.futureforcoppices.eu

rising energy demand due to the sharp population increase and the concurrent rapid development of manufacturing activity (Agnoletti 2003).

Today, the analysis of silvicultural systems according to modern 'sustainability criteria' cannot ignore the basic question of the long-lasting, primary demand-driven, role of the coppice forest. In addition, it is nearly always impossible to separate this intensive cultivation system from the manifold overlapping uses and misuses of soil and tree vegetation. In fact, coppices have not only been used for short rotation wood production, but have also been over-exploited and used for deadwood and litter removal, the collection of leafy branches for fodder, occasional intercropping following shoots harvesting, and for unregulated pasture (Piussi 2006, 2015). This means uncodified, widely-practised 'multiple use'. The archives of Tuscany farms (central Italy) highlight that two thousand years of coppicing did not reduce stools vitality and site quality in the absence of overlapping, invasive uses (Piussi and Stiavelli 1986, Piussi and Zanzi Sulli 1997).

The common judgment of non-sustainable system, in the long run has been built, therefore, on the full integration of the manifold uses on the same ground rather than on the coppice system itself (Fabbio 2010). Other external pressures (e.g. wildfires and uncontrolled grazing) or sensitive environments (e.g. steep mountain sides, shallow soils, and harsh climate conditions) have contributed to, and many times have caused, the decay of site fertility and the complete erosion of forest texture (MEDCOP 1998, Fabbio et al. 2003). This is the why there is evidence today of a vast array of conditions, from the relict cover of scattered trees to dense, well-growing coppiced forests where their management has followed the basic rules and supplementary uses have been less intensive or lasting, or site quality has supported them. Driving forces, limiting factors, and feedbacks were the determinants of the co-evolutive pattern between land use and growth medium (Fabbio 2010).

The background between the 1800's and 1900's

In the 1800's and early 1900's, coppiced forests clearly depicted the pressure exerted by the increasing population density on the available natural resources. Total forest cover at country level underwent a significant reduction in the 40 years between 1868 and 1911 when first-time coppice system prevailed over high forest. The reasons for this were the doubling of the population during this period and the industry's energy requirements which were 85% (1861) of fuelwood and charcoal

Large, vital stool in an outgrown beech forest aged 70 yrs (Tuscany).

(Agnoletti 2002). The rising price of charcoal caused social problems in the early 1900's and Italy began to import it between 1906 and 1913. At this time nine-tenths of charcoal were used for cooking food. The city of Rome alone burnt up to 90 million kg per year in the course of World War II (Hippoliti 2001). The ratio between forest resources and the population only changed in the mid-1900's when firewood and charcoal supplied 11% of the country's energy requirements compared to 85% during the previous century. At this time, the coppice system took part in the major socio-economic shift of the modern era since the first industrial revolution (Fabbio 2004).

Many factors contributed to coppice downgrading: first, the much decreased economic significance of firewood production and the related lower profitability of its harvesting; then, the less intensive practice of forestry because of the emerging societal demands other than wood production; and finally, the critical association of coppicing with an out-of-date, ecologically-incorrect forest management system. Thus, the coppice system progressively reduced its leading role and turned, to some extent, to a 'reminder of the past' (Amorini and Fabbio 1994).

The new frame of reference can be outlined as the change of the original ground hosting the common matrix of coppiced young stands into a variable texture of stand ages and structures, stand dynamics, and growing stocks. In Italy, the current

Table 1 - Coppice cover in Italy by main tree species and stand age (source: Gasparini and Tabacchi 2011, INFC *mod.*).

tree species	cover		stand age					
	ha	%	<20 years	%	20<years<40	%	>40 years	%
Fagus sylvatica	477225	13.0	7728	1.6	128513	26.9	340984	71.5
Castanea sativa	593242	16.2	91908	15.5	277709	46.8	223625	37.7
Carpinus b. e Ostrya c.	636662	17.4	85250	13.4	325034	51.1	226379	35.6
Q robur, Q. petraea, Q.pubescens	534325	14.6	54256	10.2	241590	45.2	238479	44.6
Q. cerris	675532	18.4	124999	18.5	314835	46.6	235699	34.9
Q. ilex, Q. suber	372020	10.2	27241	7.3	146679	39.4	198099	53.2
other spp.	374137	10.2	81390	21.8	151169	40.4	141578	37.8
Total	3663143	100	472772	100	1585529	100	1604843	100

distribution of coppice forests with respect to stand age (Tab.1) (Gasparini and Tabacchi 2011 *mod.*) highlights that young stands represent less than 1/3 of mature coppices, which is quite similar to that of 'ageing' stands. This composite panorama includes the stands which are still managed, the outgrown stands, and the minority proportion of stands being converted into high forests, which are mainly under public ownership and located in the upper mountain belt (Amorini and Fabbio 1992, Fabbio et al. 1998a).

Historical statistics on firewood harvesting (Hippoliti 2001, Pettenella 2002, Ciccarese et al. 2006, Pra and Pettenella 2016) (Fig.1) show a minimum exploitation in the mid-seventies, whilst the last official statistics available on domestic fellings (ISTAT 2011) are similar to 2004. According to Forest Europe (2015), the current felling rate as a percentage of Net Annual Increment in Italy is one of the lowest in Europe: 39.2% compared to 47.3% in France, 80.3% in Germany, and 55.5% in Spain (Pra and Pettenella 2016). Even if the internal consumption of firewood from forests is only a part of the total consumption of the wood biomass for energy in Italy, which is estimated to be equal to 21.20 Mt (range between 16.37 and 22.17 Mt according to Pra and Pettenella 2016) or 19 Mt (Ciccarese et al. 2012), the official statistics are heavily underestimated (Corona et al. 2007). The reasons for this are generally related to the cross-sectorial character and fragmentation of the market. The multiplicity of sources on the supply side and the presence of different sub-markets and final users on the demand side make the wood energy market complex to clearly define and quantify

Visual impact of customary, slope-oriented coppice harvesting on a mountainside (central Apennines).

Final harvesting in a coppice with standards (Turkey oak forest, Latium).

(Steierer 2007, SFC 2008 in Pra and Pettenella 2016).

The increased rotation length has been induced by several reasons: the suspension of charcoal production, the improvement in chopping tools and hauling/processing machinery (Schweier et al. 2015) and, especially, by the opportunity to harvest higher stocks per unit area. Over the last decades, the much-increased differential between manpower

Thickness of a holm oak coppice forest close to the age of rotation. Growing space occupancy is quite full (Corsica).

The high mortality rate is a typical trait of the early outgrown phase (holm oak forest, Sardinia).

Turkey oak coppice under conversion into high forest (Tuscany).

costs (x80) and firewood costs (x16), even with higher (x4) manpower productivity for processing and logging, led to doubling rotations and reaching the optimal shoot size of 10-15 cm. This trend, as highlighted by Hippoliti in 2001, is today the consolidated operational principle ruling the practice of coppice forestry.

Concurrent reasons are gaining ground today for the following reasons: the awareness of the residual stock of fossil fuels, the evidence of the climate shift in progress, the need for pro-active mitigation, i.e. that resilient and efficient forest systems have to be managed effectively, the fulfillment of emerging green economy issues (Marchetti et al. 2014). All of these call for streamlined production processes in a time of environmental change and increased bio-energy demand. This is why the coppice system has returned to play a potential, prominent role within the forestry domain.

This newly-established perspective leads, on the one hand, to the reconsideration of the legacy of the past, i.e. the fully tried techniques used for coppice systems; and on the other hand, calls for working hypotheses which are free of any subjective opinion other than ground-based evidence and the body of knowledge achieved so far.

The aim of this paper is to review the research questions which arose fifty years ago, the research pathway, and the main findings from the long-term research trials which have been established on this topic since the late sixties in Italy by the CREA - Forestry Research Centre. The main traits of the current scenario between the area managed under longer rotations, the outgrown area, the area undergoing conversion into high forest, and the role of each are outlined.

The need to handle the shift towards the suspension of harvesting/abandonment within a significant share of coppice area and to also suggest hypotheses for alternative, pro-active management has originated a series of applied research trials. These have compared the new options of coppice maintenance under the updated rules and/or conversion into high forest with the natural evolutive pattern or 'outgrown phase' in progress. These trials have contributed a better understanding of coppice system functioning above and below ground in terms of growth and re-growth ability, of dynamics and structure of the standing crop, as well as of the main drivers acting within each management choice.

The focus here is on the main tree species, i.e. the deciduous and evergreen oaks (Turkey oak, *Quercus cerris* L.; holm oak, *Quercus ilex* L.) and beech (*Fagus sylvatica* L.). Chestnut (*Castanea sativa* Mill.) was also considered in the trials but will be addressed in a next paper (Manetti, *forthcoming*) because of its peculiar traits with its set of available management options and the array of available wood assortments.

The research questions

At the time the early papers on the subject-matter were issued (Gambi 1968, Guidi 1976, Amorini and Gambi 1977) and CREA's first experimental trial was established (1969), a few basic questions were asked: (i) how could the share of coppice forests no longer being harvested be managed? (ii) what are coppiced forests' growth patterns beyond the customary rotation? (iii) what is the decay rate of stools' resprouting ability with stand ageing? (iv) what is the most suitable stand age to undertake coppice con-

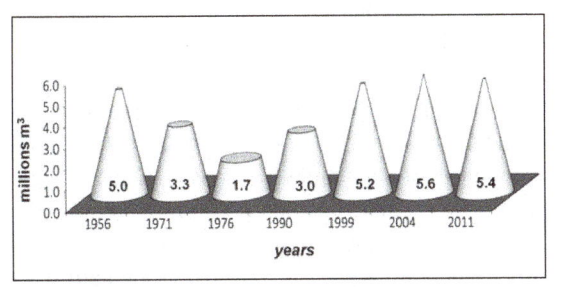

Figure 1 - Firewood harvesting over the last sixty years in Italy (2011, last official data available). (Sources: Hippoliti 2001, Pettenella 2002, Ciccarese et al. 2006 and 2012, Pra and Pettenella 2016).

version into high forest and which practices should be implemented? (v) how can economic sustainability be achieved in order to tackle any pro-active silviculture, given that profitable firewood harvesting no longer exists? (vi) how can 'standards', i.e. the trees (usually from seed) released from one up to a few coppice rotations be managed?

The consolidated management cycles, which have been improved and finely tuned throughout centuries of cultivation, could not provide any answer to the above questions due to the ruling principles at that time. Rotations in use, well-grounded on the specific growth rate, were optimal for the: (i) size of harvested assortments (brushwood, charcoal, and fuelwood); (ii) cutting tools and the hauling techniques in use; (iii) avoidance of any yield loss due to the heavy competition among shoots and stools causing natural ('regular' according to Oliver and Larson 1996) mortality beyond customary rotations. Yield tables specific to coppice forests identified the rotations in use as close to the age of mean volume increment culmination, i.e. to the age of maximum wood production (Castellani 1982, see also Bernetti 1980).

The research pathway

The first CREA trials, established in the late 1960's when the suspension of harvesting was already in progress, basically compared the fairly unknown evolutive pattern of outgrown coppices and the alternative option of pro-active silviculture for coppice conversion into high forest. All of this may be seen as the establishment, ahead of its time, of an adaptive management approach. The following tools were used: the periodical survey of mensurational parameters, stem and root branches analyses, the analyses of tree layering set up and dynamics, and the shoots' mortality rate and progress survey. The collection of sample trees at the same sites allowed the measurement of dendrotypes within the consistent size-age span and the calculation of species-specific (beech, Turkey oak, holm oak) allometric functions (Amorini et al. 1995, Brandini and Tabacchi 1996, Amorini et al. 2000, Fabbio et al. 2002, Nocetti et al. 2007). Stem and root branches analyses made it possible to plot growth patterns from the coppice rotation up to the outgrown phase. These analyses were, therefore, a contribution to understanding the functions and processes already acting within the coppice cycle.

Further trials were implemented by the CREA within; the EU-Agrimed 'Multiple use of Mediterranean forests and prevention of forest fires' (1979-82) with an Italian-French partnership (Morandini 1979, Amorini et al. 1979), the EU-AIR2 'Improvement of Mediterranean coppices' MEDCOP (1994-98) involving the five EU Med. countries (Morandini 1994, Fabbio et al. 1998a), and the regional/national projects 'LIFE-Summacop' (Grohmann et al. 2002), 'TraSFoRM' (Amorini et al. 2002), 'Ri-SelvItalia'(2002-2004) (Fabbio 2004), ARSIA-Cedui (2004-06).

Agrimed (1979-82) 'Multiple use and prevention of forest fires'. Combined thinning method: opening of parallel clear-cut strips and selective thinning of standing crop (Turkey oak coppice forest, Tuscany).

Each project contributed to the establishment of further trials and analyses (Amorini et al. 1996, 1997, 1998ab, Fabbio 1994, Fabbio et al. 1996).

A basic entry within the newly-established research design was the 'adjusted coppice system' which was revised for average rotation length, harvesting, and thinning practices, the definition of 'standards' release (number, positioning, and aggregation on the ground), and the choice of the suited phenotype (Amorini et al. 1998c, Grohmann et al. 2002, Becagli et al. 2006, Cantiani et al. 2006, Savini 2010, Savini et al. 2015).

Further analyses addressed the parameters of productivity (litterfall, leaf area index), the canopy properties and the radiative climate (Cutini 1994ab, Cutini 1997, 2006, Cutini and Hajny 2006), the ecophysiological traits (Cutini and Mascia 1998, Cutini and Benvenuti 1998), the inner microclimate (Fabbio et al. 1998b), the dynamics of tree biomass and deadwood density (Bertini et al. 2010, 2012), and the stand structure and compositional diversity (Manetti and Gugliotta 2006, Manetti et al. 2013).

The theory and practice of silviculture according to the options in progress (Fabbio and Amorini 2006, Amorini et al. 2006), genetics and stand structure (Ducci et al. 2006), harvesting operations and hauling systems (Piegai and Fabiano 2006), technological improvement of wood assortments (Berti et al. 1998), biodiversity conservation (Baragatti et al. 2006), landscape analysis (Mairota et al. 2006), and the economic sustainability of the management systems (Fagarazzi et al. 2006) were the main subjects investigated within the mentioned projects.

The final points take into account the forthcoming issue of concern, i.e. analysing the resprouting ability of outgrown coppice stools. The issue is highly important if a share of the currently abandoned coppice area will be used for coppicing again.

The revision of practices ruling the customary technique of conversion to high forest has been tackled within the LIFE-ManForCBD (2010-2015). Innovative adaptive practices consisted of lowering symmetrical competition by reducing stand evenness to better address tree canopies for future regeneration by selective thinning. A case study was carried out in a beech stand under conversion to high forest (Fabbio et al. 2014). The advance seed cutting in the same forest type is addressed by Cutini et al. (2015), whilst long-term data on litter production, leaf area index, canopy transmittance, and growth efficiency estimates are again reported from a beech trial by Chianucci et al. (2016).

Further contributions to the subject matter within the period of review are available in 'The improvement of Italian coppice forests' (Accademia Nazionale di Agricoltura 1979), 'Improvement of coppice forests in the Mediterranean region' (Morandini 1994), 'The coppice forest in Italy' (Ciancio and Nocentini eds. 2002), and 'The coppice forest. Silviculture, Regulation, Management' (Ciancio and Nocentini 2004).

The 'coppice issue' is also well-addressed by: the ongoing Cost Action FP1301 (EuroCoppice 2013 https://www.eurocoppice.uni-freiburg.de/) which aimed to develop the innovative management and multifunctional utilisation of traditional coppice forests and is an answer to future ecological, economic and social challenges in the European forestry sector; the international Conference held in Brno (Coppice 2015 http://coppice.eu/conference_en.html) where the past, present, and future of coppice forests were analysed alongside the new challenges in a changing environment; the LIFE FutureForCoppiceS (2015 http://www.futureforcoppices.eu/en/) which aimed to demonstrate the outcomes of different approaches by datasets collected from long-term experimental plots networks and improve the knowledge of Sustainable Forest Management indicators in view of the forecasted changes in key drivers and pressures.

The common goals addressed by these ongoing activities acknowledge the role of coppice system and the challenge for forestry within the newly-established economical and environmental conditions.

Main findings

Stand dynamics of outgrown coppice forests
The above ground process

In Italy, the available yield models for coppice forests date back to the 1940's up to the early 1980's (Tab.2). Predictive models set the age of 'maximum yield' or the 'age of mean annual volume increment culmination'. Scheduled rotations are quite short and vary as a function of the specific growth capacity and the site-index. The age of maximum yield often reported both for growing stock as a whole and for the firewood/brushwood component, underlines the attention paid to each harvestable assortment.

The evidence of incremental values higher than those recorded at the ages of previous rotations was provided by the repeated measurement of the standing crop volume and biomass undertaken since the establishment of the first permanent monitoring CREA sites in the late sixties (UNIF 1987). All of

Table 2 - Yield models for coppice forests in Italy (source: Castellani 1982).

main tree species	yield tables (years)	site class	stand age corresponding to the maximum yield (years)		
			growing stock	firewood	brushwood
Turkey oak	1948-49 to 1965-66	-	14-16		
	1950	I	9	12	
		II	9	12	
		III	12	12-15	
	1950	I		12	9
		II		12	9
	1966 to 1982	-		14	
holm oak	1963 to 1972	I	26-28		
		II	28	32	
		III	28		
beech	1947	I	17-18		
		II	16-22		
		III	18-23		

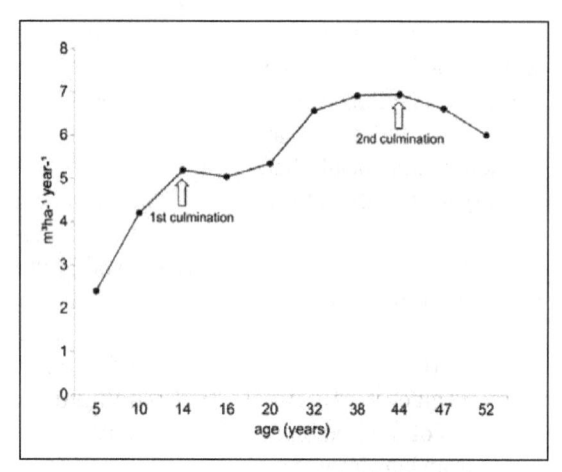

Figure 2 - Stand dynamics of an outgrown Turkey oak coppice forest: trend of mean annual volume increment.

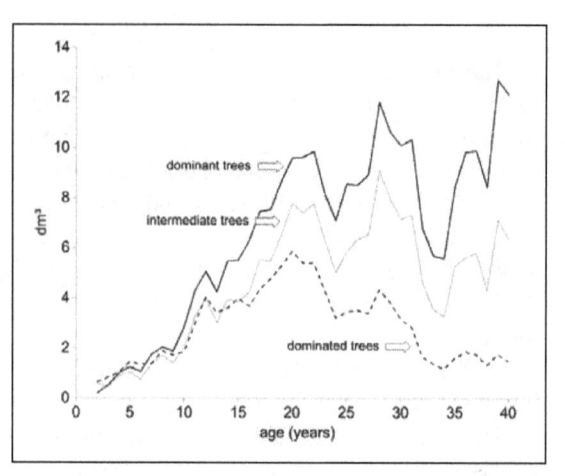

Figure 3 - Shoots dynamics of an outgrown Turkey oak coppice forest: averaged trend of current annual volume increment by the stem analysis per social ranks (same trial as in Fig. 2).

this occurred in spite of the heavier natural mortality rate recorded in between in the fully stocked outgrown coppices.

A second, higher culmination of mean annual volume increment was assessed for the first time in a 44-year-old Turkey oak coppice (Fig.2), much later than the first one (at the age of 14) as ruled by the yield models for the species (Amorini and Fabbio 1988). The same pattern was found to be common in the other stands investigated, i.e. beech and evergreen oak coppice forests (Amorini and Fabbio 1990).

This evidence clashed with previous literature and suggested further analyses. Stem analysis provided the ultimate answer to the matter. The analysis was carried out at the early-established trials on beech and Turkey oak in the Tuscan Apennines (Amorini and Fabbio 1986, 1989). The stratified tree sampling per growth layer (dominated, intermediate, and dominant) showed patterns made by synchronous current volume increment cycles in progress since the differentiation of ranks within the customary coppice rotation (Fig.3). The higher growth rate of the dominant shoots and the lower competitive ability of the not-dominant shoots over time were highlighted.

The current availability of a series of long-term monitoring trials allows the assessment of the growth dynamics in the outgrown (stored) coppice type between 44 and 75 years of age (Tab.3). The values show that growth culmination has already occurred at most sites, whilst the reduced difference of current to mean volume increment suggests that the peak is not far away at the other sites. Shade-tolerant species (holm oak, beech) show that growth culmination has not been reached yet between 60 and 75 years of age, with auto-ecology being the main driver.

What is, therefore, the meaning of the early culmination widely acknowledged by former yield models? It actually detects a first peak of growth shaped by the subsequent temporary steadiness due to the triggering of heavy competition among shoots and stools. It identifies the technical rotation suitable for small-sized wood harvesting, anticipates the occurrence of natural mortality, i.e. of any firewood production loss. It testifies the physiognomic evidence of the relative peak of growing space occupancy that takes place within the early rising stretch of the growth curve, but it is not the true culmination of stand growth from the current and mean volume increment patterns.

The unavailability of outgrown coppice forests up to the 1960's, the unfeasible checking of the temporariness of observed shoots' mortality and of the following growth recovery at the stand level, and

Table 3 - Assessment of growth pattern with reference to the age of maximum yield at the permanent sites monitored by CREA ($m.a.i.$ = mean annual volume increment; $c.a.i.$ = current annual volume increment).

main tree species	site	stand age (years)	c.a.i. $m^3ha^{-1}y^{-1}$	m.a.i. $m^3ha^{-1}y^{-1}$	m.a.i. => c.a.i.
Turkey oak	Emi1*	60	2.8	4.0	Yes
	Laz1*	50	4.3	4.2	close to
	Mar1*	50	5.6	5.9	Yes
	Sic1*	65	3.0	3.5	Yes
	Vas	47	1.8	6.6	Yes
	Cas	55	3.6	7.5	Yes
	Pop	44	1.4	3.6	Yes
holm oak	Tos1*	65	3.8	4.0	Yes
	Tos2*	70	4.8	3.6	No
	Laz2*	65	5.5	3.5	No
	Sar1*	65	4.0	4.3	Yes
	Isc	55	0.9	4.1	Yes
beech	Emi2*	60	6.3	5.4	No
	Lom3*	60	9.0	5.7	No
	Pie1*	75	6.8	4.6	No
	Cat	67	6.5	7.5	Yes

* ICP- Forests Level II plots.

the already achieved firewood size are the likely, concurrent explanatory reasons for the general acceptance of traditional yield models (Amorini and Fabbio 2009). The evidence of further positive growth patterns supports well the current shifting of coppice rotations towards higher stand ages and large-sized firewood production.

Basal area growth rates from 1.4 - 1.9% up to 2.5 - 3.1% are being recorded in the deciduous/ evergreen oaks and beech outgrown coppice plots of the ICP intensive monitoring network in Italy (Fabbio et al. 2006a). Net primary production varies from 10.2 (Turkey oak) to 11.6 Mgha^{-1}y^{-1} (beech), whilst growth efficiency varies from 2.0 to 2.6, respectively.

A further index descriptive of growth patterns is the relative space occupancy calculated as the percentage ratio of shoots' standing biomass (standards excluded) to the volume defined by mean stand height per unit area (the proxy of the age-related growing space). The two case-studies (beech and Turkey oak) reported in Fig.4 describe species-specific patterns. The shade-tolerant species shows an early drop following the age of customary harvesting (24 years), and then the tendency is for a smooth increase until the end of the observed lifespan (67 years). The pattern of light-demanding oak maintains lower values throughout and has a nearly steady increase which reaches its culmination at the age of 44.

The pro-active practice of coppice conversion into high forest (Amorini and Gambi 1977, Fabbio and Amorini 2006, Amorini et al. 2006, 2010) allowed the comparison of this option to the outgrown phase within the same trials. The early, sharp drop in density at the first thinning and the much more reduced decreases following the intermediate harvestings

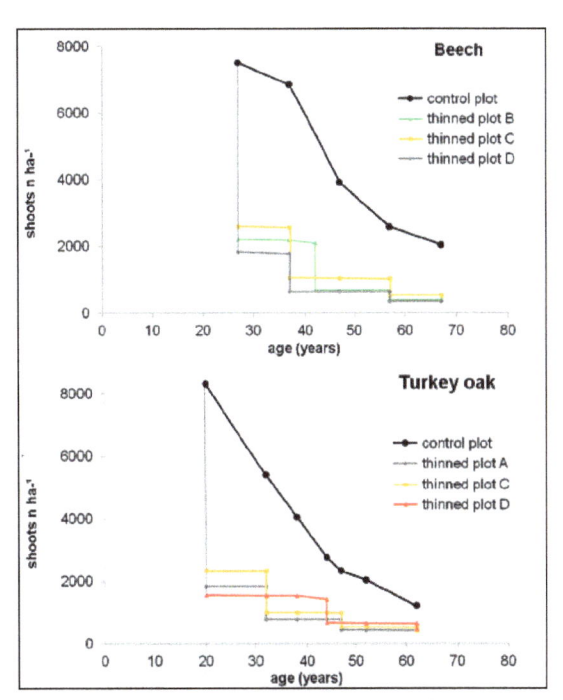

Figure 5 - Trend of shoots density at increasing stand ages at control and thinned plots in a beech and a Turkey oak outgrown coppice forest.

are compared with natural tree mortality (Fig.5). Thinning to control densities are quite similar at the last age recorded in Turkey oak, whilst a marked difference is still present in beech plots, due to the specific shade-tolerance of the latter. The growth dynamics in terms of current and mean volume increment (Fig.6) in one beech (A) and two Turkey oak (B, C) experiments points out that management changes both incremental values and their course, whilst the age of culmination control vs. thinning remains nearly unchanged at each site. Beech stand shows a delayed age of culmination (60 years) as compared to Turkey oak. A marked difference exists between the Turkey oak sites: site B, located in the pre-Apennine range at an elevation of 700 m asl, showed the mean annual volume increment culmination close to 40 years, whilst site C, located close to the Thyrrenian coast at 200 m asl, culminates earlier, at about 30 years. There is also evidence of an abrupt drop in current annual volume increment following the culmination age at control plots.

Standing volume and standing volume plus intermediate harvestings (Tab.4) describe wood allotment at control and thinned plots, respectively. Intermediate yields suggest the economic feasibility within well-accessible sites and average production levels. Patterns of periodical thinnings aimed at promoting the recovery of the high forest physiognomy and triggering the sizing of final crop trees to better address the regeneration from seed are outlined (Fig.7). Intermediate harvestings show to be not negligible, even if the current wood destination is still firewood. The mean stand dbh throughout

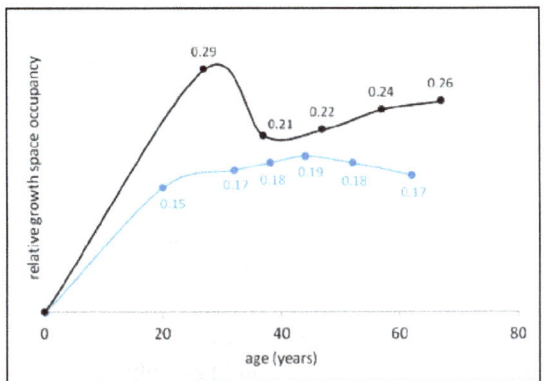

Figure 4 - Pattern of growing space occupancy by standing crop as a function of stand age in outgrown beech (black line) and Turkey oak (blue line) stands. Growing space occupancy is calculated as the relative ratio between standing crop volume and the theoretical available volume upper-bounded by the age-related mean stand height. Resprouting mass (shoots) is considered only. The small number of trees released since former cycles (the standards) at both stands makes them comparable.

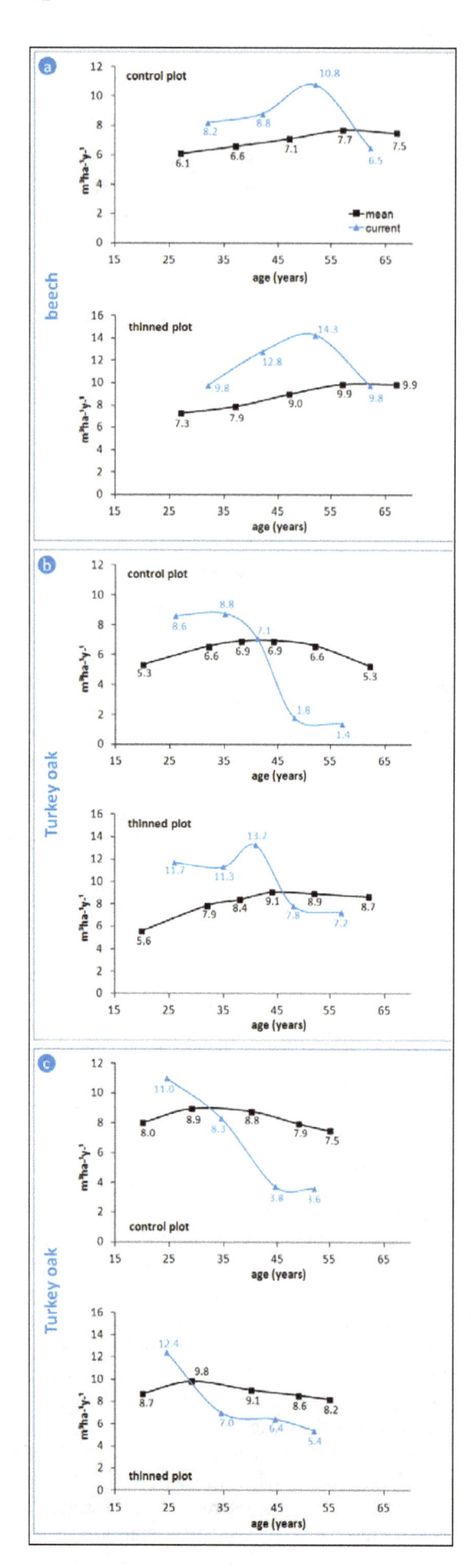

Figure 6 - Trends of current and mean volume increment of a beech (a) and Turkey oak (b, c) outgrown coppice forests at control and a thinned plot, respectively.

Table 4 - Dynamics of conversion to high forest vs. natural evolution (control plot). Current standing volume, intermediate and total yields at two permanent monitoring sites.

	BEECH (monitoring span = 27 - 67 years)		
treatment	standing volume m³ ha⁻¹	Σ thinned volumes m³ ha⁻¹	total yield m³ ha⁻¹
plots under conversion (average)	354	273 (3 thinnings)	627
control plot	505	-	505

	TURKEY OAK (monitoring span = 20 - 62 years)		
treatment	standing volume m³ ha⁻¹	Σ thinned volumes m³ ha⁻¹	total yield m³ ha⁻¹
plots under conversion (average)	287	215 (3 thinnings)	502
control plot	326	-	326

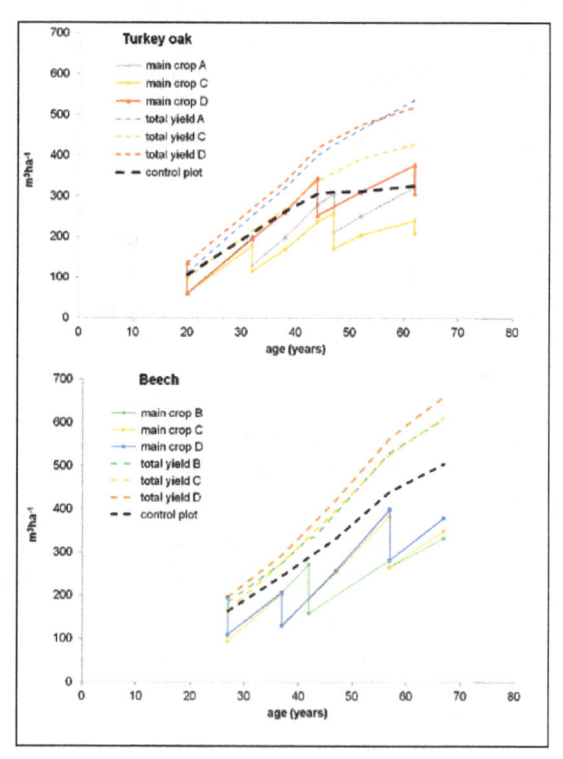

Figure 7 - Growth pattern of standing, main and total volumes at control plot and thinned plots in beech and Turkey oak outgrown coppice trials.

the explored life-span ranges from 9 to 34 cm (age 27-67 years) and from 9 to 26 cm (age 20-62 years) for beech and Turkey oak, respectively. The size of harvestable stems fits well and enhances the productivity of current handling-hauling systems.

The below ground process

The relationship between root system and above-ground biomass has a special significance in the coppice system since the prompt resprouting following the clear-cutting of stools in short time spans needs to be well supported belowground. A few hypotheses by Bernetti (1980, 1981) and Clauser (1981, 1998) regarding the development of outgrown coppice stands both in the above and underground components have produced original theoretical contributions. However, no field surveys and analyses studied the root system prior to the 1980's. A first digging trial was carried out (Fig.8-9) on a beech 'transitory crop', i.e. a coppice under conversion to high forest aged 43 years, thinned the first time at the age of 27 years, and the second time ten years later (at 37 years). The customary rotation was at 24. Therefore, the first thinning took place within the early establishment of the 'ageing period'. Two shoots were released at the first thinning and one at the second thinning on each of the sampled stools (Amorini et al. 1990).

Stem analysis was carried out for all living main (1st order) root branches (Fig.10). The age of each horizontal and vertical branch at the stool insertion, the annual radial increment, and the lengthening rate were determined. The shoot (stem) radial growth was also assessed at the height of 50 cm above ground level. Data refer to a stool living in the upper canopy layer, i.e. carrying at least one dominant shoot, which means a 'candidate' to be standing over the full conversion cycle. The results (Fig.11) apply to the time of coppice rotation (age 1-24 years) and to the transitory crop time (age 27-43 years).

- Horizontal rooting: only one living root aged before the last coppicing (1945) was detected; the new root branches sprouted as follows: +7 (age of 9), and +14 (age of 18). Three more branches sprouted before the first thinning, i.e. +17 in total (age of 27). No other branches developed over the transitory crop span (age > 27).
- Vertical rooting: no living branches aged more than the last coppicing were found. The development of new roots is slower in this sub-system +6 (age of 18), and +12 (age of 27). Only one new entry was detected within the transitory crop phase, i.e. +13 in total (age of 43).

At the age of survey (43 years), nearly all the main living root branches were developed after the last coppicing. This means that all the branches aged more, and still living during the last coppice cycle, ended their lifetime within the analysed time-span.

The resulting evidence of re-growth ability draws attention to the root system turnover as stool re-sprouting takes place after coppicing. These findings contribute a further understanding of stools' capacity to regrow several times without any depletion of their own regeneration potential. The same survey protocol, repeated a few years later in a Turkey oak ageing coppice, produced similar results.

Figure 8 - The digging trial of a beech coppice stool (1988).

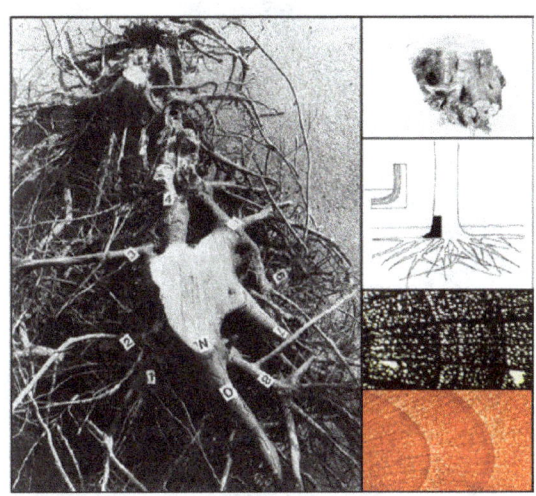

Figure 9 - Details of a beech root system.

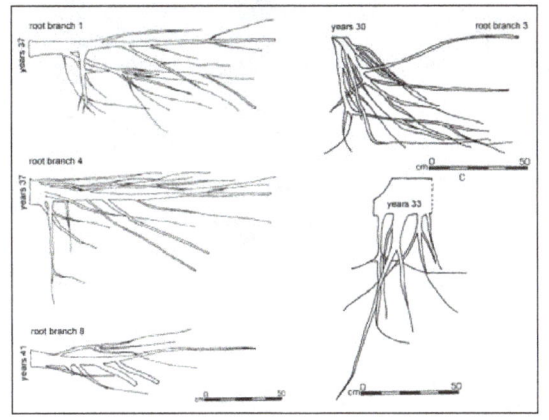

Figure 10 - Profile and age of main root branches at the insertion on the stool in a beech coppice.

An additional focus was on the development of the current radial increment at the shoot (stem) section at 50 cm above ground level and the total radial increment of the root branches measured at 10 cm from the stool insertion (Fig.12). The resulting patterns were quite similar over the full coppice cycle and also following each thinning in terms of reactive ability and growth rate. If the effect of growing space made available by the thinning on radial stem growth and crown sizing of released stems (the above ground component) was clear, there was

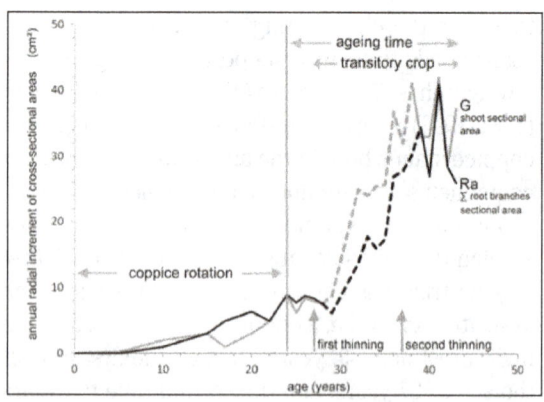

Figure 12 - Comparison between current radial increment (G) at the shoot section (height 50 cm) and total radial increment of root branches (ΣRa) at 10 cm from the stool insertion of a sampled stool in a beech coppice.

Root system mapping at digging out operations in a beech coppice stand (Tuscany).

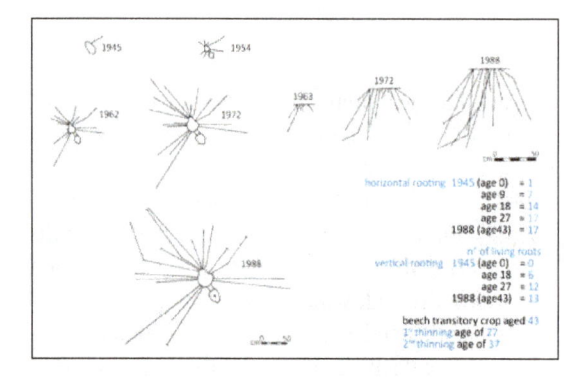

Figure 11 - Pattern of root branches development since last coppicing in a beech coppice. Age and number of living roots detected at the time of survey are reported.

no evidence that a similar, synchronous reaction takes place below ground. The adaptive ability of the new-established root system fully accomplished the development of the above ground tree biomass according to a consistent functional significance.

This finding also helps the understanding of the fulfilment of trophic autonomy by the inherent regenerative ability hastened here by thinnings, drastically anticipating the reduction of the number of shoots on the stool by natural mortality.

Other attributes of relevance

Further evidence from the same sites stress the auto-ecology of main tree species as a principal driver in the outgrown coppice's lifetime. This trait rules the quantitative outcome of growth patterns. Shoots' mortality is anticipated in light-demanding species (e.g. Turkey oak) as compared with shade-tolerant species (e.g. beech) (Fig.13a) according to the average speed of variation (Odum 1973). The different behaviour is also highlighted by the contrasting trend of the 'auto-tolerance' or 'intra-specific competitive ability' (Zeide 2005) diverging from the ages of 45-50 years when a sharp increase occurs for the oak (i.e. a higher mortality rate under similar radial increment), whilst the competitive ability remains steady for beech (Fig. 13b). Further evidence of the auto-ecology – shade tolerance in this case – as the main driver of shoots' mortality is shown by the overlapping dbh distributions of standing dead shoots in outgrown beech and holm oak coppices aged likewise but living in quite different environments (Fig. 13c).

Clear proof of Zeide's statement 'shade tolerance affects the survival of trees but not their growth' (1985, 1991, 2005) is provided by shoots' radial increments in a beech outgrown coppice forest (Fig. 14a). One-quarter of living shoots (27%) is alive but no longer able to produce new tissue and its radial

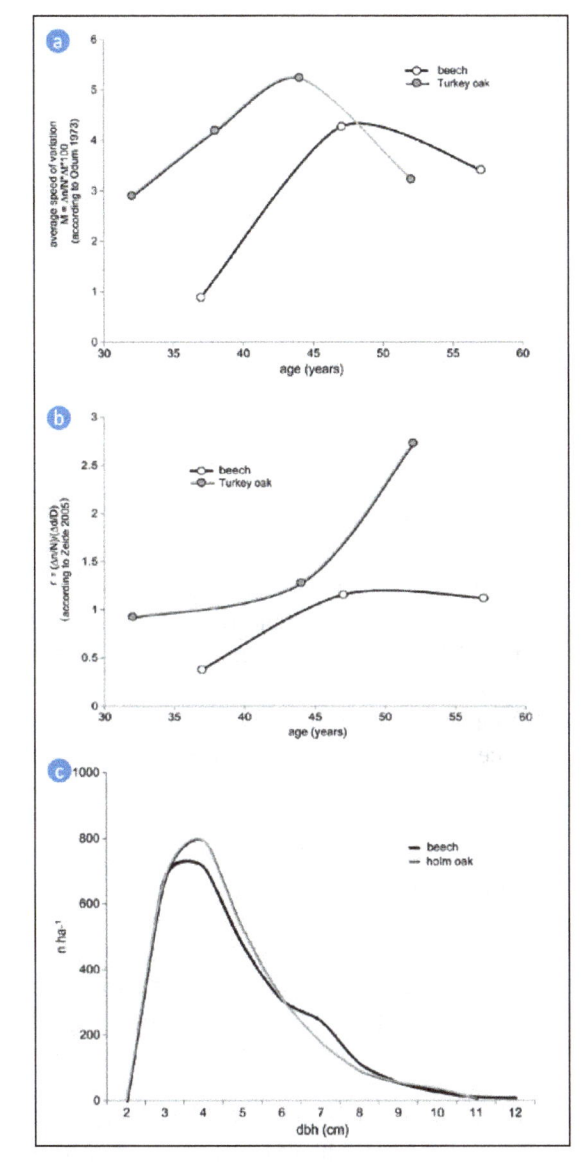

Figure 13 - (a) Average speed of variation of shoots number in a shade-tolerant (beech) and a light-demanding tree species (Turkey oak) according to Odum (1973). (b) Trend of the 'auto-tolerance' or 'intra-specific competitive ability' according to the Zeide algorithm (2005). (c) Dbh frequency distributions of standing dead shoots in outgrown beech and holm oak coppices aged likewise.

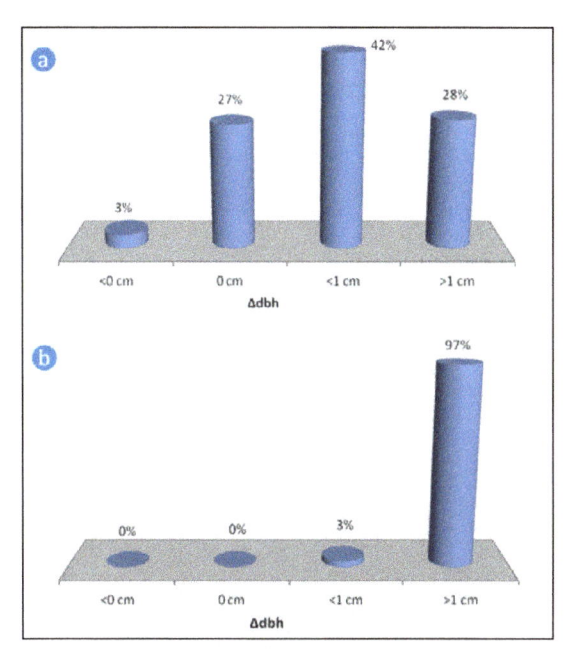

Figure 14 - Distribution of dbh growth under the natural evolutive pattern (control plot) (a) and in a thinned plot (b) within an outgrown beech coppice forest over the same life-span (age from 47 up to 57 years).

increment is zero within the not-negligible time span of ten years, whereas the standing dead population is only 3%. The quite different growth pattern within the same stand under conversion to high forest (Fig. 14b) highlights the change promoted by the repeated standing crop thinning causing a one-layered stand structure physiognomically similar to a pole stand from seed.

Tree biomass and standing/lying deadwood are complementary functional attributes along with stand age, especially in outgrown coppices. Their dynamics have a noticeable effect on deadwood accumulation on the forest floor. Tree biomass and standing/lying deadwood allocation for light-demanding (Turkey oak) and shade-tolerant (holm oak, beech) species are provided in Tab.5. The light-demanding species is characterised by the early shift of standing to lying deadwood ratio, whilst the opposite takes place for the shade-tolerant species. All of this is in spite of the quite similar mean

Table 5 - Standing shoots biomass, total, standing/lying deadwood, mean annual increments of standing biomass and deadwood according to stand age and main tree species.

main tree species	stand age	standing biomass	standing biomass mean annual increment	deadwood total	deadwood standing	deadwood lying	standing to lying deadwood ratio	deadwood mean annual increment
	years	Mg ha^{-1}	Mg ha^{-1} year^{-1}	Mg ha^{-1}	Mg ha^{-1}	Mg ha^{-1}		Mg ha^{-1} year^{-1}
Turkey oak	52	238.8	4.59	22.4	6.1	16.3	1/3	0.43
Turkey oak	55	313.0	5.69	30.0	9.8	20.2	1/2	0.55
holm oak	55	225.3	4.10	25.3	18.5	6.8	3/1	0.46
beech	57	321.6	5.64	27.7	19.5	8.2	2/1	0.49

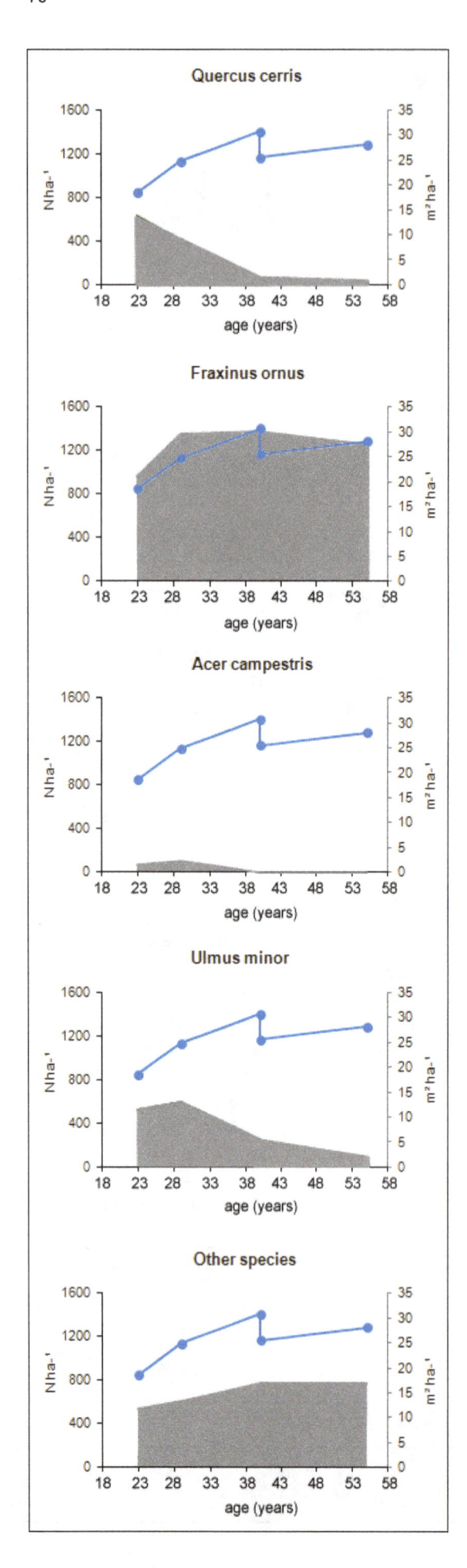

Figure 15 - Compositional diversity (number of trees per species = grey area) in the understory of a Turkey oak-dominated coppice under conversion to high forest and its change as a function of stand age and basal area in the main crop layer (blue line) used here as a proxy of upper canopy cover.

deadwood accumulation rate of about 0.5 Mgha^{-1} at all the sites (Bertini et al. 2010, 2012). This value ranges from 1/9 to 1/11 of the mean above ground biomass accumulation rate.

Also, compositional diversity follows different patterns under the same conditions according to specific ecological requirements. The presence-abundance and the time trend of complementary tree species living in the understory of a Turkey oak-dominated coppice under conversion into high forest are shown in Fig.15 (Fabbio and Amorini 2006).

Within the level II-ICP network in Italy, the outgrown coppice plots showed cases of dynamic-specific and structural stand rearrangement and among the highest values of tree richness (Fabbio et al. 2006b).

Another attribute relevant to the dynamics of coppice forests aged 40 to 60 years is the production of litterfall and leaf litter as compared with temperate-warm and temperate-cold forests (Fig.16). The ratio of leaf litter to total litter is about 70%; a typical figure for young, productive forests. In thinned stands, Leaf Area Index reduction to 4-5 optimises the Net Assimilation Rate and increases the Net Primary Production (Cutini and Hajny 2006). The amount of seed production and the frequency of mast years are species-specific traits. Turkey oak shows an annual production five times higher than beech on average (0.70 vs. 0.13 Mgha^{-1}), which is also due to the much more frequent occurrence of

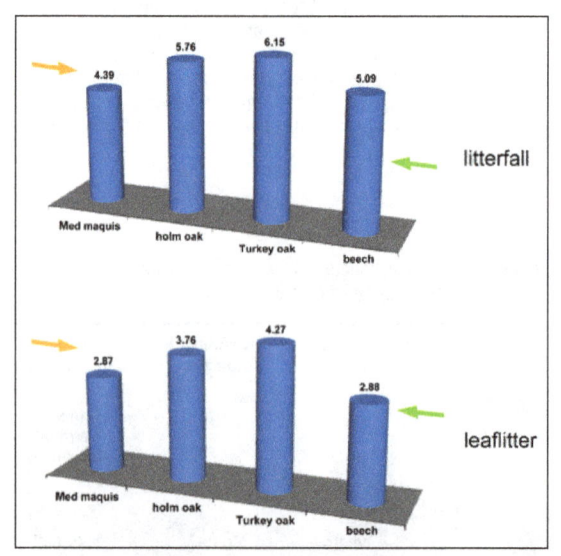

Figure 16 - Average litterfall and leaflitter production (Mgha^{-1}) in coppice stands aged 40 to 60. Arrows show the average for temperate-warm (left) and temperate-cold (right) forests (sources: Bray and Gorham 1964, O'Neil and De Angelis 1981, in Cutini and Hajny 2006).

mast years (Cutini 2000 and 2002). These attributes become of major concern at the time of stand regeneration from seed.

A paradox is finally evident if we account for the damage due to wildlife browsing on stools resprouting. The 86% of Turkey oak stools browsed the first year after coppicing did not survive over the following two years (Cantiani et al. 2006). An average reduction of standing volume up to 57% and 41% was determined six and eleven years after coppicing, respectively (Cutini et al. 2011). The issue is of even greater concern when it becomes the driver for the successful resprouting of outgrown coppice stools as reported by Pyttel (2015) for sessile oak stands aged 80-100 years. This 'modern' disturbance may be heavier than the past practice of grazing by domestic animals and is can compromise the whole wood production cycle. The issue at hand puts forward the critical question of wildlife conservation practice, i.e. protection or breeding?

The most recent research issues at the CREA trials dealt with: forest management and water use efficiency (Di Matteo et al. 2010), environment-induced specific adaptive traits and C and N pools by different compartments (Di Matteo et al. 2014ab), seed production patterns (Cutini et al. 2010, 2013), management of the final phase of conversion into high forest, i.e. the implementation of regeneration cuttings (Cutini et al. 2015), the long-term response of coppice conversion to high forest experiments (Chianucci et al. 2016) and, again, the wildlife browsing impact on stools' resprouting (Chianucci et al. 2015).

Main traits and role of the management areas established on the former coppice cover

Land use and land use change

The original coppice system area underwent a significant reduction some fifty years ago. Land use and land use change are consistent with the dynamics of social and economic context. Both originate from factual needs and when the related commodities can be usefully replaced, produced elsewhere or the customary use is no longer profitable, land use is abandoned (Del Favero 2000) or changed (Mottet et al. 2006). Other spontaneous or man-induced changes have taken place in the landscape matrix since the time of coppice abandonment on less accessible and less fertile sites. The natural afforestation of open areas (abandoned fields and rangelands) contributed to the steady increase of forest-type cover, reduced the patchy distribution of open areas and homogenized the forest cover over the last decades. The establishment of agro-forestry systems and, more recently, of short rotation forestry on lands set aside by the agricultural practice have modified the human-imprinted landscape. The newly-established tree farming created further interfaces between rural and urban areas on the plains or hilly sites no longer favourable for agriculture, but fertile enough and with good access with respect to intensive wood production. This sequence contributes recent traits of landscape dynamics. Our perception of land use is first supported and consolidated by the long-lasting direct, visual/physiognomic experience and then transferred by documental sources. This is the way a common 'heritage' value is established over generations.

If the balance between practice and its profitability and/or replaceability historically dictated land use, the modern acknowledgement of complementary societal benefits arising from its maintenance can move, to some extent, the point of balance. In this case, the community should reward the fulfilment of shared common benefits and contribute to the feasibility of the use in concern. This seems to be the basic condition for answering some questions. Reference is made to the manifold calls for coppice recovery as a heritage value or the driver of 'bio-cultural diversity' (Burgi 2015) where it was historically present, but independently of its current, profitable implementation. Also, the frequently asked question about the evidence of vegetational and faunal diversity loss (Kopecky et al. 2015, Vild et al. 2013, Mullerova 2015ab) being closely linked to the short-term opening of patchy clearings in the forest cover as in the former coppice system (Kirby 2015) should be settled within the same frame of reference.

The area managed under the coppice system

Easily-accessible areas to make profitable wood harvesting, good site fertility allowing sustained growth and closeness to the market are the basic requirements for effective management of today's coppice system. The updating of management criteria includes: (i) lengthening customary rotations, (ii) recovery of technical function for 'standards' release (i.e. reducing their number, effective selection of dendrotypes, and suitable spatial arrangement), (iii) obtaining certified productions, and (iv) search for the optimal size, shape and contiguity of clearings according to the different physical environments.

The main purpose of maintaining a coppice system is still firewood production, but there are other complementary benefits such as the contribution to the landscape mosaic texture providing specific habitats, types, and patterns of diversity (Mairota et al. 2014, Burgi 2015, Hédl et al. 2015) which is also linked to the early successional stages inside the clearings (Kirby 2015).

The emerging green economy issues (Marchetti et al. 2014), the structural change from a fossil-based economy to a bio-based economy (Corona 2014), and the increased demand for renewable energy resources may be the turning point which addresses the forthcoming role of coppice forests. Positive factors remain the basic attributes of the system, i.e. the easy management technique, the guarantee of natural regeneration, the flexibility/reversibility, and the resistance/resilience. In this regard, the adaptability of oak coppice forests to changed conditions, namely water management, is reported by Splichalova (2015), whilst the higher drought tolerance of sessile oak resproutings vs. seedlings under soil moisture limiting conditions is underlined by Holisova et al. (2015) and Pietras et al. (2016). It is also worth recalling that resprouting ability has been one of the most important keys to the building up of the paradigm of resilience and post-disturbance auto-successional nature of Mediterranean coppice forests (Espelta et al. 1999, Konstantinidis et al. 2006, in Lopez et al. 2009).

The more manageable length of stand lifespan compared with the high forest system widens the options for handling the risk and unpredictability of climate shift in the primary phase (regeneration) and throughout the forest cycle.

Further concurrent elements today are the chance to select contexts optimal to cultivation within the former coppice area, the reduced impact of historical overlapping uses, the reasonable less intensive management ruling the system, the improved knowledge achieved so far about the bio-ecological functioning (drivers, limiting factors, and feedbacks), the above and below-ground dynamics, and the growth patterns.

Final harvesting: the irregular shape of the cutting areas edge results in a lower visual impact even in case of large clearings (northern Apennines).

Special focus is being developed and consistent techniques are applied to the effective tending of valuable, even sporadic tree species within coppice stands (Pelleri, *this paper*).

Besides all its positive traits, coppice remains a low-input, high-output energy system compared to other silvicultural systems because of the natural assurance of crop regeneration. This is why, looking back but ahead too, a consistent definition of coppice may be today 'a very ancient but modern system as well' (Fabbio 2015).

Conservation and enhancement of sporadic tree species living in coppice forests: the CREA experience

Francesco Pelleri
CREA, Centro di ricerca per le foreste e il legno, Arezzo (Italy)

'Sporadic tree species' means trees living both as individuals and small groups within a forest stand; they are often able to produce quality timber, and valuable broadleaved tree species are included within this category (Mori et al. 2007). Following the ageing of unmanaged coppice stands and the customary practices of conversion into high forest, the progressive reduction of sporadic tree species is usually recorded, since most of sporadic tree species are less competitive than dominant tree species (Mori and Pelleri 2012). Because of their attributes (light-demanding, poorly competitive, reduced growth), such species are very sensitive to any practice going to change stand structure parameters. Their maintenance and increase in value runs through practices targeted to complying with their auto-ecology. The tree-oriented silviculture, an approach developed in central Europe for oak, beech and spruce high forests (Abetz 1993, Sevrin 1994, Bastien and Wilhelm 2000, Abetz and Kladtke 2002, Wilhelm 2003), especially fits both conservation and enhancement of sporadic tree species in coppice forests (Spiecker 2006, Spiecker et al. 2009, Sansone et al. 2012, Pelleri et al. 2013 and 2015, Manetti et al. 2016).

Appropriate fields of application
The economically feasible and successful practice of tree-oriented silviculture in coppice forests is based on the selection of a limited number of suitable trees as for vigour, stem quality and crown quality. Prerequisites are the presence of these dendrotypes, well-accessible sites, favourable ecological conditions. Under these assumptions, tree-oriented silviculture can ensure the persistence of these species, their natural regeneration and the increase of valuable timber production. Thinnings localized around a limited number of selected trees allows to manage the remaining standing crop as in the customary way, without reducing noticeably the firewood production of the coppice system (Sansone et al. 2012, Mori and Pelleri 2014).

This approach was recently applied within the LIFE-PProSpoT in coppice forests in central Tuscany on an overall area of 53 hectares. Ten to twenty target trees per hectare were promoted by localized crown thinning to get their free crown expansion. Trials were undertaken both in young and ageing coppice stands.

The young coppice stands

The enforcement of tree-oriented silviculture in stands aged 10-15 years allows the conservation of compositional diversity, as well as the promotion of growth pattern and timber value of the selected trees. A notable increase of radial increment from 1-2 up to 5-7 mm per year has been reached in a few years in service tree and wild service tree, whilst the growth rate has increased from 2-4 up to 8-10 mm per year in field maple. Stem diameter increment similar or higher can be achieved by field elm and wild cherry (Wilhelm e Ducos 1996, Nicolescu et al. 2009, Manetti et al. 2016, Giuliarelli et al. 2016).

The maintenance of these growth rates implies heavy crown thinnings repeated every 6-8 years to get the crown release of 2-3 m, or less intense (1-2 m of crown release) but more frequent thinnings (4-6 years).

The ageing coppice stands

Practices are aimed at favouring more the conservation and fruiting of sporadic valuable tree species, whilst less achievable is the increase in value of wood production. Light-demanding species are especially unable to react to late thinnings (service tree, European ash, wild cherry, peduncolate oak, etc.), whilst the reaction ability is more evident for shade-tolerant species (wild service tree, linden, sycamore, holly tree, etc.) even in the late life-span; these are less-sensitive to tree competition and maintain a more deep and efficient crown, just able to react also to late openings (Rasmussen 2007).

The 'standards' release

At coppicing, it is advisable to protect the young selected trees with a belt of shoots (grouped standards release) to avoid any damage due to the sudden isolation (stem quality worsening, e.g. growth of epicormic branches or stem breakage). Large-sized sporadic trees provided with well-developed crowns may be released as individuals without any significant risk. The grouping of standards may be supplemented by the customary release of individuals of the main and sporadic tree species (Mori and Pelleri 2014).

The use of the proposed tending techniques may result in the successful maintenance and improvement of wood production value as well as in the preservation of a higher compositional diversity.

Wild cherry target tree in a chestnut coppice stand aged 16 years before thinning.

Wild service target tree in a mixed coppice forest aged 15 years.

The post-cultivation area or the outgrown coppice stands

Outgrown coppice stands are widespread under marginal conditions in terms of accessibility and site quality, but they are also widespread in public-owned areas and in areas designated for nature conservation. Here, main forest functions are the soil protection or its recovery, the re-establishment of habitats similar to those former to human-imprinting, the carbon stock and sequestration (mitigation), the contribution to landscape texture, and the maintenance of specific habitats for biodiversity conservation. Small-scale silvicultural practices should be introduced to allow the maintenance of target tree species, stand structural and compositional diversity into the abandoned mosaic-like structures where protection is also a target issue (Garbarino et al. 2015, Urbinati et al. 2015). New adaptive rules allowing the coexistence of gamic and agamic regeneration in the same stand have been recently introduced in northern Italy (Motta et al. 2015).

The increased forest area under different protection levels, the enforced regulations heavily limiting or making difficult the implementation of any form of management for wood production – independently of bio-ecological conditions and site location suitable for harvesting – have contributed to the unmanaged coppice area increasing. Hence, forests also susceptible to still being under sustainable coppice system rules but included in areas protected today as a whole, are not effectively manageable (Mairota et al. 2016b). Due to this conservative position, a non-defined permanence time of stands and minimal interventions addressed to promote seed regeneration are foreseeable at now.

A point of concern is provided by the high amount of standing and lying woody necromass in these overstocked types making them very sensitive to fire. This issue has to be taken into account by managers of large, continuous forest covers within environments prone to wildfire occurrence (Corona et al. 2015).

The coppicing of aged crops provides case studies of utmost importance in order to collect grounded evidence about the long-term resprouting capacity of stools. An inventory of these cases and their analysis would be the main source of knowledge for the time being. It might provide, in addition to the very few trials established in between, the necessary expertise to again undertake coppicing under suitable conditions, to meet the renewable bio-energy demand, and reduce the unmanaged areas.

Suitable conditions mean the occurrence of: (i) geomorphological and soil attributes allowing repeated clearcutting even with doubled rotation lengths, (ii) good site quality for sustained and sustainable firewood production, and (iii) well-accessible locations allowing the reduction of harvesting costs.

The area under conversion into high forest

The main goal of the conversion of coppice into high forest is to anticipate the recovery of former high forest structure managing timely the outgrown coppice crops. The choice is consistent with sites fertile enough and stand textures sufficiently dense to get the awaited outcome within the conversion cycle. The periodic revenues from thinnings make this option economically enforceable to different sizes and types of ownership. The presence of valuable, even sporadic, tree species is an added value here.

Less relevant in terms of total area, this option has special significance in the public domain, where a share of the wide forest cover which is no longer harvested is available to a pro-active, adaptive silviculture. The decision to undertake such silviculture should be pursued according to a few, logical steps. The mountain areas, where the coppice system has been suspended earlier because of lower profitability and higher environmental sensitivity, should be taken into account first. Then, the areas which are valuable because of tree-specific composition and of scenic value, or under-targeted conservation, may be included in this option. Management rules are the tending practices applied to standing crops up to their regeneration from seed (Amorini and Fabbio 1990, 1994, 2009, Alberti et al. 2015). Adaptive strategies may be usefully added to the thinning methods in use to implement a higher canopy differentiation of standing crops already in the second half of the conversion cycle. This implies the selective tending of best phenotypes to favour individual crown development and to reduce the evenness of the one-storied stands (Fabbio et al. 2014).

The main purpose of this choice remains a more suitable balance between wood, non-wood productions, and environmental functions as in the

Harvested coppice area a few years after coppicing. The irregular shape of clear-cut makes easier the physiognomical re-establishment within surrounding texture (holm oak forest, southern Italy).

traits of the high forest system. It implies a more extended lifetime and the implementation of the intermediate set of silvicultural practices ensuring the awaited growth pattern of the selected shoots in the main crop layer as well as their health and vitality throughout the conversion cycle and up to the regeneration from seed.

Conclusive remarks

Each management type which has arisen from the common coppice matrix shows peculiar and consolidated features. All of them have become established as a result of factual macro-economic conditions and provide a range of goods and benefits. They basically run according to different management criteria, intensity, and the type of applied practices up to post-cultivation. Further inherent or practice-driven dynamics will be determined by criteria prevailing at the multiple decision-making levels and by changing scenarios as well (Millar et al. 2007, Lindner et al. 2010).

Stakeholders, planners, and managers should envisage all the available options and their possible connections on the ground as well as their complementarities in landscape planning.

They should also acknowledge the consistency of each choice according to the prevailing local function(s), the site quality, and the bio-ecological conditions at the operational scales of silviculture and forest management, i.e. from the stand up to the forest compartment level.

This rationale accomplishes and supports many, already well-achieved, statements. Among these, worthy of mention are: [...] the establishment and mosaic of the different choices builds up the organic development of land matrix and its connections (Franklin 1993); [...] intensive to extensive cultivation systems up to pure conservation tailored at the local scale may coexist and implement diverse development stages and structural diversity from stand up to landscape level (Fuhrer 2000, Farrell et al. 2000); [...] the complex and varied physical context allows post-cultivation and pro-active management approaches as well, both of them being strategic and complementary (Di Castri 1996, Teissier Du Cros 2001, Palmberg-Lerche 2001, Fabbio et al. 2003).

The forthcoming challenge will be the tuning of management strategies so they are able to make the system function effectively under the new condition(s). Two main open questions remain for coppice forests as for any other biological system living within a changing growth medium. Are we moving from a steady state to a perennial transition? Furthermore, how much of the inherent ecological buffer has been/will be eroded in between?

Acknowledgements

I am grateful to Francesco Pelleri for contributing his own specific expertise to the subject-matter.

I also want to thank here several the colleagues who have been and are working on trials on coppice forests at our Institute: Germano Gambi, Giulio Guidi, Riccardo Morandini, Emilio Amorini, Maria Chiara Manetti, Andrea Cutini, Paolo Cantiani, Francesco Pelleri, Silvano Ghetti, Vittorio Mattioli, Dino Gialli, Nevio Donati, Luigi Mencacci, Umberto Cerofolini, Mario Romani, Galeazzo Scaioli, Mauro Frattegiani, Silvia Bruschini, Claudia Benvenuti, Maurizio Piovosi, Giada Bertini, Claudia Becagli, Tessa Giannini, Luca Marchino, Walter Cresti, Eligio Bucchioni, and Leonardo Tonveronachi.

Finally, a special thank you goes to Germano Gambi and Emilio Amorini who first introduced me to the subject matter and to the CREA trials on coppice forests.

1988, the digging trial.

References

Abetz P. 1993 - *L'arbre d'avenir et son traitement sylvicole en Allemagne* [The tree for the future and silvicultural treatment in Germany]. Revue Forestiere Francaise, 45 (5): 551-560.

Abetz P., Klädtke J. 2002 - *The target tree management system.* Forstw. Cbl. 121: 73-82.

Accademia Nazionale di Agricoltura 1979 - *Il miglioramento dei cedui italiani* [The improvement of Italian coppice forests]. Bologna 401 p. [In Italian]

Agnoletti M. 2002 - *Bosco ceduo e paesaggio: processi generali e fattori locali* [Coppice woods and landscape: analysis and evaluation of a cultural factor]. In: Il bosco ceduo in Italia. (Ciancio O., Nocentini S. eds). Accademia Italiana di Scienze Forestali, Firenze: 21-62. [In Italian]

Agnoletti M. 2003 - *Note sui principali mutamenti avvenuti negli ecosistemi forestali italiani dall'Unità ad oggi* [Notes on main changes occurred in Italian forest ecosystems from the Unity up to the present time]. In: Atti III Congresso nazionale SISEF Alberi e Foreste per il nuovo millennio. (De Angelis et al. eds). Viterbo, 15-18 ottobre 2001: 127-132. [In Italian]

Alberti G., Mariotti B., Maltoni A., Tani A., Piussi P. 2015 - *Conversion of* Fagus sylvatica *coppices to high forests: results from a thirty year experiment in eastern Italian pre-Alps*. Coppice forests: past, present and future. International conference. Brno, Czech Republic, 9-11 April 2015 http://coppice.eu/conference_en.html

Amorini E., Gambi G. 1977 - *Il metodo dell'invecchiamento nella conversione dei cedui di faggio* [the method of 'ageing' in the conversion of beech coppices]. Annali Istituto Sperimentale Selvicoltura, Arezzo 8: 23-42. [In Italian]

Amorini E., Fabbio G., Gambi G. 1979 - *Sistema di diradamento del bosco ceduo per l'avviamento all'altofusto. Sperimentazione in prospettiva dell'uso multiplo con il pascolo* [Thinning method of coppice stand for conversion to high forest. Experimental trials for multiple use with grazing]. Annali Istituto Sperimentale Selvicoltura, Arezzo 10: 3-23. [In Italian]

Amorini E., Fabbio G. 1986 - *Studio auxometrico in un ceduo invecchiato e in una fustaia da polloni di faggio, sull'Appennino toscano* - Primo contributo [Growth assessment at an outgrown beech coppice and at a transitory crop in the Tuscany Apennines - First contribution]. Annali Istituto Sperimentale Selvicoltura, Arezzo 14: 283-328. [In Italian]

Amorini E., Fabbio G. 1988 - *L'avviamento all'altofusto nei cedui a prevalenza di cerro. Risultati di una prova sperimentale a 15 anni dalla sua impostazione. Primo contributo* [Conversion into high forest of Turkey oak coppices. Results of an experimental trial fifteen years later. First contribution]. Annali Istituto Sperimentale Selvicoltura, Arezzo 17: 7-101. [In Italian]

Amorini E., Fabbio G. 1989 - *L'avviamento all'altofusto nei cedui a prevalenza di cerro. Risultati di una prova sperimentale a 15 anni dalla sua impostazione. Studio auxometrico. Secondo contributo* [Conversion into high forest of Turkey oak coppices. Results of an experimental trial fifteen years later. Growth analysis. Second contribution]. Annali Istituto Sperimentale Selvicoltura, Arezzo 18: 19-70. [In Italian]

Amorini E., Fabbio G. 1990 - *Le 'vieillissement' des taillis en Italie: étude auxométrique et traitement de la futaie sur souches* [The 'outgrowing' of coppice forests in Italy: growth analysis and silviculture of the conversion into high forest]. In: Proceedings IUFRO, XIX World Congress. Montreal, August 5-11 1990. 1: 363-374.

Amorini E., Fabbio G., Frattegiani M., Manetti MC. 1990 - *L'affrancamento radicale dei polloni. Studio sugli apparati radicali in un soprassuolo avviato ad altofusto di faggio* [The turnover of root system in an outgrown beech coppice under conversion into high forest]. Annali Istituto Sperimentale Selvicoltura, Arezzo 19: 201-261. [In Italian]

Amorini E., Fabbio G. 1992 - *A rapidly changing cultivation system: the coppice*. In: Proceedings 1st European Symposium on Terrestrial Ecosystems. Forests and Woodlands. Florence, may 20-24 1991. Elsevier, London: 902-903.

Amorini E., Fabbio G. 1994 - *The coppice area in Italy. General aspects, cultivation trends and state of knowledge*. In: Proceedings Workshop Improvement of coppice forests in the Med region. Arezzo, september 24-25 1992 Annali Istituto Sperimentale Selvicoltura, Arezzo 23: 292-298.

Amorini E., Fabbio G., Tabacchi G. 1995 - *Le faggete di origine agamica: evoluzione naturale e modello colturale per l'avviamento ad alto fusto* [Ageing beech coppices: natural evolutive pattern and cultivation model for its conversion into high forest]. In: Atti Seminario Funzionalità del sistema faggeta. AISF, Firenze 16-17 Novembre: 331-345. [In Italian]

Amorini E., Bruschini S., Cutini A., Fabbio G., 1996 - *Struttura e produttività di popolamenti di leccio in Sardegna* [Structure and productivity of holm oak stands in Sardinia]. In: Atti VII Congresso nazionale Società Italiana di Ecologia, Napoli 11-14 settembre 1996: 133-137. [In Italian].

Amorini E., Cutini A., Fabbio G. 1997 - *Gestion visant la conservation des écosystèmes de chene vert résiduel en Sardaigne (Italie)* [Management vs. conservation of residual holm oak forests in Sardinia (Italy)]. In: Comptes Rendus du XI Congrès Forestier Mondial. Antalya, 13-22 Octobre 2: 171.

Amorini E., Bruschini S., Cutini A., Fabbio G., Manetti MC. 1998a - *Silvicultural treatment of holm oak (*Quercus ilex *L.) coppices in Southern Sardinia: thinning and related effects on stand structure and canopy cover*. Annali Istituto Sperimentale Selvicoltura, Arezzo 27: 167-176.

Amorini E., Bruschini S., Cutini A., Di Lorenzo MG., Fabbio G. 1998b - *Treatment of Turkey oak (*Quercus cerris *L.) coppice. Structure, biomass and silvicultural options*. Annali Istituto Sperimentale Selvicoltura, Arezzo 27: 121-129.

Amorini E., Di Lorenzo MG., Fabbio G. 1998c - *Intensity of standards release and shoots dynamics in a Turkey oak (Q. cerris L.) coppice. First contribution*. Annali Istituto Sperimentale Selvicoltura, Arezzo 27: 105-111.

Amorini E., Brandini P., Fabbio G., Tabacchi G. 2000 - *Modelli di previsione delle masse legnose e delle biomasse per i cedui di cerro della Toscana centro-meridionale* [Volume and biomass prediction models for Turkey oak coppice stands in Central and Southern Tuscany]. Annali Istituto Sperimentale Selvicoltura, Arezzo 29: 41-56. [In Italian]

Amorini E., Cantiani P., Fabbio G. 2002 - *Principali valutazioni sulla risposta degli indicatori dendrometrici e strutturali in querceti decidui dell'Umbria sottoposti a diverso trattamento selvicolturale* [Main response of mensurational and stand structure indicators in deciduous oak coppices submitted to different silvicultural practices] In: Ferretti M., Frattegiani M., Grohmann F., Savini P. (eds). Il progetto TraSFoRM, Regione Umbria. [In Italian]

Amorini E., Fabbio G., Cantiani P. 2006 - *Avviamento ad alto fusto e dinamica naturale nei cedui a prevalenza di cerro. Risultati di una prova sperimentale a 35 anni dalla sua impostazione. Il protocollo di Valsavignone (Arezzo)* [Conversion to high forest and natural pattern into ageing Quercus cerris L.- dominated coppices. Results from 35 years of monitoring. The Valsavignone site (Apennines - Tuscany)]. Annali CRA-Istituto Sperimentale Selvicoltura, Arezzo 33: 115-132. [In Italian]

Amorini E., Fabbio G. 2009 - *I boschi di origine cedua nella selvicoltura italiana: sperimentazione, ricerca, prassi operativa* [Coppice forests evolving from the suspension of management in Italy: experimental trials, applied research and operational praxis]. In: Atti III Congresso Nazionale di Selvicoltura (Ciancio O. ed.) Taormina (Me, Italy) vol. 2: 201-207. [In Italian]

Amorini E., Fabbio G., Bertini G. 2010 - *Dinamica del ceduo oltre turno e avviamento ad alto fusto dei cedui di faggio. Risultati del protocollo 'Germano Gambi' sull'Alpe di Catenaia (Arezzo)* [Stand dynamics of a beech coppice beyond the rotation age and under conversion into high forest]. Annali CRA-SEL Arezzo 36: 151-172. [In Italian]

ARSIA 2006 - *Selvicoltura sostenibile nei boschi cedui* [Sustainable practice of silviculture into coppice forests]. Progetto Arsia-Regione Toscana Annali CRA-Istituto Sperimentale Selvicoltura, Arezzo 33, 255 p. [In Italian]

Baragatti E., Frati L., Chiarucci A. 2006 - *Cambiamenti nella diversità della vegetazione in seguito a diversi tipi di matricinatura in cedui di cerro* [Changes in vegetation diversity under different silvicultural management in a *Quercus cerris* forest]. Annali CRA-Istituto Sperimentale Selvicoltura, Arezzo 33: 39-50. [In Italian]

Bastien Y, Wilhelm GJ. 2000 - *Une sylviculture d'arbres pour produire des gros bois de qualité* [A trees oriented silviculture to produce large-sized quality wood]. Revue Forestière Française, 5: 407-424.

Becagli C., Cantiani P., Fabbio G. 2006 - *Trattamento sperimentale in un ceduo composto di roverella e leccio del Chianti senese. Primi risultati* [Experimental trial in a pubescent oak and holm oak coppice with standards in the Chianti region (Siena). First results]. Annali CRA-Istituto Sperimentale Selvicoltura, Arezzo 33: 31-38. [In Italian]

Bernetti G. 1980 - *L'auxometria dei boschi cedui italiani* [A critical review of the yield tables for italian coppice stands]. L'Italia Forestale e Montana XXXV (1):1-24. [In Italian]

Bernetti G. 1981 - *Ipotesi sullo sviluppo dei boschi cedui e relative considerazioni selvicolturali e assestamentali* [An hypothesis on the growth pattern of coppice forests: silvicultural and management remarks]. Monti e Boschi XXXII (5): 61-66. [In Italian]

Berti S., Lauriola MP., Mannucci M., Ricottini G. 1998 - *Technological characterization of Turkey oak solid wood panels*. Annali Istituto Sperimentale Selvicoltura, Arezzo 27: 209-214.

Bertini G., Fabbio G., Piovosi M., Calderisi M. 2010 - *Densità di biomassa e necromassa legnosa in cedui di cerro in evoluzione naturale in Toscana* [Tree biomass and deadwood density into ageing Turkey oak coppices in Tuscany]. Forest@ (7): 88-103. doi: 10.3832/efor0620-007 [In Italian]

Bertini G., Fabbio G., Piovosi M., Calderisi M. 2012 - *Densità di biomassa e necromassa legnosa in cedui oltre turno di leccio in Sardegna e di faggio in Toscana* [Tree biomass and deadwood density into aged holm oak (Sardinia) and beech coppices (Tuscany)]. Forest@ (9): 108-129. doi: 10.3832/efor0690-009 [In Italian]

Brandini P., Tabacchi G. 1996 - *Modelli di previsione del volume e della biomassa per i polloni di leccio e di corbezzolo in boschi cedui della Sardegna meridionale* [Biomass and tree volume equations for holm oak and strawberry-tree in coppice stands of southern Sardinia]. Comunicazioni di ricerca ISAFA, 96/1: 59-69. [In Italian]

Burgi M., 2015 - *Coppicing in the past - examples of practice, context and consequences*. Coppice forests: past, present and future. International conference. Brno, Czech Republic, 9-11 April 2015 http://coppice.eu/conference_en.html

Cantiani P., Amorini E., Piovosi M. 2006 - *Effetti dell'intensità della matricinatura sulla ricostituzione della copertura e sull'accrescimento dei polloni in cedui a prevalenza di cerro* [Effect of standards density release on canopy cover recovery and shoots growth in Turkey oak coppice forests]. Annali CRA-Istituto Sperimentale Selvicoltura 33: 9-20. [In Italian]

Castellani C. 1982 - *Tavole stereometriche ed alsometriche costruite per i boschi italiani* [Tree volume and yield prediction models established for the Italian forests]. Reprint of volumes issued in 1970 and 1972 by ISAFA, Trento 277 p. [In Italian]

Chianucci F., Mattioli L., Amorini E., Giannini T., Marcon A., Chirichella R., Apollonio M., Cutini A. 2015 - *Early and long-term impacts of browsing by roe deer in oak coppiced woods along a gradient of population density*. Annals of Silvicultural Research 39 (1): 32-36. doi: 10.12899/ASR-945

Chianucci F., Salvati L., Giannini T., Chiavetta U., Corona P., Cutini A. 2016 - *Long-term response to thinning in a beech (*Fagus sylvatica* L.) coppice stand under conversion to high forest in Central Italy*. Silva Fennica, 50 (3) article id: 1549. 9 p. http://dx.doi.org /10.14214 /sf.1549.

Ciancio O., Nocentini S. (eds) 2002 - *Il bosco ceduo in Italia* [The coppice forest in Italy]. Accademia Italiana di Scienze Forestali, Firenze 678 p. [in Italian]

Ciancio O., Nocentini S. 2004 - *Il bosco ceduo. Selvicoltura, Assestamento, Gestione* [The coppice forest. Silviculture, Regulation, Management]. Accademia Italiana di Scienze Forestali, Firenze 721 p. [in Italian]

Ciccarese L., Cascio G., Cascone C. 2006 - *Biomassa legnosa da foresta e da fuori foresta* [Woody biomass from forest crops and outside forest cover]. Sherwood Compagnia delle Foreste, Arezzo 128: 5-13. [In Italian]

Ciccarese L., Crosti R., Cascone C., Cipollaro S., Ballarin Denti A., Fontanarosa E., Masiero M., Pizzuto Antinoro M., La Mela Veca DS. 2012 - *Status report of forest biomass use in the Mediterranean region*. Proforbiomed report. Case-study: Italy.

Clauser F. 1981 - *Un'ipotesi auxonomica da verificare* [A growth pattern to be verified]. Monti e Boschi XXXII (2-3): 98-98. [In Italian]

Clauser F. 1998 - *Una seconda ipotesi sullo sviluppo dei cedui verso la fustaia*. [A second hypothesis on the growth pattern of coppice towards high forest]. Monti e Boschi XLIX (3-4): 12-13. [In Italian]

Coppice 2015 - *Coppice forests: past, present and future* International conference. Brno, Czech Republic, 9-11 April 2015 http://coppice.eu/conference_en.html

Corona P. 2014 - *Forestry research to support the transition towards a bio-based economy*. Annals of Silvicultural Research 38 (2): 37-38. doi: 10.12899/ASR-1015

Corona P., Giuliarelli D., Lamonaca A., Mattioli W., Tonti D., Chirici G., Marchetti M. 2007 - *Confronto sperimentale tra superfici a ceduo tagliate a raso osservate mediante immagini satellitari ad alta risoluzione e tagliate riscontrate amministrativamente*. Forest@ 4 (3): 324-332.

Corona P., Ascoli D., Barbati A., Bovio G., Colangelo G., Elia M., Garfì V., Iovino F., Lafortezza R., Leone V., Lovreglio R., Marchetti M., Marchi E., Menguzzato G., Nocentini S., Picchio R., Portoghesi L., Puletti N., Sanesi G., Chianucci F. 2015 - *Integrated forest management to prevent wildfires under Mediterranean environments*. Annals of Silvicultural Research 39 (1): 1-22. doi: 10.12899/ASR-946

Cutini A. 1994a - *Indice di area fogliare, produzione di lettiera ed efficienza di un ceduo di cerro in conversione* [Leaf Area Index, litterfall and efficiency of a Turkey oak coppice in conversion into high forest]. Annali Istituto Sperimentale Selvicoltura Arezzo, 23: 147-166. [In Italian]

Cutini A. 1994b - *La stima del LAI con il metodo delle misure di trasmittanza in popolamenti diradati e non diradati di cerro* [The estimate of LAI from canopy transmittance in thinned and unthinned Turkey oak stands]. Annali Istituto Sperimentale Selvicoltura, Arezzo, 23: 167-181. [In Italian]

Cutini A. 1997 - *Drought effects on canopy properties and productivity in thinned and unthinned Turkey oak stands.* Plant Biosystems 131 (1): 59-65.

Cutini A., Mascia V. 1998 - *Silvicultural treatment of holm oak (Quercus ilex L.) coppices in southern Sardinia: effects of thinning on water potential, transpiration and stomatal conductance.* Annali Istituto Sperimentale Selvicoltura, Arezzo 27: 47-53.

Cutini A., Benvenuti C. 1998 - *Effects of silvicultural treatment on canopy cover and soil water content in a Quercus cerris L. coppice.* Annali Istituto Sperimentale Selvicoltura, Arezzo 27: 65-70.

Cutini A. 2000 - *Produttività e processi ecologici in popolamenti di origine agamica* [Stand productivity and ecological processes into coppice forests]. In: Atti II Congresso nazionale SISEF Applicazioni e prospettive per la ricerca forestale italiana. (Bucci G. et al. eds). Bologna, 20-22 ottobre 1999: 131-134. [In Italian]

Cutini A. 2002 - *Litterfall and Leaf Area Index at the CONECOFOR Permanent Monitoring Plots.* Journal of Limnology. CNR-ISE Verbania P. (Italy), 61 (1): 62-68.

Cutini A. 2006 - *Taglio di avviamento, ceduazione e matricinatura: effetti sulle caratteristiche della copertura forestale in cedui a prevalenza di cerro* [Coppice conversion cuts, coppicing and standards density: effects on canopy properties of Turkey oak coppice stands]. Annali CRA-Istituto Sperimentale Selvicoltura 33: 21-30. [In Italian]

Cutini A. Hajny M. 2006 - *Effetti del trattamento selvicolturale su produzione di lettiera, caratteristiche della copertura ed efficienza in un ceduo di cerro in conversione* [Effects of the silvicultural treatment on litter production, canopy characteristics and stand efficiency in a Turkey oak coppice in conversion to high forest]. Annali CRA-Istituto Sperimentale Selvicoltura 33: 133-142. [In Italian]

Cutini A., Chianucci F., Giannini T. 2010 - *Effetti del trattamento selvicolturale su caratteristiche della copertura, produzione di lettiera e di seme in cedui di faggio in conversione* [Effect of the silvicultural treatment on canopy properties, litter and seed production in beech coppices under conversion into high forest]. Annali CRA-SEL Arezzo 36: 109-124. [In Italian]

Cutini A., Bongi P., Chianucci F., Pagon N., Grignolio S., Amorini E., Apollonio M. 2011 - *Roe deer (Capreolus capreolus L.) browsing effects and use of chestnut and Turkey oak coppiced areas.* Annals of Forest Science 68 (4): 667-674. doi: 10.1007/s13595-011-0072-4

Cutini A., Chianucci F., Chirichella R., Donaggio E., Mattioli L., Apollonio M. 2013 - *Mast seeding in deciduous forests of the northern Apennines (Italy) and its influence on wild boar population dynamics.* Annals of Forest Science 70: 493-502. doi: 10.1007/s13595-013-0282-z

Cutini A., Chianucci F., Giannini T., Manetti MC., Salvati L. 2015 - *Is anticipated seed cutting an effective option to accelerate transition to high forest in European beech (Fagus sylvatica L.) coppice stands?* Annals of Forest Science 72: 631-640. doi: 10.1007/s13595-015-0476-7

Del Favero 2000 - *Gestione forestale e produzione legnosa a fini energetici.* [Forest management and wood production for energy purpose]. Sherwood (59): 5-9. [In Italian]

Di Castri F. 1996 - *Mediterranean diversity in a global economy.* In: International Symposium on Mediterranean Diversity. Rome, ENEA: 21-30.

Di Matteo G., De Angelis P., Brugnoli E., Cherubini P., Scarascia-Mugnozza G. 2010 - *Tree-ring Δ13C reveals the impact of past forest management on water-use efficiency in a Mediterranean oak coppice in Tuscany (Italy).* Annals of Forest Science, 67: 503-510. doi: 10.1051/forest/2010012

Di Matteo G., Perini L., Atzori P., De Angelis P., Mei T., Bertini G., Fabbio G., Scarascia Mugnozza G. 2014a - *Changes in foliar carbon isotope composition and seasonal stomatal conductance reveal adaptive traits in Mediterranean coppices affected by drought.* Journal of Forestry Research, 25 (4): 839-845. doi: 10.1007/s11676-014-0532-4

Di Matteo G., Tunno I., Nardi P., De Angelis P., Bertini G., Fabbio G. 2014b - *C and N concentrations in different compartments of outgrown oak coppice forests under different site conditions in Central Italy.* Annals Forest Science 71: 885-895. doi: 10.1007/s13595-014-0390-4

Ducci F., Proietti R., Cantiani P. 2006 - *Struttura genetica e sociale in un ceduo di cerro in conversione* [Genetics and stand structure of a Turkey oak coppice]. Annali CRA-Istituto Sperimentale Selvicoltura, Arezzo 33: 143-158. [In Italian]

Espelta JM., Sabaté S., Retana J. 1999 - *Resprouting dynamics.* In: Roda F., Retana J., Gracia CA., Bellot J. (eds). Ecology of the Mediterranean Evergreen Oak Forests. Ecological Studies vol. 137. Springer-Verlag, Berlin Heidelberg: 61-71.

EuroCoppice 2013 - *COST Action FP 1301*, Oct 2013-Oct 2017 https://www.euro coppice.uni-freiburg.de/

Fabbio G. 1994 - *Dinamica della popolazione arborea in un ceduo di cerro in invecchiamento.* [Dynamics of tree population in an outgrown Turkey oak coppice]. Annali Istituto Sperimentale Selvicoltura, Arezzo, 23: 41-72. [In Italian]

Fabbio G. 2004 - *Il bosco ceduo tra passato e futuro: problemi e opportunità per la selvicoltura e la gestione* [The coppice system at a turning point: a challenge for silviculture and forest management] Ri.Selv.Italia progetto 3.2 Selvicoltura, funzionalità e gestione sostenibile dei cedui nell'area appenninica e mediterranea. Doc. interno 16 p. [In Italian]

Fabbio G. 2010 - *Il ceduo tra passato e attualità: opzioni colturali e dinamica dendro-auxonomica e strutturale nei boschi di origine cedua* [Coppice system at a turning point: silvicultural options, dynamics of growth and structure into outgrown coppice forests]. In: Atti 46° Corso di Cultura in Ecologia. Gestione multifunzionale e sostenibile dei boschi cedui: criticità e prospettive. (Carraro V., Anfodillo T. eds). S. Vito di Cadore, 7-10 giugno 2010: 27-43. [in Italian]

Fabbio G. 2015 - *Shaping future coppice forestry on the legacy of the past: lesson learnt and perspectives.* Coppice forests: past, present and future. International Conference, Brno, April 9-11, 2015. http://coppice.eu/conference_en.html

Fabbio G., Manetti MC., Puxeddu M. 1996 - *La lecceta: un ecosistema 'in riserva' Uno studio condotto su aree permanenti: ipotesi di lavoro e approccio metodologico* [The holm oak stand:a relict ecosystem]. In: Atti VII Congresso nazionale Società Italiana di Ecologia. Napoli, 11-14 settembre 1996: 139-143. [In Italian]

Fabbio G., Amorini E., Cutini A. 1998a - *Towards a sustainable management of Mediterranean forest: the MEDCOP experience (1994-98).* In: Proceedings VII International Congress of Ecology, INTECOL. Florence, 19-25 July 1998. Contribution to the Symposium Perspectives in sustainable land use of marginal areas, land abandonment and restoration: 295-308.

Fabbio G., Cutini A., Mascia V. 1998b - *Silvicultural treatment of holm oak coppices (Q. ilex L.) in Southern Sardinia: effects of canopy and crop thinning on microclimate.* Annali Istituto Sperimentale Selvicoltura, Arezzo 27: 55- 63.

Fabbio G., Iovino F., Menguzzato G., Tabacchi G. 2002 - *Confronto fra modelli di previsione della biomassa arborea elaborati per cedui di leccio* [Comparison between volume and biomass equations elaborated for holm oak coppice stands]. In: Il bosco ceduo in Italia (Ciancio O., Nocentini S. eds): 469-495. [In Italian]

Fabbio G., Merlo M., Tosi V. 2003 - *Silvicultural management in maintaining biodiversity and resistance of forests in Europe-the Mediterranean region.* Journal of Environmental Management 67 (1) Special Issue: 67-76.

Fabbio G., Amorini E. 2006 - *Avviamento ad alto fusto e dinamica naturale nei cedui a prevalenza di cerro. Risultati di una prova sperimentale a 35 anni dalla sua impostazione. Il protocollo di Caselli (Pisa)* [Conversion to high forest and natural pattern into ageing Quercus cerris L. coppices. Results from 35 years of monitoring. The Caselli site (Tyrrhenian coast - Tuscany]. Annali CRA-Istituto Sperimentale Selvicoltura, Arezzo 33: 79-104. [In Italian]

Fabbio G., Bertini G., Calderisi M., Ferretti M. 2006a - *Status and trend of tree growth and mortality rate at the CONECOFOR plots, 1997-2004.* Special issue on Ecological condition of selected forest ecosystems in Italy. Annali CRA-Istituto Sperimentale Selvicoltura, Arezzo 34: 17-28.

Fabbio G., Manetti MC., Bertini G. 2006b - *Aspects of biological diversity at the CONECOFOR plots. I. Structural and species diversity of the tree community.* Special issue on Aspects of Biodiversity in selected forest ecosystems in Italy. Annali CRA-Istituto Sperimentale Selvicoltura, Arezzo 30 (2): 11-20.

Fabbio G., Cantiani P., Ferretti F., Chiavetta U., Bertini G., Becagli C., Di Salvatore U., Bernardini V., Tomaiuolo M., Matteucci G., De Cinti B. 2014 - *Adaptive silviculture to face up to the new challenges: the ManForCBD experience.* In: Proceedings 2nd International Congress of Silviculture. Florence, November 26-29, 2014 vol. I: 531-538 http://dx.doi.org/10.4129/2cis-gf-ada

Fagarazzi C., Fabbri LC., Fratini R., Riccioli F. 2006 - *Sostenibilità economica delle utilizzazioni dei boschi cedui di quercia nel territorio toscano* [Economical sustainability of oak coppice harvesting in Tuscany]. Annali CRA-Istituto Sperimentale Selvicoltura, Arezzo 33: 63-78. [In Italian]

Farrell EP., Fuhrer E., Ryan D., Andersson F., Huttl R., Piussi P. 2000 - *European forest ecosystems: building the future on the legacy of the past.* Forest Ecology and Management 132: 5-20.

Forest Europe 2011 - *Conference Proceedings.* Ministerial Conference on the Protection of Forests in Europe, Oslo 14-16 June 2011.

Forest Europe 2015 - *State of Europe's Forests 2015.* Ministerial Conference on the Protection of Forests in Europe, Liaison Unit Madrid, Madrid.

Franklin JF. 1993 - *Preserving biodiversity: species, ecosystems, or landscapes?* Ecological Applications 3 (2): 202-205.

Fuhrer E. 2000 - *Forest functions, ecosystem stability and management.* Forest Ecology and Management 132: 29-38.

FutureForCoppiceS 2015 - *Shaping future forestry for sustainable coppices in southern Europe: the legacy of past management trials.* 2015-2018 http://www.futurefor coppices.eu/en/

Gambi G. 1968 - *Le conversioni dei cedui in altofusto sull'Appennino Tosco-Emiliano* [The conversion of coppice forests in the Apennines]. Annali Accademia nazionale Agricoltura Bologna, 3a serie 78: 1-49. [In Italian]

Garbarino M., Allegrezza M., Ciucci V., Ottaviani C., Renzaglia F., Tesei G., Vitali A., Urbinati C. 2015 - *Legacies of past coppicing on the structure and vegetation diversity of beech forests in central Apennines, Italy.* Coppice forests: past, present and future. International Conference, Brno, April 9-11, 2015. http://coppice.eu/conference_en.html

Gasparini P., Tabacchi G. 2011 - *Inventario Nazionale delle Foreste e dei Serbatoi forestali di Carbonio INFC 2005. Metodi e Risultati* [National Inventory of Forests and forest Carbon Sinks. Methods and Results]. MiPAAF-CFS-CRA-MPF. Edagricole, Milano, 653 p. [In Italian]

Grohmann F., Savini P., Frattegiani M. 2002 - *La matricinatura per gruppi. L'esperienza del progetto SUMMACOP.* Sherwood. Foreste e Alberi oggi 80: 25-32. [In Italian]

Guidi G. 1976 - *Primi risultati di una prova di conversione in un ceduo matricinato di cerro* [First results of the conversion into high forest of a Turkey oak coppice with standards]. Annali Istituto Sperimentale Selvicoltura, Arezzo 6: 255-278. [In Italian]

Hédl R., Chudomelova M., Kolar J., Kopecky M., Mullerova J., Szabo P. 2015 - *Historical legacy of coppice systems in herbaceous vegetation of central European forests.* Coppice forests: past, present and future. International conference. Brno, Czech Republic, 9-11 April 2015 http://coppice.eu/conference_en.html

Hippoliti G. 2001 - *Sul governo a ceduo in Italia (XIX-XX secolo)* [On the coppice system in Italy (XIX-XX c.)]. In: Storia e risorse forestali (Agnoletti M. ed.) AISF, Firenze: 353-374. [In Italian]

Holisova P., Pietras J., Darenova E., Novosadova K., Pokorny R. 2015 - *Comparison of assimilation parameters of coppice and non-coppiced sessile oak.* Coppice forests: past, present and future. International Conference, Brno, April 9-11, 2015. http://coppice.eu/conference_en.html

ISTAT 2011 - *Serie storica utilizzazioni boschive.* Istituto Nazionale Statistica, Roma. www.istat.it

Kirby K. 2015 - *Coppice woods. temporal and spatial diversity creating rich wildlife assemblages.* Coppice forests: past, present and future. International Conference, Brno, April 9-11, 2015. http://coppice.eu/conference_en.html

Konstantinidis P., Tsiourlis G., Kofis P. 2006 - *Effect of fire season, aspect and pre-fire plant size on the growth of Arbutus unedo L. (strawberry tree) resprouts.* Forest Ecology and Management 225 (1-3): 359-367.

Kopecky M., Hedl R., Szabo P. 2013 - *Non-random extinctions dominate plant community changes in abandoned coppices.* Journal Applied Ecology 50:79-87. doi:10.1111/1365-2664.12010

Lindner M., Maroschek M., Netherer S., Kremer A., Barbati A., Gonzalo JG., Seidl R., Delzon S., Corona P., Kolstrom M., Lexer ML., Marchetti M. 2010 - *Climate change impacts, adaptive capacity, and vulnerability of European forest ecosystems.* Forest Ecology and Management 259: 698-709.

Lopez BC., Gracia CA., Sabaté S., Keenan T. 2009 - *Assessing the resilience of Mediterranean holm oaks to disturbance using selective thinnings*. Acta Oecologica 35: 849-854.

Mairota P., Manetti MC., Amorini E., Pelleri F., Terradura M., Frattegiani M., Savini P., Grohmann F., Mori P., Terzuolo PG., Piussi P. 2016a - *Opportunities for coppice management at the landscape level: the Italian experience*. iForest (early view). doi: 10.3832/ifor1865-009 [online 2016-08-04]

Mairota P., Buckley P., Suchomel C., Heinsoo K., Verheyen K., Hédl R., Terzuolo PG., Sindaco R., Carpanelli A. 2016b - *Integrating conservation objectives into forest management: coppice management and forest habitats in Natura 2000 sites*. iForest 9: 560-568. doi:10.3832/ifor1867-009 [online 2016-05-12]

Mairota P., Manetti MC., Amorini E., Pelleri F., Terradura M., Frattegiani M., Savini P., Grohmann F., Mori P., Piussi P. 2014 - *Socio-economic and environmental challenges of responsible coppice management: Italian examples*. COST Action FP 1301 EuroCoppice. International event People and Coppice. November 3-5, University of Greenwich, UK.

Mairota P., Tellini Florenzano G., Piussi P. 2006 - *Valutazione della funzione paesaggistica delle fustaie transitorie di cerro nel territorio delle Colline Metallifere* [Forest management and biodiversity conservation: landscape ecological analysis of wooded lands in southern Tuscany (Italy)]. Annali CRA-Istituto Sperimentale Selvicoltura, Arezzo 33: 187-244. [In Italian]

Manetti MC., Gugliotta OI. 2006 - *Effetto del trattamento di avviamento ad altofusto sulla diversità specifica e strutturale delle specie legnose in un ceduo di cerro* [Impact of the conversion into high forest on tree specific and structural diversity in a Turkey oak coppice]. Annali CRA-Istituto Sperimentale Selvicoltura, Arezzo 33: 105-114. [In Italian]

Manetti MC., Giannini T., Chianucci F., Casula A., Cutini A. 2013 - *Cambiamenti strutturali ed ecologici in cedui di leccio in Sardegna a 25 anni dal taglio di avviamento ad altofusto* [Stand structure and ecological changes in holm oak coppices 25 years later the opening of thinning operations for the conversion into high forest]. Annals of Silvicultural Research 37 (1): 22-28. [In Italian] doi: 10.12899/ASR-770

Manetti MC., Becagli C., Sansone D., Pelleri F. 2016 - *Tree-oriented silviculture: a new approach for coppice stands*. iForest (early view) doi: 10.3832/ifor1827-009 [online2016-08-04]

Marchetti M., Vizzarri M., Lasserre B., Sallustio L., Tavone A. 2014 - *Natural capital and bioeconomy: challenges and opportunities for forestry*. Annals of Silvicultural Research 38 (2): 37-38. doi: 10.12899/ASR-1013

MEDCOP 1998 - *Improvement of Mediterranean coppice forests*. Special issue Annali Istituto Sperimentale Selvicoltura 27, Arezzo, 217 p. (Morandini R., Fabbio G., Cutini A., Manetti MC. eds).

Millar CI., Stephenson NL., Stephens S. 2007 - *Climate change and forest of the future: managing in the face of uncertainty*. Ecological Applications 17 (8): 2145-2151.

Morandini R. 1979 - AGRIMED Internal document, 12 p.

Morandini R. (ed.) 1994 - *Improvement of coppice forests in the Mediterranean region*. Proceedings of the workshop held in Arezzo, september 24-25, 1992. Annali Istituto Sperimentale Selvicoltura, Arezzo 23: 257-333.

Mori P., Bruschini S., Buresti E., Giulietti V., Grifoni F., Pelleri F., Ravagni S., Berti S., Crivellaro A. 2007 - *La selvicoltura delle specie sporadiche in Toscana* [The silviculture of sporadic tree species]. Supporti tecnici alla Legge Regionale Forestale della Toscana, 3. ARSIA Firenze 355 p. http://www.regione.toscana.it/documents/10180/13328713/4_Manuale-specie-sporadicheToscana.pdf/aa46b002-3609-4681-89e9-69b0043f01bb [in Italian]

Mori P., Pelleri F. 2012 - *PProSpoT: un Life+ per le specie arboree sporadiche* [PProSpoT: a LIFE+ project for the sporadic tree species]. Sherwood, foreste e alberi oggi, 179: 7-11. [in Italian] http://www.pprospot.it

Mori P., Pelleri F. (eds) 2014 - *Selvicoltura per le specie arboree sporadiche: Manuale tecnico per la selvicoltura d'albero proposta dal progetto LIFE+ PProSpoT* [Silviculture for sporadic tree species: Technical handbook for tree silviculture proposed in the LIFE+ PProSpoT]. Compagnia delle Foreste Arezzo 144 p. [in Italian] http://www.pprospot.it

Motta R., Berretti R., Meloni F., Nosenzo A., Terzuolo PG., Vacchiano G. 2015 - *Past, present and future of the coppice silvicultural system in the italian North-West*. Coppice forests: past, present and future. International Conference, Brno, April 9-11, 2015. http://coppice.eu/conference_en.html

Mottet A., Ladet S., Coque N., Gibon A. 2006 - *Agricultural land-use change and its drivers in mountain landscapes: a case study in the Pyrenees*. Agriculture Ecosystems & Environment 114 (2-4): 296-310.

Mullerova J., Hedl R., Szabo P. 2015a - *Coppice abandonment and its implications for species diversity in forest vegetation*. Forest Ecology and Management 343: 88-100. doi:10.1016/j.foreco.2015.02.003.

Mullerova J., Szabo P., Hédl R., Dorner P., Veverkova A. 2015b - *The rise and fall of coppicing - historical processes and consequences for nature conservation*. 'Coppice forests: past, present and future'. International Conference, Brno, April 9-11, 2015. http://coppice.eu/conference_en.html

Nicolescu VN., Hochbichler E., Coello Gomez J., Ravagni S., Giulietti V. 2009 - *Ecology and silviculture of wild service tree (Sorbus torminalis (L.) Crantz): a literature review*. Die Bodenkultur, 60 (3): 35-44.

Nocetti M., Bertini G., Fabbio G. Tabacchi G. 2007 - *Equazioni di previsione della fitomassa arborea per i soprassuoli di cerro in avviamento ad altofusto in Toscana* [Equations for the prediction of tree phytomass in Quercus cerris stands in Tuscany, Italy]. Forest@ 4 (2): 204-212. [In Italian]

Odum EP. 1973 - *Principi di Ecologia*. Piccin Ed. Padova, 584 p.

Oliver CD., Larson BC. 1996 - *Forest stand dynamics*. In: Biological Resource Management Series. (WM. Getz ed.). Mc Graw-Hill, New York 467 p.

Palmberg-Lerche C. 2001 - *Conservation of forest biological diversity and forest genetic resources*. In: Forest Genetic Resources 28 Rome, FAO.

Pelleri F., Sansone D., Bianchetto E., Bidini C., Sichi A. 2013 - *Selvicoltura d'albero in fustaie di faggio: valorizzazione delle specie sporadiche e coltivazione della specie dominante* [Tree-oriented silviculture in European beech high forests: silvicultural practices aimed both at enhancing sporadic species and at managing the dominant species]. Sherwood, foreste e alberi oggi, 190: 43-47. [English version] http://www.pprospot.it/english-products.htm

Pelleri F., Sansone D., Fabbio G., Mori P. 2015 - *Sporadic tree species management for preserving biodiversity and increasing economic stands value: the PProSpoT experience.* 'Coppice forests: past, present and future'. International Conference, Brno, April 9-11, 2015. http://coppice.eu/conference_en.html

Pettenella D. 2002 - *Fattori di inerzia nelle forme di gestione e nuovi sviluppi del mercato per i boschi cedui* [Problems and perspectives for market development of coppice forest in Italy]. In: Il bosco ceduo in Italia. (Ciancio O., Nocentini S. eds) AISF, Firenze: 541-560 [In Italian]

Piegai F., Fabiano F. 2006 - *Il lavoro per la raccolta di legna da ardere da cedui e da avviamenti ad alto-fusto* [Harvesting of firewood at coppice clearcutting and at first thinning for coppice conversion into high forest]. Annali CRA-Istituto Sperimentale Selvicoltura, Arezzo 33: 51-62 [In Italian]

Pietras J., Stojanović M., Knott R., Pokorný R. 2016 - *Oak sprouts grow better than seedlings under drought stress.* iForest 9: 529-5357. doi: 10.3832/ifor1823-009 [online 2016-03-17]

Piussi P. 1980 - *Il trattamento a ceduo di alcuni boschi toscani dal XVI al XX secolo* [The coppice system in Tuscany from 16th to 20th c.]. Dendronatura 1 (2): 8-15 [In Italian]

Piussi P. 1982 - *Utilizzazione del bosco e trasformazione del paesaggio: il caso di Monte Falcone (XVII-XIX secolo)* [The use of forests and landscape transformation]. Quaderni storici XVII, Il Mulino, Bologna, 49 (1): 84-107 [In Italian]

Piussi P., Stiavelli S. 1986 - *Dal documento al terreno. Archeologia del bosco delle Pianora (colline delle Cerbaie, Pisa)* [From the documents to the ground. Archaeology of the Pianora woods (Pisa, Italy)]. Quaderni storici XXI, Il Mulino, Bologna, 62 (2): 445-466. [In Italian]

Piussi P., Zanzi Sulli A. 1997 - *Selvicoltura e storia forestale* [Silviculture and history of forests]. Annali AISF, Firenze 46: 25-42. [In Italian]

Piussi P. 2006 - *Close to nature forestry criteria and coppice management.* In: Nature-based forestry in Central Europe. (Diaci J. ed.), University of Lubiana Biotechnical Faculty: 27-30.

Piussi P. 2015 - *Coppice management and nutrition.* Coppice forests: past, present and future. International Conference, Brno, April 9-11, 2015. http://coppice.eu/conference_en. html

Pra A., Pettenella D. 2016 - *Consumption of wood biomass for energy in italy: a strategic role based on weak knowledge.* L' Italia Forestale e Montana/ Italian Journal of Forest and Mountain Environments 71 (1): 49-62.

Pyttel P., Fischer U., Bauhus J. 2015 - *The effect of harvesting on stump mortality and re-sprouting in aged oak coppice forests.* Coppice forests: past, present and future. International Conference, Brno, April 9-11, 2015. http://coppice.eu/conference_en. html

Rasmussen KK. 2007 - *Dendro-ecological analysis of a rare subcanopy tree: effect of climate, latitude, habitat condition and forest history.* Dendroecologia, 25: 3-17.

Sansone D., Bianchetto E., Bidini C., Ravagni S., Nitti D., Samola A., Pelleri F. 2012 - *Selvicoltura d'albero nei cedui giovani: interventi di valorizzazione delle specie sporadiche nell'ambito del Progetto LIFE+ PProSpoT* [Tree-oriented silviculture in young coppices. Silvicultural practices to enhance sporadic species: the LIFE+PPRoSpoT project experience]. Sherwood foreste e alberi oggi, 185: 5-10. [English version] http://www.pprospot.it/english-products.htm

Savini P. 2010 - *Nuove tecniche di intervento nei boschi cedui* [New intervention techniques in coppice woodland]. In: Atti 46° Corso di Cultura in Ecologia. Gestione multifunzionale e sostenibile dei boschi cedui: criticità e prospettive. (Carraro V., Anfodillo T. eds) S. Vito di Cadore, 7-10 giugno 2010: 73-84. [in Italian]

Savini P., Cantiani P., Frattegiani M., Pedrazzoli M., Prieto D., Terradura M. 2015 - *Innovative coppice management in Umbria: coppice with groups of standards.* Coppice forests: past, present and future. International Conference, Brno, April 9-11, 2015. http://coppice.eu/conference_en. html

Scarascia-Mugnozza G., Oswald H., Piussi P., Radoglou K. 2000 - *Forests of the Mediterranean region: gaps in knowledge and research needs* Forest Ecology and Management 132: 97-109.

Schweier J., Spinelli R., Magagnotti N., Becker G. 2015 - *Mechanized coppice harvesting with new small-scale feller-bunchers. Results from harvesting trials with newly manufactured felling heads in Italy.* Biomass and Bioenergy, vol. 72, (1): 85-94.

Sevrin E. 1994 - *Chênes sessile et pédunculé.* [Sessile oak and Penduculate oak]. Institut pour le Développement Forestiere, Paris, 96 p.

Spiecker H. 2006 - *Minority tree species: a challenge for a multi-purpose forestry.* In: Nature based forestry in central Europe. Alternative to industrial forestry and strict preservation. Studia Forestalia Slovenica, 126: 47-59.

Spiecker H., Hein S., Makkonen-Spiecker K., Thies M. 2009 - *Valuable broadleaved forests in Europe.* EFI Research Report 22, 256 p.

Splichalova M. 2015 - *Aspects of oak (Quercus sp.) management in Spain and its application.* Coppice forests: past, present and future. International Conference, Brno, April 9-11, 2015. http:// coppice.eu/conference_en. html

Szabo P., Mullerova J., Suchankova S., Kotacka M. 2015 - *Coppice woodlands since the Middle Ages: spatial modelling based on archival sources.* Coppice forests: past, present and future. International Conference, Brno, April 9-11, 2015. http://coppice.eu/conference_en. html

Teissier du Cros E. 2001 - *Conserving forest genetic resources: objectives, research, networks.* In: Forest Genetic Resources Management and Conservation. INRA DIC, Paris: 4-5.

UN-ECE/FAO 2000 - *Forest resources of Europe, CIS, North America, Australia, Japan and New Zealand* (TBFRA-2000). ECE7TIM/SP/17, Geneva, 466 p.

UNIF 1987 - *La conversione dei boschi cedui in altofusto. Stato attuale delle ricerche* [Coppice conversion into high forest. Current research progress]. In: Atti convegno: Parchi e Riserve naturali nella gestione territoriale. (E. Giordano ed.): 1-8. Università della Tuscia, Viterbo. [in Italian]

Urbinati C., Iorio G., Agnoloni S., Garbarino M., Vitali A. 2015 - *Beech forests in central Apennines: adaptive management for structure and functions in transition.* Coppice forests: past, present and future. International Conference, Brno, April 9-11, 2015. http://coppice.eu/conference_en.html

Vild O., Roleček J., Hedl R., Kopecky M., Utinek D. 2013 - *Experimental restoration of coppice with standards: response of understorey vegetation from the conservation perspective.* Forest Ecology and Management 310: 234-241. doi: 10.1016/j.foreco.2013.07.056

Wilhelm GJ., Ducos Y. 1996 - *Suggestions pour le traitement de l'alisier torminal en mélange dans les futaies feuillues sur substrats argileux du Nord-Est de la France* [Suggestions for the silvicultural treatment of wild service tree scattered in the broadleaved high forests on clay soils in the North-East of France]. Rev. For. Fr., 2: 137-143.

Wilhelm GJ. 2003 - *Qualification-grossissement: la stratégie sylvicole de Rhénanie-Palatinat* [Qualification and sizing: the silviculture strategy of the Rhénanie-Palatinat]. Rend Des-Vous techniques, Office National des Forêt, 1: 4-9.

Zeide B. 1985 - *Tolerance and self-tolerance of trees*. Forest Ecology and Manag. 13: 149-166.

Zeide B. 1991 - *Self-thinning and stand density*. Forest Science 37: 517-523.

Zeide B. 2005 - *How to measure stand density*. Trees - Structure and Function 19: 1-14.

Explore inhabitants' perceptions of wildfire and mitigation behaviours in the *Cerrado* biome, a fire-prone area of Brazil

Giovanni Santopuoli [1*], Jader Nunes Cachoeira [2], Marco Marchetti [1], Marcelo Ribeiro Viola [3], Marcos Giongo [2]

Abstract - Fire represents an important natural feature of Brazilian landscape, especially in the *Cerrado* biome. The Cerrado is the economic livelihood of thousands of people from rural areas in Brazil. It is one of the most important hotspots of biodiversity in the world but also it is a fire-prone area thanks to the high flammability index of the vegetation. Residents and native people of this environment use fire very frequently. The majority of wildfires are caused by humans, though there are some aggravating natural factors affecting the risk, intensity and severity of wildfires. Since residents are continuously involved in fire suppression activities, understanding their perceptions is important for the decision makers who must assess the local capacity to preserve natural resources. This study explores perceptions about wildfire risk and fire mitigation behaviours within three municipalities of the state of Tocantins (Brazil). The study demonstrates that survey participants perceived wildfire risk as rather high, although the perceptions were complex and conflicting among interviewees. A wide range of confused perceptions regarding fire ignition and heterogeneous points of view have emerged from the survey. However, the residence of interviewees and their educational attainment result in variables that significantly affect the inhabitants' perceptions.

Keywords - Perceptions; fire risk; mitigation behaviors; Cerrado; Brazil

Introduction

Although fire is an important ecological factor, it is also a natural disturbance and one of the most significant threats to forest and savannah ecosystems worldwide. It prevents the effectiveness of forest ecosystems to provide the ecosystem services and socio-economic benefits. This is of particular relevance in Latin America, especially in Brazil's Cerrado biome, which is characterized by vegetation with a high flammability index (Klink and Machado 2005). Although lightning represents the principal natural ignition origin (Ramos-Neto and Pivello 2000), the most common causes of wildfire in Brazil are anthropogenic (Pereira Jr. et al. 2014). Anthropogenic ignition is often related to improper use of fire in agricultural and livestock practices, carelessness in fishing and hunting, cigarettes butts discarded along roadsides and the burning of trash (Klink and Machado 2005, Mistry and Bizerril 2011, Pivello 2011).

Over recent history, population growth and the expansion of agricultural areas (Klink et al. 2002, Phalan et al. 2013, Welch et al. 2013) have increased the use of fire. Most landowners use fire as a tool when they are attempting to replace natural vegetation with crop cultures or pastures, to perform shifting (slash-and-burn) cultivation, and also to stimulate the regrowth of grasses to feed cattle during the dry season (Klink and Machado 2005). Moreover, native populations augment fire frequency and fire risk through traditional uses, such as using fire in rituals, for signals, to kill or drive away pests and snakes, to eliminate waste, to slash-and-burn and to attract or drive game during hunting (Hecht 2009, Mistry et al. 2005, Welch et al. 2013).

Therefore, the Cerrado biome is continuously under strong fire peril, which threatens one of the most important hotspots of biodiversity in the world (Myers et al. 2000). The situation is exacerbated by climate change, which increases the fire risk by creating warmer and drier conditions (Maezumi et al. 2015, Ribeiro, JF and Walter 2008). Likewise, international agreements aimed at reducing the carbon emissions from deforestation and forest degradation in the Amazonian rainforest have allowed the replacement of savannah ecosystems with forest plantations. Thus, climate change mitigation, economic pressure and human carelessness all contribute to the spread of wildfire within the Cerrado.

Assessing the relationships between socio-ecological systems is of paramount importance in order

---

[1] University of Molise, Dipartimento di Bioscienze e Territorio, Campobasso (Italy)
[2] Universidade Federal do Tocantins, Centro de Monitoramento Ambiental e Manejo do Fogo (Brazil)
[3] Universidade Federal de Lavras, Soil and Water Engineering Group (Brazil)
* giovanni.santopuoli@unimol.it

to discover the key drivers that determine changes in forest ecosystems (Ferrara et al. 2016). Similarly, the socio-economic context (Kosmas et al. 2016) of fire-prone areas could exacerbate negative fire-related phenomena such as land degradation and depletion of natural resources, with subsequently loss of biodiversity and other ecosystem services (Sallustio et al. 2015, Vizzarri et al. 2015).

Within the vegetation of the Cerrado, it is very common to find morphological and physiological adaptations to frequent fires, such as twisted trees, thick fruit skins and corky bark (Pivello 2011, Simon and Pennington 2012). Their presence confirms the role of fire in governing forest dynamics (Hoffmann et al. 2009, Pivello and Coutinho 1996). For example, fire causes top kill and often destroys the above ground biomass of saplings and smaller trees (Hoffmann et al. 2009). However, it also alters the forest composition, promotes regeneration and increases timber production (Certini 2005). Although mature trees often survive, fire stress affects their growth and fosters susceptibility to other stressors (Odhiambo et al. 2014). In particular, high fire frequency impairs the ability of trees to reach adult stages, accumulate bark and reach sufficient height to avoid top kill (Batalha et al. 2011, Hoffmann et al. 2003). Similarly, by altering soil properties (D. M. Silva et al. 2013), high fire frequency modifies the Cerrado physiognomies, reducing biodiversity (Bond et al. 2005, Lehmann et al. 2011, I. A. Silva and Batalha 2008).

In combatting wildfire, residents can play a crucial role in the conservation of natural resources by adopting risk mitigation strategies which have developed over the years and which currently represent the social memory regarding the ability of a local community to manage and cope with fire issues (Wilson et al. 2017). Such mitigation behaviour, also known as community resilience (Kelly et al. 2015), is the consequence of several dynamic factors (Champ et al. 2013), which interact with each other and influence people's perception of the risk. The relationship between risk perception and mitigation action has been a subject of studies across several disciplines (Beringer 2000, Dondo Bühler et al. 2013, Gounaridis et al. 2014, McCaffrey et al. 2013). Moreover, the exploration of stakeholders' perceptions has been increasingly adopted in the management and governance of forest resources (Pastorella et al. 2016, Santopuoli et al. 2016, Santopuoli et al. 2012). The main objective of this study is to explore local perceptions of wildfire risk and the mitigation behaviours in a fire-prone area of Brazil. Furthermore, since many authors agree that indigenous populations use fire in almost all of their traditional, ritual, cultural and daily activities, an additional aim of the study is to compare residents' perceptions among three municipalities that are contiguous with Bananal Island, where two important native populations live (Valente et al. 2013).

Methods

Study area

The present study was conducted in three municipalities (Dueré, Formoso do Araguaia and Lagoa da Confusão) located in the south-western part of the Tocantins of central Brazil (Fig. 1). Located along the western border of the state, Bananal Island represents the most important hot spot of biodiversity conservation in Tocantins, with Araguaia National Park in the north and the reservations of the Xavante and Javaés indigenous groups in the South (Valente et al. 2013). Particularly significant is the presence of Xavante group, which historically has been famous for its use of fire for management (Pivello 2011). The vegetation is of the Cerrado type, which represents one of the most important types of savannah in terms of species richness and level of endemism, (Forzza et al. 2012, Simon and Pennington 2012). Nevertheless, intensive human use has cleared more than 30% of the Cerrado biome, predominantly for pasture, intensive monoculture -e.g., soybean, rice and maize (Phalan et al. 2013) - and more recently for forest plantations of species such as eucalyptus and pine (Ceccon and Miramontes 2008, Klink et al. 2002).

According to the Koppen classification, the climate is seasonal: the wet season is from October to March and the dry season from April to September (Klink and Machado 2005). The annual mean temperatures range from 22°C to 27°C and the annual rainfall from 1'300 to 1'900 mm (Alvares et al. 2013). In this fire dependent/influenced ecosystem (Hard-

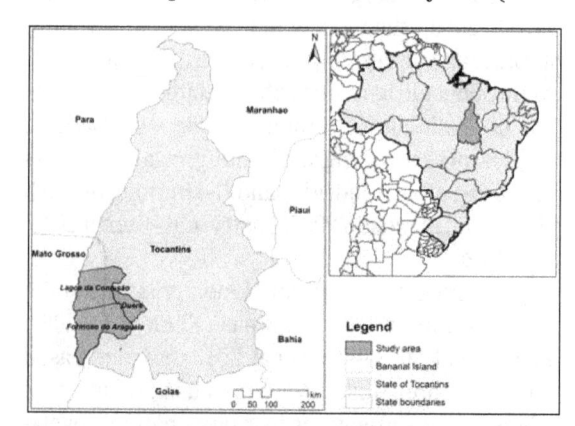

Figure 1 - Study area. The position of the three municipalities within the state of Tocantins and the border of the Bananal Island indigenous area on the left. The total area is 27,412 km², (3,424 km2 in Dueré, 13,423 km² in Formoso do Araguaia and 10,564 km² in Lagoa da Confusão), within which forest cover represents the 70% of the total area. The position of Tocantins within Brazil is on the right.

esty, Jeff and Myers, Ron and Fulks 2005), fire is an essential factor in conserving the native animal and plant species and maintaining ecological processes (Pivello 2011). The National Institute for Space Research (INPE 2015), reports that from 2003 to 2011, the annual average number of fires in Tocantins was 11'682, corresponding to almost 32'000 km^2 (11.5% of total surface) burned each year (Cachoeira 2015).

According to the Brazilian Statistics and Geography Institute database (Instituto Brasileiro de Geografia e Estatística), in 2014 the total human population was 35'352 (4'720 in Dueré, 18'773 in Formoso do Araguaia and 11'859 in Lagoa da Confusão).

Survey method

Data on the perception of fire were collected from February to September of 2014, during the development of the Forest Fire Prevention Plans (FFPPs) for the three municipalities, e.g. "*Plano Operativo de Prevenção e Combate aos Incêndios Florestais do município de Dueré*" (Giongo et al. 2014). In-person interviews were conducted with individuals over 16 years of age who inhabited the three municipalities of the area studied.

Respondents were mainly surveyed in private agroforestry enterprises and seldom in public locations such as local municipal offices, restaurants and market places. The main difficulty encountered for collecting information was the distance between forestry businesses, which affected the number of respondents. However, the final sample size consisted of 116 interviews, and considering the total population of the study area (35'352 inhabitants), the sampling error is estimated as ± 9.08% at the confidence level of 95%.

The interviewees were selected according to the combination of theoretical and snowball sampling for structured interviews. This approach allows collection of the maximum variety of concepts, ideas and practical experiences (López-Santiago et al. 2014, Paveglio et al. 2017). This strategy is designed to achieve both consistency and representativeness of the topic studied (Corbin and Strauss 1990, Strauss and Corbin 1994). It is important to highlight that the theoretical sampling strategy provides greater understanding of the representativeness of a slice of life such as individual acquaintances and practical experiences rather than prioritizing representativeness of population (Patton 1990, Strauss and Corbin 1994). The snowball sampling is a complimentary strategy that uses chain referral for the identification of additional informants from an initial sample (Paveglio et al. 2017).

The interviews were conducted through a semi-structured questionnaire based on the Ministerial guidance provided for developing the FFPPs (IBAMA 2009). The questionnaire contained 38 questions, including both open-ended and closed questions. There were 24 open-ended questions on socio-demographic characteristics such as age, race, gender and education, and open-ended questions were also used for explanations of the choices made with the closed questions. The 14 closed questions (Tab. 1) were designed to investigate the opinions of interviewees regarding two aspects:

(i) The use of fire for daily activities and its impact on the environment;
(ii) Knowledge about fire suppression techniques, their implementation and the interviewees personal capability.

The closed questions included two types of questions. The first consisted of 'Yes/No' answers, while the second consisted of giving a weight to the choices highlighted during the interview, ranking them from 1 "less important" to 9 "very important".

Table 1 - Questions used to deliver the in-person interviews. The first group of questions was designed to investigate the daily use of fire among residents. The second group of questions addressed the mitigation behaviors of interviewees.

	Questions	Explanation
Wildfire risk perceptions	Fire opinion	To give a preference (positive or negative) about the usefulness of fire for land management, adding a justification of the choice.
	Fire use	To describe whether residents use fire for daily activities.
	Activities that use fire	To give at least one example of an activity (rural or domestic) for which fire is necessary and quantify its usefulness from 1 to 9.
	Problem for municipality	To evaluate whether fire represents a problem for the municipality where the interviewee live and to give the weight of its relevance from 1 to 9.
	Causes of fire ignition	To list the main causes of fire ignition, giving a weight to the frequency of each cause on a scale from 1 to 9.
	Risk for urban and rural areas	To indicate whether rural or urban areas are more affected by fire, giving weight to its importance on a scale from 1 to 9.
	Action to reduce risk of fire	To mention at least one potential action useful for reducing fire ignition or limiting fire impacts. For each action mentioned, the interviewee weighed its effectiveness from 1 to 9.
Fire mitigation behavior	Training course	To indicate whether the interviewee has taken part in a training course on firefighting.
	Voluntary activity	To indicate whether the interviewee has taken part in firefighting activities as a volunteer.
	Prescribed fire	To indicate whether the interviewee is familiar with techniques of prescribed fire.
	Example of prescribed fire	Give at least one example of prescribed fire techniques.
	Authorization for prescribed fire	To indicate whether the interviewee is familiar with institutions that authorize prescribed fires.
	Current fire suppressors	To indicate who currently takes part in fire suppression, giving a weight to the importance of each on a scale from 1 to 9.
	Expected fire suppressor	To list who should be expected to take part in fire suppression, giving the weight to the importance of each group on a scale from 1 to 9.

In this study, we use the term "wildfire risk perception" to mean understanding that fire can occur and damage natural resources, crops, domestic animals and human infrastructure. "Mitigation behaviours" are strategies that individual inhabitants are able to adopt or consider useful to firefighting, to reduce wildfire occurrence and to limit the impact on the environment.

Data analysis

Statistical analyses were carried out separately for: (i) classifying the social structure and the frequency of answers among municipality; (ii) testing whether the different socio-demographic variables affect the inhabitants' opinions; and (iii) the identification of common trends among municipalities.

Statistical analyses included an independent-sample t-test (p-value <0.05) and the Principal Component Analysis (PCA), using SPSS v 15.0.0 (2006). The t-test assessed whether the different socio demographic variables affect local opinions about wildfire risk and mitigation behaviour. By contrast, the PCA allows to identify the overall trend of investigated population about the awareness of risk perception and the preparedness to manage fire. Starting from the original variables considered in this study, the PCA identified the Principal Components (PCs) that allow us to assess the closeness and differences of perceptions among the three municipalities. Once the PCs were extracted, we displayed the score factors of the first two PCs through a scatterplot in order to show the variability among the interviewees as well as among the three municipalities. For the statistical analysis, the input data for the answers to the first type of closed questions ('Yes/No') was set as a value of 9 for 'Yes' and 1 for 'No', and we ranked the answers for the second type of closed questions from 1 to 9. "No opinion" and missing answers were ranked as 0. The open-ended questions were used to enrich the interpretation of the statistical findings.

Results

Social structure of population studied

The participatory approach involved a total of 116 local inhabitants, 25.0% female and 75.0% male, of the three municipalities (Tab. 2). Their ages ranged between 16 and 87 years with a median age of 42, and most of the interviewees (31.9%) were between 31 and 45. The people contacted from Formoso do Araguaia were younger than those in the other municipalities, with an average age of 36. Interviewees from Dueré were slightly older, with a median age of 46 years.

Most of interviewees were of mixed race, 56.0% of the overall total. This percentage reached 64.1% in Formoso do Araguaia probably due to the proximity of the indigenous reserves. White respondents (22.4%) were more frequent than black ones (18.1%). Finally, although Bananal Island was excluded from the study area, four native people (3.4% of total) took part in the interviews as well.

In terms of education, roughly a third of the interviewees (33.6%) declared that they had completed high school and 12.1% had more advanced education. Nevertheless, many interviewees (30.2%) did not complete the elementary school or were illiterate (7.8%), mainly in Lagoa da Confusão and in Dueré.

Table 2 - Population and social structure of the studied municipalities in the state of Tocantins. Elementary school inc. stands for elementary school incomplete.

Parameters	Municipalities						Total	
	Dueré		Formoso do Araguaia		Lagoa da Confusão			
	Num.	%	Num.	%	Num.	%	Num.	%
Num. Interview	35	30.17	39	33.62	42	36.21	116	100
Age								
16-30 years	7	20.00	17	43.59	9	21.43	33	28.45
31-45 years	11	31.43	10	25.64	16	38.10	37	31.90
46-60 years	12	34.29	11	28.21	8	19.05	31	26.72
61-75 years	4	11.43	1	2.56	8	19.05	13	11.21
76-87 years	1	2.86			1	2.38	2	1.72
Race								
Mixed race	19	54.29	25	64.10	21	50.00	65	56.03
White	7	20.00	9	23.08	10	23.81	26	22.41
Black	7	20.00	4	10.26	10	23.81	21	18.10
Indigenous	2	5.71	1	2.56	1	2.38	4	3.45
Gender								
Male	25	71.43	27	69.23	35	83.33	87	75.00
Female	10	28.57	12	30.77	7	16.67	29	25.00
Educational qualification								
Illiterate	2	5.71	-	-	7	16.67	9	7.76
Elementary school inc.	10	28.57	8	20.51	17	40.48	35	30.17
Elementary school	5	14.29	11	28.21	3	7.14	19	16.38
High school	12	34.29	18	46.15	9	21.43	39	33.62
Higher educational	6	17.14	2	5.13	6	14.29	14	12.07

Risk perception

The larger part of the interviewees (94.0%) agreed that fire represents a negative phenomenon, particularly for the rural environment (Tab. 3). Most explained their negative opinions about fire in terms of their awareness of the resulting environmental degradation (78.9%). According to the respondents, environmental degradation means degradation of soil, plants and animals, which alters the annual agricultural and hunting production. However, they also include the economic aspects such as the destruction of fences and infrastructure. The second most important explanation stressed by interviewees was the effect of fire on climate change. In particular, some explicitly referred to temperature increase and carbon dioxide emission, while others more generally referred to the emission of gases in the atmosphere with strong repercussions on human health. This was especially true of surveys from the municipality of Dueré, where the education levels among interviewees were higher. Finally, the lack of local firefighters was also noted. Though only one person mentioned this deficiency, it is significant given the high flammable index of Cerrado biome as well as the important role that fire plays for land management in these rural environments.

The study revealed that 50.0% of the interviewees used fire in their own activities, particularly in the municipality of Dueré. Although the overall use of fire in the study area was mostly for slash-and-burn practices (55.2%), in Dueré fire was used mostly for burning domestic waste. Furthermore, people also use fire for producing charcoal (5.2%) for cooking in their homes, along the river or in the field during the fishing and hunting expeditions (3.4%). Finally, people used fire for industrial activities (1.7%) such as soldering or ceramics and for bonfires (1.7%).

Most of interviewees (83.6%) perceived wildfire as a serious problem for the municipalities. They also highlighted three main types of damage: (i) environmental stress like destruction of plants, animals and soil; (ii) the adverse health effects in exposed humans; and (iii) the lack of preparedness of people to prevent and control the fires.

Dueré is the municipality where most interviewees (32 out of 35) described fire as a problem for the municipality and, additionally, it was the municipality with the most use of fire. Furthermore, the high rate of landowners and employees involved in the suppression activities, as a consequence of the lack of firefighters, underlines their perception of greater risk. The inhabitants were better acquainted with mitigation actions, even though a general lack of awareness about the causes of fire ignition remains one significant limitation highlighted by interviewees.

There was a wide range of confused perceptions about fire ignition, and heterogeneous points of view. Interviewees highlighted nine different causes of fire ignition, most of anthropogenic origin. The most often cited were cigarette butts (24.1%), intentional ignition (19.8%), mismanaged intentional fires in daily activities (18.9%), and agricultural practices (15.5%). The complexity of the inhabitants' perceptions was underscored by the other minor causes identified by interviewees and by those who "do not know".

Results showed that interviewees from Formoso do Araguaia considered the carelessness in the use of fire to be the most frequent cause of fire ignition. They also identified a prevalence of arsonists, and believed that improving the awareness of fire use among inhabitants represents the best solution for reducing fire impacts. Interviewees from Lagoa da Confusão highlighted the widest range of causes of fire ignition, including short-circuits, even though the cigarette butts and agricultural practices were the most commonly identified. Fire risk was barely noted, and it was mainly linked to the use of fire in the agricultural practices. The actions that could be useful for reducing fires include improving awareness and strictly limiting the use of fire to those activities for which fire is indispensable. In Dueré, interviewees declared that improving the awareness among residents is the most important solution for reducing fire risk.

Mitigation behaviours

Almost all the interviewees (98.2%) said they had taken part in fire suppression activities as a volunteer (Tab. 4). Nevertheless, only 30.2% of the interviewees had taken part in at least one training course on fire prevention and firefighting. However, several interviewees (24.1%) did not know what a prescribed fire is, nor did they know the procedure for its implementation. This was particularly evident in Formoso do Araguaia, where this value was 43.6% and the majority of the remaining interviewees (63.6%), who said they did know what prescribed fire is, did not mention any examples of prescribed fire. Furthermore, 31.8% recognized the firebreaks as a prerequisite to undertake prescribed fire and 4.5% mentioned backfires as a prescribed fire technique.

In contrast, interviewees from the other two municipalities were more confident about mitigation actions. In particular, in Lagoa da Confusão almost 98.0% of the interviewees were aware of at least one prescribed fire technique. They reported firebreaks to be the most frequent example of prescribed fire. Many interviewees (17.1%) associate prescribed fire with elements such as the season of the year and the time of day, saying that prescribed fires should

Table 3 - Perception of fire risk within three municipalities in the state of Tocantins.

Parameters	Municipalities						Total	
	Dueré		Formoso do Araguaia		Lagoa da Confusão			
	Num.	%	Num.	%	Num.	%	Num.	%
Fire opinion								
Negative	34	97.14	38	97.44	37	88.10	109	93.97
Positive/negative	-	-	1	2.56	3	7.14	4	3.45
Positive	1	2.86	-	-	2	4.76	3	2.59
Explanation of negative opinion								
Environmental degradation	23	67.65	33	84.62	34	82.93	90	78.95
Climate change	9	26.47	6	15.38	7	17.07	22	19.30
No local firefighters	1	2.94	-	-	-	-	1	0.88
Do not know	1	2.94	-	-	-	-	1	0.88
Personal use of fire								
Yes	23	65.71	13	33.33	22	52.38	58	50.00
No	12	34.29	26	66.67	20	47.62	58	50.00
Activities which use fire								
Slash and Burn	8	34.78	8	61.54	16	72.73	32	55.17
Burning domestic waste	13	56.52	3	23.08	2	9.09	18	31.03
Charcoal production	-	-	1	7.69	2	9.09	3	5.17
Cooking	1	4.35	-	-	1	4.55	2	3.45
Industrial activity	-	-	-	-	1	4.55	1	1.72
Prescribed fire	1	4.35	-	-	-	-	1	1.72
Bonfire	-	-	1	7.69	-	-	1	1.72
Problem for municipality								
Yes	32	91.43	29	74.36	36	85.71	97	83.62
No	3	8.57	10	25.64	6	14.29	19	16.38
Causes of fire ignition								
Cigarettes	10	28.57	5	12.82	13	30.95	28	24.14
Intentional	9	25.71	9	23.08	5	11.90	23	19.83
Unconscious	7	20.00	14	35.90	1	2.38	22	18.97
Slash-and-burn	6	17.14	3	7.69	9	21.43	18	15.52
Do not know	-	-	5	12.82	1	2.38	6	5.17
Glass bottom of bottles	1	2.86	-	-	4	9.52	5	4.31
Indigenous	-	-	1	2.56	4	9.52	5	4.31
Hunting/fishery	1	2.86	-	-	2	4.76	3	2.59
Short-circuit	-	-	-	-	3	7.14	3	2.59
High temperature	1	2.86	2	5.13	-	-	3	2.59
Environment at risk								
Urban environment	3	8.57	3	7.69	4	9.52	10	8.62
Both	3	8.57	-	0.00	15	35.71	18	15.52
Rural environment	29	82.86	36	92.31	23	54.76	88	75.86
How to reduce fire risk								
Awareness	21	60.00	18	46.15	22	52.38	61	52.59
Surveillance	4	11.43	5	12.82	6	14.29	15	12.93
Not use fire	3	8.57	5	12.82	7	16.67	15	12.93
Other	-	-	7	17.95	1	2.38	8	6.90
Do not know	3	8.57	2	5.13	2	4.76	7	6.03
Firebreak	2	5.71	-	-	3	7.14	5	4.31
Prevention	2	5.71	2	5.13	1	2.38	5	4.31

be scheduled at night and during the rainy season. Only 4.9% of interviewees mentioned backfire and 2.4% associated prescribed fire with large numbers of people involved in the suppression of fire.

Finally, most of interviewees in Dueré (71.4%) said they were aware of at least one prescribed fire technique. Firebreaks and equipment, such as fire swatters, backpack fire pumps and drip torches were the most frequent examples of prescribed fire tools, 60.0% and 12.0%, respectively. It was interesting to note that some of interviewees recognized the necessity of expertise (8.0%) in order to implement prescribed fires. Nevertheless, 12.0% of the interviewees were not able to provide any examples of prescribed fire.

Results showed that interviewees from Lagoa da Confusão were more familiar with the procedure for implementing prescribed fire. They were also aware of the institutions responsible for the prescribed fire authorization. In contrast, most of the interviewees from Formoso do Araguaia did not answer (43.6%) or did not know (20.5%) about this issue. Similarly, in Dueré, 28.6% of interviewees did not answer and 34.3% did not know.

The results showed that usually the main fire suppressors are landowners, employees and neighbors (40.5%). However, this is mainly evident in Dueré and Lagoa da Confusão, while in Formoso do Araguaia, the main fire suppressor is the National Centre for Fire Prevention (PREVFOGO), as stated by 59.0% of interviewees. Overall, often the people involved in fire suppression are volunteers (10.3%) who occasionally travel to the area where the fire occurs, followed by trained individuals and rural workers (9.5%). Finally, 6.9% of the interviewees said they did not know the answer to this question, especially in Formoso do Araguaia (15.4%).

Most of the interviewees (34.5%) believed that national public institutions and their firefighters are the most important of those expected to suppress fires. Furthermore, they believed that the party responsible for ignition (25.0%), landowners (16.4%), and volunteers (13.8%) should participate in the fire suppression to speed the effort. Only 9.5% of the interviewees consider that trained people should participate in the fire suppression activities.

Table 4 - Personal opinions about fire mitigation strategies that interviewees consider useful to reduce fire impacts.

Parameters	Municipalities						Total	
	Duéré		Formoso do Araguaia		Lagoa da Confusão			
	Num.	%	Num.	%	Num.	%	Num.	%
Training course								
No	25	71.43	27	69.23	29	69.05	81	69.83
Yes	10	28.57	12	30.77	13	30.95	35	30.17
Voluntary activity								
Yes	29	97.75	32	97.63	39	99.15	100	98.25
No	6	2.25	7	2.37	3	0.85	16	1.75
Prescribed fire								
Yes	25	71.43	22	56.41	41	97.62	88	75.86
No	10	28.57	17	43.59	1	2.38	28	24.14
Prescribed fire example								
Firebreak	15	60.00	7	31.82	31	75.61	53	60.23
Any example	3	12.00	14	63.64	-	-	17	19.32
Season/time	-	-	-	-	7	17.07	7	7.95
Backfire	1	4.00	1	4.55	2	4.88	4	4.55
Equipment	3	12.00	-	-	-	-	3	3.41
Number of persons	1	4.00	-	-	1	2.44	2	2.27
Expertise	2	8.00	-	-	-	-	2	2.27
Authorization for prescribed fire								
Yes	13	37.14	14	35.90	27	64.29	54	46.55
Do not know	12	34.29	8	20.51	14	33.33	34	29.31
No Answer	10	28.57	17	43.59	1	2.38	28	24.14
Current fire suppressors								
Landowners/employees	23	65.71	2	5.13	22	52.38	47	40.52
Prevfogo	5	14.29	23	58.97	9	21.43	37	31.90
Volunteers	3	8.57	7	17.95	2	4.76	12	10.34
Trained/employees	3	8.57	-	-	8	19.05	11	9.48
Do not know	1	2.86	6	15.38	1	2.38	8	6.90
Arsonist	-	-	1	2.56	-	-	1	0.86
Expected fire suppressor								
Firefighter	8	22.86	20	51.28	12	28.57	40	34.48
Arsonist	7	20.00	11	28.21	11	26.19	29	25.00
Landowners	8	22.86	-	-	11	26.19	19	16.38
Volunteers	8	22.86	4	10.26	4	9.52	16	13.79
Trained	4	11.43	4	10.26	3	7.14	11	9.48
Do not know	-	-	-	-	1	2.38	1	0.86

Independent-sample t-test

The independent-samples t-test found few significant differences among interviewees (Fig. 2). Opinions did not differ by gender. Age and race had a stronger effect, though this was perhaps due to the sample sizes of certain age classes and races (e.g., only two people within the age range 76-87, and only four indigenous people took part in the interviews). The residence of interviewees and their education were the only variables that were associated with significant differences in responses.

Overall, the t-tests revealed more differences in mitigation behaviours than in wildfire risk perceptions. This demonstrates the widespread perceptions of wildfire risk among interviewees and highlights the heterogeneity of perceptions about the mitigation behaviours. Although most of interviewees had taken part in the suppression activities, they did not have adequate knowledge for wildfire prevention. Usually they acted on experiences and beliefs of the owner once a wildfire was ignited. Results confirmed that interviewees from Formoso do Araguaia were less familiar with mitigation behaviours (Tab. 5). In particular, there were significant differences about prescribed fire and prescribed fire examples between Formoso do Araguaia and Lagoa da Confusão.

Furthermore, the results confirmed that high school and higher educational attainment positively affect awareness of wildfire mitigation behaviours (Tab. 6), even if the differences are not highly significant.

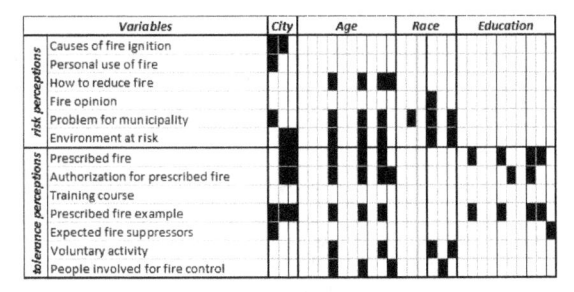

Figure 2 - Independent t-test. The figure shows the variables that presented statistically significant differences (black square), p-value <0.05. Gender is absent because it did not present any significant differences.

Principal Component Analysis

Since some variables did not have any significant correlations, the PCA combined only seven of the original variables, extracting four PCs that explained

Table 5 - Pairwise comparisons among municipalities (p-value is <0.05). The independent variables shown are only those that displayed significant differences. In the table, df stands for "Degrees of Freedom", M for "Mean" and SD for "Standard deviation". M and SD are given for the two municipalities of pairwise comparison.

	Duaré - Formoso do Araguaia						
	p-value	t	df	M	SD	M	SD
Causes of fire ignition	0.038	2.12	63.51	7.31	1.78	6.13	2.94
Personal use of fire	0.005	2.9	71.01	6.26	3.85	3.67	3.82
Problem for municipality	0.05	1.99	65.49	8.31	2.27	6.95	3.54
Prescribed fire example	0.001	3.63	68.56	6.09	4.06	2.82	3.61
Expected fire suppressors	0.014	2.52	71.45	5.40	2.69	7	2.75
	Duaré - Lagoa da Confução						
	p-value	t	df	M	SD	M	SD
Causes of fire ignition	0.048	2.01	73.35	7.31	1.78	6.33	2.49
Environment at risk	0.05	1.99	74.36	7.97	2.44	6.81	2.68
Prescribed fire	0.002	3.23	40.47	6.71	3.67	3.48	3.74
Authorization for prescribed fire	0.011	2.63	70.72	3.69	4.16	6.12	3.91
Prescribed fire example	0.001	3.75	40.63	6.09	4.06	8.79	1.39
	Formoso do Araguaia - Lagoa da Confução						
	p-value	t	df	M	SD	M	SD
Environment at risk	0.005	2.92	77.50	8.38	2.16	6.81	2.68
Prescribed fire	0.000	4.91	44.63	5.51	4.02	8.81	1.23
Authorization for prescribed fire	0.004	2.95	77.19	3.44	4.23	6.12	3.91
Prescribed fire example	0.000	9.67	48.30	2.82	3.61	8.79	1.39

Table 6 - Pairwise comparisons among educational attainment categories (p-value is <0.05). The independent variables shown are the only ones that displayed significant differences between fire mitigation behaviors. In the table, df stand for "Degrees of Freedom", M for "Mean" and SD for "Standard deviation". M and SD are given for the educational qualifications of the groups in each pairwise comparison.

	Prescribed fire						
Grouping variables	p-value	t	df	M	SD	M	SD
Illiterate - Elementary	0.017	2.56	23.10	8.11	2.67	4.79	4.10
Elementary incomplete - Elementary	0.023	2.39	30.45	7.40	3.25	4.79	4.10
Elementary - High	0.038	2.17	30.54	4.79	4.10	7.15	3.41
Elementary - Higher educational	0.003	3.30	28.37	4.79	4.10	8.43	2.14
	Authorization for prescribed fire						
Grouping variables	p-value	t	df	M	SD	M	SD
Elementary incomplete - High	0.048	2.02	71.84	3.54	4.02	5.49	4.28
Elementary - High	0.042	2.10	36.37	3	4.20	5.49	4.28
	Prescribed fire example						
Grouping variables	p-value	t	df	M	SD	M	SD
Illiterate - Elementary	0.003	3.30	21.43	8	3	3.42	4.18
Elementary incomplete - Elementary	0.012	2.64	35.63	6.51	3.99	3.42	4.18
Elementary - High	0.039	2.15	34.43	3.42	4.18	5.90	4.01
Elementary - Higher educational	0.010	2.73	30.93	3.42	4.18	6.93	3.2
	Expected fire suppressors						
Grouping variables	p-value	t	df	M	SD	M	SD
High - Higher educational	0.049	2.08	23.43	6.69	2.66	5	2.6

76.2% of the total variance. In order to represent the common trends of perceptions about wildfire risk and mitigation behaviours in the three municipalities, we used the factor scores obtained from the first two PCs to draw a scatterplot (Fig. 3). These two explained 20.5% (PC1) and 11.5% (PC2) of the variability. The variables that scored the highest values in PC1 were the authorization for prescribed fire and its implementation, reflecting the mitigation behaviours. The variables that yielded the highest values for PC2 were training courses, a current role as fire suppressor and the environment at risk reflecting the perceived risk.

In the scatterplot, the position of the symbols represents the interviewees' answers according to the two PC scores. Symbols on the far right reflect greater awareness of prescribed fire techniques and their implementation. The symbols at the top of the score plot reflect a perception of greater risk.

Interviewees from Formoso do Araguaia had

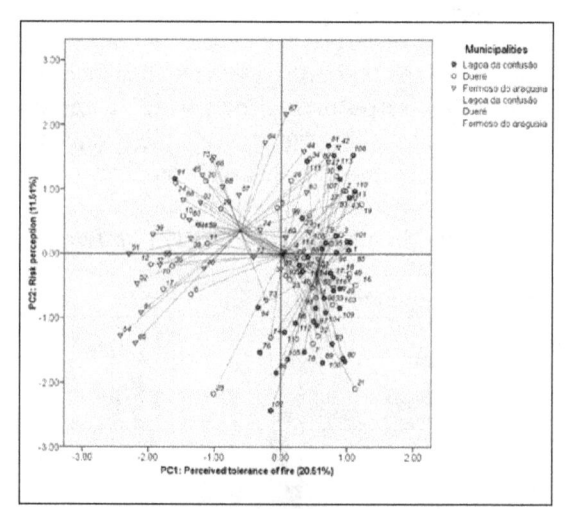

Figure 3 - PCA score plot. The y-axis represents the PC1 "mitigation behaviors" while the x-axis represents the PC2 "risk perception" among the citizens. The star indicates the municipalities of the Lagoa da Confusão, the circle Dueré and the triangle Formoso do Araguaia.

the most variability in mitigation behaviour, ranging from -2.65 (the lowest value) to 1.32, with more than 50.0% of negative score values positioned, for this reason, on the left side of the score plot. In contrast, in Dueré, the score values ranged from -2.34 to 1.32, but most of them presented positive values, reflecting the inhabitants' familiarity with mitigation behaviours. Finally, in Lagoa da Confusão, the score plot ranged from -1.17 to 1.32. Since most of them were positive, the interviewees from Lagoa da Confusão were overall the most aware of fire mitigation.

Risk perception increases along the vertical axis of the score plot. Results showed that those who participated in training courses had higher perceived risk of fire. In addition, interviewees who recognized the wildfire risk for the rural environment and those who declared the PREVFOGO as the most important current suppressor of fire are positioned at the top of the score plot. Interviewees from Dueré showed more variability in their scores, ranging from -2.15 (the lowest value) to 1.77. Within the municipality of Formoso do Araguaia, the results were largely positive, highlighting the higher risk perception by local inhabitants, with scores that ranged from -1.73 to 2.07. Finally, in Lagoa da Confusão, the variability was lowest, ranging from -1.98 to 1.36 with a negative trend.

The centroids of the score values for the three municipalities revealed the overall perceptions about wildfire risk and the mitigation behaviours. Formoso do Araguaia showed the highest level of risk perception and had the least capability in mitigation behaviours. This is likely due to the strong presence of firefighters who represented the most frequent suppressors of fire and to the high perception of fire risk for the rural environment. On the contrary, Lagoa da Confusão showed more familiarity with mitigation actions and perceived less risk. Finally, in Dueré the risk perception and the mitigation awareness were both intermediate amongst those of the other municipalities.

Discussion and conclusion

This study demonstrates that survey participants perceived wildfire risk as rather high, although the perceptions were complex and conflicting among interviewees. Interviewees felt that fire is a destructive phenomenon, but at the same time, they also highlighted its functionality in many activities. The most common use of fire was for traditional practices in agriculture and animal breeding (Klink and Machado 2005, Mistry 1998, Pereira Jr. et al. 2014, Pivello 2011), though the mismanagement of these practices represents one of the main causes of fires in this area. These findings stress the importance of disseminating knowledge about fire prevention and fire management among local inhabitants in limiting fire damage (Berkes 2004, Eriksen 2007, Mistry and Bizerril 2011). This represents a crucial challenge for policy decision makers because limiting fire impacts can strongly contribute to the reduction of land degradation (Kosmas et al. 2016) and halting the loss of biodiversity.

Furthermore, the interviewees perceived fire as a problem for the municipalities, not only in ecological aspects, such as soil degradation and biodiversity loss (Bond et al. 2005, Lehmann et al. 2011, I. A. Silva and Batalha 2008), but also for the administrative challenges of fire control. Although the second and third actions of the National Action Plan, developed by the State Committee for Forest Fire Fighting and Control of Fires, promote the mobilization, prevention, control and fighting of illegal fires by civilian individuals, this study reveals that only 30.2% of interviewees have taken part in training courses, and they rarely took part in suppression activities. Conversely, landowners, employees and rural inhabitants are the most frequent suppressors. Significant improvements would be needed to improve fire prevention activities and reduce the risk of fire. For example, promoting cooperation between trained people and landowners in order to schedule and implement prescribed fire would be useful. Likewise, it is important to enhance awareness regarding the causes of fire ignition, not only due to the main origin being Anthropogenic, but also in order to identify the main drivers of socio-ecological changes (Ferrara et al. 2016) and their impacts on the natural resources. The number of ignition causes listed by local respondents reflects, on one hand, the diversity of perceptions, and on the other hand a lack of knowledge among respondents. Cigarette butts were commonly identified as the main cause, but previous research has demonstrated the low probability of their causing fire ignition, given that they would require many favourable conditions such as wind and specific road surfaces (Xanthopoulos et al. 2006). Similarly, they referred to the bottom of a glass bottle, stating that bottles function as magnifying glasses and thereby enable fire ignition. Although this is another widespread opinion, experimental studies have demonstrated the low probability of fire ignition by glass fragments (Wittich and Müller 2009). Often respondents referred to the irresponsible use of fire, as well. Finally, there is a common bias that considers native people responsible for fires due to the pervasive use of fire in traditional customs. In contrast, some authors argue that indigenous practices reduce fire intensity because they maintain lower levels of fuel (Welch et al. 2013). However, the present study confirms that interviewees with high school diplomas recognized the high risk of fire in the area studied. They strongly agreed with the proposition that more awareness is necessary in order to reduce fire impacts. Although the social memory of local communities represents an important cultural heritage (Wilson et al. 2017), new efforts are necessary to improve the effectiveness of training activities and limiting fire impacts on land degradation and loss of ecosystem services. The social perceptions about the expected ecosystem services could represent an important starting point for a deeper evaluation of inhabitant behaviours.

Regarding fire mitigation behaviours, this study demonstrates that, although almost all the interviewees have taken part in fire suppression activities, most of them did not have an adequate level of competence. The participation in training activities was quite low as was the awareness of topics such as prescribed fire and its implementation. Most of the examples given for prescribed fire were rather vague and confusing. Most referred to the firebreaks as prescribed fire techniques. Only a few interviewees explicitly mentioned backfire as an example of prescribed fire, and even fewer recognized expertise as a prerequisite of implementing prescribed fires. Similarly, interviewees deemed the equipment and personnel sufficient for prescribed fire implementation. In a few cases, interviewees mentioned the burning season and mainly referred to tie of day and avoiding high temperatures (i.e. during the night and in November). Although they were not aware of the vegetative cover responses and phenology (Grace et al. 2006, Santos et al. 2003), early fire (May–June) favours woody plants, while later fire (September–October) favours grassy plants (Pivello and Coutinho 1996). These findings demonstrate the lack of awareness about forest fire control techniques and firefighting among local inhabitants, but also demonstrate the local experience in several voluntary efforts in fire suppression activities.

Finally, this study shows the variability of perceptions about wildfire issues among inhabitants in the three municipalities. Overall, interviewees that perceived higher risk were those from Formoso do Araguaia, followed by residents of Dueré and finally by those from Lagoa da Confusão. In contrast, interviewees from Formoso do Araguaia were less familiar with fire mitigation, followed by those from Dueré and then by residents from Lagoa da Confusão. Although fires were recognized as often induced by humans, inhabitants' perceptions were mostly focused on accidental rather than intentional fire. A general lack of awareness about the wildfire risk and fire mitigation behaviours has emerged from the study. Since the study area is fire dependent (Hardesty, Jeff and Myers, Ron and Fulks 2005, Pivello 2011), improving awareness about fire management represents the most suitable solution to reduce fire impacts. The role of local inhabitants is important in order to help strike the appropriate balance between developing and conserving natural resources and managing undesired fires. Based on their educational backgrounds, training in fire prevention, control and firefighting represent the most important tools for improving awareness. In conclusion, the study demonstrates that assessing inhabitant's perceptions offers a strong contribution to the evaluation of areas that are critically exposed to fire impacts due to the lack of preparedness of local inhabitants. Since fire issues are very common in Brazil as in other prone areas, this study can be further exploited to evaluate the perceptions among citizens elsewhere. The approach used highlights aspects of weakness in training activities and suggests an improvement of policy makers' efforts to overcome these challenges. Significant improvements could be achieved by a more interdisciplinary approach in order to increase the effectiveness of perceptions evaluation.

Acknowledgement

We are grateful to the voluntary interviewees of the municipalities of the Dueré, Formoso do Araguaia and Lagoa da Confusão. Many thanks to Allan D. P. da Silva from the Federal University of Tocantins for his contribution to the data collection. We gratefully acknowledge financial support from CAPES-Brazil, the Ministry of Education of Brazil, project 88881.062168/2014-01.

References

Alvares CA., Stape JL., Sentelhas PC., De Moraes Gonçalves JL., Sparovek G. 2013 - *Köppen's climate classification map for Brazil*. 22 (6): 711-28. doi:10.1127/0941-2948/2013/0507

Batalha MA., Silva IA., Cianciaruso MV, França H., de Carvalho GH. 2011 - *Phylogeny, traits, environment, and space in cerrado plant communities at Emas National Park (Brazil)*. Flora: Morphology, Distribution, Functional Ecology of Plants 206 (11): 949-56. doi:10.1016/j.flora.2011.07.004

Beringer J. 2000 - *Community fire safety at the urban rural interface: the bushfire risk*. Fire Safety Journal 35 (1): 1-23. doi:10.1016/S0379-7112(00)00014-X

Berkes F. 2004 - *Rethinking community-based conservation*. Conservation Biology 18 (3): 621-30. doi:10.1111/j.1523-1739.2004.00077.x

Bond WJ., Woodward FI., Midgley GF. 2005 - *The global distribution of ecosystems in a world without fire*. New Phytologist 165 (2): 525-38. doi:10.1111/j.1469-8137.2004. 01252.x

Cachoeira JN. 2015 - *Caracterização das Queimadas e Incêndios Florestais no Estado do Tocantins no Período de 2003 a 2011*. Universidade Federal do Tocantins.

Ceccon E., Miramontes O. 2008 - *Reversing deforestation? Bioenergy and society in two Brazilian models.* Ecological Economics 67 (2): 311-17. doi:10.1016/j.ecolecon.2007.12.008

Certini G. 2005 - *Effects of fire on properties of forest soils: A review.* Oecologia 143 (1): 1-10. doi:10.1007/s00442-004-1788-8

Champ PA., Donovan GH., Barth CM. 2013 - *Living in a tinderbox: Wildfire risk perceptions and mitigating behaviours.* International Journal of Wildland Fire 22 (6): 832-40. doi:10.1071/WF12093

Corbin JM., Strauss A. 1990 - *Grounded theory research: Procedures, canons, and evaluative criteria.* Qualitative Sociology 13 (1): 3-21. doi:10.1007/BF00988593

Dondo Bühler M., de Torres Curth M., Garibaldi LA. 2013 - *Demography and socioeconomic vulnerability influence fire occurrence in Bariloche (Argentina).* Landscape and Urban Planning 110 (1): 64-73. doi:10.1016/j.landurbplan.2012.10.006

Eriksen C. 2007 - *Why do they burn the "bush"? Fire, rural livelihoods, and conservation in Zambia.* Geographical Journal 173 (3): 242-56. doi:10.1111/j.1475-4959.2007.00239.x

Ferrara A., Kelly C., Wilson GA., Nolè A., Mancino G., Bajocco S., Salvati L. 2016 - *Shaping the role of "fast" and "slow" drivers of change in forest-shrubland socio-ecological systems.* Journal of Environmental Management 169: 155-66. doi:http://doi.org/10.1016/j.jenvman.2015.12.027

Forzza RC., Baumgratz JFA., Bicudo CEM., Canhos DAL., Carvalho Jr. AA., Coelho MAN., Costa AF., Costa DP., Hopkins MG., Leitman PM., Lohmann LG., Lughadha EN., Maia LC., Martinelli G., Menezes M., Morim MP., Peixoto AL., Pirani JR., Prado J. 2012 - *New brazilian floristic list highlights conservation challenges.* BioScience 62 (1): 39-45. doi:10.1525/bio.2012.62.1.8

Giongo M., Batista AC., Cachoeira JN., Pereira AD., Viola MR., da Silva DB., Santopuoli G., Barilli J., Patriota JN., Sousa Pereira IM., de Souza Junior MR. 2014 - *Plano Operativo de Prevenção e Combate aos Incêndios Florestais do município de Dueré (TO).* Gurupi (TO)

Gounaridis D., Zaimes GN., Koukoulas S. 2014 - *Quantifying spatio-temporal patterns of forest fragmentation in Hymettus Mountain, Greece.* Computers, Environment and Urban Systems 46: 35-44.

Grace J., José JS., Meir P., Miranda HS., Montes RA. 2006 - *Productivity and carbon fluxes of tropical savannas.* Journal of Biogeography 33 (3): 387-400. doi:10.1111/j.1365-2699.2005.01448.x

Hardesty J., Myers R., Fulks W. 2005 - *Fire, ecosystems, and people: a preliminary assessment of fire as a global conservation issue.* The Nature Conservation 22 (4): 78-87.

Hecht SB. 2009 - *Kayapó savanna management: Fire, soils, and forest islands in a threatened biome.* Amazonian Dark Earths: Wim Sombroek's Vision: 143-162 p. doi:10.1007/978-1-4020-9031-8_7

Hoffmann WA., Adasme R., Haridasan M., De Carvalho MT., Geiger EL., Pereira MAB., Gotsch SG., Franco AC. 2009 - *Tree topkill, not mortality, governs the dynamics of savanna-forest boundaries under frequent fire in central Brazil.* Ecology 90 (5): 1326-37. doi:10.1890/08-0741.1

Hoffmann WA., Orthen B., Vargas Do Nascimento PK. 2003 - *Comparative fire ecology of tropical savanna and forest trees.* Functional Ecology 17 (6): 720-26. doi:10.1111/j.1365-2435.2003.00796.x

IBAMA 2009 - *Roteiro Metodológico para a Elaboração de Planos Operativos de Prevenção e Combate aos Incêndios Florestais.* Brasilia: 43 p. https://brigadaro selynunes.files.wordpress.com/2013/08/roteiro-metodolc3b3gico.pdf

INPE 2015 - *Monitoramento dos Focos Ativos por estado: TOCANTINS - Brasil.* http://www.inpe.br/queimadas/estatisticas_estado.php?estado=TO&nomeEstado=TOCANTINS. Accessed 25 June 2015

Kelly C., Ferrara A., Wilson GA., Ripullone F., Nolè A., Harmer N., Salvati L. 2015 - *Community resilience and land degradation in forest and shrubland socio-ecological systems: Evidence from Gorgoglione, Basilicata, Italy.* Land Use Policy 46: 11-20. doi:http://doi.org/10.1016/j.landusepol.2015.01.026

Klink CA., Machado RB. 2005 - *Conservation of the Brazilian Cerrado.* Conservation Biology 19 (3): 707-13. doi:10.1111/j.1523-1739.2005.00702.x

Klink C., Moreira AG. 2002 - *Past and current human occupation, and land use.* In The cerrados of Brazil: ecology and natural history of a neotropical savanna. Columbia University Press New York, New York, USA

Kosmas C., Karamesouti M., Kounalaki K., Detsis V., Vassiliou P., Salvati L. 2016 - *Land degradation and long-term changes in agro-pastoral systems: An empirical analysis of ecological resilience in Asteroussia - Crete (Greece).* CATENA 147: 196-204. doi:http://doi.org/10.1016/j.catena.2016.07.018

Lehmann CER., Archibald SA., Hoffmann WA., Bond WJ. 2011 - *Deciphering the distribution of the savanna biome.* New Phytologist 191 (1): 197-209. doi:10.1111/j.1469-8137.2011.03689.x

López-Santiago CA., Oteros-Rozas E., Martín-López B., Plieninger T., Martín EG., González JA. 2014 - *Using visual stimuli to explore the social perceptions of ecosystem services in cultural landscapes: The case of transhumance in Mediterranean Spain.* Ecology and Society 19 (2). doi:10.5751/ES-06401-190227

Maezumi SY., Power MJ., Mayle FE., McLauchlan KK., Iriarte J. 2015 - *Effects of past climate variability on fire and vegetation in the cerrãdo savanna of the Huanchaca Mesetta, NE Bolivia.* Climate of the Past 11 (6): 835-53. doi:10.5194/cp-11-835-2015

McCaffrey S., Toman E., Stidham M., Shindler B. 2013 - *Social science research related to wildfire management: An overview of recent findings and future research needs.* International Journal of Wildland Fire 22 (1): 15-24. doi:10.1071/WF11115

Mistry J. 1998 - *Decision making for fire use among farmers in savannas: An exploratory study in the Distrito Federal, central Brazil.* Journal of Environmental Management 54 (4): 321-34. doi:10.1006/jema.1998.0239

Mistry J., Berardi A., Andrade V., Krahô T., Krahô P., Leonardos O. 2005 - *Indigenous fire management in the cerrado of Brazil: The case of the Krahô of Tocantíns.* Human Ecology 33 (3): 365-86. doi:10.1007/s10745-005-4143-8

Mistry J., Bizerril M. 2011 - *Why it is important to understand the relationship between people, fire and protected areas.* Biodiversidade Brasileira 1: 40-49. doi:10.2307/302397

Myers N., Mittermeler RA., Mittermeler CG., Da Fonseca GAB., Kent J. 2000 - *Biodiversity hotspots for conservation priorities.* Nature 403 (6772): 853-58. doi:10.1038/35002501

Odhiambo B., Meincken M., Seifert T. 2014 - *The protective role of bark against fire damage: A comparative study on selected introduced and indigenous tree species in the Western Cape, South Africa*. Trees - Structure and Function 28 (2): 555-65. doi:10. 1007/s00468-013-0971-0

Pastorella F., Giacovelli G., Maesano M., Paletto A., Vivona S., Veltri A., Pellicone G., Mugnozza GS. 2016 - *Social perception of forest multifunctionality in southern Italy: The case of Calabria Region*. Journal of Forest Science 62 (8): 366-79. doi:10.17221 /45/2016-JFS

Patton MQ. 1990 - *Qualitative evaluation and research methods*. 2nd ed. SAGE Publications, inc: 532 p.

Paveglio TB., Nielsen-Pincus M., Abrams J., Moseley C. 2017 - *Advancing characterization of social diversity in the wildland-urban interface: An indicator approach for wildfire management*. Landscape and Urban Planning 160: 115-26. doi:http://doi.org/10.1016/j.landurbplan.2016.12.013

Pereira Jr. AC., Oliveira SLJ., Pereira JMC., Turkman MAA. 2014 - *Modelling fire frequency in a Cerrado savanna protected area*. PLoS ONE 9 (7). doi:10. 1371/journal .pone.0102380

Phalan B., Bertzky M., Butchart SHM., Donald PF., Scharlemann JPW., Stattersfield AJ., Balmford A. 2013 - *Crop Expansion and Conservation Priorities in Tropical Countries*. PLoS ONE 8 (1). doi:10.1371/journal.pone.0051759

Pivello VR. 2011 - *The use of fire in the cerrado and Amazonian rainforests of Brazil: Past and present*. Fire Ecology 7 (1): 24-39. doi:10.4996/fireecology.0701024

Pivello VR., Coutinho LM. 1996 - *A qualitative successional model to assist in the management of Brazilian cerrados*. Forest Ecology and Management 87 (1-3): 127-38. doi:10.1016/S0378-1127(96)03829-7

Ramos-Neto MB., Pivello VR. 2000 - *Lightning fires in a Brazilian Savanna National Park: Rethinking management strategies*. Environmental Management 26 (6): 675-84. doi:10.1007/s002670010124

Ribeiro JF. Walter B. 2008 - *As principais fitofisionomias do bioma cerrado in: Sano, SM; Almeida, SP; Ribeiro, JF Cerrado: Ecologia e flora*. Brasilia, Embrapa Informação Tecnológica: 406 p.

Sallustio L., Quatrini V., Geneletti D., Corona P., Marchetti M. 2015 - *Assessing land take by urban development and its impact on carbon storage: Findings from two case studies in Italy*. Environmental Impact Assessment Review 54: 80-90. doi:http://dx. doi.org/10.1016/j.eiar.2015.05.006

Santopuoli G., Ferranti F., Marchetti M. 2016 - *Implementing Criteria and Indicators for Sustainable Forest Management in a Decentralized Setting: Italy as a Case Study*. Journal of Environmental Policy and Planning 18 (2): 177-96. doi:10.1080/ 1523908X.2015.1065718

Santopuoli G., Requardt A., Marchetti M. 2012 - *Application of indicators network analysis to support local forest management plan development: A case study in Molise, Italy*. iForest 5: 31-37.

Santos AJB., Silva GTDA., Miranda HS., Miranda AC., Lloyd J. 2003 - *Effects of fire on surface carbon, energy and water vapour fluxes over campo sujo savanna in central Brazil*. Functional Ecology 17 (6): 711-19. doi:10.1111/j.1365-2435.2003.00790.x

Silva DM., Batalha MA., Cianciaruso MV. 2013 - *Influence of fire history and soil properties on plant species richness and functional diversity in a neotropical savanna*. Acta Botanica Brasilica 27 (3): 490-97. doi:10.1590/S0102-33062013000300005

Silva IA., Batalha MA. 2008 - *Species convergence into lifeforms in a hyperseasonal cerrado in central Brazil*. Brazilian Journal of Biology 68 (2): 329-39. doi:10.1590 / S1519-69842008000200014

Simon MF., Pennington T. 2012 - *Evidence for adaptation to fire regimes in the tropical savannas of the Brazilian Cerrado*. International Journal of Plant Sciences 173 (6): 711-23. doi:10.1086/665973

SPSS - Statistical Package for Social Science Inc. 2006 - Chicago.

Strauss A., Corbin J. 1994 - *Grounded theory methodology*. Handbook of qualitative research 17: 273-85.

Valente CR., Latrubesse EM., Ferreira LG. 2013 - *Relationships among vegetation, geomorphology and hydrology in the Bananal Island tropical wetlands, Araguaia River basin, Central Brazil*. Journal of South American Earth Sciences 46: 150-60. doi:10.1016/j.jsames.2012.12.003

Vizzarri M., Tognetti R., Marchetti M. 2015 - *Forest Ecosystem Services: Issues and Challenges for Biodiversity, Conservation, and Management in Italy*. Forests (6): 1810-38. doi:10.3390/f6061810

Welch JR., Brondízio ES., Hetrick SS., Coimbra Jr. CEA. 2013 - *Indigenous burning as conservation practice: Neotropical savanna recovery amid agribusiness deforestation in Central Brazil*. PLoS ONE 8 (12). doi:10.1371/journal.pone.0081226

Wilson GA., Kelly CL., Briassoulis H., Ferrara A., Quaranta G., Salvia R., Detsis V., Curfs M., Cerda A., El-Aich A., Liu H., Kosmas C., Alados CL., Imeson A., Landgrebe-Trinkunaite R., Salvati L., Naumann S., Danwen H., Iosifides T. 2017 - *Social Memory and the Resilience of Communities Affected by Land Degradation*. Land Degradation and Development 28 (2): 383-400. doi:10.1002/ldr.2669

Wittich KP., Müller T. 2009 - *An experiment to test the potential for glass fragments to ignite wildland fuels*. International Journal of Wildland Fire 18 (7): 885-91. doi:10. 1071/WF08069

Xanthopoulos G., Ghosn D., Kazakis G. 2006 - *Investigation of the wind speed threshold above which discarded cigarettes are likely to be moved by the wind*. International Journal of Wildland Fire 15 (4): 567-76. doi:10.1071/WF05080

Figuring the features of the Roman Campagna: recent landscape structural

Luca Salvati[1*], Lorenza Gasparella[2], Michele Munafò[3], Raoul Romano[4], Anna Barbati[2]

Abstract - This article evaluates the impact of urban expansion on landscape composition and structure of a landscape icon, the Roman Campagna, central Italy, during the last 30 years. Landscape attributes were assessed between 1974, when the distinguishing features of Roman Campagna are still widespread, and 2008, following decades of urban decentralization and urban sprawl. Changes in landscape structure were explored by spatial pattern analysis to detect how structural changes in landscape components can modify land structure and landscape profile. Non-parametric correlation statistics and factor analysis showed that the distinctive features of the Roman Campagna landscape are now blurred. A generalized landscape mix was generated by the juxtaposition of different land-use, reflected in a negative relationship between changes in surface area and patchiness found in natural and agricultural uses of land. Adaptation measures for preserving peri-urban agriculture in a changing landscape were finally discussed.

Keywords - landscape icon; Morphological Spatial Pattern Analysis (MSPA); fragmentation; peri-urban agriculture; Roman Campagna

Introduction

The low-lying landscape surrounding Rome (Latium, central Italy) has been shaped by humans for thousands of years, like other areas with ancient colonization in the northern Mediterranean basin. The area known as Roman Campagna was used for agriculture since Roman time and then abandoned during the Middle Ages, due to malaria and water shortage for farming needs. The Campagna became a famous landscape icon in Europe during the 18th and 19th centuries and an excursion into the Roman countryside was a must for travellers in the *Grand Tour* (Fig. 1).

The Campagna was reclaimed in the 19th and 20th centuries: efforts to make land more productive marked the beginning of its decline and the loss of agricultural land-use for a large part of the area. Starting with the 1950s, the Rome's expansion took over large parts of the Campagna, all around the city. However, until the 1970s the distinguishing features of Roman Campagna were still widespread: extensive arable land and pastures in lowlands, areas suitable for crops and sheep farming, and vineyards laid in a concentric semi-circle to the top of the Alban hills, where extensive cropping is spatially mixed with residential urban expansion.

Following a homogeneous path of urban decentralization and sprawl common to many European cities since the 1980s (Kasanko et al. 2006, Longhi and Musolesi 2007, Turok and Mykhnenko 2007, Schneider and Woodcock 2008, Munafò et al. 2013, Kazemzadeh-Zow et al. 2016), urban areas have greatly expanded in Rome's countryside (Fig. 2). The high rate of cropland conversion to urban uses is the most visible alteration of fringe landscapes,

Figure 1 - Cole Thomas, Roman Campagna, c. 1843, Wadsworth Athenaeum (photo in public domain).

---

[1] Consiglio per la ricerca in agricoltura e l'analisi dell'economia agraria, Centro di ricerca per le foreste e il legno, Arezzo (Italy)
[2] University of Tuscia, Dipartimento per l'innovazione nei sistemi Biologici, Agroalimentari e Forestali (UNITUS-DIBAF), Viterbo (Italy)
[3] Istituto Superiore per la Protezione e la Ricerca Ambientale, Roma (Italy)
[4] Consiglio per la ricerca in agricoltura e l'analisi dell'economia agraria, Research Centre for Agricultural Policies and Bioeconomiy, Roma (Italy)
*luca.salvati@crea.gov.it

Figure 2 - 2012 Roman Campagna ©[Gasparella L.]

resulting in mixed land-use, the unwatched territory of the so called "edgelands" (Shoard 2002). Land-use spatial polarization, ecosystem deterioration, and loss in biodiversity are the main threats for the sustainability of rural landscape experiencing urban expansion (Cakir et al. 2008, Mavrakis et al. 2015; Colantoni et al. 2016, Salvati et al. 2017).

Paradoxical as it may sound, the ecological and functional-productive values of the countryside around towns in Europe are less recognized and appreciated by an urbanized society, than landscape amenities provided by a heterogeneous and complex agricultural land-use (Zasada 2011). In this framework, the alteration of the structure of the traditional agricultural landscape surrounding cities and, particularly, its interplay with the dynamics of urban development requires specific attention in a broader academic discourse on landscape transformations (Salvati et al. 2013, Ferrara et al. 2014, Serra et al. 2015, Smiraglia et al. 2015).

Based on these considerations, the present study aims at figuring out changes occurred in the landscape structure of Roman Campagna over an extended period of intense urban development (1974-2008). Other studies have approached this issue by landscape-level estimation of percolation and proximity indices (Metzger and Muller 1996; Metzger and Decamps 1997; Pili et al. 2017). We propose, instead, a novel application of the Morphological Spatial Pattern Analysis (Soille and Vogt 2009), which is conceived for the diagnosis of the spatial patterns of a given land cover class and the classification of their individual components. We apply MSPA to figure out to what extent observed land cover changes altered the inner structure of the Roman Campagna landscape.

The specific aims of this paper are therefore (i) to identify the most important landscape transformations during a period of rapid urbanization, (ii) to quantify changes in the structure of the rural landscape by using Morphological Spatial Pattern Analysis and, finally, (iii) to verify spatial segregation or association patterns between two (or more) land-use classes. Results may inform rural policies and spatial planning, with the aim to take account of specific peri-urban agriculture types when tailoring environmental and land management incentives.

Materials and Methods

Study area

The investigated area covers a regular, squared landscape scene of 3,000 km^2 in the Nuts-3 district of Rome that consists of 50% lowlands, 40% hills, and 10% mountains. This landscape scene includes the so-called Roman Campagna district (in Italian, "Campagna Romana"), corresponding to the territorial units of Sabatini Mountains (Rome province), Rome Countryside and parts of Tiber Valley and Rome Coastline, as illustrated in Fig. 3 using a

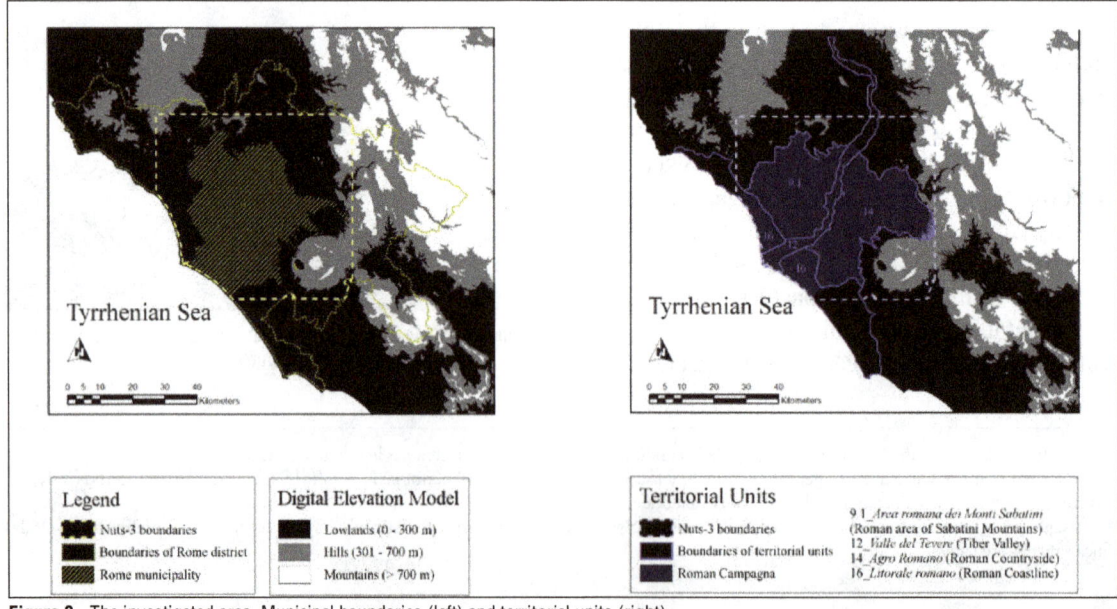

Figure 3 - The investigated area. Municipal boundaries (left) and territorial units (right).

specific district code (see also Savo et al. 2012 for a description of "Agro Romano"). Rome municipality occupies the 76% of Roman Campagna. Although nowadays urban areas cover a significant (and still increasing) almost 20% of the district land, the landscape matrix still consists of a mosaic of rural land-use. Rome is, namely, the most densely populated and the largest farming municipality of Italy at the same time (Salvati 2013). According to earlier studies (see Salvati et al. 2012 and references therein), compact growth occurred in Rome mainly between the early 1950s and the 1980s while peri-urban development was observed only in the following decades (Quatrini et al. 2015). In the 'compact growth' urban wave, population grew in the urban area at a higher rate compared to the peri-urban area (Munafò et al. 2010). The difference in population density between the two areas was high and the ratio between peri-urban and urban population increased slightly from 27% to 35% (Savo et al. 2012). During the 'dispersed growth' wave, population declined in the urban area while rose in peri-urban areas at a relatively high rate (1.5% per year).

Land cover maps

Two digital land cover maps (scale 1:25 000) were used as input data sources for this study: (i) the 'Agricultural and forest map of Rome region' produced by the district authority of Rome in 1974 (hereafter LCM74) and based on the elementary datasets developed by the Italian Geographical Military (IGM) and (ii) a land cover map derived from photo-interpretation of digital ortho-images released from the Italian National Geoportal (Italian Ministry for Environment, Land and Sea) with a 0.5 meters pixel related to 2008 (hereafter LCM08). The LCM74 map had a minimum mapping unit smaller (0.375 ha) than the LCM08 one (1.56 ha).

To harmonize spatial data for land cover change classification, the polygons of the LCM74 map with area smaller than the minimum mapping unit of the LCM08, were merged with the neighboring polygon with the largest area. Both maps were reclassified according to a land cover classification system based on 8 classes compatible with Corine Land Cover (EEA 2007). An additional class includes water bodies and other minor land-uses.

Landscape analysis

Mathematical morphology is a framework for analyzing the shape and form of objects (Soille 2003), that recently has been used in landscape ecology and environmental geography applications (Soille and Vogt 2009). The Morphological Spatial Pattern Analysis (MSPA) implements a series of image processing routines to identify features that are relevant to the diagnosis of the structural connectivity of land cover classes. The individual classes identified by MSPA are: (i) core areas, (ii) islets, (iii) perforations, (iv) edges, (v) loops, (vi) bridges, and (vii) branches. These categories cover a wide range of spatial elements mainly used, so far, to classify and map the structural connectivity of forest habitats (Elbakidze et al. 2011, Saura et al. 2011) and green infrastructures (Kuttner et al. 2013).

In order to identify and quantify variations in the spatial pattern of the Roman Campagna landscape between 1974 and 2008, we applied MSPA by means of the GUIDOS software (http://forest.jrc.ec.europa.eu/download/software/guidos/, Version 2.3, JRC, Ispra) to the 8 land-use classes mapped in the study area: (i) arable land, (ii) heterogeneous agricultural areas, (iii) vineyards, (iv) olive groves, (v) pastures, (vi) forests and semi-natural areas, (vii) continuous urban fabric, and (viii) discontinuous urban fabric.

Raster binary maps of each land-use class for the years 1974 and 2008 were processed to segment each given land use type into the mutually exclusive structural MSPA categories earlier described. MSPA classification routine starts by identifying core areas, based on user-defined rules for criteria linking connectivity and edge width (Soille and Vogt 2009). Connectivity was set for a node pixel to its adjacent neighboring pixels by considering 8 neighbors (a pixel border and a pixel corner in common). Edge width, that is the thickness of the pixels at the boundary of a core area, was set 100 m large. According to Vogt et al. (2009), this value for edge thickness is considered suitable for core-edge discrimination within a vast range of animal and plant species.

Statistical analysis

Variations in land-use and landscape structure observed between 1974 and 2008 in the study area were explored by parametric (Pearson) and non-parametric (Spearman) correlation coefficients testing for significance at $p < 0.05$. Pearson and Spearman correlation analysis were used with the aim to identify both linear and non-linear relationships among variables. Pair-wise correlations between the changes in the seven landscape structure categories observed between 1974 and 2008 for investigated classes were also carried out using Pearson and Spearman correlation tests testing for significance at $p < 0.05$.

To explore the evolution of landscape structure in Rome, a Multiway Factor Analysis (MFA) was applied to the matrix composed of the percentages of 7 GUIDOS landscape structure categories observed in Rome for the years under study (1974, 2008). The MFA is a generalization of the Principal

Table 1 - Variations in land use classes during 1974-2008 in Rome district.

1974					2008					
	Arable land	Heterogeneous agricultural areas	Vineyards	Olive groves	Pastures	Forests and semi-natural areas	Continuous urban fabric	Discontinuous urban fabric	Others land uses	Total
1 Arable land	**33.89**	0.60	0.31	0.48	2.87	2.30	3.33	4.09	1.29	49.15
2 Heterogeneous agricultural areas	1.25	**0.63**	0.16	0.63	0.38	0.27	0.62	0.78	0.23	4.95
3 Vineyards	1.01	0.32	**1.10**	0.22	0.16	0.07	0.23	0.54	0.17	3.81
4 Olive groves	0.36	0.07	0.04	**0.74**	0.06	0.10	0.04	0.09	0.05	1.56
5 Pastures	5.04	0.16	0.10	0.12	**3.32**	4.30	1.25	1.93	0.64	18.36
6 Forests and semi-natural areas	1.00	0.07	0.04	0.10	0.37	**4.20**	0.12	0.14	0.07	6.10
7 Continuous urban fabric	0.41	0.04	0.01	0.01	0.55	0.15	**9.03**	1.66	0.59	12.45
8 Discontinuous urban fabric	0.06	0.01	0.00	0.01	0.05	0.05	0.16	**1.00**	0.03	1.36
9 Others land uses	0.11	0.02	0.01	0.02	0.07	0.11	0.23	0.14	**1.54**	2.26
Total	43.13	1.91	1.77	2.33	7.82	11.55	15.79	11.09	4.61	**1.00**

Component Analysis (PCA) whose goal is to analyze variables collected on the same set of observations. The general objectives of MFA are (i) to analyze the relationship between the different data sets, (ii) to combine them into a common structure called 'compromise' which is then analysed via PCA to reveal the common structure between the observations and, finally, (iii) to project each of the original data sets into the compromise to analyse communalities and discrepancies (Lavit et al. 1994). Points placed close each other in the factorial plane, generated by the two main MFA axes (Coppi and Bolasco 1989), indicate spatial association, while points placed far each other indicate spatial segregation (Salvati et al. 2012).

Results

Land-use Changes in Rome (1974-2008)

To develop a comprehensive understanding of land use change dynamics in Roman Campagna district over nearly 35 years, we analysed land-use flows between urban and rural areas (Tab. 1). Looking at land-use classes transitions from 1974 to 2008 the most remarkable land-use dynamics are related to (i) land uptake by built up areas from arable lands, pastures and heterogeneous agricultural areas; discontinuous urban fabric consumed nearly 7.6% of the study area, formerly occupied by rural lands in 1974, while continuous urban fabric took nearly 5.6%; (ii) withdrawal of farming with woodland and semi-natural areas creation mainly on pastures and arable land (nearly 6% of the study area) and internal conversion in farmlands, notably conversion of heterogeneous agricultural lands into arable land occurring in more than 1% of the area. Some transition cases (e.g. urban fabric converted into pastures and forests and semi-natural areas), that could seem false detected cases, can be explained by the significant and rapid changes especially in productive areas that have resulted in underutilized land colonized by vegetation (Lafortezza et al. 2004).

Landscape structure dynamics

Changes in the spatial pattern of single land-use classes in the period from 1974 to 2008 are summarized in Tab. 2, which reports the change in landscape structure, as percentage of the study area, together with the observed surface variation for each land-use class. A pattern of change common to all land-use classes is the decrease in the presence of landscape components in the form of 'core areas' and the increase of land fragments in the form of 'islets'.

Arable land, heterogeneous agricultural land, pastures and vineyards have been significantly fragmented, loosing respectively 14.4%, 18.2%, 34.1%, 5.6% of the area occupied by 'core patches' in 1974.

Table 2 - Variations in land use and landscape structure during 1974-2008 in Rome district.

Land use	% change (1974 - 2008)							
	Total surface	Core	Islet	Perforation	Edge	Loop	Bridge	Branch
Arable land	-6.02	-14.44	2.42	-0.02	-3.04	1.38	11.42	2.29
Heterogeneous agricultural areas	-3.04	-18.20	53.41	0.00	-21.24	-0.62	-3.47	-9.88
Vineyards	-2.04	-34.12	27.70	-1.68	-14.57	0.32	20.13	2.21
Olive groves	0.77	-5.60	16.87	0.00	-11.99	-0.52	4.30	-3.06
Pastures	-10.54	-22.80	35.00	-0.12	-13.53	1.24	-0.55	0.75
Forests an semi-natural areas	5.45	2.29	0.65	0.14	-4.51	2.49	0.97	-2.30
Continuous urban fabric	3.34	-6.39	0.77	-1.21	-0.44	1.96	4.45	0.27
Discontinuous urban fabric	9.73	-12.60	15.06	0.10	-17.26	4.76	6.49	3.45
Pearson correlation test	-	0.55	-0.46	0.05	0.08	0.61	0.02	0.11
Spearman rank correlation test	-	0.67	-0.60	0.50	0.05	0.62	0.19	0.07

Also the increased area of olive groves occurred by addition of 'islets' (nearly 16.9%) and associated connectors structures (4.3%), while the area of core patches decreased by 5.6%. The increase of forest and semi-natural areas was mainly related to the closing of clearings with an increase of 'core areas' (nearly 2.3%) and connectors structures emanating from the same core connected component (nearly 2.5%) and, consequently, a reduction of 'edge' areas (nearly -4.5%). At the same time, urban development had a severe impact on landscape composition leading to (i) a remarkable land uptake by fragments of residential settlements in the class 'discontinuous urban fabric', amounting to nearly 15% of the study area, and (ii) an increase of linear infrastructure (loop and branches) connecting discontinuous urban fabric (11.3% of the study area) and continuous urban fabric (+6.4% of the study area).

Correlation between land-use changes and landscape structure

Even if significant correlations between percent surface area of each cover class and morphological spatial pattern classes percent area were not found at the landscape level, pair-wise correlations among structural classes (Tab. 3) indicate a negative non-parametric relationship among 'islet' surface area and 'core', 'edge', 'loop'. 'Bridge' showed a positive, linear correlation with 'perforation' but also with 'branch' surface area using both linear Pearson correlation and non-parametric Spearman correlation.

Summarizing landscape patterns by way of the MFA

The MFA extracted two axes explaining respectively 43.9% and 22.9% of the total variance for a total of 66.8% cumulated variance. Factor loadings are shown in Table 4 by year. The structural landscape classes mostly associated, during all the study period, to the MFA first axis were 'core',

Table 4 - Loadings of the Multiway Factor Analysis by year and landscape structure category.

Variable	1974		2008	
	Axis 1	Axis 2	Axis 1	Axis 2
Core	-0.92	0.27	-0.79	-0.11
Islet	0.75	-0.45	0.94	-0.06
Perforation	-0.83	0.13	-0.73	-0.37
Edge	0.20	0.95	-0.67	0.53
Loop	0.08	-0.88	-0.25	-0.40
Bridge	-0.64	-0.59	-0.84	-0.07
Branch	0.93	-0.01	0.08	0.37
Explained variance	43.93		22.88	
Cumulated variance		66.81		

'perforation', 'edge', 'loop', and 'bridge'. 'Islet' and 'branch', instead, were associated with axis 2 (Fig. 4). The first axis illustrates an urban-rural gradient discriminating different settlement forms (continuous vs discontinuous settlements). The second axis segregates urban land-use classes (positive values) from natural and semi-natural land use classes (negative values).

Different trends in morphological spatial pattern transformation for different land-use classes clearly stand out from point positions into the factorial plan that relate variables (i.e. structural landscape classes) and observations (i.e. land-use classes). The coordinates of structural classes points indicate a strong segregation between 'core' and 'bridge' for continuous urban fabric and 'core' and 'branch', 'islet' and 'perforation' for discontinuous urban fabric, while a spatial segregation between structural classes are not clearly recognizable for farmland (Fig. 5).

Discussion

The physical pattern of low-density expansion, typical of urban sprawl (EEA 2006), is clearly figured out, under the investigated case study. A remarkable growth of patchy and scattered areas was observed for all land-use classes (Tab. 2). Occlusion and shredding of portions of open land, interconnected in earlier time periods, highlights a trend towards fragmentation of the peri-urban agricultural land of the Roman Campagna. The processes of soil sealing that affected Rome's plain and the neighbouring areas in recent decades is certainly the main but not the only cause of the progressive patchiness. Urban development further away from the urban fringe was accelerated by the growth of linear infrastructures in response to needs of transportation and mobility. This is well captured by the increase of linear connectors in discontinuous urban fabric areas. This trend was observed in almost all other land-use classes determining a juxtaposition of different land uses and a generalized landscape mix (Zambon et al. 2015).

Table 3 - Pair-wise correlation between the changes in the seven landscape structure categories observed during 1974-2008 in Rome district (n = 8 land use classes); bold indicates significant correlation at p < 0.05.

Variable	Islet	Perforation	Edge	Loop	Bridge	Branch
Pearson correlation coefficient						
Core	-0.61	0.51	0.52	0.30	-0.50	-0.20
Islet		0.03	-0.87	-0.52	-0.28	-0.56
Perforation			-0.16	0.19	**-0.64**	-0.30
Edge				0.21	0.15	0.36
Loop					0.04	0.61
Bridge						**0.65**
Spearman rank correlation coefficient						
Core	**-0.74**	0.65	0.50	0.43	-0.17	-0.31
Islet		-0.28	**-0.76**	**-0.76**	-0.33	-0.21
Perforation			-0.19	0.32	-0.40	-0.20
Edge				0.36	0.24	0.05
Loop					0.31	0.60
Bridge						**0.74**

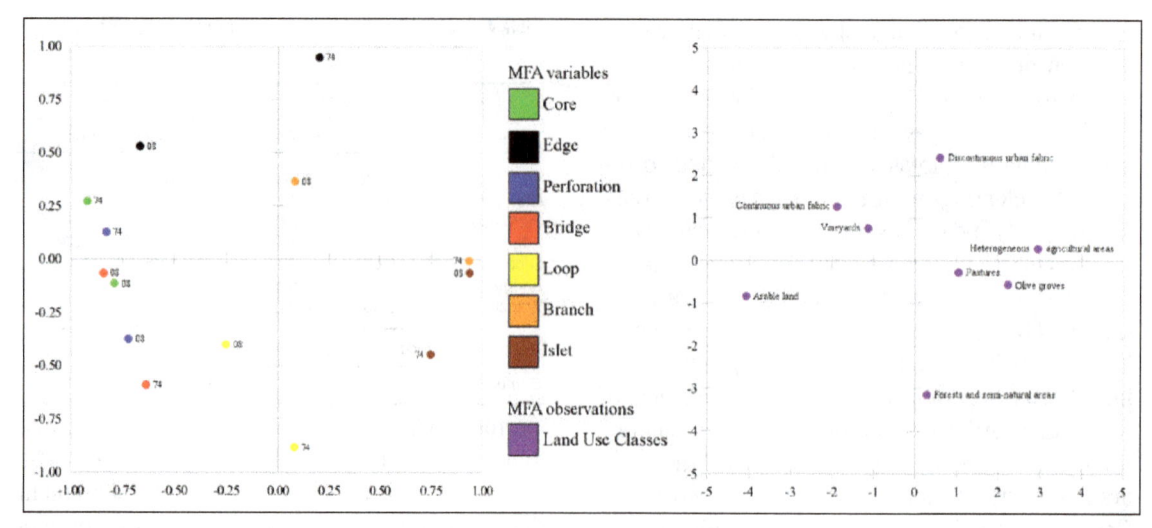

Figure 4 - MFA correlation map of variable (left) and MFA observation factor score plot (right).

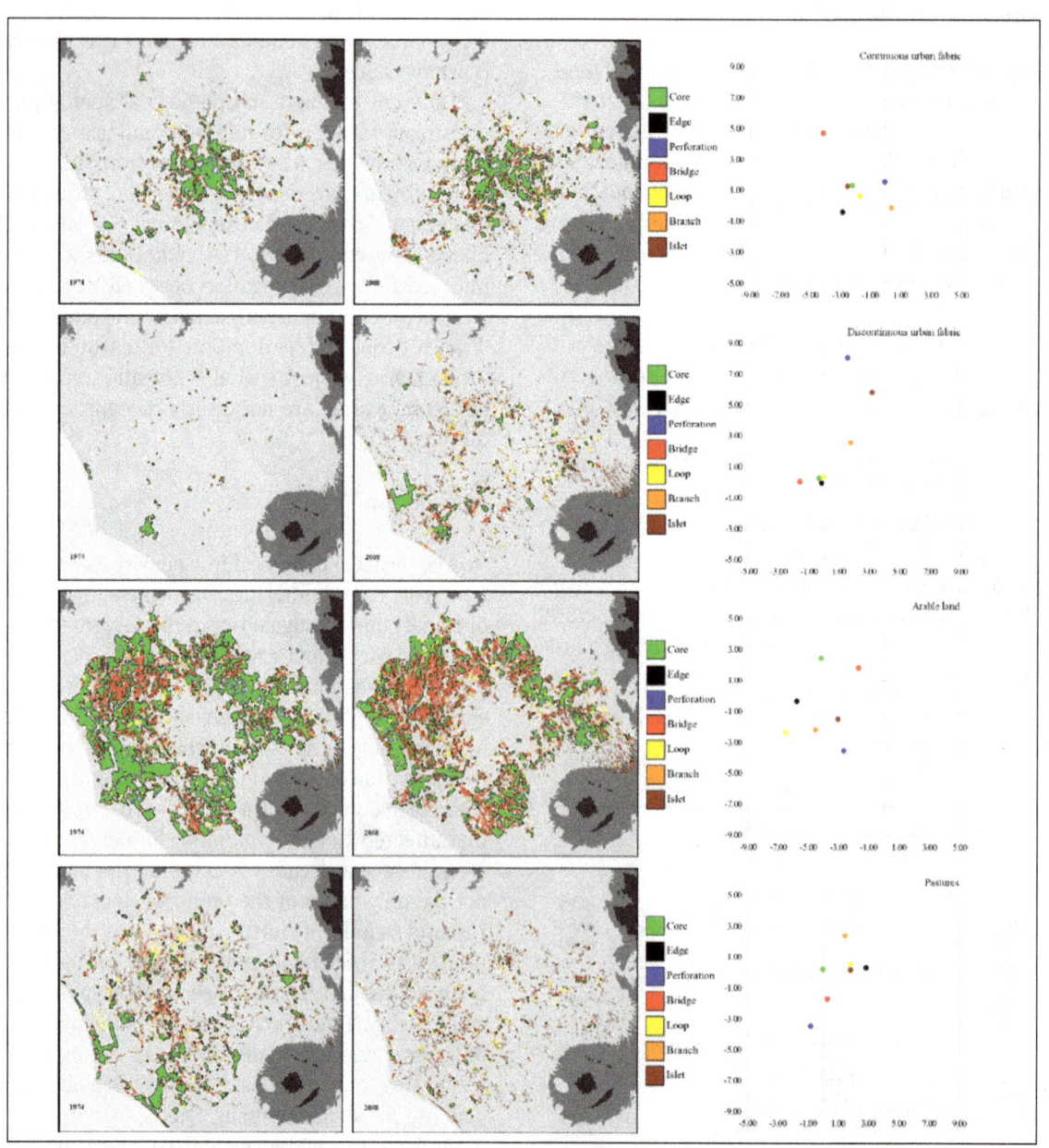

Figure 5 - MSPA pattern classes in 1974 and 2008 for continuous urban fabric, discontinuous urban fabric, arable land and pastures and the corresponding MFA factorial plan.

Landscape fragmentation of farmland, tracked by the decline in the surface of arable land, heterogeneous agricultural areas and pastures, and of core areas therein, is the combined effect of urbanization, agricultural intensification and land abandonment. This corroborates the data presented by Salvati et al. (2012). While arable land is still the main land-use, traditional farming (pastures and heterogeneous agricultural areas) decreased significantly throughout the study area. The significant increase of forest and semi-natural areas, typical of many landscapes in Italy (Corona et al. 2008, 2012), showed a peculiar spatial pattern (creation of 'islets' and 'core areas' with associated connectors) caused by the withdrawal of farming in arable land and heterogeneous agricultural areas (Barbati et al. 2013, Biasi et al. 2015, Ferrara et al. 2015). However, only absolute values of land-use classes gain or loss are not enough to explain landscape transformation. Each land-use class changes its morphological structure in different way, as shown by MFA analysis.

Continuous urban fabric landscape structure tends to the consolidation (spatial association for almost all MSPA classes into MFA factorial plan) while discontinuous urban fabric morphological pattern tends to expansion and spatial segregation between 'core' and 'islet' patches. Islet patches expands the wildland-urban interface, with the consequent significant increase of both wildfire risk and danger (Moreira et al. 2011). Arable land and pastures patterns of change are more complex. In the former case, a progressive erosion of cropland influenced negatively the expansion of connective fragments; in the latter case, a process of transformation of large patches into small patches was observed. Disappearance of connection patches is particularly relevant in processes of landscape fragmentation.

Conclusions

Land fragmentation is the most impressive change due to peri-urban development in Rome as well as in other Mediterranean cities (Chorianopoulos et al. 2010, Munafò et al. 2010). This kind of knowledge can be effectively used to steer socioeconomic policies towards sustainable urban growth (Salvati and Ferrara 2013, Smiraglia et al. 2014, Tombolini and Salvati 2014, Quatrini et al. 2015). Rome's municipality has made some efforts to protect peri-urban rural landscapes: the designation of protected areas has had positive effects for conservation and expansion of core forest areas (Barbati et al. 2013, Colantoni et al. 2015, Zitti et al. 2015). Rome is still surrounded by a large agricultural region, historically linked with the city. The present study shows that a multi-temporal MSPA can effectively highlight key changes in the landscape structure of a traditional and iconic rural landscape, like the Roman Campagna. This knowledge is a relevant base to inform integrated and co-operative spatial planning across the urban gradient, with the final objective to strengthen and modernize the role of farming activities in peri-urban areas, responding to pressures and opportunities derived from proximity to urban areas. Farming practices that promote conservation of heterogeneous and small-scale farming structure, punctuated with natural elements, are prioritized (Zasada 2011).

These "new" demands (i.e. conservation of cultural rural landscape and recreation in addition to food production) linked to peri-urban agriculture are not yet acknowledged by the city planning and land management tools. The General Rome City Plan, approved in March 2006, aims to support the peri-urban agriculture, but the lack of regional or national laws regulating by year and municipality the maximum surface of agricultural land which can be converted to space for housing or commercial development is a major bottleneck that negatively impacts on the surrounding landscape, as observed for other urban regions of the Mediterranean (Paul and Tonts 2005, Christopoulou et al. 2007, Jomaa et al. 2008, Cimini et al. 2013). Furthermore, policies of urban containment provide only a prerequisite for the preservation of open spaces in the Roman Campagna. Rural development programs (e.g. environmental and land management incentives) need to be locally tailored to face the gradual simplification and fragmentation of peri-urban landscapes. This approach can be applied to updated digital land cover maps, with the final aim to assess the impact of integrated planning measures.

References

Barbati A., Corona P., Salvati L., Gasparella L. 2013 - *Natural forest expansion into suburban countryside: gained ground for a green infrastructure?* Urban Forests and Urban Greening 12: 36-43.

Biasi R., Colantoni A., Ferrara C., Ranalli F., Salvati L. 2015 - *In-between Sprawl and Fires: long-term Forest Expansion and Settlement Dynamics at the Wildland-Urban Interface in Rome, Italy*. International Journal of Sustainable Development and World Ecology 22 (6): 467-475.

Cakir G., Ün C., Baskent E.Z., Köse S., Sivrikaya F., Keleş S. 2008 - *Evaluating urbanization, fragmentation and land use/land cover change pattern in Istanbul city, Turkey from 1971 to 2002*. Land Degradation and Development 19: 663-675.

Ceccarelli T., Bajocco S., Perini L., Salvati L. 2014 - *Urbanisation and Land Take of High Quality Agricultural Soils: Exploring Long-term Land Use Changes and Land Capability in Northern Italy*. International Journal of Environmental Research 8 (1): 181-192.

Chorianopoulos I., Pagonis T., Koukoulas S., Drymoniti S. 2010 - *Planning, competitiveness and sprawl in the Mediterranean city: The case of Athens.* Cities 27: 249-259.

Christopoulou O., Polyzos S., Minetos D. 2007 - *Peri-urban and urban forests in Greece: obstacle or advantage to urban development?* Journal of Environmental Management 18: 382-395.

Cimini D., Tomao A., Mattioli W., Barbati A., Corona P. 2013 - *Assessing impact of forest cover change dynamics on high nature value farmland in Mediterranean mountain landscape.* Annals of Silvicultural Research 37 (1): 29-37.

Colantoni A., Mavrakis A., Sorgi T., Salvati L. 2015 - *Towards a 'polycentric' landscape? Reconnecting fragments into an integrated network of coastal forests in Rome.* Rendiconti Accademia Nazionale dei Lincei 26 (3): 615-624.

Colantoni A., Grigoriadis E., Sateriano A., Venanzoni G., Salvati L. 2016 - *Cities as selective land predators? A Lesson on Urban Growth, (Un)effective planning and Sprawl Containment.* Science of the Total Environment 545-546: 329-339.

Coppi R., Bolasco S. 1989 - *Multiway data analysis.* North Holland, Amsterdam.

Corona P., Calvani P., Mugnozza Scarascia G., Pompei E. 2008 - *Modelling natural forest expansion on a landscape level by multinomial logistic regression.* Plant Biosystems 142: 509-517.

Corona P., Barbati A., Tomao A., Bertani R., Valentini R., Marchetti M., Fattorini L., Perugini L. 2012 - *Land use inventory as framework for environmental accounting: an application in Italy.* iForest 5: 204-209.

Elbakidze M., Angelstam P., Andersson K., Nordberg M., Pautov Y. 2011 - *How does forest certification contribute to boreal biodiversity conservation? Standards and outcomes in Sweden and NW Russia.* Forest Ecology and Management 262: 1983-1995.

European Environment Agency 2006 - *Urban sprawl in Europe – The ignored challenge.* EEA Report no. 10. Office for Official Publications of the European Communities, Luxembourg.

European Environment Agency 2007 - CLC2006 technical guidelines. EEA Technical report no 17. Office for Official Publications of the European Communities, Luxembourg.

Ferrara A., Salvati L., Sabbi A., Colantoni A. 2014 - *Urbanization, Soil Quality and Rural Areas: Towards a Spatial Mismatch?* Science of the Total Environment 478: 116-122.

Ferrara A., Salvati L., Sateriano A., Carlucci M., Gitas I., Biasi R. 2015 - *Unraveling The 'Stable' Landscape: a Multi-factor Analysis of Unchanged agricultural and Forest land (1987-2007) in a Rapidly-expanding Urban Region.* Urban Ecosystems 19 (2): 835-848.

Jomaa I., Auda Y., Abi Saleh B., Hamzé M., Safi S. 2008 - *Landscape spatial dynamics over 38 years under natural and anthropogenic pressures in Mount Lebanon.* Landscape and Urban Planning 87: 67-75.

Kasanko M., Barredo J.I., Lavalle C., McCormick N., Demicheli L., Sagris V., Brezger A. 2006 - *Are European Cities Becoming Dispersed? A Comparative Analysis of Fifteen European Urban Areas.* Landscape and Urban Planning 77: 111-130.

Kazemzadeh-Zow A., Zanganeh Shahraki S., Salvati L., Neisani Samani N. 2016 - *A Spatial Zoning Approach to Calibrate and Validate Urban Growth Models.* International Journal of Geographic Information Systems 31 (4): 763-782.

Kuttner M., Hainz-Renetzeder C., Hermann A., Wrbka T. 2013 - Borders without barriers – *Structural functionality and green infrastructure in the Austrian–Hungarian transboundary region of Lake Neusiedl.* Ecological Indicators 31: 59-72.

Lafortezza R., Sanesi G., Pace B., Corry R.C., Brown R.D. 2004 - *Planning for the rehabilitation of brownfield sites: a landscape ecological perspective.* In: Donati A., Rossi C., Brebbia C.A. editors. Brownfield Sites II. Assessment, Rehabilitation and Development. WIT Press, Southampton, pp. 21–30.

Lavit C., Escoufier Y., Sabatier R., Traissac P. 1994 - *The ACT (STATIS) method.* Computational Statistical and Data Analysis 18: 97–119.

Longhi C., Musolesi A. 2007 - *European cities in the process of economic integration: towards structural convergence.* Annals of Regional Science 41: 333-351.

Mavrakis A., Papavasileiou C., Salvati L. 2015. *Towards (Un)sustainable Urban Growth? Climate aridity, land-use changes and local communities in the industrial area of Thriasio Plain, Greece.* Journal of Arid Environment 121: 1-6.

Metzger J.P., Muller E. 1996 - *Characterizing the complexity of landscape boundaries by remote sensing.* Landscape Ecology 11: 65–77.

Metzger J.P., Decamps H. 1997 - The structural connectivity threshold: an hypothesis in conservation biology at the landscape scale. Acta Oecologica 18: 1–12.

Moreira F., Viedma O., Arianoutsou M., Curt T., Koutsias N., Rigolot E., Barbati A., Corona P., Vaz P., Xanthopoulos G., Mouillot F., Bilgili E. 2011 - *Landscape-wildfire interactions in southern Europe: implications for landscape management.* Journal of Environmental Management 92: 2389-2402.

Munafò M., Norero C., Salvati L. 2010 - *Soil sealing in the growing city: a survey in Rome, Italy.* Scottish Geographical Journal 126 (3): 153-161.

Munafò M., Salvati L., Zitti M. 2013 - *Estimating soil sealing at country scale – Italy as a case study.* Ecological Indicators 26: 36-43.

Paul V., Tonts M. 2005 - *Containing urban sprawl: trends in land use and spatial planning in the Metropolitan Region of Barcelona.* Journal of Environmental Planning and Management 48: 7-35.

Pili S., Grigoriadis E., Carlucci M., Clemente M., Salvati L. 2017 - *Towards Sustainable Growth? A Multi-criteria Assessment of (Changing) Urban Forms.* Ecological Indicators 76: 71-80.

Ploeg van der J.D. 2008 - *The New Peasantries. Struggles for Autonomy and Sustainability in an Era of Empire and Globalization.* Earthscan, London.

Quatrini V., Barbati A., Carbone F., Giuliarelli D., Russo D., Corona P. 2015 - *Monitoring land take by point sampling: Pace and dynamics of urban expansion in the Metropolitan City of Rome.* Landscape and Urban Planning 143: 126-133.

Salvati L. 2013 - *Urban expansion and high-quality soil consumption - an inevitable spiral?* Cities 31: 349-356.

Salvati L., Ferrara C. 2013 - *Do changes in vegetation quality precede urban sprawl?* Area 45 (3): 365-375.

Salvati L., Perini L., Bajocco S., Sabbi A. 2012 - *Climate aridity and land use change: a regional-scale analysis.* Geographical Research 50 (2): 193-203.

Salvati L., Quatrini V., Barbati A., Tomao A., Mavrakis A., Serra P., Sabbi A., Merlini P., Corona P. 2017 - *Soil occupation efficiency and landscape conservation in four Mediterranean urban regions.* Urban Forestry and Urban Greening 20: 419-427.

Salvati L., Zitti M., Sateriano A. 2013 - *Changes in the City Vertical Profile as an Indicator of Sprawl: Evidence from a Mediterranean Region.* Habitat International 38: 119-125.

Saura S., Vogt P., Velázquez J., Hernando A., Tejera R. 2011 - *Key structural forest connectors can be identified by combining landscape spatial pattern and network analyses.* Forest Ecology and Management 262: 150-160.

Savo V., De Zuliani E., Salvati L., Perini L., Caneva G. 2012 - *Long-term changes in precipitation and temperature patterns and their possible impact on vegetation (Tolfa-Cerite area, central Italy).* Applied Ecology and Environmental Research 10 (3): 243-266.

Schneider A., Woodcock C.E. 2008 - *Compact, dispersed, fragmented, extensive? A comparison of urban growth in twenty-five global cities using remotely sensed data, pattern metrics and census information.* Urban Studies 45: 659-692.

Serra P., Vera A., Tulla A.F., Salvati L. 2015 - *Beyond urban-rural dichotomy: exploring socioeconomic and land-use processes of change in Spain (1991-2011).* Applied Geography 55: 71-81.

Shoard M. 2002 - *Edgelands.* In: Jenkins J. Editor. Remarking the Landscape: The Changing Face of Britain. London: Profile Books.: 117-146.

Smiraglia D., Rinaldo S., Ceccarelli T., Bajocco S., Salvati L., Perini L. 2014 - *A Cost-Effective Approach For Improving The Quality of Soil Sealing Change Detection From Landsat Imagery.* European Journal of Remote Sensing 47: 805-809.

Smiraglia D., Ceccarelli T., Bajocco S., Perini L., Salvati L. 2015 - *Unraveling landscape complexity: land use/land cover changes and landscape pattern dynamics (1954-2008) in an agro-forest region of northern Italy.* Environmental Management 56 (4): 916-932.

Soille P. 2003 - *Morphological image analysis: principles and applications.* Springer-Verlag, Berlin.

Soille P., Vogt P. 2009 - *Morphological segmentation of binary patterns.* Pattern Recognition Letters 30: 456-459.

Tombolini I., Salvati L. 2014 - *A Diachronic Classification of Peri-urban Forest Land based on Vulnerability to Desertification.* International Journal of Environmental Research 8 (2): 279-284.

Turok I., Mykhnenko V. 2007 - *The trajectories of European cities, 1960-2005.* Cities 24: 165-182.

Vogt P., Ferrari J.R., Todd R., Lookingbill R.H., Gardner K.H., Ostapowicz K. 2009 - *Mapping Functional Connectivity.* Ecological Indicators 9: 64-71.

Zambon I., Sabbi A., Shuetze T., Salvati L. 2015 - *Exploring Forest 'FringeScapes': Urban Growth, Society and Swimming Pools as a Sprawl Landmark in Coastal Rome.* Rendiconti Accademia Nazionale dei Lincei 26 (2): 159-168.

Zasada I. 2011 - *Multifunctional peri-urban agriculture—A review of societal demands and the provision of goods and services by farming.* Land Use Policy 28: 639-648.

Zitti M., Ferrara C., Perini L., Carlucci M., Salvati L. 2015 - *Long-term Urban Growth and Land-use Efficiency in Southern Europe: Implications for Sustainable Land Management.* Sustainability 7 (3): 3359-338

An approach to public involvement in forest landscape planning in Italy: a case study and its evaluation

Isabella De Meo [1], Fabrizio Ferretti [2*], Alessandro Paletto [2], Maria Giulia Cantiani [3]

Abstract - In Italy, in the last decade, there have been both new social requests and an ever-increasing sensitivity towards the multiplicity of values attributed to forests. This has led to a profound revision of the structure of forest planning. This paper illustrates the planning system, characterized by a hierarchical approach, focusing on the upper level, that is Forest Landscape Management Plan (FLMP). At this level of planning, attention to the different needs and targets expressed by the population is considered of strategic importance and thus requires a participative attitude. In the first part of the paper the authors show the approach currently used in forest landscape planning, through a case study carried out in a rural area of the Apennine mountains, focusing on the method established for the process of participation. In the second part, after describing the methodology followed to identify a set of criteria for success, the quality of participation in the case study is analyzed.

Keywords - Forest Landscape Management Planning; public participation; evaluation; success criteria

Introduction

While the social dimension has been developing as an integral part of sustainability, there has been a gradual increase in the involvement of local communities in the decision-making process regarding environmental matters (FAO-ECE-ILO 2000, Appelstrand 2002, European Commission 2003, Lee and Abbot 2003).

As for the forestry sector, the adoption of participatory planning has been seen from the outset as an instrument and an opportunity to take into consideration social sustainability in order to enhance sustainable forest management (FAO-ECE-ILO 2000, Kangas et al. 2006, Ananda 2007). This reflects a clear shift towards a post-productivist approach to natural resources management (Appelstrand 2002, Farcy and Devillez 2005, Cantiani et al. 2013) and shows an ever-increasing need for taking into account the multiple uses and multiple values of forests (Farrel et al. 2000, O'Brien 2003, Leskinen 2006, Schmithüsen 2007).

In Italy, such a need has led, in the last decade, to a profound revision of the very structure of forest planning (Cantiani et al. 2010, Ferretti et al. 2011, Paletto et al. 2011, Paletto et al. 2015a) which is now based on a hierarchical approach (Ferretti et al. 2011, Paletto et al. 2015a). It introduces the Forest Landscape Management Plan (FLMP), a higher level to the existing traditional Forest Unit Management Plan (FUMP), which pertains to single ownership. The FLMP includes all non-urban and non-agricultural land, mainly forests and pastures, referring to a homogeneous area from a geomorphological point of view, irrespective of ownership boundaries.

The theoretical framework has been provided thanks to the activities of workgroups made up of researchers and practitioners. Their work has been carried out within a long term national research project promoted by the Ministry for Agriculture and Forestry Policies, together with most Regional Agriculture and Forestry Administrations (Ferretti et al. 2011, Paletto et al. 2011).

The methodology of forest landscape planning involves a series of interdependent phases, according to a logical procedure summarized in Fig. 1.

The landscape scale was deemed the most suitable for considering long term general interests, such as soil protection, nature and landscape conservation, while taking into account local community needs (Bettelini et al. 2000, Cantiani 2012).

The FLMP was thought of as an instrument entrusted with two tasks: providing management guidelines for the subordinate FUMPs and integrating and coordinating with other types of plans or

---

[1] Consiglio per la ricerca in agricoltura e l'analisi dell'economia agraria Research Centre for Agriculture and Environment, Firenze (Italy)
[2] Consiglio per la ricerca in agricoltura e l'analisi dell'economia agraria Research Centre of Forestry and Wood, Arezzo (Italy)
[3] Università degli Studi di Trento, Department of Civil, Environmental, and Mechanical Engineering, Trento (Italy)
*fabrizio.ferretti@crea.gov.it

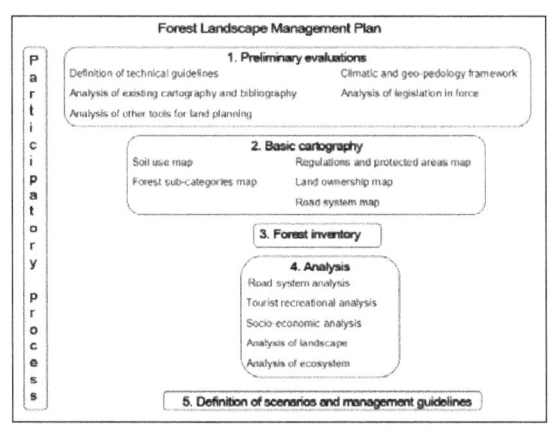

Figure 1 - Structure of a Forest Landscape Management Plan.

projects existing in the same area.

At this level of planning, care about the different needs and targets expressed by the population was considered of strategic importance.

With regard to this, a workgroup on participation in forest planning, where the authors were directly involved, was set up within the above-mentioned research project. Based on foregoing experience from several case studies and on a careful analysis of the literature a methodological approach to participation in forest landscape planning was outlined.

The main questions and concerns of the workgroup revolved around the following issues:

a) feasibility of public involvement in the decision-making process to build up a planning process well-rooted in the socio-economic context (Cantiani 2012);
b) identification of the most advisable level of involvement (Bettelini et al. 2000);
c) understanding of suitable means to reach and involve the stakeholders in the highly rural contexts typical of the Italian mountains. Here, generally, the actors more directly in charge of the management of the land have only a marginal role in the local social network (Paletto et al. 2012, Cantiani et al. 2013);
d) provision of opportunities for enhancing people's awareness of the values of their own territory (De Meo et al. 2011);
e) design of a flexible procedure, easily adaptable and reproducible in other rural contexts (Cantiani 2012, Paletto et al. 2015a).

In this paper, among the case studies realised throughout the research project, we refer to the one carried out in a hilly and mountainous district of Southern Italy, the *Comunità Montana* Collina Materana; the Comunità Montana is the Italian administrative body that coordinates the municipalities located in the mountainous areas and is responsible for administration and economic development. This was actually the first Forest Landscape Management Plan carried out in Southern Italy and one of the first ever realised, on this scale, in the entire country. According to the project philosophy, the main purpose of the plan was that of defining medium/long term natural resources management strategies, able to guide a sustainable and harmonious development of the area.

In the first part of the paper we describe the case study (the FLMP of the Comunità Montana Collina Materana), focusing on the methodological approach established for the process of participation. In the second part, we analyse the quality of participation in our case study, after describing the approach followed to identify a group of success criteria, deemed particularly relevant in relation to our concerns.

Materials and methods

The planning context

The Comunità Montana Collina Materana (40°29'30" N; 16°09'0" E) is located in the Basilicata Region and occupies a surface of 60'784 ha (Fig. 2). This case is one of a typical rural area, with few industrial activities and generally poor infrastructure. The population density is low (19.8 inhabitants/km^2) in comparison to other regions of Italy and to the national density (201 inhabitants/km^2). The primary sector plays an essential role in the economic structure of the Comunità Montana, involving 24% of the active population (national average about 8%). Agricultural activities, which are mostly extensive, also include the cultivation of high quality products, such as durum wheat, used for the production of "pasta".

The area covered by forest is 22'221 ha, corresponding to 36.5% of the territory and the main forest types present are: forests of Turkey oak (*Quercus cerris* L.), downy oak (*Quercus pubescens* Willd.), Holm oak (*Quercus ilex* L.), Hungarian oak

Figure 2 - Basilicata Region and Comunità Montana Collina Materana.

Figure 3 - The typical gully landscape of the Materana district.

(*Quercus farnetto* Ten.) and reforestation of Aleppo pine (*Pinus halepensis* Mill.). The large diversity of forest types is due to the great variability in morphology, altitude, and lithology of the area (Fig. 3).

About 5% of the territory falls within a protected area (Regional Natural Park of Gallipoli Cognato). 36% of forest land is public property (Municipalities, Regional Park) and 64% is privately owned, often forming part of a larger agro-forest enterprise.

The surface of pastureland is only 272 ha, despite the importance of husbandry, which relies on 1'500 heads of cattle (without considering smaller livestock) (Argenti et al. 2008).

Forest land is still very important today for the economy of the local community, mainly in relation to firewood production and the supply of pasture resources. Forest management, is strongly characterized and influenced by grazing in the forests (Fig. 4). This has been a common practice since the Middle Ages, as it was the case in large areas of Europe (Rotherham 2007), and has played a major role in the socio-economic organisation of Collina Materana. It has, in fact, always helped to ensure the survival of the population when the conditions were not favourable to forage production in pastures and meadows, due to the Mediterranean climate. At the same time, however, grazing in the forest has also posed serious constraints on forest management, interfering with other functions of the forests, in particular that of protection. Indeed, the continuous overgrazing, which causes unfavourable conditions for the vegetation, may result in the reduction of species, a decline in wood production, soil compaction and damage caused by animal tracks. It goes without saying that it may be one of the main sources of conflict in this area.

The FLMP of the Comunità Montana Collina Materana was carried out between 2006 and 2007, coherent with the theoretical framework of reference (Argenti et al. 2008). The plan came into force in 2009 and was due to last 20 years. Since it was the first experimental plan, special attention was paid to the development and testing of the participatory approach, which should serve as a model for the following planning activity (Cantiani 2012).

The participatory process

From the beginning of planning, the participatory process took place along with the other planning activities. This process consisted of a series of steps: a preliminary evaluation, the establishment of a "participatory support group", the definition of the participation method, a stakeholder analysis and the first stage of consultation, the SWOT analysis and the second stage of consultation.

The participatory process was coordinated and followed in all its steps by the authors in person.

The participatory process: preliminary evaluation

In order to verify the real applicability of the participation process and to structure it properly, several meetings were organised between the responsible parties of the Plan (National Institute of Agricultural Economics – INEA - and the Basilicata Region) and the planning team. The objective was to assess the human and financial resources available, as well as the commitment required to activate and nurture the process. In this regard it was necessary to clearly evaluate the timing of the various phases, bearing in mind the specific socio-cultural context, too.

Figure 4 - Cattle-grazing in the Collina Materana forests.

The participatory process: establishment of the "participatory support group"

A crucial aspect of the participatory approach adopted was the setting up of a participatory support group. This was meant to guide and accompany the entire participation process, it being in charge of defining aims and strategic choices of participation (Cantiani 2012). In particular, it was in charge of deciding the most appropriate method of participation for the specific context and it identified and contacted all stakeholders, assessing their degree of influence in the process. It was also responsible for providing feedback to participants while assessing the effectiveness of the approach taken at the end of each step in the process.

In this case study the support group was formed by:
a) the person responsible for planning, a freelance technician in charge of the Plan with the responsibility for coordinating the data inventory, data processing and formulation of silvicultural guidelines;
b) the person responsible for participation (in this case one of the authors), with experience in forestry participatory processes;
c) two institutional participants (representatives from the Region of Basilicata and from INEA);
d) an actuator, responsible for logistic and secretarial aspects;
e) two local referees, well-known and respected persons from the local community with a profound knowledge of the territory and whose task was to collaborate in the analysis of the socio-economic context and to ease interactions with local actors.

The participatory process: definition of the participation method and start of the process

The choice of the level of participation is extremely important, since different levels correspond to different degrees of participants' involvement, which then lead to different possibilities in influencing the decision-making process (Chess 2000). Each case has to be evaluated individually, taking into careful consideration the specific objectives of the planning process and the socio-economic and cultural peculiarities of the local context (Paletto et al. 2015a).

In our case, the participatory support group opted for the activation of a consultative approach.

Consultation is a method by which the public is informed and then its needs, interests and opinions are heard. No guarantee is given that public demands will really affect final decisions. However, feedback is provided regarding the level of acknowledgement and inclusion of people's expectations in the decision-making process (Linder et al. 1992, Bettelini et al. 2000, Buchy and Hoverman 2000, IAP2 2007, Cantiani 2012).

In our case, the consultation was carried out at two different levels and with different objectives.

The first stage of consultation was mainly aimed at: i) understanding the expectations and needs of people directly involved in land management; ii) gathering local knowledge; iii) identifying any conflict.

The second stage of consultation was carried out at a more technical level and was directed at stakeholders who had specific competence regarding the matter in hand.

Particular importance has been devoted to information, with the purpose of raising the public's interest in the forthcoming planning process and, at the same time, fostering awareness of the functions and values of forests. In this case study institutional actors were informed of the planning process through written communication and a public meeting. On this occasion, the participatory plan process was officially considered to have begun. Thereafter, information was extended to the public at large, through the use of leaflets posted at the Comunità Montana centre and the municipality headquarters. These leaflets provided a useful tool in reaching large number of people. The meaning of the plan and the role of participation were illustrated in eye-catching graphics and clear language.

The participatory process: stakeholder analysis and first-stage consultation

The stakeholder analysis is a complex but important step (Ananda and Herath 2003, Candrea and Bouriaud 2009), since it allows the identification, characterization and classification of the stakeholders, with the objective of involving them in future decision-making processes. It obviously requires a great deal of work (Paletto et al. 2015b).

In our study case the stakeholder identification was an iterative process based on the principles of snowball sampling (or referral sampling): starting from the institutional actors, other previously unknown representative parties were identified (Harrison and Qureshi 2000, Hislop 2004). This type of sampling is advantageous since the costs and the size of the sample can be controlled. The limit is represented by the fact that distortions can be generated if the group formed in the beginning is not representative of the different categories involved (Hair et al. 2000).

In the Comunità Montana Collina Materana, as it often happens in a small rural area, almost all the institutional actors showed widespread knowledge of the territory. They therefore were crucial for the identification of other stakeholders in the area. In total, 63 stakeholders were identified who were then subdivided into several categories of interest, as shown in Table 1. Particular attention was given to the farmers, as the relationship between pasture and forest is one of the most critical elements in the system (De Meo et al. 2011).

Table 1 - Stakeholders involved in the consultation (* the institutional actors).

Categories of actor	Number
Municipalities*	7
Forest Bureau (Comunità Montana) *	1
State Forestry Corps*	4
No-profit associations	4
Tourist activities	5
Farmers	27
Forest enterprises	11
Forest owners	4

In the first stage of consultation, the participants were involved in the process through face-to-face interviews during which they responded to semi-structured questionnaires. The aim of these questionnaires was to elicit needs and expectations, to highlight problems and opportunities, and to gather suggestions on the basis of hypotheses concerning the future development of the territory under FLMP.

The questions were based on the following topics: the values and main functions attributed to the forest; the potentiality and critical aspects of the forestry sector; the relationship between livestock farming and forest management; the value attributed to the landscape and perception of landscape change; the bond between population and its home territory and the relationship between people and institutions.

The participants were firstly contacted by telephone, interviewees were met wherever they felt most comfortable. During the interview, people were given the opportunity to expand the conversation and to deepen issues considered particularly relevant or tricky. This often led to the collection of unexpected and interesting information.

The interview schedule was the result of several discussions and reviews between researchers, technicians and experts with a deep knowledge of the area, in order to obtain a tool that would serve to combine the clarity of language, the completeness of the information sought and the effectiveness of the questions raised.

The participatory process: SWOT analysis and the second phase of consultation

The data obtained by the interviews were analyzed and summarized by means of the Strengths, Weaknesses, Opportunities and Threats (SWOT) analysis.

In forest planning, the SWOT analysis, used in an

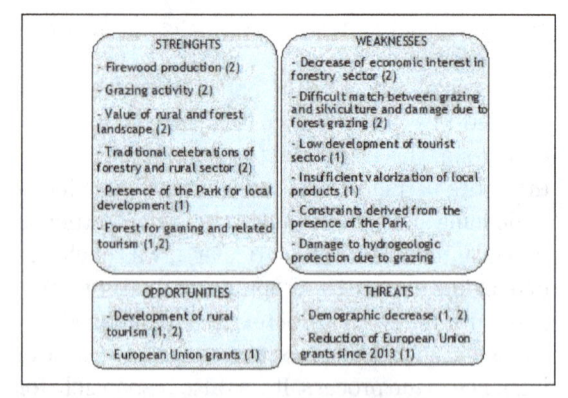

Figure 5 - Summary matrix of SWOT analysis. (1) Data from the literature, (2) Data from interviews.

ex-ante phase, is a method of analysis suitable for integrating a programme or a plan within the real context in which it is implemented. In doing so, factors, both the internal and external to the system, are considered in a systematic way (Kurttila et al., 2000).

The factors characterizing the forests of the Comunità Montana were summarized in a matrix. At first, they were marked differently, according to whether they were the result of the bibliographic survey or directly from the interviews (Fig. 5). In this respect, it is worth noticing that some of these factors, while being considered elements of strength or opportunities in the bibliography concerning the area, were actually regarded as critical by those interviewed. This is the case, for example, of the presence of a protected area such as the Regional Park. In the subsequent evaluation, the same weight was, however, assigned, regardless of the source. This phase of the SWOT analysis was of fundamental importance in providing an initial list of the main functions fulfilled by the forest ecosystems of the area.

The qualitative information obtained from the first consultation stage and synthetized by means of the SWOT analysis was then integrated with the data of the forest inventory and the technical information derived from other stages of the FLMP (Fig. 1, Phases 1 and 3).

In particular, the functions that were acknowledged as a priority during the inventory phase (Table 2) were closely related to the findings of the SWOT analysis so that different alternative land development scenarios could be suggested, each one characterised by different objectives and strategies and supported by a publicly participated GIS (PPGIS) (De Meo et al. 2013). The management proposals, corresponding to the different scenarios, were focused on the internal forces of the area and represented some possible alternatives in order to respond to the threats of external factors (Tab. 3).

These proposals, synthesized in a clear and simple working document, were submitted, in the

Table 2 - Percentage (%) forest type per function in hectares.

Function/Forest type	Turkey oak	Downy oak	Holm oak	Hungarian oak	Aleppo pine	Others	Total
Landscape and biodiversity	15.5	8.2	2.1	2.1	0.0	1.0	28.8
Leisure	1.0	1.0	0.0	0.0	0.0	0.0	2.1
Production	16.5	15.5	2.1	1.0	0.0	1.0	36.1
Protection	11.3	15.5	3.1	1.0	1.0	1.0	33.0
Total	44.3	40.2	7.2	4.1	1.0	3.1	100.0

Table 3 - Synthetic management proposals.

Regulation of the relationship between grazing and forest
Need to develop the tourist-recreational potentialities of the area
Valorisation of the production function, especially firewood
Valorisation of hydro-geological protection function, particularly in relation to geo-morphological characteristics of the territory

second round of consultation, to the institutional actors and some key stakeholders. The latter were identified by the support group taking into account their representativeness and their technical competence. The actors, involved in working groups, were invited to discuss the management proposals, in order to bring about the most widely shared version of the Plan, ready to be presented to the decision makers in charge of the final decision.

Methodology for the evaluation of the participatory process

A few years after the approval of the plan, we felt the need for an assessment of the participatory process which had been carried out. This effort was deemed a requisite in order to inform future participative planning at the landscape level in mountainous rural areas of the country. In particular, we wanted to reflect on the success and failures of the methodological approach chosen and decide on both the feasibility of reproducing it and any likely improvements.

This is an important issue, considering that interest towards FLMP in Italy is currently growing, also due to the fact that in many regions today any kind of project for local development must be set within the frame of higher level planning, such as the FLMP, in order to obtain either national or European funding, which is generally channelled through the regions to the local communities. For these reasons we decided, despite the plan being in an initial phase of implementation, to undertake an evaluation process, which may be considered, with regard to the timing, an ex-post summative evaluation (Blackstock et al. 2007), where our attention was mainly focused on the process, rather than on the outcome.

Success is a multi-dimensional, complex concept and the measure of it depends heavily on motivation and the perspective adopted in the participation approach. It also has to take into account the local governance context (Blackstock et al. 2007, Faehnle and Tyrväinen 2013).

In our case, the process being our main concern, we paid particular attention to both the normative and the substantive rationale. From a normative perspective, people's empowerment deriving from participation represents a measure of success, whereas substantive reasons call for the need to encompass a multiplicity of voices, concerns and values (Fiorino 1989, Blackstock et al. 2007, Menzel et al. 2012). In this particular phase of evaluation we were less interested in the instrumental rationale, which focuses on participation as a means to facilitate implementation and avoid conflicts.

Identification of success criteria

As a first step towards the development of an evaluation framework, the theoretical and empirical literature was scrutinised in order to select success criteria suitable for our case study.

Our analysis ranged over the specific literature on participative forest planning (Shindler and Neburka 1997, Tuler and Webler 1999, Buchy and Hoverman 2000, Webler et al. 2001, Saarikoski et al. 2010, Menzel et al. 2012, Robson and Rosenthal 2014), on participative natural resources management and environmental decision making (McCool and Guthrie 2001, Olsson et al. 2004, Blackstock et al. 2007, Lockwood 2010, Faehnle and Tyrväinen 2013), but also the more general literature on quality of participation (Innes and Booher 1999, Rowe and Frewer 2000, Asthana et al. 2002, Brinkerhoff 2002).

A few criteria, though often cited in the literature, have been considered unsuitable for our scale and timing and, for this reason, disregarded. This is the case, for example, of the criterion "Conflict resolution among competing interests" (Robson and Rosenthal 2014) which, in our context, has been deemed appraisable only over a longer lapse of time. All the same, other criteria were considered first, but then abandoned because too narrow or too specific for the context of the research in which they had been utilised. Besides, as we noticed many cases of blurring or superimposing, we merged some criteria. After this preparatory work, we finally set a preliminary list of success criteria. Since we looked for an evaluation framework well rooted in the local governance context and meaningful for the stakeholders, we decided to submit this list, for scrutiny and discussion, to the same group of stakeholders who had been involved in the second stage of consultation.

The involvement of local actors in the evaluation can be undertaken in different ways. As time was held to be the main constraint, we invited people to be engaged, for one day only, in activities carried out within small focus groups, followed by a plenary discussion in the evening. The focus group technique is gaining more and more interest among researchers, planners and evaluators. A planned discussion carried out in small groups of participants, in a relaxed atmosphere, is considered a good method to analyse and obtain in-depth comprehension of complex issues (van Asselt and Rijkens-Klomp 2002).

The initial list was questioned and reshaped. Some criteria were rejected because considered too abstract, too vague or unsuitable for the local context. This is the case, for example, of "Legitimacy" and "Fairness", which have been regarded as concepts that are too blurred. Both issues, indeed, are seen in the literature as quite closely interconnected and, at the same time, particularly related to the outcomes of the process (Webler et al. 2001, Saarikoski et al. 2010).

As a result, we obtained a final shared list, shorter than the previous one, consisting of eight criteria that are reported in alphabetical order in Table 4.

Evaluation according to the success criteria identified

The evaluation was carried out following two main tracks: an "expert" perspective and a "participated" one.

Firstly, we made a keen analysis of a large amount of documentation:
a) planning documentation;
b) documents related to the participation process, such as reports, minutes of meetings, field notes, feedback of the support group etc.;
c) other documentation somehow related to the FLMP, such as conference presentations, media reports etc.

Then we examined the projects and plans (such as the forest unit management plan) realised within the Comunità Montana, following the FLMP's guidelines. We also considered material related to a larger area of interest than forestry if considered relevant.

As only a few years had elapsed since the approval of the plan, we could not rely on much documentation. However, we found this exercise very useful and we think that, generally speaking, good lessons can be learnt thanks to such an approach.

By the end of this phase, we had gained a good insight into participation performance, but we still needed to collect the participants' perceptions and experiences regarding this.

A whole day was committed to involving local actors in evaluation. The same institutional actors and key stakeholders, already involved in building the evaluation framework, were invited to discuss issues in focus groups (differently composed than in the previous case) and then in a final plenary session.

An external observer was invited to attend this

Table 4 - List of success criteria.

EVALUATION CRITERIA	CRITERIA DESCRIPTION	
ACCESSIBILITY	Timely information is available to all participants and any kind of resources and facilities necessary to support participation are provided throughout the entire process (Asthana et al. 2002, Menzel et al. 2012, Saarikoski et al. 2010, Tuler and Webler 1999).	
CHALLENGING STATUS QUO AND FOSTERING CREATIVE THINKING	Participation encourages questioning the status quo and stimulates the imagination of alternative future scenarios (Innes and Booher 1999, Menzel et al.2012, Olsson et al. 2004).	
COST-BENEFIT	From the organisational perspective: COST EFFICIENCY	The accrued costs for organising participation must be balanced throughout the process (Blackstock et al. 2007, Faehnle and Tyrväinen 2013, Rowe and Frewer 2000).
	From the participants perspective: PARTICIPATION "WORTH THE EFFORT"	Perceived costs must not outweigh perceived benefits, especially when time is the main cost variable (Cheng and Mattor 2006, Faehnle and Tyrväinen 2013).
INCLUSIVENESS	All the stakeholders and interest groups willing to participate are involved in planning; a broad range of the population of the affected public is present (Blackstock et al. 2007, Buchy and Hoverman 2000, Cantiani 2012, Lockwood 2010, McCool and Guthrie 2001, Rowe and Frewer 2000, Saarikoski at al. 2010).	
INTERACTIVENESS	Participation is dialogical, based on a constructive long lasting face-to-face interaction (Saarikoski at al. 2010, Shindler and Neburka 1997, Tuler and Webler 1999)	
KNOWLEDGE INTEGRATION	Participation improves the knowledge and value base of planning because of the utilisation of experiential information (Cantiani 2012, Faehnle and Tyrväinen 2013, Saarikoski at al. 2010, Blackstock et al. 2007).	
SOCIAL LEARNING	Participation changes individual values and behaviour, thus influencing collective culture and norms (Blackstock et al. 2007, McCool and Guthrie 2001, Faehnle and Tyrväinen 2013).	
TRANSPARENCY	The participants can understand what is going on and how decisions are made and, at the same time, external observers can audit the process (Blackstock et al. 2007, Brinkerhoff 2002, Lockwood 2010, Menzel et al. 2012, Rowe and Frewer 2000).	

last meeting. His remarks were useful when we finally compared the expert evaluation to the participated one, in order to integrate and merge the results to arrive at a concluding assessment.

Discussions

Accessibility

When dealing with the notion of accessibility to the process of participation, the various authors often refer, in turn, to different issues. These issues may correspond to different criteria, such as availability of early and timely information (Saarikosky et al. 2010b, Faehnle and Tyrväinen 2013), adequacy, quality and quantity of information (Menzel et al. 2012, Blackstock et al. 2007), provision of adequate resources (Rowe and Frewer 2000, Asthana et al. 2002), access to policy makers and leaders (Blackstock et al. 2007), the circumstance of physically getting people to be present and involved (Tuler and Webler 1999, Menzel et al. 2012).

In our case, it was deemed that all these elements could be profitably summarized in one single criterion, accessibility. This is indeed closely related to another important criterion, that of inclusiveness, and also, following Tuler and Webler (1999), to the concept of fairness.

In the perspective of accessibility, the process has been evaluated as satisfactory, thanks also to the procedure expressly thought for and tailor made for rural areas.

Challenging status quo and fostering creative thinking

With regard to this criterion, a unanimous positive opinion of the results of participation in our case study was expressed.

Actually, the process promoted reflection and constructive discussions, often questioning the traditional forms of management and envisioning alternatives of development capable of overcoming the weaknesses intrinsic to the local socio-ecological system. Thanks to knowledge building and social learning, which were enhanced by the participation process, possible scenarios for future management have been designed and interesting solutions have been found. These were later acknowledged in the drafts of the plan, contributing in a substantial way to the realisation of the management guidelines.

Two challenging issues, in particular, have profitably stimulated creative thinking:
a) The age-old conflict between pasture and forest, i.e. between farmers and foresters. In this regard, possible areas of overlapping and new management strategies have been identified, in order to make grazing activity in wooded lands reconcilable with the existence of vital, viable forests. As a matter of fact, the consequences of climate change are already manifest in the Mediterranean region and are expected to become more and more severe in the near future, with longer periods of drought. In such periods the forest's contribution to the production of palatable, nutritious forage is particularly valuable and must be carefully considered (De Meo et al. 2011);
b) The development of eco-friendly tourism. In Italy, the Apennines are much less exploited for tourism than the Alps and their potential in this respect is mostly unknown or little recognised even by the residents themselves. Thanks to the participation process, the multifunctional landscape that characterises the area has finally been regarded with new interest in relation to the development of activities connected to rural tourism. In particular, the supply of natural and healthy food, typical of the area, appears to be bound to gain more and more importance in a time when special attention is being paid to the production of high quality food as an element of sustainability. Talking about creative thinking, we can definitely say that in general, beyond our case study, in periods characterised by great changes such as we are experiencing right now, one of the main results of participation is indeed that of showing the way forward to different approaches and innovative solutions when looking at problems.

Cost-benefit

Measuring the cost-effectiveness of participation is a difficult but necessary task, which must be accomplished especially when dealing with an experimental phase of planning. Evaluation in this respect, in fact, may help to avoid wasting public and stakeholder resources in future planning processes.

Participation necessarily entails participants' commitment, accrued costs and more time for planning and should not be taken lightly either from the organisational or the participants' perspective. From the organisational perspective, only if the quality of the decisions is concretely improved, the participation efforts prove to be reasonable in terms of cost efficiency. Though sometimes neglected in favour of the organisational perspective, the participants' standpoint must be taken into careful consideration, too. The perceived costs, especially in terms of time required for the involvement, should not outweigh the perceived benefits, otherwise people might no longer be willing to participate. In other words, participation must be "worth the effort" (Faehnle and Tyrväinen 2013, p. 336).

In our case study, the participative approach has been acknowledged as very advantageous from both perspectives, despite consuming time (115 man-days) and money. In the planners' opinion, participation provided very useful information and made it possible to shape more appropriate planning strategies. Forest technicians of the Comunità Montana deemed particularly convenient the spatialization of information, by means of the PPGIS, in order to reflect on real or presumed conflicts. Many stakeholders appreciated the fact that a wide range of possible solutions could be considered because of participation. In the words of a farmer: "It is only thanks to the fact that we (the category) have been listened, that the plan can now take into consideration the possibility of sending our animals into the woods to graze and we can discuss the way to do it and also its limits."

A general empowerment of participants has finally been acknowledged as a positive effect of participation, and this is thought to favour future implementation of the plan.

Inclusiveness

Inclusiveness is largely acknowledged as a critical requisite for an effective participatory planning process. Especially when planning in rural areas, it is not easy to involve those stakeholders from the primary sector who have generally a marginal role in the social system, despite being directly in charge of the management of local natural resources and landscape. In this case, two main consequences may become apparent:
a) a loss of valuable experiential information during the elaboration of the plan;
b) possible conflicts arising during the implementation stage.

Strictly connected to the issue of inclusiveness is the need for a broad representation of the various views and interests in the planning process (McCool and Guthrie 2001). Actually, a fair and balanced representation is hard to attain and requires a great effort in the phase of designing the participation process.

Being aware of this problem, in our case study we tried our very best to give different voices the chance to be heard and to represent different interests appropriately, focusing in particular on both the key and primary stakeholders (Paletto et al. 2015a). Even if the public at large was not our main target, we tried to open the process up as much as possible and to also reach citizens who are not directly affected but potentially interested, by trying to distribute timely, clear information.

As for the stakeholders' involvement, the process has been evaluated as successful, mainly due to the approach taken in the stakeholders' analysis and the work carried out by the support group. Within the latter, the role of the two local referees was regarded as very helpful in interacting with the stakeholders and in assessing and balancing their power. The first stage of consultation, carried out by means of face-to-face interviews, was particularly appreciated for the reason that it accomplished the outreach task well.

In contrast, a greater effort to include the general public has been deemed necessary. For this purpose, appropriate tools should be studied when designing future participation processes in FLMPs. Particular attention should be paid to addressing women and young people. In fact, in communities of mountainous areas especially in the south of the country, women are inclined to exclude themselves from a public and visible social debate, whereas young adults are less and less interested in forest or agriculture related professions, and are increasingly willing to out-migrate.

Interactiveness

In the present case study, participation was implemented through consultation. Actually, there was a disregard for the use of participatory methods that directly involve citizens in identifying objectives and strategies of the plan in deliberative spaces. These methods, in fact, are generally more expensive in terms of time and energy and, above all, require from the population a keen interest in participation and a willingness to work in groups (Linder et al. 1992), which is uncommon in the geographical context investigated.

In the literature, beginning with the classic paper by Arnstein (1969), the consultation process is generally imputed with strong limitations, considered ineffective and sometimes even counterproductive.

In our opinion, these negative aspects are in reality more attributable to the way in which the consultation is implemented rather than to the method itself (Bettelini et al. 2000, Cantiani 2012, Paletto et al. 2015a). As a matter of facts, in the past, the consultation process has been associated with the decision-making of public bodies, which is characterized by very formal protocols with the sole purpose of either complying with a law or legitimizing decisions already taken by the administration. This fact has often resulted in belated involvement of the population, a procedure with partial clarity and a highly technical content, with the use of language poorly understood by most people and a complete absence of constructive integration.

In our case study, in the first stage of consultation the stakeholders were involved through a dialogical attitude, stimulating a constructive discussion between interviewer and interviewee. In the second consultation stage the interactiveness within the working group was continually encouraged and kept alive.

Finally, with regard to the criterion of interactiveness, the opinion that emerged during the evaluation was generally positive. Consultation is considered a method suitable for the local socio-cultural context and thanks to the way it has been structured, capable of enhancing not only official moments of exchange, but also informal social interaction, in a relaxed climate of trust and reciprocal understanding.

The institutional actors, in particular, said they were glad for the opportunity to coordinate better across different sectors, due to sustained interaction.

Knowledge integration

If the main aim of participation is that of improving the content of planning, as in our case, knowledge building is to be considered a critical ingredient in a successful process. Especially when planning in geographical contexts such as ours, experiential information is as valuable as the technical and scientific kind, and complementary to it. Local people are source of knowledge deriving from cultural heritage or from their personal experience and capacity to interpret the relationship between human beings and the environment in complex socio-ecological systems (Raymond et al. 2010).

Actually, knowledge integration may be an important surplus value, strictly connected to other criteria, such as the cost-benefit of participation and social learning. From a planning perspective, knowledge integration means improving not only knowledge, but also the value base, which cannot be considered separately (Faehnle and Tyrväinen 2013).

In our case, from the beginning, we understood that we could not manage without the experiential information of foresters and farmers and, for this reason, we based the first stage of consultation on a systematic action of reaching out. Both analysing the documents and listening to the opinions deriving from the focus-group activities, clearly emerged the enormous contribution to the solution of problems obtained from the first stage of consultation.

Finally, evaluation showed that, in both stages of consultation, knowledge integration was greatly enhanced. In particular, by some institutional actors it was remarked that the use of PPGIS proved to be a very helpful tool while working on the drafts of the plan in order to detect the areas of existing or latent conflicts, thus facilitating the identification of possible solutions.

Social learning

Learning is a typical "two-way or interactive concept" (McCool and Guthrie 2001, p. 317). Social learning can be enhanced, in strict connection with knowledge integration if participation is carried out with an approach that stimulates back and forth discussion and a reflective attitude.

A particular effect of social learning is the empowerment that originates within the local community, thanks to sensitization efforts and the deriving awareness of the functions and values of the ecosystems present in the area. In the participants' eyes in our case study, this issue has been especially stressed. As one institutional actor pointed out: "Participation helped me to reflect on the values of my area. For example: before, I had never considered the landscape of the Comunità Montana as beautiful, nor had I thought that somebody from outside could wish to come here on holidays".

Following the evaluation, this criterion can actually be considered largely fulfilled in our case study.

Transparency

Transparency is generally acknowledged as an important requisite for a genuine, fair participation process.

It is nevertheless true that it is not easy to evaluate it, due to the complex structure of a participation process in forest planning on the one hand, and to the subjective nature of the criterion itself on the other.

If transparency means that throughout the entire process "established channels for continuous dialogue and information sharing" exist and "timely response to information requests" (Brinkerhof 2002, p. 222) is provided, the criterion has to be considered fully satisfied in our case study.

When shifting attention onto why and how the decisions have been made, however, things are more complicated. A greater effort has been deemed necessary in the future designing of participation, in transmitting information in this regard in a more direct form, accessible also to non institutional or expert actors.

Conclusions

One of the main outcomes of landscape planning is that it definitely contributes to a sustainable development of the area. What, however, often happens is that the implementation of a plan is disregarded or even sometimes boycotted by some local actors.

For this reason, participation is more and more frequently called upon, in order to set up a planning process which is well grounded in the local context and thus more effective. Quite often, though, disappointment about participation results and a kind of frustration may show up during or at the end of the process. This is mainly due to the fact that initial expectations are too high, both in the planning organisation and on the part of participants.

These considerations prompted us to reflect on targets and the effects of participation in our case study, involving local actors in the evaluation stage.

Considering the timing of evaluation, it is too early to argue over tangible outcomes. Only in the long term will it be apparent if stakeholders' sensitization and empowerment, activated by participation, are kept alive, contributing to the implementation phase and if the institutional actors are able to mobilise resources, bring networks into play and adapt to changing conditions, in order to shape a sustainable development in tune with people's expectations.

On the basis of the evaluation carried out so far, the participation process illustrated for the FLMP of Collina Materana can be considered satisfactory, although future improvements are deemed necessary. It goes without saying that when the implementation phase is advanced, it may be necessary to take into consideration other criteria and develop suitable qualitative and quantitative indicators to measure performances.

We hope that the case study itself and the framework set up for its evaluation might be useful for anyone who decides to undertake planning processes through a participative approach.

Our experience actually showed the importance of concretely integrating participation into the planning process. The procedure adopted for this aim, flexible and divided into phases, allowed to incorporate the findings of participation into the goals and strategies of the plan, with a reasonable commitment of financial resources and time.

The framework tested for the evaluation proved to be effective and not too costly in terms of either time or money. Since it is quite flexible, it could easily be adapted to other contexts, identifying specific criteria and indicators, tailor made for local needs.

Acknowledgments

We are grateful to all those people, too many to be mentioned individually here – institutional actors, stakeholders, ordinary citizens – who believed in our work and actively collaborated in the participative planning process and in its evaluation.

The authors contributed equally to this work.

Funding

This project was funded by the MIPAF (Ministry for Agriculture and Forestry Policies) National Research Project Ri.Selv.Italia – Sub-task 4.2 – "Geographic informative system for forest management", the Region of Basilicata and the Ministry of Environment, within the frame of PON-ATAS (National Operative Program on technical assistance and systemic action), measure I.2 "Soil conservation", Action 2.2 "Support for highly complex issues".

References

Ananda J. 2007 - *Implementing Participatory Decision Making in Forest Planning.* Environmental Management 39: 534-544.

Ananda J., Herath G. 2003 - *Incorporating stakeholder values into regional forest planning: a value function approach.* Ecological Economics 45: 75-90.

Appelstrand M. 2002 - *Participation and societal values: the challenge for lawmakers and policy practitioners.* Forest Policy and Economics 4: 281-290.

Argenti G., Bellotti G., Bernetti J., Bianchetto E., Cantiani MG., Cantiani P., Costantini G., De Meo I., Ferretti F., Frattegiani M. 2008 - *Piano Forestale Territoriale di Indirizzo della Comunità Montana Collina Materana.* Vol.1, Vol. 2. INEA, Basilicata, Italy.

Arnstein SR. 1969 - *A Ladder of Citizen Participation.* Journal of the American Planning Association 35 (4): 216-224.

Asthana S., Richardson S., Halliday J. 2002 - *Partnership working in public policy provision: a framework for evaluation.* Society and Policy Administration 36 (7): 780-795.

Bettelini D., Cantiani MG., Mariotta S. 2000 - *Experiences in participatory planning in designated areas: the Bavona Valley in Switzerland.* Forestry 73 (2): 187-198.

Blackstock KL., Kelly GJ., Horsey BL. 2007 - *Developing and applying a framework to evaluate participatory research for sustainability.* Ecological Economics 60: 726-742.

Brinkerhoff JM. 2002 - *Assessing and improving partnership relationships and outcomes: a proposed framework.* Evaluation and Program Planning 25: 215-231.

Buchy M, Hoverman S. 2000 - *Understanding public participation in forest planning: a review.* Forest Policy and Economics 1:15-25.

Candrea AN., Bouriaud L. 2009 - *A stakeholders' analysis of potential sustainable tourism development strategies in Piatra Craiului National Park.* Annals of Forest Research, 52 (1): 191-198.

Cantiani MG. 2012 - *Forest planning and public participation: a possible methodological approach.* iForest 5:72-82. doi: 10.3832/ifor0602-009

Cantiani MG., De Meo I., Paletto A. 2013 - *What do Human Values Suggest about Forest Planning? An International Review Focusing on the Alpine Region.* International Journal of Humanities and Social Science 1: 228-243.

Cantiani P., De Meo I., Ferretti F., Paletto A. 2010 - *Forest functions evaluation to support forest landscape management planning.* Forest Ideas 16 (1): 44–51.

Chess C. 2000 - *Evaluating Environmental Public Participation: Methodological Questions.* Journal of Environmental Planning and Management 43 (6): 769-784.

De Meo I., Cantiani MG., Ferretti F., Paletto A. 2011 - *Stakeholders' Perception as Support for Forest Landscape Planning*. International Journal of Ecology. doi:10.1155/2011/685708

De Meo I., Ferretti F., Frattegiani M., Lora C., Paletto A. 2013 - *Public participation GIS to support a bottom-up approach in forest landscape planning*. iForest 6: 347-352.

European Commission. 2003 - *Natura 2000 and forests "Challenges and opportunities" interpretation guide*. Office for Official Publications of the European Communities. Available from: http://ec.europa.eu/environment/nature/info/pubs/docs/nat2000/n2kforest_en.pdf.

Faehnle M., Tyrväinen L. 2013 - *A framework for evaluating and designing collaborative planning*. Land Use Policy 34: 332-341.

FAO-ECE-ILO 2000 - *Public participation in forestry in Europe and North America. Report of the FAO/ECE/ILO Joint Committee Team of Specialists on Participation in Forestry*. Working paper 163. Sectorial activities department. International labour office. Ginevra.

Farcy C., Devillez F. 2005 - *New orientations of forest management planning from an historical perspective of the relations between Man and Nature*. Forest Policy and Economics 7: 85-95.

Farrell EP., Führer E., Ryan D., Anderson F., Hüttl R., Piussi P. 2000 - *European forest ecosystems: building the future on the legacy of the past*. Forest Ecology and Management 132: 5-20.

Ferretti F., Di Bari C., De Meo I., Cantiani P., Bianchi M. 2011 - *ProgettoBosco: a Data-Driven Decision Support System for forest planning*. International Journal of Mathematical and Computational Forestry & Natural-Resource Sciences (MCFNS) 3 (1): 27–35.

Fiorino DJ. 1989 - *Environmental risk and democratic process: a critical review*. Columbia Journal of Environmental Law 14: 501-547.

Hair JF., Bush RP., Ortinau DJ. 2000 - *Marketing Research: A Practical Approach for the Millennium*. McGraw-Hill. Boston.

Harrison SR., Qureshi ME. 2000 - *Choice of stakeholder groups in multicriteria decision models*. Natural Resources Forum 24:1-19.

Hislop M. 2004 - *Involving people in forestry: a toolbox for public involvement in forest and woodland planning*. Forestry Commission. London

IAP2 2007 - *IAP2 Spectrum of Public Participation*. Available from: http://www.iap2.org/associations/4748/files/spectrum.pdf.

Innes JE., Booher DE. 1999 - *Consensus building and complex adaptive systems – A framework for evaluating collaborative planning*. Journal of the American Planning Association 65: 412-423.

Kangas A., Laukkanen S., Kangas J. 2006 - *Social choice theory and its applications in sustainable forest management – a review*. Forest Policy and Economics 9: 77-92.

Kurttila M., Pesonen M., Kangas J., Kajanus M. 2000 - *Utilizing the analytic hierarchy process (AHP) in SWOT analysis - a hybrid method and its application to a forest-certification case*. Forest Policy and Economics 1: 41-52.

Lee M., Abbot C. 2003 - *The Usual Suspects? Public Participation under the Aarhus Convention*. Modern Law Review 1: 80-108.

Leskinen LA. 2006 - *Adaptation of the regional forestry administration to national forest, climate change and rural development policies in Finland*. Small-scale Forest Economics, Management and Policy 2: 231-247.

Linder W., Lanfranchi P., Schnyder D., Vatter A. 1992 - *Procédures et modèles de participation: propositions pour une politique de participation de la Confédération selon l'art. 4 LAT. (Models of participation: suggestions for a participation policy in Switzerland, following the article four LAT.)* Office fédéral de l'aménagement du territoire (Swiss Federal Bureau for Land Planning). Berne (Switzerland).

Lockwood M. 2010 - *Good governance for terrestrial protected areas: a framework, principles and performance outcomes*. Journal of Environmental Management 91 (3): 754-766.

McCool SF., Guthrie K. 2001 - *Mapping the dimensions of successful public participation in messy natural resources management situations*. Society and Natural Resources 14 (4): 309-323.

Menzel S., Nordstrom EM., Buchecker M., Marques A., Saarikoski H., Kangas A. 2012 - *Decision support systems in forest management: requirements from a participatory planning perspective*. European Journal of Forest Research 131: 1367-1379. doi:10.1007/s10342-012-0604-y

O'Brien EA. 2003 - *Human values and their importance to the development of forestry policy in Britain: a literature review*. Forestry 76 (1): 3-17.

Olsson P., Folke C., Berkes F. 2004 - *Adaptive Co-management for Building Resilience in Social-Ecological Systems*. Environmental Management 34 (1): 75-90.

Paletto A., Ferretti F., Cantiani P., De Meo I. 2011 - *Multifunctional approach in forest management land plan: an application in Southern Italy*. Forest System 2: 66-80.

Paletto A., Ferretti F., De Meo I. 2012 - *The Role of Social Networks in Forest Landscape Planning*. Forest Policy and Economics 15: 132–139.

Paletto A., Cantiani MG., De Meo I. 2015a - *Public Participation in Forest Landscape Management Planning (FLMP) in Italy*. Journal of Sustainable Forestry 34 (5): 465-483.

Paletto A, Hamunen K, De Meo I. 2015b - *Social Network Analysis to Support Stakeholder Analysis in Participatory Forest Planning*. Society and Natural Resources 28 (10): 1108-1125.

Raymond CM., Fazey I, Reed MS., Stringer LC., Robinson GM., Evely AC. 2010 - *Integrating local and scientific knowledge for environmental management*. Journal of Environmental Management 91: 1766-1777.

Robson M., Rosenthal J. 2014 - *Evaluating the effectiveness of stakeholder advisory committee participation in forest management planning in Ontario*. Canada. Forestry Chronicle 90 (3): 361-370.

Rotherham ID. 2007 - *The implications of perceptions and cultural knowledge loss for the management of wooded landscapes: a UK case-study*. Forest Ecology and Management 249: 100-115.

Rowe G., Frewr LJ. 2000 - *Public participation methods: A framework for evaluation*. Science, Technology & Human Values: 25 (1): 3-29.

Saarikoski H., Tikkanen J., Leskinen LA. 2010 - *Public participation in practice – Assessing public participation in the preparation of regional forest programs in Northern Finland.* Forest Policy and Economics 12: 349-356.

Schmithüsen F. 2007 - *Multifunctional forestry practices as a land use strategy to meet increasing private and public demands in modern societies.* Journal of Forest Science 53 (6): 290-298.

Shindler B., Neburka J. 1997 - *Public participation in Forest Planning: 8 Attributes of Success.* Journal of Forestry 95 (1): 17-19.

Tuler S., Webler T. 1999 - *Voices from the Forest: What Participants Expect of a Public Participation Process.* Society and Natural Resources 12 (5): 437-453.

Van Asselt MBA., Rijkens-Klomp N. 2002 - *A look in the mirror: reflection on participation in Integrated Assessment from a methodological perspective.* Global Environmental Change 12: 167-184.

Webler T., Tuler S., Krueger R. 2001 - *What is a good public participation process? Five perspectives from the public.* Environmental Management 27 (3): 435-450.

Geospatial analysis of woodland fire occurrence and recurrence in Italy

Leone Davide Mancini[1*], Anna Barbati[1], Piermaria Corona[1]

Abstract - This technical note aims to exemplify the potential of annual time series of wildfire geodatasets to quantify fire occurrence and recurrence amongst different woodland types at large scale, under an international forestry perspective. The study covers a time series of areas affected by wildfire between 2007 and 2014 in Italy. A GIS operation of geometric intersection was carried out between time-series, geo-referred data of burned areas and synchronic Corine Land Cover maps. Mediterranean pine forest, high maquis, transitional woodland-shrub and high oro-Mediterranean pine forest are the woodland types most preferred in terms of fire occurrence and recurrence. Large fires and megafires hold a significant share of total burned area. An unexpected finding is the huge impact of fires in wildland-urban-interface areas. The proposed analysis provides spatial information that is central to any approach to fire management at large scale. Research findings provide support that can be used e.g. for positive and normative advancements, basically prioritization of fire prevention, suppression measures, economic incentive allocation, and sustainable land management in peri-urban districts.

Keywords - Wildfire, woodland type, geodataset, large fires, wildland-urban interface

Introduction

Wildfires are major disturbances in Mediterranean ecosystems worldwide. Wildfires affect natural resources and ecosystem services (e.g. tourism, local development, soil protection, biomass production, carbon sequestration, habitat provision), but also strategic policies and governance concerning the protection and enhancement of forest landscapes (Carmo et al. 2011, Kolström et al. 2011, Corona et al. 2015). Over the last few decades, the massive abandonment of, and the reduced land-use pressure on, rural lands have determined extensive woodland and shrubland re-colonization and fuel accumulation in many areas of Mediterranean Europe (Moreira et al. 2009, Moreira et al. 2011).

In regions such as the Mediterranean basin, almost all wildland fires are human-caused (Ganteaume et al. 2013). The number and size of wildfires is significantly influenced by the type, patterns and conditions of vegetation and fuels. For instance, the fuels produced by the main tree species that build up the forest physiognomy primarily drive its "baseline" flammability level (Xanthopulos et al. 2012, Corona et al. 2014), i.e. the relative ease by which a fuel will ignite and burn with a flame (Stacey at al. 2012, Fares at al. 2017). Xanthopoulos et al. (2012) proposed an assessment framework to assign a baseline flammability index to vegetation classes of the European Forest Types classification (Barbati et al. 2007, Barbati et al. 2014). Certain woodland types (e.g. shrublands or conifers) are at much greater risk of being affected by fire than other types (i.e. greater fire proneness or fire incidence), because of differences in forest management, vegetation structure, fuel load composition and proximity to human activities, all intended as causes of fires ignition (Moreira et al. 2011, Ganteaume et al. 2013, Barros and Pereira 2014, Oliveira et al. 2014, Pereira et al. 2014). In addition, the growing Wildland Urban Interface (WUI, e.g. Lampin-Maillet et al. 2010) poses new challenges to fire management, in terms of increase of fire ignition by human activities (*risk*), and of threats to local communities and properties (*vulnerability*).

A proper quantification of fire occurrence and recurrence amongst different woodland types is therefore essential to quantify need and priorities in fire management (Vazquez et al. 2015). The fire proneness of land cover types in Mediterranean Europe has been investigated in several studies (e.g. Nunes et al. 2005, Bajocco and Ricotta 2008, Moreira et al. 2009, Carmo et al. 2011, Marques et al. 2011, Barros and Pereira 2014; Oliveira et al. 2014, Pereira et al. 2014, Salvati et al. 2015). Earlier studies have highlighted the need to connect burned

---

[1] University of Tuscia, Dipartimento per l'innovazione nei sistemi Biologici, Agroalimentari e Forestali (UNITUS-DIBAF), Viterbo (Italy)
* leone.mancini@unitus.it

areas' perimeters to fire affected land cover types, in order to gain an understanding of differences in fire proneness amongst woodland types (e.g. Gonzalez et al. 2006, Silva et al. 2009, Corona et al. 2014, Oliveira et al. 2014). However, at least to our knowledge, no published study has addressed fire recurrence at country or regional scales in Mediterranean woodlands using annual time series of the wildfires perimeters, despite the most critical effects on ecosystem resilience are produced by recurrent fire events with short return periods (Moreira et al. 2011, Ricotta and Di Vito 2014, Barbati et al. 2015).

Wildfire geo-databases (i.e. geometry and topology of burnt areas or ignition point data) have a high potential to qualify fire occurrence and recurrence amongst different woodland types over large areas, under an international forestry perspective. This kind of information is currently available in many Countries of Mediterranean Europe (Chiriacò et al. 2013). In Italy, the National Forest Service (Corpo Forestale dello Stato, CFS) and the Forest Services of autonomous Regions have provided aggregated data on wildfires since 1970. A vector geodataset of the areas affected by wildfires (forest and rural areas), recorded by ground-based GPS surveys, is currently available for the years 2007-2014. The dataset currently covers 17 of the 20 EU-NUTS2 level administrative units of Italy (i.e. Sardinia, Sicily and all the Regions with ordinary statute), totaling 95% of the area burned during this period.

This technical note examines Italy as a case study to exemplify the potential of very detailed time series analysis of wildfire geodatasets for addressing questions that are central to any approach to fire management at large scale: which woodland types are most prone to fire recurrence? Which woodland types are most affected by WUI fires, large fires and megafires? Detailed evidence-based answers to such questions have practical application for fire management prioritization at country and local levels, and even for stimulating legislative, planning and operational advances in fire protection.

Materials and methods

A Corine Land Cover (CLC) complete database has been chosen as a reference land cover map (CLC 2006; CLC 2012; EEA 2007). CLC is an European database in vector format available for the reference years 1990, 2000, 2006 and 2012; the minimum mapping unit is 25 ha and mapping technology is based on computer assisted visual interpretation of satellites images. CLC fourth hierarchical level, introduced in Italy since 2000, maps forest and semi-natural areas with a higher thematic detail compared to the standard CLC (ISPRA 2010), distinguishing, namely, the different physiognomies of woodland types reported in Table 1.

A GIS operation of geometric intersection was carried out between 2007-2011 wildfire perimeters and the CLC 2006 geodataset and between 2012-2014 wildfire perimeters and the CLC 2012 geodataset. The same method, using CLC 2006 only, was applied to identify repeatedly burned areas in the period 2007-2014.

Following San-Miguel-Ayanz et al. (2013), in this work we have considered large fires those larger than 100 hectares and megafires those larger than 500 hectares. Moreover, in order to locate WUI fires, a buffer zone of 200 meters from urban settlements has been delineated, according to the distance defined at national level for human settlements protection by the Italian Civil Protection Department (Presidenza Consiglio Ministri 2007). A similar definition of WUI has been adopted by other Mediterranean countries (e.g. Modugno et al. 2016).

Burned Area Selection Ratio (BASR) index has been chosen to measure wildfires incidence by woodland type. BASR is defined as the ratio between percentage of burned area of a given woodland type compared to the total burned woodland area and the percentage of the area of the woodland type compared to the total woodland area:

$$BASR = \frac{A_b/T_b}{A_r/T_r}$$

where A_b is the burned area of the considered woodland type, T_b is the total burned woodland area, A_r is the area of the considered woodland type and T_r is the total woodland area.

BASR is founded on the principle of resource selection function (for previous relevant applications to forest fires, see e.g. Bajocco and Ricotta 2008, and Pezzati et al. 2009). If BASR for a given woodland type equals 1, then that type burned in proportion to its wide spreadness on the territory. If BASR is higher than 1, the type burned more often than expected due to chance, thus indicating fire preference. Conversely, if BASR is lower than 1, the type burned less than expected (i.e. fire avoidance).

BASR has been also used here for characterizing large fires and megafires and for fire recurrence analysis. In the first two cases, numerator consists of the ratio between burned area of woodland type and total woodland burned area of a given kind of fires. In the case of fire recurrence, numerator consists of the ratio between area affected by repeated burning of a given woodland type and the total area affected by repeated burning in woodlands.

Results

Since 1970, in Italy 46% of wildfires has occurred in the woodlands, with an annual mean surface of 48 700 ha. Annual mean number of fires has been 8700, with a decrease of about one third since the year 2000 compared to previous decades. The most critical situations were recorded in 1985 for the number of fires (18,664) and in 2007 for the wooded burned area (116,602 ha).

During the study period, the total burned area was 669 375 ha, corresponding to 2.4% of the land in Italy. The most fire-affected land cover class is scrub and/or herbaceous vegetation associations (35%), while forests represent 22% of the burned area (Fig.1).

Between 2007 and 2014 fires have affected all woodland types. According to Mann Kendall non-parametric test (Kendall 1975), statistically significant decreasing trends ($p<0.05$) have occurred in most woodland types as concern total burned area, number of fires and average area of fires. Exotic broad-leaved forest and Mediterranean pine forest do not show any trend for total burned area and number of fires.

The incidence of fires varies substantially across woodland types (Tab.1). In absolute terms, transitional woodland-shrub and deciduous oak forest have recorded the highest values both in terms of number of wildfires (5 685 and 6 843, respectively) and of total burned area (50 989 and 44 544 hectares, respectively). Larch and arolla pine forest and high oro-Mediterranean pine forest present the highest value of average burned area for fire (10.5 and 11.3 hectares, respectively).

BASR ranges from 0.04 to 3.46 (Fig.2). The rate of fire incidence in deciduous oak, hygrophilous and broad-leaved evergreen forest types is balanced with their land coverage, with a BASR value close to 1, while wildfire shows a strong preference for Mediterranean pine forest (BASR 3.46) and, to a lesser extent, high maquis, transitional woodland-shrub and high oro-Mediterranean pine forest (BASR around 2). Spruce/fir and larch and arolla pine forest

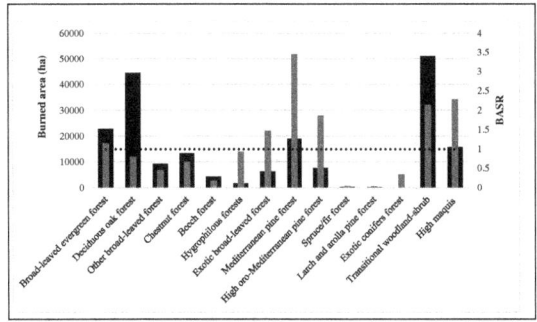

Figure 2 - Burned area (black histogram) and BASR values of burned area (grey histogram) by woodland type; the dashed line represents BASR=1.

Table 1 - Number of wildfires, total burned area, average burned area and contribution by large fires (>100 ha) and megafires (>500 ha) by woodland type in the period 2007-2014.

Woodland type	Number of fires (No.)	Total burned area (ha)	Average burned area (ha)	Large fires area (ha)	Megafires area (ha)
Broad-leaved evergreen forest	2 868	22 880	8.0	12 748	7 575
Deciduous oak forest	6 843	44 544	6.5	19 017	6 170
Other broad-leaved forest	1 405	9 254	6.6	3 761	1 287
Chestnut forest	2 189	13 412	6.1	4 118	758
Beech forest	542	4 284	7.9	2 199	418
Hygrophilous forests	307	1 667	5.4	419	5
Exotic broad-leaved forest	786	6 349	8.1	2 411	286
Mediterranean pine forest	1 892	18 975	10.0	11 589	5 855
High oro-Mediterranean pine forest	677	7 642	11.3	4 384	2 311
Spruce/fir forest	61	330	5.4	237	0
Larch and arolla pine forest	14	147	10.5	130	0
Exotic conifers forest	19	67	3.5	0	0
Transitional woodland-shrub	5 685	50 989	9.0	21 337	6 869
High maquis	1 534	15 765	10.3	9 598	5 899

are the least affected types, in both absolute and relative terms.

During the study period, 1023 large fires (i.e. fire size >100 ha), corresponding to 317 996 hectares, have occurred in Italy and 115 of these are megafires (i.e. fire size >500 ha), corresponding to 142 946 hectares. Although large fires account for 2.3% of total fires, they represent the 47 % of total burned area. Similarly, megafires represent 0.25% of total fires, but 21 % of total burned area. Table 1 and Figure 3 highlight a breakdown by woodland types of the total area affected by large fires and megafires with respect to total burned area. Transitional woodland-shrub, deciduous oak forest and broad-leaved evergreen forest are the woodland types most affected by extensive fires. In Mediterranean pine

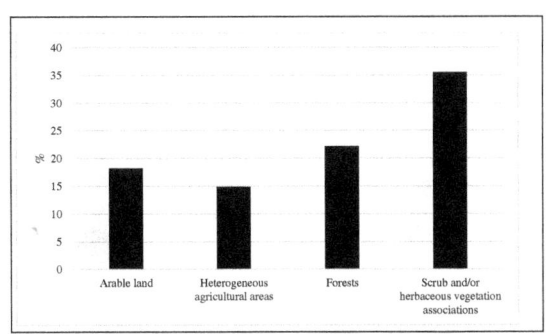

Figure 1 - Share of total burned area by land cover class in the period 2007-2014.

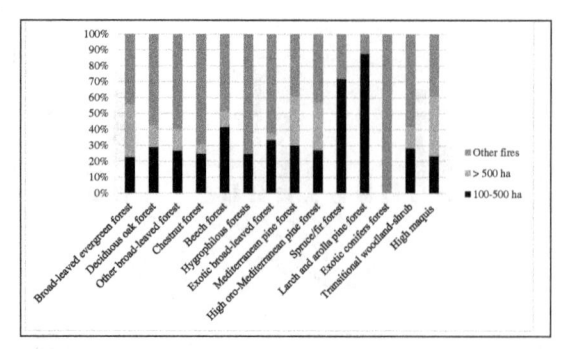

Figure 3 - Contribution of large fires and megafires to the total burned area by woodland type.

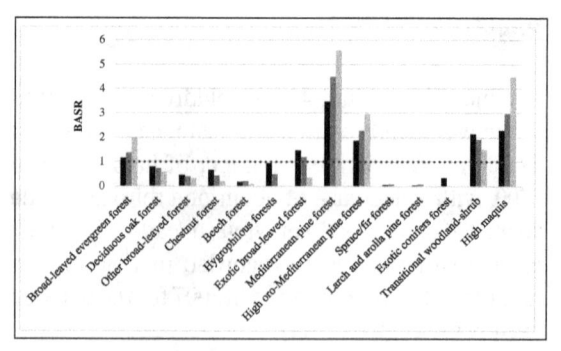

Figure 4 - BASR values of burned area (black histogram), BASR values of large fires (dark grey histogram) and BASR values of megafires (light grey histogram) by woodland type; the dashed line represents BASR=1.

forests, high maquis, spruce/fir forests and larch and arolla pine forests the contribution of large and megafires exceeds 60 % of total burned area (Fig. 3). A comparison amongst the BASR values referred to total fires, large fires and megafires is shown in Figure 4. In the case of Mediterranean pine forest, high oro-Mediterranean pine forest and high maquis and transitional woodland-shrub all such values are greater than 1, and even much higher in the case of megafires and large fires, except for transitional woodland-shrub.

Overall, 5 411 woodland fire events, for a total of 143 850 hectares, have affected WUI space. These events represent 21% of total burned area and 12% of fire events number. Broad-leaved evergreen forest, transitional woodland-shrub, deciduous oak forest and high maquis have the highest values in terms of burned area (Tab. 2). The contribution to the total burned area is higher than 20% in high maquis, spruce/fir forest, broad-leaved evergreen forest and Mediterranean pine forest types.

As shown in Figure 5, fire recurrence is a phenomenon that affects primarily scrub and/or herbaceous vegetation types (44%), but also a significant proportion of forests (18%). During the 8 years covered by the time series, 19 396 hectares of woodlands have been affected by at least two fire events (9% of the woodlands burned area), of which 2 038 hectares burned more than twice (0.95% of woodland burned area). Larch and arolla pine forest type, spruce/fir forest and exotic conifers forest types are negligibly affected by fire recurrence, while transitional woodland-shrub is the most affected type (6 137 hectares), followed by deciduous oak forests (4 169 hectares), in the case of forest areas burned twice, and by Mediterranean pine forests (393 hectares) in the case of areas with more than two fire events (Tab. 2). The contribution of repeatedly burned area to total burned area is greater than 10% in transitional woodland-shrub, Mediterranean pine forests and hygrophilous forests. Mediterranean pine forest presents the highest values of BASR (i.e. 3.4), in terms of repeatedly burned area (Fig. 6), followed by transitional woodland-shrub and

Table 2 - Burned areas in the Wildland-Urban Interface and fire recurrence by woodland type.

Woodland types	WUI fires		Fire recurrence		
	Area (ha)	Share of total burned area (%)	Repeatedly burned area (ha)	Share of total burned (%)	More than 2 fire events (ha)
Broad-leaved evergreen forest	6 476	28.3	1 344	5.9	120
Deciduous oak forest	5 413	12.2	4 169	9.4	300
Other broad-leaved forest	878	9.5	641	6.9	59
Chestnut forest	1 904	14.2	1 159	8.6	105
Beech forest	452	10.6	83	1.9	3
Hygrophilous forests	96	5.8	206	12.3	30
Exotic broad-leaved forest	306	4.8	754	11.9	54
Mediterranean pine forest	4 830	25.5	1 736	9.1	393
High oro-Mediterranean pine forest	815	10.7	575	7.5	45
Spruce/fir forest	111	33.5	9	2.9	1
Larch and arolla pine forest	6	4.3	0	0.0	0
Exotic conifers forest	0	0.0	6	9.1	0
Transitional woodland-shrub	9 619	18.9	6 137	12.0	766
High maquis	5 059	32.1	1 105	7.0	78

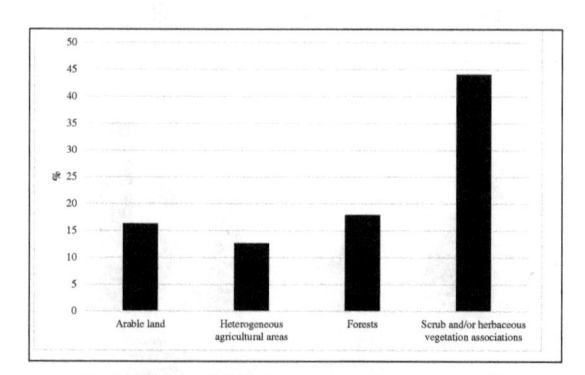

Figure 5 - Share of total repeatedly burned area by land cover class in the period 2007-2014.

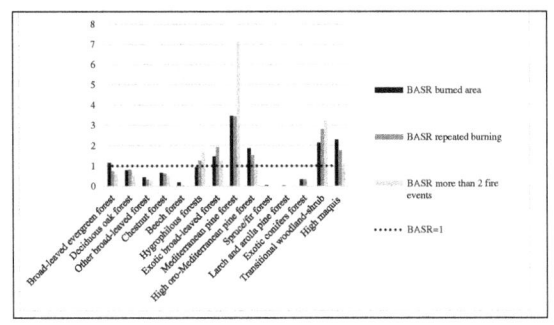

Figure 6 - BASR values of total burned area (black histogram), BASR values of total repeatedly burned area (dark grey histogram) and of repeatedly burned area with more than 2 fire events (light grey histogram) by woodland type; the dashed line represents BASR=1.

exotic broad-leaved forest (respectively, BASR = 2.8 and BASR = 1.9). In the case of area with more than two fire events Mediterranean pine forests are by far the most affected (BASR = 7.2), followed by transitional woodland-shrub (BASR = 3.2). 75 % of large fires and 86% of megafires have been affected by fire recurrence. Repeated burning affects 12% and 9% of the areas burned by large fires and megafires, respectively.

Discussion

Our study demonstrates how time series analysis of wildfires perimeters can help to figure out differences in fire occurrence and recurrence across different woodland types, over relatively large areas. Sinergy between administrative data (wildfire cadastre) and land cover maps (Corine Land Cover initiative) is particularly relevant in this case for identification of vulnerable areas to forest fires. In Italy, Mediterranean pine forest is the most preferred by fire, followed by high maquis, transitional woodland-shrub and high oro-Mediterranean pine forest. Fire selectivity is in line with previous research findings indicating that Mediterranean coniferous forests show the highest fire incidence compared with other woodland types (e.g. Gonzalez et al. 2006, Silva et al. 2009, Barros and Pereira 2014). The same types, with the addition of broad-leaved evergreen forest, have also the highest fire incidence for large fires. Thus, for these woodland type fire incidence does not change with fire size.

Fire recurrence affects Mediterranean pine forest too, followed by transitional woodland-shrub, high maquis, and high oro-Mediterranean pine forest. The high incidence of recurrent fire events in Mediterranean pine forest can lead to severe effects in terms of soil and vegetation degradation processes. Despite the high post-fire resilience, regeneration of Mediterranean pines may fail when time lag between fires is so short (Pausas et al. 2009, De las Heras et al. 2012).

Even the high fire recurrence in transitional shrub-woodland requires a careful land management, especially where such woodland type represents a stage of forest succession, as most often happens in Italy. Repeated fire disturbance may hamper or even reverse vegetation dynamics, and ultimately leads to land degradation.

In the considered period, there is an overall decreasing trend of total burned area, number of fires and average burned area. Our findings agree with the results by Turco et al. (2016) in the case of burned area. As concerns the trend of fire number, findings from previous studies are relatively controversial: Turco et al. (2016) ascribe these differences to the different methods, datasets and studied periods. The significant declining trend here observed can be mainly explained by improvements of fire prevention and suppression measures, even under a fire management governance perspective (Corona et al. 2015), as well as by progresses in public awareness and education occurred in the last decade.

Not surprisingly, large fires and megafires hold a significant share of total burned area, as observed over all the Mediterranean Europe (San-Miguel-Ayanz et al. 2013). On the other hand fire recurrence does not seem to affect such areas with an incidence higher than that of other burned areas.

A perhaps unexpected finding is the huge impact of fires in WUI areas, as concern both their with respect to total burned area (21%) and their average size, that is 1.7-fold larger than the average size of all fires. Our outcome is in accordance with Modugno et al. (2016) that showed a higher frequency of large fires in WUI areas in Italy as in other Mediterranean countries (e.g. Spain, France). This calls upon more and more fire management to come to terms with fire prevention in order to more effectively tackle such complex issue in the WUI, e.g. prioritally allocating there fire suppression resources, planning fuel modification treatments and educational programs.

Conclusions

The analytical approach here applied can be easily replicated under an international forestry perspective, e.g. in other countries where forest type maps and annual time series of wildfire perimeters are available on large scales. This technical note has shown examples on how to unlock the power of such tools whose information potential becomes even more higher as the time window of available data becomes longer.

The spatial analyses allow quantitative assessment of the phenomenon, providing information (e.g. BASR) that can be used e.g. for advancements in research (e.g. risk models), prioritization of fire prevention, suppression measures, economic incentive allocation, and landscape and peri-urban planning, as shown for WUI which has proven to have an unexpectedly huge impact in Italy. Ultimately, such analyses support basic knowledge for evidence-based understanding and management of fires, with special reference to those with short return period which are one of the main socio-ecological factors influencing the state and dynamics of Mediterranean woodlands (Filotas et al. 2014, Moreira et al. 2011).

Acknowledgements

We fully acknowledge the provision of burned areas geodaset by Italian Forest Service (Corpo Forestale dello Stato), Regional Forest Services of Sardinia (Corpo forestale e di vigilanza ambientale) and Sicilia (Corpo Forestale della Regione Sicili-ana). This work was partially supported by the Project MedWildFireLab ("Global Change Impacts on Wildland Fire Behaviour and Uses in Mediterranean Forest Ecosystems, towards a « wall less » Mediterranean Wildland Fire Laboratory") funded by ERANET FORESTERRA.

References

Bajocco S., Ricotta C. 2008 - *Evidence of selective burning in Sardinia (Italy): which land-cover classes do wildfires prefer?* Landscape Ecology 23 (2): 241-248.

Barbati A., Corona P., D'Amato E., Cartisano R. 2015 - *Is Landscape a Driver of Short-term Wildfire Recurrence?* Landscape Research 40 (1): 99-108.

Barbati A., Corona P., Marchetti M. 2007 - *A forest typology for monitoring sustainable forest management: the case of European forest types*. Plant Biosystems 141(1): 93-103.

Barbati A., Marchetti M., Chirici G., Corona P. - 2014. *European forest types and forest Europe SFM indicators: tools for monitoring progress on forest biodiversity conservation*. Forest Ecology and Management 321: 145-157.

Barros AM., Pereira JM. 2014 - *Wildfire selectivity for land cover type: does size matter?* PloS One 9 (1): e84760.

Carmo M., Moreira F., Casimiro P., Vaz P. 2011 - *Land use and topography influences on wildfire occurrence in northern Portugal*. Landscape and Urban Planning 100 (1): 169-176.

CFS 2014 - *Rapporto annuale incendi boschivi* [Annual report on forest fires]. Corpo Forestale dello Stato, Rome, Italy.

Chiriacò MV., Perugini L., Cimini D., D'Amato E., Valentini R., Bovio G., Corona P., Barbati A. 2013 - *Comparison of approaches for reporting forest fire-related biomass loss and greenhouse gas emissions in southern Europe*. International Journal of Wildland Fire 22 (6): 730-738.

CLC 2006 - Corine Land Cover 2006. http://www.sinanet.isprambiente.it/it/sia-ispra/download-mais/corine-land-cover/corine-land-cover-2006-iv-livello/view.

CLC 2012 - Corine Land Cover 2012. http://www.sinanet.isprambiente.it/it/sia-ispra/download-mais/corine-land-cover/corine-land-cover-2012-iv-livello/view.

Corona P., Ascoli D., Barbati A., Bovio G., Colangelo G., Elia M., Garfì V., Iovino F., Lafortezza R., Leone V. - 2015. *Integrated forest management to prevent wildfires under mediterranean environments*. Annals of Silvicultural Research 39 (1): 24-45.

Corona P., Ferrari B., Cartisano R., Barbati A. 2014 - *Calibration assessment of forest flammability potential in Italy*. iForest - Biogeosciences and Forestry 7 (5): 300.

De Las Heras J., Moya D., Vega, JA., Daskalakou E., Vallejo R., Grigoriadis N., Tsitsoni T., Baeza J., Valdecantos A., Fernandez C., Espelta J., Fernandes P. 2012 - *Post-Fire Management of Serotinous Pine Forests*. In: "*Post-Fire Management and Restoration of Southern European Forests*". Moreira F., Arianoutsou M., Corona P., De Las Heras, J. (Ed.). pp: 79-92. Springer, Netherlands.

EEA 2007 - CLC2006 technical guidelines. EEA Technical report No 17/2007, pp 66. http://www.eea.europa.eu/publications//technical_report_2007_17.

Fares S., Bajocco S., Salvati L., Camarretta N., Dupuis JL., Xanthopoulos G., Guijarro M., Madrigal J., Hernando C., Corona, P. 2017 - *Characterizing potential wildland fire fuel in live vegetation in the Mediterranean region*. Annals of Forest Science, 74: 1.

Filotas E., Parrott L., Burton PJ., Chazdon RL., Coates KD., Coll L., Haeussler S., Martin K., Nocentini S., Puettmann KJ., Putz FE., Simard SW., Messier C. 2014 - *Viewing forests through the lens of complex systems science*. Ecosphere 5(1): 1-23.

Ganteaume A., Camia A., Jappiot M., San-Miguel-Ayanz J., Long-Fournel M., Lampin C. 2013 - *A review of the main driving factors of forest fire ignition over Europe*. Environmental Management 51 (3): 651-662.

Gonzalez JR., Palahi M., Trasobares A., Pukkala T. 2006 - *A fire probability model for forest stands in Catalonia (northeast Spain)*. Annals of Forest Science 63 (2): 169-176.

Kendall MG. 1975 - *Rank correlation methods*, 4th edn. Charles Griffin, London.

Kolström M., Lindner M., Vilén T., Maroschek M., Seidl R., Lexer MJ., Netherer S., Kremer A., Delzon S., Barbati A., Marchetti M., Corona P. 2011 - *Reviewing the science and implementation of climate change adaptation measures in European forestry*. Forests 2 (4): 961-982.

Lampin-Maillet C., Jappiot M., Long M., Bouillon C., Morge D., Ferrier JP. 2010 - *Mapping wildland-urban interfaces at large scales integrating housing density and vegetation aggregation for fire prevention in the South of France*. Journal of Environmental Management 91 (3): 732-741.

Marques S., Borges JG., Garcia-Gonzalo J., Moreira F., Carreiras JMB., Oliveira MM., Cantarinha A., Botequim B., Pereira JMC. 2011 - *Characterization of wildfires in Portugal*. European Journal of Forest Research 130 (5): 775-784.

Modugno S., Balzter H., Cole B., Borrelli P. 2016 - *Mapping regional patterns of large forest fires in Wildland–Urban Interface areas in Europe*. Journal of Environmental Management 172: 112-126.

Moreira F., Vaz P., Catry F., Silva JS. 2009 - *Regional variations in wildfire susceptibility of land-cover types in Portugal: implications for landscape management to minimize fire hazard*. International Journal of Wildland Fire 18 (5): 563-574.

Moreira F., Viedma O., Arianoutsou M., Curt T., Koutsias N., Rigolot E., 2011 - *Landscape–wildfire interactions in southern Europe: implications for landscape management*. Journal of Environmental Management 92 (10): 2389-2402.

Nunes MC., Vasconcelos MJ., Pereira JM., Dasgupta N., Alldredge RJ., Rego FC. 2005 - Land cover type and fire in Portugal: do fires burn land cover selectively?. Landscape Ecology 20 (6): 661-673.

Oliveira S., Moreira F., Boca R., San-Miguel-Ayanz J., Pereira JM. 2014 - *Assessment of fire selectivity in relation to land cover and topography: a comparison between Southern European countries*. International Journal of Wildland Fire 23 (5): 620-630.

Pausas JG., Llovet J., Rodrigo A., Vallejo R. 2009 - *Are wildfires a disaster in the Mediterranean basin?–A review*. International Journal of Wildland Fire 17 (6): 713-723.

Pereira MG., Aranha J., Amraoui M. 2014 - *Land cover fire proneness in Europe*. Forest Systems 23 (3): 598-610.

Pezzatti GB., Bajocco S., Torriani D., Conedera M. 2009 - *Selective burning of forest vegetation in Canton Ticino (southern Switzerland)*. Plant biosystems 143 (3): 609-620.

Presidenza del Consiglio dei Ministri 2007 - *Manuale operativo per la predisposizione di un piano comunale o intercomunale di protezione civile*. O.P.C.M., N. 3606, Roma.

Ricotta C., Di Vito S. 2014 - *Modeling the landscape drivers of fire recurrence in Sardinia (Italy)*. Environmental Management 53 (6): 1077-1084.

Salvati L., Ferrara A., Mancino G., Kelly C., Chianucci F., Corona P. 2015 - *A multidimensional statistical framework to explore seasonal profile, severity and land-use preferences of wildfires in a Mediterranean country*. International Forestry Review 17 (4): 485-497.

San-Miguel-Ayanz J., Moreno JM., Camia A. 2013 - *Analysis of large fires in European Mediterranean landscapes: lessons learned and perspectives*. Forest Ecology and Management 294: 11-22.

Silva JS., Moreira F., Vaz P., Catry F., Godinho-Ferreira P. 2009 - *Assessing the relative fire proneness of different forest types in Portugal*. Plant Biosystems 143 (3): 597-608.

Stacey R., Gibson S., Hedley P. 2012 - *European Glossary for wildfires and forest fires*. EUFOFINET Project.

Turco M., Bedia J., Di Liberto F., Fiorucci P., Von Hardenberg J., Koutsias N., Llasat M., Xystrakis F., Provenzale A. 2016 - *Decreasing Fires in Mediterranean Europe*. PLoS one 11 (3): e0150663.

Vázquez A., Climent JM., Casais L., Quintana JR. 2015 - *Current and future estimates for the fire frequency and the fire rotation period in the main woodland types of peninsular Spain: a case-study approach*. Forest Systems 24 (2): e 031.

Xanthopoulos G., Calfapietra C., Fernandes P. 2012 - *Fire hazard and flammability of European forest types*. In: "Post-fire management and restoration of southern European forests". Moreira F., Arianoutsou M., Corona P., De Las Heras J. (Ed.). pp: 79-92. Springer, Netherlands.

Single-entry volume table for *Pinus brutia* in a planted peri-urban forest

Kyriaki Kitikidou[1*], Elias Milios[1], Kalliopi Radoglou[1]

Abstract - Brutia pine is a Mediterranean tree species of high ecological value, widely planted for soil protection, windbreaks and timber, both in its native area and elsewhere in the Mediterranean region. However, there is not yet enough information relating its growth dynamics and yield. The aim of this study was to evaluate the volume of Pinus brutia in a planted peri-urban forest (reforested area) in Greece. A single-entry, individual tree volume model has been developed using data from 18 permanent experimental plots, in the context of a research project regarding recovery of degraded coniferous forests..

Keywords - Pinus brutia; volume equation; volume estimation; volume table

Introduction

Brutia pine, also known as red pine, Turkish pine and Calabrian pine, is a widely distributed species, native to the eastern Mediterranean / western Asia regions. *Pinus brutia* is a fast-growing conifer, often associated with the related Aleppo pine (*Pinus halepensis*), extensively used in reforestations, in many degraded areas in Greece.

The wood of *Pinus brutia*, though resinous, can be successfully sawn and has higher density than *Pinus radiata* or *Pinus pinaster* (Raymond et al. 2004), while appropriate silviculture and genetic improvement can increase its growth rates and augment merchantable volume through improved stem form and branching (Arnold et al. 2005).

Pinus brutia is a drought-tolerant species, with fire resistant cones allowing it to successfully colonize dry, abandoned and burnt areas, particularly adapted to dry and cold sites and shallow, calcareous soils (Arnold et al. 2005). It has a remarkable adaptation to recurrent and severe fires, because it is an obligate seeder (Keeley et al. 2011), so knowledge about post-fire growth is useful for assessing not only current management practices, but also growth rate changes under climate warming (Sugihara et al. 2006). Climate affects forest fire regimes, in the short term, because weather rules fire ignition and propagation, and in the long term, because climate determines primary productivity, thus potential fuel and global fire patterns (Dale et al. 2000 and 2001, Urbieta et al. 2015). Moreover, research on the species yield dynamics and its impact on climate change via carbon storage, can contribute to the implementation and development of management strategies for climate change mitigation. Forest management practices should focus on maximizing increments, not stocks, in order to be more efficient under different climate scenarios. Volume dynamics are related to biomass increments, which should be maximized instead of standing biomass, since many regions in Europe have already high carbon stocks in forests (Kindermann et al. 2013).

The most usual way of estimating yield is through the use of volume tables (volume equations). Secondary variables such as diameter and height are used when it becomes difficult to measure volume directly in the field. Volume equations relating the tree volumes and auxiliary variables are applied to quicken this process. In developing volume tables, two variables can be used as dependent: individual tree volume or stand volume. Individual tree-based tables predict volume per tree, whereas stand volume tables predict volume per unit area (Philip 1994). The individual tree volume tables are further divided into three categories: local/single-entry, standard/double-entry, and form class/multiple-entry volume tables (Husch et al. 1982).

Local volume tables estimate tree volume using only the diameter at breast height, while standard tables are using diameter at breast height and height. Local volume tables are supposed to be restricted to

---

[1] Democritus University (Greece)
*kkitikid@fmenr.duth.gr

a local area. However, the terms "local" and "standard" do not in any way connote that one is better than the other. Both table categories are normally developed for a single species and specific region. The main difference between them is that local volume tables don't consider the height-diameter relationship. When this relationship is known, then a double-entry volume equation can be transformed to a single-entry volume equation (Husch et al. 1982). The form-class/multiple entry tables are different from the previous two, as they provide the volume in terms of some measure of form in addition to diameter and height. Examples of such form are the Girard form class and the absolute form quotient (Spurr 1952, Husch et al. 1982, Avery and Burkhart 1994).

Despite the ecological importance of Brutia pine, there is still a knowledge gap about the growth and yield properties of the species. This research is a preliminary investigation to assessing the productivity potential of *Pinus brutia*, providing a basis for future studies. The purpose of this study is to develop an individual tree volume table for the Brutia pine as simple as possible, i.e. a single-entry volume equation.

Materials and Methods

Study area

The peri-urban forest of Xanthi (41° 09' 27.33" N - 24° 54' 09.80" E) is located northern of Xanthi city, in northeastern Greece, and it covers an area of 2.366,137 ha (Theodoridis 2016) (Fig. 1).

The topography of the area varies, due to many hills, gorges and streams. Slopes vary from 5% to 80%, while the minimum altitude is 100 and the maximum 630 m above the sea level (information from Forest Service).

In 1936, planting activities began and took place periodically up to 2007, even though most of them were made till 1973. In 2006, the forest was designated as protective. The main species that were used for reforestation were mainly *Pinus brutia*, and secondarily *Pinus pinaster*, *Pinus pinea*, *Pinus nigra*, and *Cupressus spp*. A few broadleaves such as *Robinia pseudoacakia* were used as well (Theodoridis 2016).

According to the meteorological data from the closest meteorological station to the peri-urban forest (in the city of Xanthi), the mean annual temperature is 15.5 °C and the mean annual precipitation is 675 mm. The xerothermic period lasts from July till the middle of October (Papaioannou 2008). The soil is classified as alkaline with poor humus (Theodoridis 2016).

Experimental plots and data used

The most reliable sources of data for the estimation and modeling of growth and yield are the Permanent Sample Plots - PSPs. PSPs are classified into two groups: passive monitoring PSPs and experimental PSPs. The major difference between the two groups lies in the scope of their use; passive PSPs are used for monitoring only existing conditions, whereas the experimental plots are used for monitoring treatments like varying intensities of thinning (Alder and Synott 1992, Vanclay et al. 1995).

In the context of the LIFE14 CCM/IT/000905 project entitled "recovery of degraded coniferous FOrests for environmental sustainability, REStoration and climate change MITigation" (FORESMIT), 18 circular experimental PSPs with 13 m radius were placed in the study area in February 2016. The following measurements are used in the present work:
- diameter at breast height (d) of each tree, with caliper, in cm
- total height (h) of each tree, with Haglöf Vertex laser hypsometer, in m
- form height (fh) of the trees with $d \geq 15$ cm, with Bitterlich's Spiegel relaskop (first measurement with the relaskop at breast height).

The total volume v (m³) of each tree with $d \geq 15$ cm was derived following the formula (Van Laar and Akça 1997):

$$v = \frac{\pi}{4} d^2 1,3 + \frac{\pi}{4} d^2 fh$$

For each tree with d<15 cm its volume was calculated as a cylinder:

$$v = \frac{\pi}{4} d^2 h$$

Tree volume estimation

The mean tree method is one method for estimating stand volume and yield – there are various others. This method, and that of volume tables, shall be briefly discussed in this section.

The mean tree method of stand volume estimation

In the simplest of terms, in this method, the stand volume is obtained by carefully measuring the tree of mean volume and multiplying this volume

Figure 1 - Plot centers in the peri-urban forest of Xanthi, northeastern Greece.

by the total number of trees in the stand or plot (Spurr 1952).

The usual way of doing this is by getting the average volume of sub-sampled trees in each plot as the mean tree volume. The volume per hectare and the volume of each plot are calculated using this value and the number of trees.

This method involves two stages of sampling, with the sub-sampled trees being the second stage sample. For a precise estimate of the mean tree volume, the minimum sub-sampled size ought to be about 20 trees per plot (Philip 1994). This same postulation proposes pooling of the sub-sampled trees in all plots, to obtain a pooled tree of mean volume. However, this proposal comes with a warning: a serious bias can come out, if different plots provide different numbers of trees in the sub-sample and have different sized trees.

A common issue with this method is that the sub-sampled size is normally small, especially when there is a need for felling the sub-sampled trees to get detailed measurements. A substitute to this approach is founded on the assumption that the tree of mean volume is the one with the mean basal area (Spurr 1952, Crow 1971). Though this substitute approach offers some fairly positive results, a fallacy has been observed in the assumption (Spurr 1952). In this case, the mean tree is the tree that has a diameter approximately equal to the quadratic mean diameter of a sample of trees from the target stand.

Following this step the (mean) tree must be isolated in order to properly obtain the volume. After this, the plot volume estimate can be obtained by multiplying the total basal area of the plot by the ratio of the volume to the basal area of the mean tree (Schreuder et al. 1993).

Stand volume estimation using volume tables

The most usual way of estimating yield is through the use of volume tables. Secondary variables, such as diameter and height, are used when it becomes impossible to measure the individual tree volume in the field (Murchison 1984). In preparing volume tables, two variables can be used: single tree or stand volume. Single tree-based tables predict volume per tree, whereas stand volume tables predict volume per unit area (Philip 1994).

The single tree volume tables are further divided into three (3): local/single entry, standard/double entry, and form class/multiple entry volume tables (Husch et al. 1982).

Local volume tables present tree volume in terms of only the diameter at breast height (dbh). Tables that are restricted to a local area fall in this division. However, the terms "local" and "standard" do not in any way connote that one is greater than the other. Both table categories are normally prepared for single species or a group of species and specific localities. The main difference between these two divisions is that local volume tables don't generally consider the total height-dbh relationship. When the relationship is considered, then a standard volume table is the result (Husch et al. 1982).

The form-class/multiple entry tables are different from the previous two in that they provide the volume in terms of some measure of form in addition to dbh and total height. Examples of such form are the Girard form class and the absolute form quotient (Spurr 1952, Husch et al. 1982, Avery and Burkhart 1994).

One of the most common problems encountered in constructing volume tables is heteroscedasticity of residuals. Cunia (1964) proposed a solution to this problem. The proposed solution is through the use of weighted least squares when constructing the tables.

There are basically three methods that can be used for preparing a single tree volume table. The graphical method is the oldest and requires less mathematical techniques (Spurr 1952). The downside to this method is that it is prone to errors and subjectivity (Philip 1994, Spur 1952).

The next method is the alignment chart method for correcting curve linearity in multiple regression equations (Spurr 1952). A usual drawback of this method is the fact that prepared base charts are required – which are rarely available. Also, the charts cannot be read accurately as they are prone to errors associated to changes in paper dimensions (Spurr 1952).

A more modern and better method is the group of regression methods (Husch et al. 1982). Here, mathematical models and functions are used for preparing the tables. The advantage of this approach is the improved accuracy of the estimates. This method is applied in the present study.

Results

Data (v-d scatterplot) suggest that volume increases as diameter increases, following a trend that could be either linear or curve, with a constant term, as shown in Fig. 2. This was the reason for testing the following ten regression models ([1] to [10]) for fitting (Arlinghaus 1994):

Linear	$\hat{v} = b_0 + b_1 d$	[1]
Logarithmic	$\hat{v} = b_0 + b_1 \ln d$	[2]
Inverse	$\hat{v} = b_0 + \dfrac{b_1}{d}$	[3]
Quadratic	$\hat{v} = b_0 + b_1 d + b_2 d^2$	[4]
Cubic	$\hat{v} = b_0 + b_1 d + b_2 d^2 + b_3 d^3$	[5]
Compound	$\hat{v} = b_0 \, b_1^d$	[6]

Power $\quad \hat{v} = b_0\, d^{b_1}$ [7]

S-curve $\quad \hat{v} = e^{b_0 + \frac{b_1}{d}}$ [8]

Growth $\quad \hat{v} = e^{b_0 + b_1 t}$ [9]

Logistic $\quad \hat{v} = \dfrac{1}{\dfrac{1}{u} + b_0\, b_1^{t}}$

where u = upper boundary value = max h rounded up = 5.00 [10]

where:
$\hat{v}$: estimated volume (m³)
d: diameter at breast height (cm)
b_i (i = 1,2): regression coefficients.

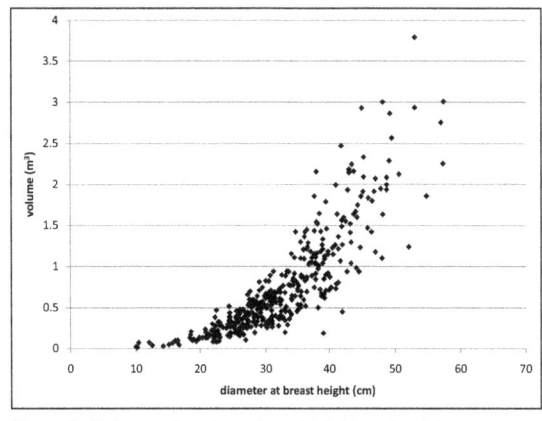

Figure 2 - Volume - diameter at breast height scatterplot.

Table 1 - Comparison criteria for tested regression models.

No	Criterion	Formula	Optimum value
1	Absolute mean error	$\dfrac{\sum_{i=1}^{n} \lvert v_i - \hat{v}_i \rvert}{n}$	0
2	Standard error of the estimate	$\sqrt{\dfrac{\sum_{i=1}^{n} (v_i - \hat{v}_i)^2}{n - p}}$	min
3	Coefficient of determination R^2	$1 - \sum_{i=1}^{n}(v_i - \hat{v}_i)^2 \Big/ \sum_{i=1}^{n}(v_i - \bar{v})^2$	1
4	Root of the mean squared error	$\sqrt{\dfrac{\sum_{i=1}^{n}(v_i - \hat{v}_i)^2}{n}}$	min
5	Sum of squared errors	$\sum_{i=1}^{n}(v_i - \hat{v}_i)^2$	0

where:
v: measured volume (m³)
$\hat{v}$: estimated volume (m³)
$\bar{v}$: average measured volume (m³)
$\bar{\hat{v}}$: average estimated volume (m³)
p: number of regression coefficients
n: number of observations (404 trees).

We tested the assumptions for the Least Squares Method, in order to fit regression equations to data, i.e.: autocorrelation, homoscedasticity and normality of residuals. Five criteria were used for model comparison (Draper and Smith 1998) (Tab. 1). Firstly, we checked the significance of regression coefficients; then we calculated the comparison criteria and selected the best regression model for volume estimation.

A summary of the statistics for the measured and calculated variables are given in Tab. 2.

Regression coefficients of all models were significant ($p<0.05$), except for the cubic model [5],

Table 2 - Summary statistics of individual tree variables.

Variable	Mean	Standard deviation	Minimum	Maximum
diameter at breast height d (cm)	32.19	8.37	10.00	57.40
total height h (m)	19.74	4.58	2.80	31.40
form height fh	6.39	3.21	.26	17.73
form factor $f = \dfrac{fh}{h}$	0.3185	0.1327	0.0147	0.7802
volume v (m³)	0.7464	0.6172	0.0148	3.7946

which had $p=0.097$ for the coefficient b_3 (all other coefficients had $p<0.05$). Therefore, the cubic model was excluded from further assessment. Comparison criteria values for the nine remaining models are given in Tab. 3 (best values for each criterion are highlighted).

Table 3 - Values for comparison criteria for tested regression models.

Criterion	1	2	3	4	5
Optimum	0	min	1	min	0
Model					
[1] Linear	0.229	0.324	0.741	0.323	42.229
[2] Logarithmic	0.269	0.384	0.637	0.383	59.275
[3] Inverse	0.326	0.463	0.473	0.462	86.119
[4] Quadratic	**0.192**	**0.291**	**0.792**	**0.290**	**34.030**
[6] Compound	0.233	0.432	0.540	0.431	75.117
[7] Power	0.197	0.298	0.781	0.298	35.780
[8] S-curve	0.234	0.379	0.647	0.378	57.674
[9] Growth	0.233	0.432	0.540	0.431	75.117
[10] Logistic	0.199	0.301	0.777	0.300	36.328

The quadratic model clearly excels over the other regression models. The selected volume-diameter model for *Pinus brutia* is $\hat{v}= 0.201\ 0.032d+0.001d^2$, with $R^2=0.792$ and standard error of the estimate=0.291.

Regarding the assumptions for the Least Squares Method, in residuals autocorrelation check, by applying the Durbin-Watson test, DW value was equal to 1.494, which is a value fairly within the confidence interval [1.5,2.5]; therefore, residuals are considered non-autocorrelated. Homoscedasticity was checked with the Kruskal-Wallis test ($p=0.058>0.05$). Finally, normality of residuals was checked with the Q-Q plot (Fig. 3); points are fairly close to the normal line.

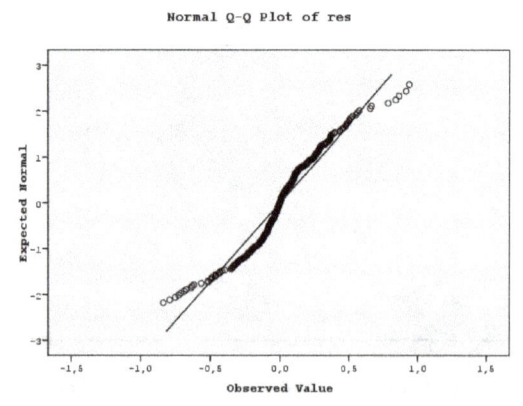

Figure 3 - Normality Q-Q plot of the residuals of the volume table.

Discussion

With this work, we have developed a single-entry equation for individual tree volume estimation, using a large sample size (404 trees from 18 permanent sample plots). Comparing this volume table with the equation of Özçelik et al. (2010) for *Pinus brutia* created for Burdur in Turkey, for the stands of the Bucak Forest Enterprise (Fig. 4), we observe that the peri-urban forest in Xanthi has higher individual tree volume, with the same diameter, than that of Turkey. In Fig. 4, the curve of Özçelik et al. (2010) $\hat{v} = 0.428753 d^{2.054628} h^{0.843735}$ was drawn using actual pairs of d and h from the database of the present work.

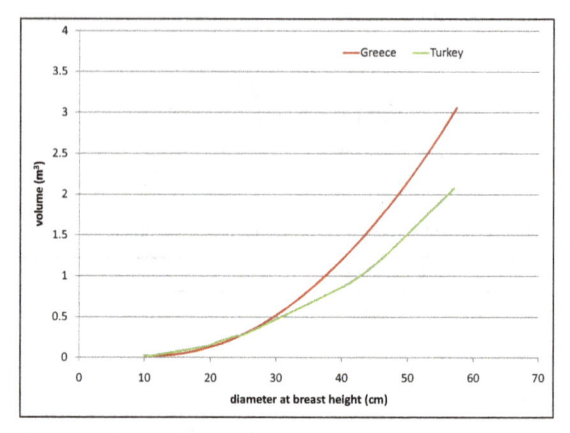

Figure 4 - Comparison of the two volume curves for *Pinus brutia* from Greece and Turkey.

In both areas stands were even-aged. Additional future research, based on climatic, geopedological, and stand structural conditions in Burdur, is essential before extracting conclusions regarding differences between the volume tables of Xanthi and Bucak.

Acknowledgments

This research was funded by the LIFE14 CCM/IT/000905 project entitled: recovery of degraded coniferous FOrests for environmental sustainability, REStoration and climate change MITigation (FORESMIT).

References

Alder D., Synott T. 1992 - *Permanent sample plot techniques for mixed tropical forests*. Tropical forestry paper No 25. Oxford Forestry Institute, UK, 124 p.

Arlinghaus S. 1994 - *Practical handbook of curve fitting (1st ed.)*. CRC Press, Boca Raton, USA. 272 p.

Arnold R., Bush D., Stackpole D. 2005 - *Genetic variation and tree improvement*. In: "New Forests: Wood production and environmental services". Sadanandan Nambiar E., Ferguson I. Eds, Csiro Publishing, Collingwood, Australia: 25-50.

Avery T., Burkhart H. 1994 - *Forest measurements (4th ed)*. McGraw-Hill Book Co. New York. 408 p.

Crow, T. 1971 - *Estimation of biomass in an even aged stand - regression and the "mean treen methods techniques*. pp.35-50. In: Young, H. (ed.) Forest biomass studies. XVTH IUFRO Congress. Univ. Of Florida.

Cunia, T. 1964 - *Weighted least squares method and the construction of volume tables*. Forest Science 10: 180-191.

Dale V., Joyce L., McNulty S., Neilson R. 2000 - *The interplay between climate change, forests, and disturbances*. Science of the Total Environment 262: 201-204.

Dale V., Joyce L., McNulty S., Neilson R., Ayres M., Flannigan M., Hanson P., Irland L., Lugo A., Peterson C., Simberloff D., Swanson F., Stocks B., Wotton M. 2001 - *Climate change and forest disturbances*. BioScience 51: 723-734.

Draper N., Smith H. 1998 - *Applied regression analysis (3rd edition)*. John Wiley and Sons Incorporation, London. 736 p.

Husch B., Miller C., Beers T. 1982 - *Forest mensuration (3rd ed)*. John Wiley and Sons Incorporation, New York, USA. 402 p.

Keeley J., Pausas J., Rundel P., Bond W., Bradstock R. 2011. *Fire as an evolutionary pressure shaping plant traits*. Trends in Plant Science 16: 406-411.

Kindermann G., Schörghuber S., Linkosalo T., Sanchez A., Rammer W., Rupert Seidl R., Lexer M. 2013 - *Potential stocks and increments of woody biomass in the European Union under different management and climate scenarios*. Carbon Balance Management 8(2). 20 p.

Murchison, H. G. 1984 - *Efficiency of multi-phase and multi-stage sampling for tree heights in forest inventory*. Ph.D Thesis. Univ. of Minnesota, St. Paul. MN. 158 p.

Özçelik R., Diamantopoulou M., Brooks J., Wiant H. 2010 - *Estimating tree bole volume using artificial neural network models for four species in Turkey*. Journal of Environmental Management 91(3): 742-753.

Papaioannou G. 2008. *The torrential environment of Kosynthos river*. MSc thesis, Democritus University of Thrace, Department of Forestry and Management of the Environment and Natural Resources, Greece.

Philip M. 1994. - *Measuring trees and forests (2nd ed)*. CAB International, Wallingford, UK. 31 p.

Raymond C., Dickson R., Rowell D., Blakemore P., Clark N., Williams M., Freischmidt G., Joe B. 2004 - *Wood and fiber properties of dryland conifers*. Rural Industries Research and Development Corporation (RIRDC) Publication No 04/099, New South Whales, Australia, 69 p.

Schreuder, T., Gregoire, T., Wood. G. 1993 - *Sampling methods for muliresource inventory*. John Wley and Sons, New York. 446 p.

Sugihara N., van Wagtendonk J., Fites-Kaufman J. 2006 - *Fire as ecological process*. In: "Fire in California's ecosystems". Sugihara N., van Wagtendonk J., Shaffer K., Fites-Kaufman J., Andrea E., Eds, University of California Press, Berkeley, USA: 58-74.

Smith D., Larson B., Kelty M., Ashton P. Mark S. 1997 - *The practice of silviculture: Applied forest ecology (9th edition)*. John Willey & Sons, Incorporated, New York, USA. 560 p.

Spurr S. 1952 - *Forest inventory*. The Ronald Press Co, New York, USA. 475 p.

Theodoridis P. 2016 - *Management study of the public forestry division of Xanthi-Geraka-Kimmerion*. Forest Services Xanthi-Stavroupoli, 2017-2026, vol.1, pp. 23-34. Xanthi Forest Directorate, Xanthi, Greece.

Urbieta I., Zavala G., Bedia J., Gutiérrez J., Miguel-Ayanz J., Camia A., Keeley J., Moreno J. 2015 - *Fire activity as a function of fire–weather seasonal severity and antecedent climate across spatial scales in southern Europe and Pacific western USA*. Environmental Research Letters 10: 114013.

Van Laar A., Akça A. 1997 - *Forest Mensuration*. Cuvillier Verlag, Göttingen, Germany. 418 p.

Vanclay J., Skovsgaard J., Hansen P. 1995. *Assessing the quality of permanent sample plot data for growth modelling in plantations*. Forest Ecology and Management 71: 177-186.

Use of Sentinel-2 for forest classification in Mediterranean environments

Nicola Puletti[1], Francesco Chianucci[2], Cristiano Castaldi[1*]

Abstract - Spatially-explicit information on forest composition provides valuable information to fulfil scientific, ecological and management objectives and to monitor multiple changes in forest ecosystems. The recently developed Sentinel-2 (S2) satellite imagery holds great potential for improving the classification of forest types at medium-large scales due to the concurrent availability of multispectral bands with high spatial resolution and quick revisit time. In this study, we tested the ability of S2 for forest type mapping in a Mediterranean environment. Three operational S2 images covering different phenological periods (winter, spring, summer) were processed and analyzed. Ten 10 m and 20 m bands available from S2 and four vegetation indices (VIs) were used to evaluate the ability of S2 to discriminate forest categories (conifer, broadleaved and mixed forests) and four forest types (beech forests; mixed spruce-fir forests; chestnut forests; mixed oak forests). We found that a single S2 image acquired in summer cannot discriminate neither the considered forest categories nor the forest types and therefore multitemporal images collected at different phenological periods are required. The best configuration yielded an accuracy > 83% in all considered forest types. We conclude that S2 can represent an effective option for repeated forest monitoring and mapping.

Keywords - Forest Classification; European Forest Types; Multispectral satellite imagery; Jeffries-Matusita (J-M) distance test; Random Forest

Introduction

Classification of forest categories and types is strongly required for addressing a wide range of ecological questions related to the determination of forest classes and/or successional stages (Laurin et al., 2013), rate of afforestation/deforestation (Hirose et al., 2016; Omruuzun et al., 2015), functional composition (Laurin et al., 2016) and global environmental changes (Trumbore et al., 2015). All these kinds of application require very fine mapping and monitoring of forest types, which have so far been limited by the spectral, spatial and temporal resolution available from current satellite open access data (e.g., Landsat, MODIS).

Previous studies using satellite multispectral sensors indicate that the visible and near-infrared wavelength regions are important for forest classification (Immitzer et al., 2012; Moore and Bauer, 1990; Waser et al., 2014). However, very few studies using multispectral sensors have evaluated the importance of red-edge bands for forest classification (e.g., Adelabu et al., 2013). As alternative to satellite remote sensing, unmanned aerial vehicles have recently gained increasing attention to obtain detailed information at local scale and at flexible temporal resolution, but their large scale applications in forestry are still at an experimental stage (Chianucci et al., 2016). Accurate discrimination of forest types is also essential for sustainable forest management and planning (Barbati et al., 2014), for estimating carbon stock (Noguiera et al., 2005) and for modelling the distribution of species and communities (Foody et al., 2003). Therefore, there is an increasing demand in both open access high quality data and quick turnaround series from remote sensing sensors for accurate mapping and monitoring of forest environments. This is particularly relevant for Mediterranean forests, which are characterized by high level of complexity (e.g., large number of species and variable canopy densities), which can complicate the discrimination of forest types from optical satellite imagery (Bajocco et al. 2013; Maselli et al., 2009; Pignatti et al., 2009). For example, previous studies indicated that Mediterranean forest types are often characterized by very high canopy density (Leaf Area Index, LAI > 5 $m^2 m^{-2}$, Chianucci, 2016; Chianucci et al., 2014; Chianucci and Cutini, 2013; Cinnirella et al., 2002; Thimonier et al., 2010), which can limit the retrieval of optical information from satellite data. Indeed, optical measures often saturate at leaf area index values of about 5 (Thenkabail et al., 2000), while vegetation indices using near-infrared (NIR) bands may saturate at

---

[1] CREA Research Centre for Forestry and Wood, Arezzo, Italy
[2] CREA Research Centre for Agriculture and Environment, Rome, Italy
*cristiano.castaldi@crea.gov.it

lower values (Davi et al., 2006; Turner et al., 1999). In addition, Mediterranean forests exhibits different phenological patterns according to forest categories and types, and therefore accurate temporal resolution data are strongly required for discriminating different forest types in these environments.

The recent Sentinel-2 (S2) mission, which started June 2015, holds great potential for the fine classification and monitoring of forest types on large scales (Baillarin et al., 2012). Even if S2 does not carry on a hyperspectral sensor, it was specifically conceived for vegetation sensing purposes and offers innovative features for environmental remote sensing (Immitzer et al., 2016). S2 can combine high spatial resolution, wide coverage and quick revisit time (about 5 days), which offers unprecedented opportunities for fine discrimination of land-cover classes. S2 carries a multispectral sensor with 13 bands, from 0.443 to 2.190 µm. The visible R, G, B and the NIR bands are available at a 10 m spatial resolution, highly suitable for application in vegetation canopies. Four red-edge bands at 20 m spatial resolution are also available and are particularly suited for chlorophyll content analysis and to parametrize ecophysiological large-scale models. Despite its potential, few studies have evaluated the ability of S2 in forest mapping and monitoring (Nelson, 2017). Immitzer et al. (2016) used actual S2 data for forest mapping, but they used pre-operational data without radiometric and geometric corrections, which hampers the comparison with other datasets. Indeed, the pre-operational data often showed artifacts which limit the consistency of the remotely-sensed information available from S2 (Immitzer et al., 2016; Vaiopoulos and Karantzalos, 2016). In addition, because of the relatively recent release of S2, most previous studies were based on simulated S2 data (e.g. Hill, 2013; Laurin et al., 2016).

The main objective of this study was to evaluate the capability of S2 operational data (i.e. after the correction from ESA) in classifying both forest categories (pure coniferous forests, pure broadleaves forests, and mixed forests) and European Forest Types (EFT; Barbati et al 2014) in a Mediterranean environment, which has not been possible before due to limited pre-operational S2 data availability. Because of the different phenological patterns of Mediterranean forests, we also compared the use of multitemporal data versus single time data for the discrimination of forest categories and types from S2.

2. Material and methods

2.1. Test site

The study was carried out in an extensive forest area (about 470 km^2) located in the Eastern part of the Tuscany Region (Figure 1). A Region Of Interest (ROI) made of 1,061 stands distributed over three forest compartments (3,960 ha)was created. The forest types that covered the study area, according to the European classification (Barbati et al., 2014), are the Apennine-Corsican mountainous beech forests (EFT code 7.3), the Thermophilous deciduous forests dominated by chestnut (EFT code 8.7), the Subalpine and mountainous spruce and mountainous mixed spruce-silver fir forest (EFT code 3.2), and the Turkey oak, Hungarian oak and Sessile oak forest (EFT code 8.2), which cover 13.4%, 35.3%, 8.9%, and 38.1% of the forest surface in the AOI, respectively.

2.2. S2 products collection and pre-processing

Sentinel-2 features 13 spectral bands with 10, 20 and 60 m spatial resolution (Table 1) at 12 bit radiometric resolution. For the remainder of the analysis, we focused only on 10 m and 20 m bands (Table 1). The three 60 m spatial resolution bands were not used in this study because they are primarily relevant for atmospheric corrections. In this work, three S2 products were downloaded as Level-1C Top-of-Atmosphere reflectance products from the Scientific Hub (https://scihub.copernicus.eu): one relative to winter (January, 2017, product code "S2A MSIL1C 20170104T101402 N0204 R022 T32TQP 20170104T101405"), one to spring (March,

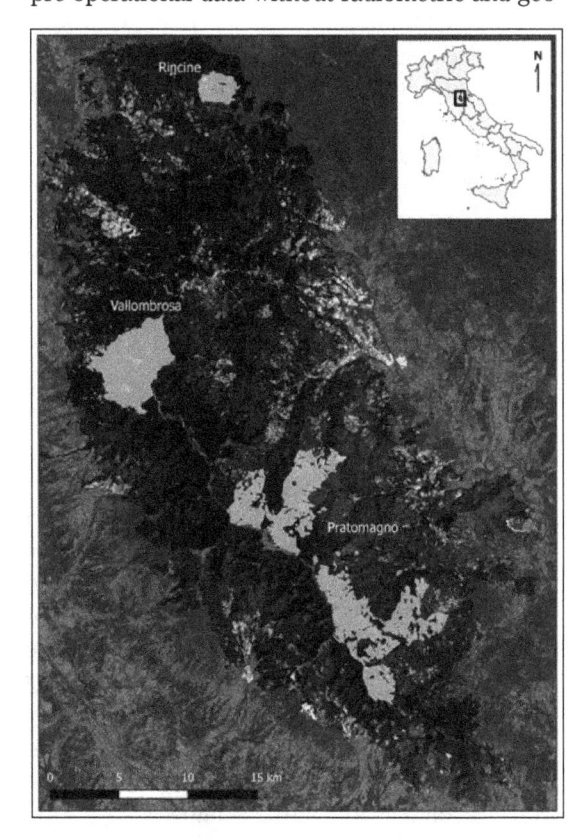

Figure 1 - True colour composition of the entire study area from Sentinel-2 imagery. The Regions of Interest (ROI) have been labelled in white.

Table 1 - Spectral bands available from Sentinel-2. Only bands with finer spatial resolution (i.e. 10 m and 20 m) have been used in this work.

Sentinel-2 Bands	Central Wavelength (μm)	Spatial Resolution m
Band 1 - Coastal aerosol	0.443	60
Band 2 - Blue	0.490	10
Band 3 - Green	0.560	10
Band 4 - Red	0.665	10
Band 5 - Red Edge	0.705	20
Band 6 - Red Edge	0.740	20
Band 7 - Red Edge	0.783	20
Band 8 - NIR	0.842	10
Band 8A - Red Edge	0.865	20
Band 9 - Water vapour	0.945	60
Band 10 - SWIR - Cirrus	1.375	60
Band 11 - SWIR	1.610	20
Band 12 - SWIR	2.190	20

2017, product code "S2A MSIL1C 20170315T101021 N0204 R022 T32TQP 20170315T101214") and one to summer (June, 2017, product code "S2A MSIL1C 20170613T101031 N0205 R022 T32TQP 20170613T101608").

The products were resampled at a resolution of 10 m by the Sentinel Application Platform (SNAP), available at the ESA website (http://step.esa.int/main/toolboxes/snap). Finally, the 10 bands were imported in ENVI software, stacked and cropped over the area of interest.

2.3. Model assessment

The separability between forest categories (pure coniferous forests, pure broadleaves forests, and mixed forests) and forest types (EFTs) was evaluated. Firstly, we define the classification ability of winter, spring and summer products separately using ten S2 bands resampled at 10 m. Vegetation indices were also included in a second step of the analysis; considering the used spectral bands, we computed the Normalized Difference Vegetation (NDVI), the Simple Ratio (SRI), the red-edge Normalized Difference Vegetation (RENDVI) and the Anthocyanin Reflectance Index 1 (ARI1) indices (Table 2).

As suggested from other experiences (Puletti et al., 2016; Laurin et al., 2016), a preliminary analysis on ROIs separability was performed by Jeffries-Matusita (J-M) distance test applied to the validation set (see Table 3). The value of the J-M measurement ranges from 0 to 2.0 and indicates how the selected ROI pairs are statistically separated: values above 1.8 indicate a statistically good separability (Richards and Jia, 1999). As a third step, we repeated the separability analysis by combining (layer-stack) the three S2 images, to explore the capability of S2 to improve the forest type classification using multitemporal information. All J-M analyses were performed in ENVI software.

The best configuration obtained from J-M distance test was used to classify the S2 products using the Random Forest method (Breiman, 2001). This method requires two input parameters: the number of predictor variables performing the data partitioning at each node and the total number of trees to be grown in the model run. For categorical classification based on the Random Forest model, the number of predictor variables was set as the square root of the number of predictor within the dataset used in the study (Liaw and Wiener, 2002).

2.4. Model evaluation

The ROI pixels were randomly divided into training and validation sets (Table 3 and Table 4), using a proportion of 70% and 30% respectively. The classification analysis was performed using the 'randomForest' package in R (R Core Team, 2017). The Random Forest model was built over the training set; the Overall Accuracy (OA), Producer Accuracy (PA), User Accuracy (UA) and Kappa coefficient of the classification were computed using information from the contingency matrix, obtained applying the model to the entire validation set (Congalton, 1991).

Table 2 - The vegetation indices calculated from S2 imagery. ρ refers to the reflectance value of the S2 band considered. For the identification of the band number used, see Table

Vegetation Index	Formula
NDVI	$(\rho_{842}-\rho_{665})/(\rho_{842}+\rho_{665})$
SRI	ρ_{842}/ρ_{665}
RENDVI	$(\rho_{740}-\rho_{705})/(\rho_{740}+\rho_{705})$
ARI1	$(1/\rho_{560})-(1/\rho_{705})$

Table 3 - Number of 10x10 m pixels used in this study as ROI for forest category classification, distinguished by training and validation sets.

Forest Compartment	Forest category	Training set (number)	Validation set (number)	Total number of pixels
Pratomagno	coniferous	24,684	10,579	35,263
	broadleaves	75,230	32,242	107,472
	mixed	78,437	33,616	112,053
Rincine	coniferous	8,862	3,798	12,660
	broadleaves	2,770	1,187	3,957
	mixed	7,244	3,105	10,349
Vallombrosa	coniferous	25,901	11,100	37,001
	broadleaves	17,003	7,287	24,290
	mixed	37,028	15,869	52,897
	Total	277,159	118,783	395,942

Table 4 - Number of 10x10 m pixels used in this study as ROI for forest type classification (EFT), distinguished by training and validation sets. EFT 3.2: Subalpine and mountainous spruce and mountainous mixed spruce-silver fir forest; EFT 7.3: Apennine-Corsican mountainous beech forest; EFT 8.2: Turkey oak, Hungarian oak and Sessile oak forest; EFT 8.7: Chestnut forest.

	Training set (number)	Validation set (number)	Total number of pixels
EFT 3.2	29,336	12,572	41,908
EFT 7.3	85,850	36,793	122,643
EFT 8.2	6,944	2,976	9,920
EFT 8.7	49,935	21,401	71,336
Total	172,065	73,742	245,807

Table 5 - J-M scores for pairs of forest groups (Con-pure conifers; Broad-pure broadleaves; Mix-mixed) using single date and multitemporal S2 imagery.

Input	Pair	Winter	Spring	Summer	Multitemporal
10 bands	Mix vs Con	0.67	0.56	0.47	1.14
	Mix vs Broad	1.30	1.52	0.89	1.87
	Broad vs Con	1.71	1.83	1.59	1.98
10 bands + NDVI	Mix vs Con	0.94	1.77	0.59	1.95
	Mix vs Broad	1.32	1.94	1.09	1.97
	Broad vs Con	1.80	1.99	1.71	1.99
10 bands + SRI	Mix vs Con	0.94	1.77	0.52	1.92
	Mix vs Broad	1.31	1.82	0.91	1.94
	Broad vs Con	1.80	1.99	1.61	1.99
10 bands + RENDVI	Mix vs Con	0.93	1.80	0.58	1.95
	Mix vs Broad	1.32	1.95	1.04	1.98
	Broad vs Con	1.80	1.99	1.69	1.99
10 bands + ARI1	Mix vs Con	0.73	0.77	0.52	1.30
	Mix vs Broad	1.33	1.54	0.98	1.88
	Broad vs Con	1.77	1.87	1.60	1.98
10 bands + 4 VIs	Mix vs Con	1.21	1.96	0.69	2.00
	Mix vs Broad	1.54	2.00	1.25	2.00
	Broad vs Con	1.93	2.00	1.78	2.00

2.5 Map production

The validated model was applied to the entire study area (470 Km^2) and 10 m spatial resolution maps for both forest categories and EFTs have been obtained.

3. Results

3.1. J-M test

Results from spring image indicated that the single image 10 bands discriminated well between broadleaved and needleleaved forests, regardless of the phenological acquisition period, but did not differentiate mixed forests (Table 5). Including the vegetation indices only slightly increased the separability between these classes, which was significant only for spring image using RENDVI. When exploiting the multitemporal information of S2, the J-M separability markedly improved compared with

Table 6 - J-M scores for pairs of selected EFT. EFT 3.2: Subalpine and mountainous spruce and mountainous mixed spruce-silver fir forest; EFT 7.3: Apennine-Corsican mountainous beech forest; EFT 8.2: Turkey oak, Hungarian oak and Sessile oak forest; EFT 8.7: chestnut forest.

Input	Pair	J-M score
10 bands (spring) + 10 bands (summer)	EFT 3.2 / EFT 7.3	1.93
	EFT 3.2 / EFT 8.2	1.96
	EFT 3.2 / EFT 8.7	1.78
	EFT 7.3 / EFT 8.2	1.98
	EFT 7.3 / EFT 8.7	1.92
	EFT 8.2 / EFT 8.7	1.63
10 bands + RENDVI (spring) + 10 bands + RENDVI (summer)	EFT 3.2 / EFT 7.3	2.00
	EFT 3.2 / EFT 8.2	1.98
	EFT 3.2 / EFT 8.7	1.88
	EFT 7.3 / EFT 8.2	2.00
	EFT 7.3 / EFT 8.7	1.99
	EFT 8.2 / EFT 8.7	1.73

the single date analysis (Table 5). The solely combination of bands did not allow the discrimination of the forest categories (J-M value of 1.14 for mixed vs coniferous, see Table 5), and inclusion of VIs was therefore required (Table 5).

As described in Figure 2, the same procedure has been adopted for EFTs classification (Table 6). In this case, due to results obtained in the first

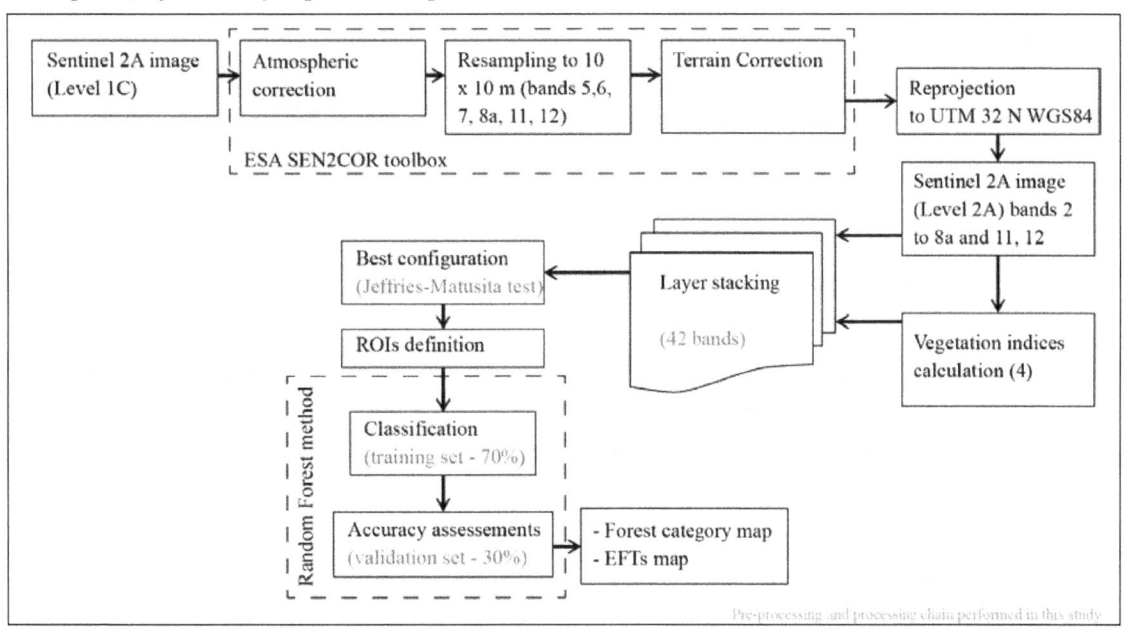

Figure 2 - Flowchart of the Random Forest methodology implemented in the study.

Table 7 - Confusion matrix of the best configuration result (see Table 5) for forest group classification, expressed as number of pixels of the validation set.

pred	true Pure broadleaves	Pure coniferous	Mixed
Pure broadleaves	33,825	526	3,205
Pure coniferous	416	17,383	3,814
Mixed	6,475	7,568	45,571

Table 8 - Confusion matrix of the best configuration result (see Table 6) for EFT classification, expressed as number of pixels of the validation set. EFT 3.2: Subalpine and mountainous spruce and mountainous mixed spruce-silver fir forest; EFT 7.3: Apennine-Corsican mountainous beech forest; EFT 8.2: Turkey oak, Hungarian oak and Sessile oak forest; EFT 8.7: Chestnut forest.

pred	true EFT 3.2	EFT 7.3	EFT 8.2	EFT 8.7
EFT 3.2	10,919	943	33	476
EFT 7.3	949	34,944	95	776
EFT 8.2	15	26	2,498	120
EFT 8.7	689	880	350	20,029

Table 9 - Accuracy by the Random Forest classifier applied to the validation set of forest groups. CE: Commission Error; OE: Omission Error; PA: Producer Accuracy; UA: User Accuracy.

EFT	CE (%)	OE (%)	PA (%)	UA (%)
Broadleaves	8.43	12.78	91.47	87.22
Coniferous	15.10	22.26	84.90	77.74
Mixed	16.94	10.49	83.06	89.51

Table 10 - Accuracy by the Random Forest classifier applied to the validation set of EFT. CE: Commission Error; OE: Omission Error; PA: Producer Accuracy; UA: User Accuracy. EFT 3.2: Subalpine and mountainous spruce and mountainous mixed spruce-silver fir forest; EFT 7.3: Apennine-Corsican mountainous beech forest; EFT 8.2: Turkey oak, Hungarian oak and Sessile oak forest; EFT 8.7: Chestnut forest.

EFT	CE (%)	OE (%)	PA (%)	UA (%)
EFT 3.2	11.74	13.15	88.26	86.85
EFT 7.3	4.95	5.03	95.05	94.97
EFT 8.2	6.05	16.06	93.95	83.94
EFT 8.7	8.74	6.41	91.26	93.59

step of analysis, only spring and summer images and relative RENDVI have been used to get best results based on J-M scores. Results indicated that the single image bands discriminated well between all considered EFT with exception for forests dominated by oaks and chestnut (Table 6). The inclusion of RENDVI improves the separability between these EFTs.

The best configuration, respectively made of 33 layers (10 bands for winter, 10 for spring, 10 for summer, and the 3 RENDVI) and 22 layers (10 for spring, 10 for summer, and the 2 RENDVI) for forest categories and EFTs, have been separately used as input variables in Random Forest.

3.2. Random forest classification

The confusion matrix and accuracy results are reported in Table 7 and in Table 8. For forest categories, an overall accuracy of 86.2% and a Kappa coefficient of 86.1% have been obtained (Table 9).

For EFTs, an overall accuracy of 92.7% and a Kappa coefficient of 92.6% have been obtained. All the classes reveal comparable producer accuracy, although slightly lower user accuracy was observed in Subalpine and mountainous spruce and mountainous mixed spruce-silver fir forest (Table 10). However, both the PA and UA are above 83% in all the classes. A ranking of variables indicated that the summer bands in Blue, Red-Edge wavelength are the most important for classification (Figure 3).

The validated models were applied to the entire study area and both forest categories and EFTs maps have been produced (Figure 4).

4. Discussion and conclusion

In this study, we demonstrated the effective per-

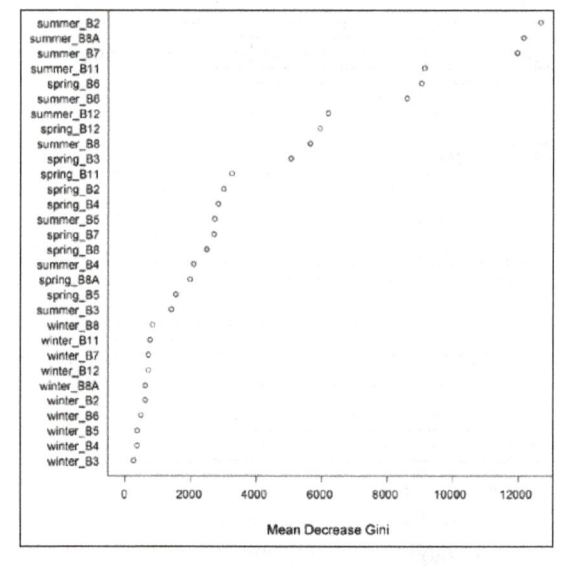

Figure 3 - Random Forest variable importance of S2 bands for EFT classification.

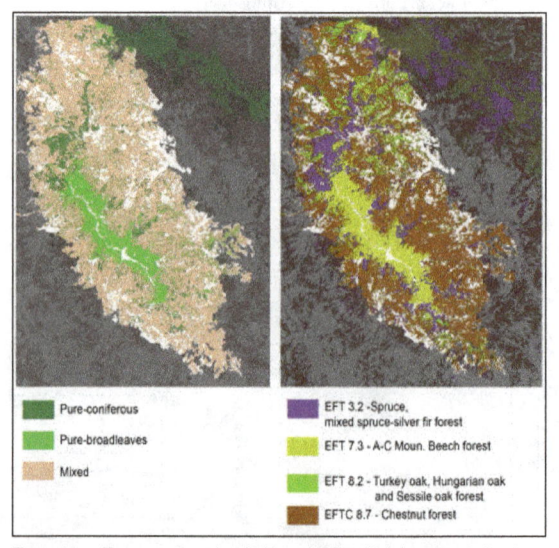

Figure 4 - Forest categories (left) and EFTs (right) maps derived from models prediction over the study area.

formance of S2 in forest mapping based on real operational data under Mediterranean forest environments. We obtained accurate discrimination of EFTs using a single summer image and VIs. We attributed the results mainly to the high spatial resolution available from the 10 m S2 bands and the capability of S2 to include red-edge bands. Indeed, previous studies indicated that using narrow-bands located in the red edge can overcome the well-known problem of saturation of NIR-based vegetation indices (Mutanga et al., 2004). Thus, the availability of four red-edge bands in S2 holds great potential to improve the applicability of optical remote sensing of forests compared with past satellite data (Sellers, 1985; Todd et al., 1998; Gao et al., 2000; Thenkabail et al., 2000).

At another level, we observed that single image data was not able to significantly discriminate forests categories nor forest types, unless VIs are included in the analysis. As expected, the inclusion of multitemporal data gave best results in the classification.

The use of ESA corrected images (L1C level) allows robust S2 product derivation which is prerequisite for standardizing the protocol of image processing and allows comparability among future studies involving operational S2 data.

We conclude that S2 data have been proved to be suitable for routine, medium to large scale mapping and monitoring of forest changes due to the combination of high spatial resolution and quick revisit time.

Acknowledgments

The study was supported by the Project "AL-ForLab" (PON03PE_00024_1) co-funded by the Italian Operational Programme for Research and Competitiveness (PON R,C) 2007-2013, through the European Regional Development Fund (ERDF) and national resource (Revolving Fund - Cohesion Action Plan (CAP) MIUR).

References

Adelabu, S., Mutanga, O., Adam, E., Cho, M.A., 2013. *Exploiting machine learning algo- rithms for tree species classification in a semiarid woodland using RapidEye image*. J. Appl. Remote. Sens. 7 (1) (073480–1–073480–13).

Baillarin S.J., Meygret A., Dechoz C., Petrucci B., Lacherade S., Tremas T., Isola C., Martimort P. Spoto F. 2012 - *Sentinel-2 level 1 products and image processing performances*. IEEE International Geoscience and Remote Sensing Symposium, 7003-7006.

Bajocco S., Raparelli E., Patriarca F., Di Matteo G., Nardi P., Perini L., Salvati L., Mugnozza G.S.,2013 - *Exploring forest infrastructures equipment through multivariate analysis: complementarities, gaps and overlaps in the Mediterranean basin*. Annals of Silvicultural Research 37(1):1-6. doi: 10.12899/asr-774.

Barbati A., Marchetti M., Chirici G., Corona P. 2014 - *European forest types and forest Europe SFM indicators: tools for monitoring progress on forest biodiversity conservation*. Forest Ecology and Management 321: 145-157.

Bartholomé E., Belward A.S. 2005 - *GLC2000: A new approach to global land cover mapping from Earth observation data*. International Journal of Remote Sensing 26: 1959–1977.

Breiman L. 2001 - *Random forests*. Machine Learning, 45(1): 5–32.

Chianucci F. 2016 – *A note on estimation canopy cover from digital cover and hemispherical photography*. Silva Fennica 50, doi: 10.14214/sf.1518

Chianucci F., Cutini A. 2013 - *Estimation of canopy properties in deciduous forests with digital hemispherical and cover photography*. Agricultural and Forest Meteorology 168: 130-139.

Chianucci F., Disperati L., Guzzi D., Bianchini D., Nardino V., Lastri C., Rindinella A., Corona P. 2016 - *Estimation of canopy attributes in beech forests using true colour digital images from a small fixed-wing UAV*. International Journal of Applied Earth Observation and Geoinformation 47: 60-68.

Chianucci, F., Puletti, N., Venturi, E., Cutini, A. and Chiavetta, U. 2014. *Photographic assessment of overstory and understory leaf area index in beech forests under different management regimes in Central Italy*. Forestry Studies, 61(1), pp.27-34.

Cinnirella S., Magnani F., Saracino A., Borghetti M. 2002 - *Response of a mature Pinus laricio plantation to a three-year restriction of water supply: structural and functional acclimation to drought*. Tree Physiology 22(1): 21-30.

Congalton R. G. 1991 - *A review of assessing the accuracy of classifications of remotely sensed data*. Remote Sensing of Environment 37: 35–46.

Davi H., Soudani K., Deckx T., Dufrene E., Le Dantec V., Francois, C. 2006 - *Estimation of forest leaf area index from SPOT imagery using NDVI distribution over forest stands*. International Journal of Remote Sensing 27: 885–902.

Fassnacht F.E., Latifi, H., Stereńczak K., Modzelewska A., Lefsky M., Waser L.T., Straub C., Ghosh A. 2016 - *Review of studies on tree species classification from remotely sensed data*. Remote Sensing of Environment 186: 64-87.

Feddema J.J., Oleson K.W., Bonan G.B., Mearns L.O., Buja L.E., Meehl G.A., Washington W.M. 2005 - *The importance of land-cover change in simulating future climates*. Science 310: 1674–1678.

Foody G. M., Boyd D. S., Cutler M. E. 2003 - *Predictive relations of tropical forest biomass from Landsat TM data and their transferability between regions*. Remote sensing of environment 85(4): 463-474.

Gao X, Huete A.R., Ni W., Miura, T. 2000 - *Optical-biophysical relationships of vegetation spectra without background contamination*. Remote Sensing of Environment 74: 609–620,

Hansen M.C., DeFries R.S., Townshend J.R., Sohlberg R. 2000 - *Global land cover classification at 1 km spatial resolution using a classification tree approach*. International Journal of Remote Sensing, 21(6-7): 1331-1364.

Hill M.J. 2013 - *Vegetation index suites as indicators of vegetation state in grassland and savanna: An analysis with simulated SENTINEL 2 data for a North American transect*. Remote Sensing of Environment 137: 94-111.

Hirose K., Osaki, M., Takeda, T., Kashimura O., Ohki, T., Segah H., Gao Y., Evri, M. 2016 - *Contribution of Hyperspectral Applications to Tropical Peatland Ecosystem Monitoring*. Tropical Peatland Ecosystems (pp. 421-431). Springer Japan.

Immitzer, M., Atzberger, C., Koukal, T., 2012. *Tree species classification with random Forest using very high spatial resolution 8-Band WorldView-2 satellite data*. Remote Sens. 4 (9), 2661–2693.

Immitzer M., Vuolo F., Atzberger C. 2016 - *First experience with sentinel-2 data for crop and tree species classifications in Central Europe*. Remote Sensing 8(3): 166.

Laurin G.V., Liesenberg V., Chen, Q., Guerriero, L. Del Frate, F., Bartolini, A., Coomes D., Wilebore, B., Lindsell, J., Valentini, R. 2013 - *Optical and SAR sensor synergies for forest and land cover mapping in a tropical site in West Africa*. International Journal of Applied Earth Observation and Geoinformation 21: 7-16.

Laurin G.V., Puletti N., Hawthorne W, Liesenberg V., Corona P. Papale D., Chen Q., Valentini R. 2016 - *Discrimination of tropical forest types, dominant species, and mapping of functional guilds by hyperspectral and simulated multispectral Sentinel-2 data*. Remote Sensing of Environment 176: 163-176.

Liaw A., Wiener M. 2002 - *Classification and regression by randomForest*. R news 2, no. 3:18-22.

Loveland, T.R., Merchant J.W., Brown J.F., Ohlen D.O., Reed B.C., Olson P., Hutchinson J. 1995 - *Map supplement: Seasonal land-cover regions of the United States*. Annals of the American Association of Geographers 85: 339–355.

Maselli F., Moriondo M., Chiesi M., Chirici G., Puletti N., Barbati A., Corona P. 2009 - *Evaluating the effects of environmental changes on the Gross Primary Production of Italian forests*. Remote Sensing, 1(4): 1108-1124.

Moore, M.M., Bauer, M.E., 1990. *Classification of forest vegetation in North-Central Min- nesota using Landsat multispectral scanner and thematic mapper data*. For. Sci. 36 (2), 330–342.

Mutanga O., Skidmore A.K. 2004 - *Narrow band vegetation indices overcome the saturation problem in biomass estimation*. International Journal of Remote Sensing 25: 3999-4014.

Nogueira E.M., Yanai A.M., Fonseca F.O., Fearnside P.M. 2015 - *Carbon stock loss from deforestation through 2013 in Brazilian Amazonia*. Global change biology 21: 1271-1292.

Omruuzun F., Baskurt D.O., Daglayan H., Cetin Y.Y. 2015 - *Utilizing hyperspectral remote sensing imagery for afforestation planning of partially covered areas*. In SPIE Remote Sensing (pp. 96432N-96432N). International Society for Optics and Photonics.

Petropoulos G.P., Kalaitzidis C., Vadrevu K.P. 2012 - *Support vector machines and object-based classification for obtaining land-use/cover cartography from Hyperion hyperspectral imagery*. Computers, Geosciences, 41: 99-107.

Pignatti S., Cavalli R.M., Cuomo V., Fusilli L., Pascucci S., Poscolieri M., Santini F. 2009 - *Evaluating Hyperion capability for land cover mapping in a fragmented ecosystem: Pollino National Park, Italy*. Remote Sensing of Environment 113(3): 622-634.

Puletti N., Camarretta N., Corona P. 2016 - *Evaluating EO1-Hyperion capability for mapping conifer and broadleaved forests*. European Journal of Remote Sensing 49: 157-169.

R Core Team 2017 - *R: A language and environment for statistical computing*. R Foundation for Statistical Computing, Vienna, Austria. URL https://www.R-project.org/.

Richards J.A., Jia X. 1999 - *Remote sensing digital imaging analysis: an introduction*. (3rd ed.). Berlin, Springer.

Sellers P.J. 1985 - *Canopy reflectance, photosynthesis and transpiration*. International Journal of Remote Sensing 6: 1335–1372.

Sellers P.J., Dickinson R.E., Randall D.A., Betts A.K., Hall F.G., Berry J.A., Collatz G.J., Denning A.S., Mooney H.A., Nobre C.A., Sato N. 1997 - *Modeling the exchanges of energy, water, and carbon between continents and the atmosphere*. Science 275: 502–509.

Sophie B., Pierre D. 2009 - *GlobCover 2009 Products Description and Validation Report* European Space Agency: Paris, France, 201

Thenkabail P.S, Smith R.B., De Pauw E. 2000 - *Hyperspectral vegetation indices and their relationships with agricultural crop characteristics*. Remote Sensing of Environment 71: 158–182.

Thimonier A., Sedivy I., Schleppi P. 2010 - *Estimating leaf area index in different types of mature forest stands in Switzerland: a comparison of methods*. European Journal of Forest Research 129: 543-562.

Todd S.W, Hoffer R.M., Milchunas D.G. 1998 - *Biomass estimation on grazed and ungrazed rangelands using spectral indices*. International Journal of Remote sensing, 19: 427–438.

Trumbore S., Brando P., Hartmann H. 2015 - *Forest health and global change*. Science 349: 814-818.

Turner P.D., Cohen W.B., Kennedy R.E., Fassnacht K.S., Riggs J.M. 1999 - *Relationships between leaf area index and Landsat TM spectral vegetation indices across three temperate zone sites*. Remote Sensing of Environment 70: 52–68.

Vaiopoulos A.D., Karantzalos K. 2016 - *Pansharpening on the Narrow Vnir and SWIR Spectral Bands of SENTINEL-2*. ISPRS-International Archives of the Photogrammetry, Remote Sensing and Spatial Information Sciences. 723-730.

Vizzarri M., Chiavetta U., Chirici G., Garfì V., Bastrup-Birk A., Marchetti M. 2014 - *Comparing multisource harmonized forest types mapping: A case study from central Italy*. iForest 8: 59-66. doi: 10.3832/ifor1133-007

Waser, L.T., Küchler, M., Jütte, K., Stampfer, T., 2014. *Evaluating the potential of World- View-2 data to classify tree species and different levels of ash mortality*. Remote Sens. 6 (5), 4515–4545.

The role of dominant tree cover and silvicultural practices on the post-fire recovery of Mediterranean afforestations

Ilaria Cutino[1*], Salvatore Pasta[2], Concetta Valeria Maggiore[3], Emilio Badalamenti[3], Tommaso La Mantia[3]

Abstract - Fire is one of the major disturbance factors in Mediterranean-type ecosystems, where since long time man has deeply modified the natural fire regime. To know how woody species recover after fire is of prominent importance for understanding vegetation dynamics, as well as for the management of Mediterranean plantations, especially where broadleaved and coniferous trees coexist. Our research was carried out at Monte Petroso (Sicily), within an historical afforestation intervention in the Mediterranean basin. We assessed the post-fire response of mixed oaks and oak-pine afforestations within six experimental plots (two plots per homogeneous sector) differing in dominant tree species (*Quercus ilex* or *Pinus pinea*), time since last wildfire (1954 or 1982), and post-fire management (understory cleaning and removal of dead biomass or no management). Dendrometric surveys and phytosociological relevés were carried out to characterize the tree layers, the regeneration by woody species plus *Ampelodesmos mauritanicus*, as well as plant species richness. Our field surveys have confirmed a notably high resilience to fire by Mediterranean woody species, regardless of post-fire management practices. The dominant tree species played a significant role as *Quercus ilex* seems to foster stand development and the regeneration dynamics in the understory, especially that of *Quercus pubescens*. By contrast, *Pinus pinea* seems to slow down the regeneration by woody species, especially at higher stand density. Post-fire management practices seemed to favor mantle shrubs (*Prunetalia spinosae*) and grassland species (*Hyparrhenietalia hirtae*), while negatively affecting shrub species (*Cisto-Ericetalia multiflorae*). In presence of sufficient propagules of native woody species, the option of no management after fire has to be considered. The results of our research may be useful to improve the management of fire-prone Mediterranean plantations, taking into account the differences in plant strategies to cope with fire, as well as the dominant canopy.

Keywords - *Quercus* spp.; *Pinus pinea*; regeneration; forest management; vegetation dynamics

Introduction

Fire plays a prominent role in shaping Mediterranean plant communities (Bengtsson et al. 2000, Lloret et al. 2002, García-Jiménez et al. 2017). Its millennial selective pressure induced in Mediterranean plants a considerable resilience to it (Buhk et al. 2006, Keeley et al. 2011), i.e. a great ability to return to a pre-disturbance state (Hanes 1971, Trabaud and Galtié 1996, Moya et al. 2011). Mediterranean woody species may adopt two main strategies to cope with fire: sprouting and seeding (Keeley and Zedler 1978, Calvo et al. 2003). Species such as *Cistus* spp. and *Rhamnus alaternus* L. are seeders, as they increase germination after fire passage, thus implying the existence of a resistant soil seed bank (Crosti et al. 2006). Other species such as thermophilous oaks (*Quercus coccifera* L., *Quercus ilex* L. and *Quercus pubescens* Willd.) and *Erica* spp. are resprouters, responding to fire through the emission of new shoots, either by roots, by stumps or by damaged stems (Buhk et al. 2006, Pausas 2006, Moreira et al. 2013). Despite both strategies may coexist in the same woody species, one strategy tends to prevail over the other (Kazanis and Arianoutsou 2004). The Mediterranean conifers do not resprout after a disturbance event, thus exclusively relying on sexual reproduction and seed traits (Díaz-Delgado et al. 2002, Pausas et al. 2008, García-Jiménez et al. 2017). Serotinous pines such as *Pinus halepensis* Mill. and *Pinus pinaster* Aiton are particularly favoured by fire (Moreira et al. 2012), while non-serotinous pines such as *Pinus nigra* J.F. Arnold and *Pinus sylvestris* L. lack of specific adaptations to fire (Retana et al. 2012). *Pinus pinea* L., despite not bearing serotinous cones, yet it has hard-coated seeds and thick bark giving it a good resistance to fire passage and some post-fire natural regeneration (Escudero et al. 1999). The response of both seeders and resprouters is severely hindered when fire frequency exceeds some threshold, and large differences among these functional groups, as well as among woody species exist (Pausas 2006). *Pinus* spp. are known to be more negatively affected by and less resilient to short-interval recurrent fires than *Quercus* spp. (Pausas et al. 2008, Vallejo and Alloza 2012). Within oaks, broadleaved species are characterized by slower recovery time after fire than

---

[1] CREA Difesa e Certificazione, Firenze (Italy)
[2] Departement de Biologie, Université de Fribourg, Fribourg (Switzerland)
[3] Department of Agricultural, Food and Forest Sciences, Università degli Studi di Palermo, Palermo (Italy)
*ilaria.cutino@crea.gov.it

evergreen species, requiring even more than 50-60 years (Trabaud and Galtié 1996, Moreira et al. 2011). Evergreen-sclerophyllous species have an inherent higher ability to overcome post-fire limiting stressful conditions (e.g.: water stress due to irradiance) (Espelta et al. 2012).

The effective recovery after fire is also strongly dependent on wildfire characteristics, such as type, frequency, intensity and seasonality, which all play a crucial role on forest stand dynamics and tree species response (Bellingham and Sparrow 2000, Corona et al. 2015). The frequency of the disturbance has large effects on landscape traits (Trabaud and Galtié 1996, Moreira et al. 2011, Guiomar et al. 2015), as well as on the structure and composition of plant communities (Pausas 2006, Espelta et al. 2008). Wildfires with moderate intensity and low frequency may exert a positive effect on biodiversity in the Mediterranean basin (Maggiore et al. 2005), and may also accelerate the renaturalization process of plantations, fostering the development of complex, more species-rich and stable forest stands, dominated by native broadleaved species (Curt et al. 2009, Badalamenti et al. 2017, 2018). By contrast, the landscape homogenization, the reduction of the resilience of forest stands, and a prevalence of shrub species over tree species generally result from increased fire frequencies (Díaz-Delgado et al. 2002, Moreira et al. 2011). Since ancient times man has changed the natural fire regime in the Mediterranean basin, increasing its frequency, intensity and potential fuel loads (Moreira et al. 2012, San-Miguel-Ayanz et al. 2013). Within Mediterranean plantations the lack of specific fuels treatments or of understory management has significantly enhanced the potential fuel load and continuity, increasing potential fire severity (Corona et al. 2015). As increasingly frequent and intense fire events are expected in the Mediterranean basin (Moreira et al. 2013), to know how woody species respond to fire is of primary importance to understand the evolution of Mediterranean vegetation, as well as for improving the management of fire-prone Mediterranean plantations. After the massive afforestation interventions carried out throughout the Mediterranean basin during the last century, large areas are now covered with artificial plantations (Pausas et al. 2004, Maetzke et al. 2017), mainly dominated by introduced Pines and Eucalypts (Moreira et al. 2013, Rühl et al. 2015). Such tree species were mainly chosen with the aim to increase forest cover, to provide effective defence against soil erosion and to stabilize unstable and degraded slopes (Pausas et al. 2004, Corona et al. 2009, Rühl et al. 2015), due to their pioneer character (e.g.: initial fast growth rates), limited ecological requirements as well as easy propagation. Native woody species were seldom used. The large preference accorded to non-native woody species has led to the formation of simplified forest stands in the Mediterranean basin, with low overall biodiversity (Granados et al. 2016). However, renaturalization processes by native woody species have been increasingly observed within many Mediterranean afforestations, that were gradually converted into multilayered and complex mixed pine-oak forests (Sheffer 2012). As concerns fire, this has further complicated the overall picture. In fact, as said before, pines and oaks adopt different strategies to cope with fire (Granados et al. 2016). In recent years, the knowledge of post-fire vegetation dynamics and management in the Mediterranean basin has greatly increased (Buhk et al. 2006, Moreira et al. 2012). Many recent studies have underlined the role played by post-fire management practices on the recovery ability of different biological communities (Bros et al. 2011, Rollan and Real 2011), including plants (Moreira et al. 2012). However, field investigations about the effects of fire on plant species richness and regeneration by woody species of mixed plantations are relatively scarce in the Mediterranean basin (Curt et al. 2009). Mediterranean plantations are frequently affected by wildfires and their response to such disturbance factor is of crucial importance for an effective management. Furthermore, such simplified forest ecosystems include both seeders and resprouters, so that their after fire is quite unpredictable and needs to be experimentally tested (Bellingham and Sparrow 2000). To increase our knowledge of such fire-prone and quite heterogeneous Mediterranean forest ecosystems, we assessed the long-term post-fire response of mixed plantations in areas differing in post-fire treatments (understory management vs. no management), time since last wildfire occurrence (1954 or 1982) and dominant tree species (*Quercus ilex* or *Pinus pinea*). More in detail, we aimed at assessing the role of dominant tree species and post-fire forest interventions in post-fire recovery of woody species in Mediterranean afforestations. We believe our results contribute in increasing knowledge about post-fire vegetation response in the Mediterranean basin in order to better manage afforested areas prone to frequent wildfires.

Material and methods

Study site

The study was carried out at "*Monte Petroso*", (663 m a.s.l.) located about 3 Km from Palermo city, on the Western sector of Palermo plain (Sicily) (Fig. 1). Dolomitic limestone is the most common outcropping rock (Abate et al. 1982), while the dominant soil association includes Lithic Xerorthens,

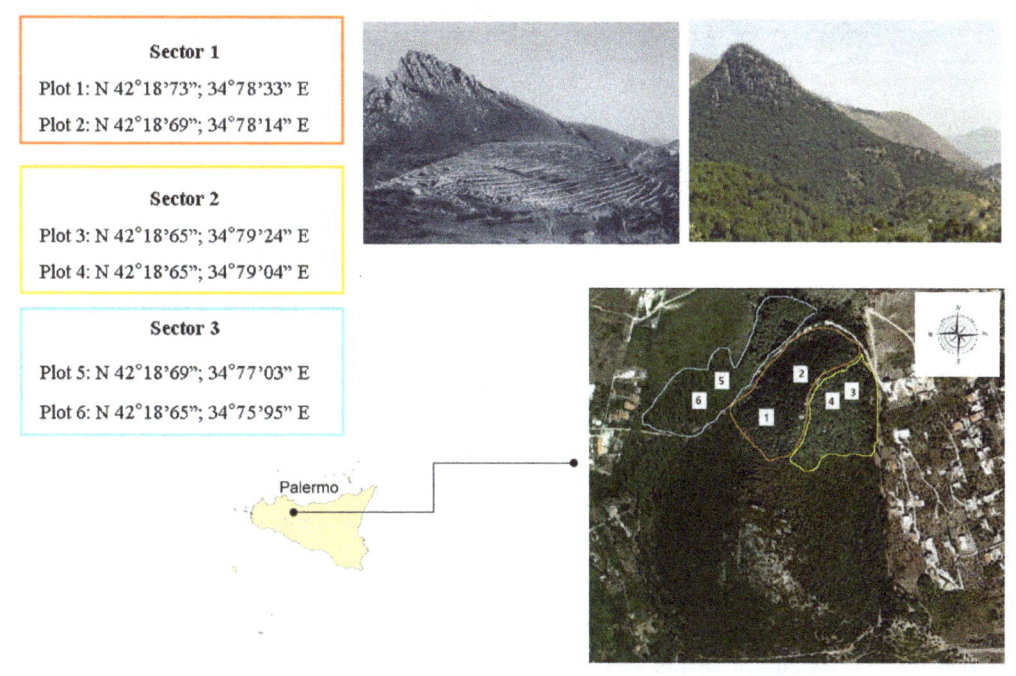

Figure 1 - Location of the study area (Mt. Petroso). Numbers from 1 to 6 show the location of the sampling plots. The picture on the center of upper part was taken in 1934 and shows the SW slopes, while the picture on the upper right corner shows the S-SW slopes.

Rock Outcrop, Typic and Lithic Rhodoxeralfs (Fierotti 1988). According to Rivas-Martínez (2008), the study area falls within the thermo-Mediterranean bioclimatic belt.

The reforestation of Mt. Petroso, extending on approx. 54 hectares (Troía et al. 2011), is a very well documented case study in the Mediterranean. The afforestation techniques adopted have a remarkable historical value (AA. VV. 1943), representing a paradigmatic case for the whole biogeographic area. A photo taken in 1934, illustrating the ongoing forest interventions, is reported under the word "rimboschimento" (= afforestation) within an Italian Encyclopaedia (AA. VV. 1943). All the slopes of Monte Petroso were reforested with native oak species (*Quercus ilex*, *Q. pubescens* and *Q. suber*) and manna ash (*Fraxinus ornus* L.), while locally non-native conifers (*Pinus pinea*, *P. halepensis* and *Cupressus sempervirens* L.) were mainly planted at low-altitude slopes. Mixed forest stands and afforestations are found between 500 and 600 m a.s.l., while above 600 m the slopes are covered by native tree species, which survived on rocky cliffs and crevices. Further details on the vascular flora and vegetation of Mt. Petroso are given by Troìa et al. (2011).

Selection and characteristics of the sampling plots

Before investigating the vegetation response, a preliminary analysis was carried out by means of aerial photos coming from the archives of the Italian Military Geographical Institute, dating back to October 1954 and June 1968, and from the archives of the Agriculture and Forest Office of Sicilian regional government, dating back to June 1978, June 1987 and October 1997. Information provided by photo interpretation suggested us to split the overall study area (Mt. Petroso) into three homogeneous sectors, according to the main forest types and time since last wildfire. Then, through dedicated field surveys, we validated the classification of the three main sectors obtained from photointerpretation. The first sector includes the northern and northwestern slopes which suffered wildfire in 1982. The second sector includes the northeastern slopes, burned in 1954 but spared by flames in 1982. In both sectors a mixed oaks forest with *Quercus ilex* as dominant tree species is present. The third sector includes stone pine forests subject to the fire of 1982, which destroyed the shrub and the herb layers, without significantly affecting the dominant tree layer. Two plots have been established within each sector, for a total of six sampling plots (Table 1). It was possible to consider small sample size (2 plots x 3 sectors) due to the relatively homogeneous conditions of the forest stands within each sector. The investigated plots mainly differ in terms of dominant tree species (*Quercus ilex* or *Pinus pinea*), time since last wildfire (1954 or 1982), and post-fire management practices, including understory cleaning and removal of dead biomass every two years in managed plots and no intervention in unmanaged plots. The aerial photos were also used to verify that no other wildfire has occurred in the investigated plots after 1954 and 1982, respectively. As we learned from interviews to local foresters and forest workers,

Table 1 - Overview of the main physical and silvicultural characteristics of sampling plots.

Parameter	Sector (Plot)					
	1 (1)	1 (2)	2 (3)	2 (4)	3 (5)	3 (6)
Altitude (m a.s.l.)	525	525	485	525	480	500
Slope (°)	27	16	25	30	0	6
Aspect	NW	N	NE	NE	flat	SE
Stone outcrop (%)	< 30	< 30	< 30	< 30	< 5	< 30
Rock outcrop (%)	~50	~50	~50	~50	< 5	< 5
Year since last wildfire	1982	1982	1954	1954	1982	1982
Dominating tree species	Q. ilex	Q. ilex	Q. ilex	Q. ilex	P. pinea	P. pinea
After fire understory management	every two years	none	every two years	none	every two years	none

post-fire management practices started soon after the fire passage, both in 1954 and 1982, and were regularly carried out every two years until spring 2005, when the field surveys were carried out.

Sampling method, dendrometric, regeneration and floristic surveys

Forest resources were sampled using a simplified version of MNTFR (Monitoring of Non-Timber Forest Resources), as proposed by Rühl et al. (2005). Field surveys were carried out within a circular sampling plot of about 450 m^2 (radius = 12 m). Plots from 1 to 4 are dominated by *Quercus* spp., mainly holm oak, whereas stone pine is the dominant species within plots 5 and 6. For each tree species, the following dendrometric parameters have been assessed: density, diameter at breast height (DBH), total height and crown height, crown projection along the four cardinal directions, plant position according to a Cartesian-coordinate system, whose centre is one of the sampling plot borders (Table 2).

Phytosociological relevés according to Braun-Blanquet (1964) were performed within a circular subplot having the same centre of the main plot and an area of about 200 m^2 (radius = 8 m). The plot size is in line with other studies on the response of Mediterranean vegetation after fire (e.g.: Álvarez et al. 2009, Tessler et al. 2016). Plant nomenclature follows Pignatti (1982), Conti et al. (2005) and Raimondo et al. (2010). Regeneration and shrub layer were assessed in the same subplots, assessing density, basal diameter and total height (Table 3). *Ampelodesmos mauritanicus* (Poir.) T. Dur. & Schinz (hereafter *Ampelodesmos*), a widespread native perennial grass, was also included in the field investigations for its recognized importance for understanding Mediterranean vegetation dynamics after fire (Incerti et al. 2013). For this grass species we considered the tussock basal diameter, as assessed in other grass species (e.g.: Yu et al. 2011). In order to provide a spatial representation of the vegetation structure within the selected plots, the software SVS (Stand Visualisation System) 3.30 (McGaughey 1997) was used. From a phytosociological point of view, the investigated vegetation is referred to the class *Quercetea ilicis* Br.-Bl. ex A. & O. de Bolòs 1947, order *Quercetalia ilicis* Br.-Bl. ex Molinier 1934 em. Rivas-Mart. 1975, and alliance *Quercion ilicis* Br.-Bl. ex Molinier 1934 em. Brullo, Di Martino & Marcenò 1977 (Brullo et al. 2009).

Results

Dendrometric surveys

Dendrometric data for the dominant tree layer are provided in Table 2, whereas data for regeneration layers, including trees, as well as shrubs plus *Ampelodesmos* are provided in Tables 3 and 4, respectively. The structural representation of the investigated plots is reported in figure 2.

Table 2 - Dendrometric parameters of the dominant tree species within the investigated plots.

Sector (Plot)	Woody species	Frequency (%)	Stumps (N ha^{-1})	Coppice shoots (N ha^{-1})	Stems (N ha^{-1})	DBH (cm)	Total height (m)	Crown height (m)	Crown cover (m^2)	Basal area (m^2 ha^{-1})
1 (1)	Quercus ilex[2]	89.9	995	3'736	-	7.17	6.65	2.86	2.36	16.11
	Quercus pubescens[2]	8.0	133	332	-	9.56	6.95	4.28	2.48	2.60
	Quercus suber[2]	2.1	-	-	88	8.50	5.05	2.58	0.85	0.60
1 (2)	Fraxinus ornus	20.0	177	1'238	-	3.75	3.79	1.51	0.44	1.60
	Quercus ilex[2]	64.3	796	3'979	-	4.12	2.74	0.75	1.11	7.83
	Quercus pubescens[2]	8.6	88	531	-	14.67	4.80	1.77	6.36	9.66
2 (3)	Quercus ilex[2]	72.4	442	1'857	-	8.81	7.66	2.28	3.20	13.20
	Quercus pubescens[2]	6.9	-	-	177	28.00	7.90	2.80	23.35	10.90
	Quercus suber[2]	20.7	88	531	-	21.67	6.18	2.13	12.80	22.80
2 (4)	Quercus ilex[2]	97.1	1'680	5'924	-	4.13	3.49	0.99	2.45	22.94
	Quercus pubescens[2]	2.9	-	-	177	15.00	8.75	4.00	10.47	3.14
3 (5)	Pinus pinea[1]	6.7	-	-	531	36.67	17.35	14.68	62.57	57.42
	Quercus ilex[2]	91.1	2'299	7'250	-	3.51	3.28	0.82	1.15	8.55
3 (6)	Pinus pinea[1]	41.2	-	-	619	27.21	13.93	8.93	31.45	36.33
	Quercus ilex[2]	58.8	265	884	-	3.55	3.37	0.83	2.05	0.95

[1]Only planted; [2]Mostly planted

Table 3 - Density and dendrometric parameters of the tree regeneration layer.

Sector (Plot)	Species	Density (N ha^{-1})	Basal diameter (cm)	Height (m)
1 (1)	Quercus ilex	2'403	0.75	0.42
	Quercus pubescens	420	0.79	0.23
	Quercus suber	44	0.70	0.30
	Rhamnus alaternus	66	2.55	0.50
1 (2)	Quercus ilex	1'238	2.25	1.39
	Quercus pubescens	1'149	1.02	0.95
	Fraxinus ornus	88	0.80	0.80
	Rhamnus alaternus	619	2.86	1.49
2 (3)	Quercus ilex	19'983	1.39	0.84
	Quercus pubescens	2'299	0.23	0.08
	Rhamnus alaternus	265	1.45	0.57
2 (4)	Quercus ilex	6'366	0.83	0.76
	Quercus pubescens	88	0.50	0.40
3 (5)	Quercus ilex	2'653	0.34	0.28
	Quercus pubescens	796	0.28	0.18
3 (6)	Quercus ilex	1'592	0.73	0.86
	Quercus pubescens	88	0.50	0.60
	Rhamnus alaternus	707	1.58	0.57

Table 4 - Density and dendrometric parameters of the shrub layer plus Ampelodesmos.

Sector (Plot)	Species	Density (N ha^{-1})	Basal diameter (cm)	Height (m)
1 (1)	Ampelodesmos mauritanicus	1'172	8.60	0.30
	Calicotome infesta	44	8.60	0.30
	Erica multiflora	155	3.10	0.29
1 (2)	Cistus creticus	2'387	1.78	0.40
	Emerus major subsp. major	2'299	5.00	1.22
	Spartium junceum	531	4.13	1.38
	Teucrium flavum	354	1.98	0.50
2 (3)	Ampelodesmos mauritanicus	3'095	10.20	0.44
	Calicotome infesta	442	2.70	0.74
	Cistus salviifolius	88	2.30	0.40
	Spartium junceum	442	4.82	1.20
	Teucrium flavum	177	1.65	0.40
2 (4)	Ampelodesmos mauritanicus	2'829	8.30	0.74
	Calicotome infesta	265	2.85	0.87
	Cistus salviifolius	88	2.30	0.40
	Teucrium flavum	1'061	1.95	0.67
	Teucrium fruticans	1'592	1.45	0.41
3 (5)	Ampelodesmos mauritanicus	3'979	5.35	0.30
	Cistus creticus	88	2.00	1.57
	Crataegus monogyna	531	2.58	1.50
3 (6)	Ampelodesmos mauritanicus	1'680	4.86	0.30
	Erica multiflora	4'244	2.00	0.80
	Teucrium flavum	884	2.00	0.80

Plot 1 is dominated by holm oak (frequency approaching 90%), with only few downy oak and cork oak individuals occurring in the tree layer. During 1982, a fire completely destroyed the tree layer. About 1'000 stumps with more than 3'700 coppice shoots per hectare of holm oak were found. Mean DBH for all tree species was 8.4 cm, while mean height was 6.2 m. The total basal area was 19.3 m^2 ha^{-1}. Holm oak regeneration was particularly abundant, exceeding 2'400 individuals per hectare. Such high value has also to be attributed to the limited occurrence and poor development of shrubs, which have been seriously damaged by the continuous understory cleaning practices. Only *Calicotome infesta* and *Erica multiflora* L. occur in the shrub layer both, all represented by a few individuals. *Ampelodesmos* regeneration is quite abundant, exceeding 1'000 individuals per hectare.

Plot 2 is still dominated by *Quercus ilex* (frequency of more than 60%), but *Q. pubescens* and *Fraxinus ornus* are also present in the plot, both with well established natural regeneration. However, about 800 stumps with almost 4'000 coppice shoots per hectare of holm oak were found. More than 1'200 coppice shoots of *Fraxinus* also occur *Q. pubescens* is less common in the dominant layer, yet it showed a natural regeneration as abundant as that of *Q. ilex*, both with more than 1'000 individuals per hectare. Mean DBH for all tree species was 7.5 cm, while mean height was 3.8 m. The total basal area was 19.1 m^2 ha^{-1}. However, the largest trees belong to *Q. pubescens*, this species reaching a mean DBH of almost 15 cm. The area is almost impenetrable because it is largely covered by shrubs such as *Cistus creticus* L. and *Emerus major* Mill. subsp. *major*, both quite abundant in terms of density and with large mean size.

Plot 3 is formed by a mixed oak forest stand dominated by *Quercus ilex* (frequency of more than 70%), followed by *Q. suber* (frequency of more than 20%) and *Q. pubescens* (frequency of about 7%), reaching a total tree cover of 80%. Holm oak regeneration was partially originated from the seeds produced by mature plants (density of almost 20'000 individuals per hectare), and partially from coppice shoots after coppicing (density of more than 1'800 coppice shoots per hectare). *Q. pubescens* showed a lower regeneration potential, with seedlings often

Figure 2 - Structural representation of the vegetation structure within the investigated plots.

not still established and with reduced size. Mean DBH was quite different among tree species, ranging from 8.8 cm by holm oak to 28.0 cm by cork oak. Downy oak reached a mean DBH of 21.7 cm. Mean height for all tree species was 7.2 m, while the total basal area was about 47 m^2 ha^{-1}. Despite the occurrence of four shrubs, the understory is largely covered by the dominant grass *Ampelodesmos*, exceeding 3'000 individuals per hectare.

Plot 4 is strongly dominated by *Quercus ilex* (frequency of about 97%), with only few *Q. pubescens* individuals. The tree regeneration layer is mainly formed by holm oak individuals, belonging to two different growth stages. The largest seedlings are concentrated along the edge of a terrace, where there are some burned stumps. The smallest seedlings can hardly stand out from the shrub layer. Because of the abundance of sprouts there is a strong competition among coppice shoots, being more than 6'000 individuals per hectare. Natural regeneration is widespread despite being mainly concentrated under the stone terraces and still characterized by poor development. Mean DBH ranged from 4.1 cm to 15.0 cm and mean height ranged from 3.5 m to 8.8 m in holm oak and downy oak, respectively. The total basal area was 26.1 m^2 ha^{-1}. Despite the tree layer reached a noteworthy cover of 70%, the shrub layer is well represented and mainly formed by *Teucrium flavum* L. and *Teucrium fruticans* L., together reaching a total cover of about 40% and both exceeding 1'000 individuals per hectare. *Ampelodesmos* is abundant as well, with more than 2'800 individuals per hectare.

Plot 5 shows a different structure of vegetation. Two main tree layers may be recognized: the dominant layer is formed by larger *Pinus pinea* individuals, which cover most of the plot area (density: 531 stems ha^{-1}), despite representing less than 7% of all the tree individuals. *Pinus* reached a mean DBH of 36.7 cm and a mean height of 17.4 m. After fire passage, no sign of natural regeneration by stone pine was observed, suggesting a very low regeneration performance. The dominated layer is composed by smaller holm oak individuals, having a mean DBH of 3.5 cm and a mean height of 3.3 m. The total basal area was 66.0 m^2 ha^{-1}. Despite the tree cover is 80%, it is possible to distinguish a third woody layer, formed by two shrubs (*Cistus creticus* and *Crataegus monogyna* Jacq.) and by a very recent regeneration by holm oak (more frequently) and downy oak. The understory is largely covered by the dominant grass *Ampelodesmos*, reaching almost 4'000 individuals per hectare.

Plot 6 is a stone pine forest with a more closed canopy, and a holm oak dominated layer. *Pinus pinea* individuals cover most of the plot area (density: 619 stems ha^{-1}), and reached a mean DBH of 27.2 cm and a mean height of 13.9 m. After fire passage, natural regeneration by stone pine was almost absent. The dominated layer is composed by smaller holm oak individuals, having a mean DBH of 3.6 cm and a mean height of 3.4 m. The total basal area was 37.3 m^2 ha^{-1}. The tree regeneration layer is mainly composed by holm oak, reaching about 1'600 individuals per hectare. The regeneration of holm oak, downy oak and *Rhamnus alaternus* has a gamic origin. The understory layer is mostly formed by an *Erica multiflora* L. open maquis with a cover of 60% and more than 4'000 individuals per hectare. The regeneration of *Ampelodesmos* is also abundant.

Species richness

A synthesis of the results of the phytosociological relevés is reported in Table 5 (full data are available in Annex). Overall, total species richness ranged from 17 (Sector 2, Plot 4) to 38 species (Sector 1, Plot 2), with a mean value of about 25 species. Within the first sector, a quite higher plant species richness was found in the unmanaged plot (38 vs 29), with a corresponding higher frequency and cover of pre-forest and forest woody species. Within the second sector, species richness was higher in the managed plot (27 vs 17), mainly due to the contribution of herbs, referred to the order *Hyparrhenietalia hirtae*, and of mantle shrubs, referred to the order *Prunetalia spinosae*. Both plots host the same number of pre-forest and forest woody species. The last sector includes the plots with the lowest number of species, both hosting 21 plant species, and more than half of them are herbs or shrubs. The managed plot hosts a slightly higher number of pre-forest and forest woody species.

Table 5 - Synthesis of the phytosociological relevés carried out within the investigated plots.

	Sector (Plot)					
	1 (1)	1 (2)	2 (3)	2 (4)	3 (5)	3 (6)
Tree layer						
Cover (%)	50	60	80	70	80	70
Mean height (cm)	622	378	725	612	1032	865
Shrub layer						
Cover (%)	40	60	30	60	40	60
Mean height (cm)	30	85	60	65	170	120
Herb layer						
Cover (%)	40	50	30	20	30	15
Mean height (cm)	10	15	15	10	10	15
Total taxa	29	38	27	17	21	21
Planted species	0	0	0	0	1	1
char. *Quercetalia ilicis*	8	10	5	5	5	4
char. *Quercetalia calliprini*	4	7	5	5	3	2
char. *Cisto-Ericetalia multiflorae*	3	3	2	2	0	6
char. *Prunetalia spinosae*	2	3	2	0	3	0
char. *Hyparrhenietalia hirtae*	7	6	7	2	7	6
plants linked to rocky/stony habitats	1	4	2	2	1	1
other herbs and grasses	4	5	4	1	1	1

Discussion

The post-fire response of Mediterranean plantations, assessed in terms of dominant tree canopy, woody species regeneration and plant species richness, was found to be affected by the dominant tree species and post-fire management strategies. Such aspects should be taken into account to understand the mechanisms underlying vegetation dynamics, as well as seedling regeneration and recruitment patterns in fire-prone Mediterranean afforested areas. Overall, field observations have confirmed the notable resilience by Mediterranean woody vegetation to fire.

Comparison plots burned in 1982

The comparison between plots 1-2 and 5-6, sharing the post-fire recovery time, allowed us to highlight the differential role of post-fire regeneration ability of dominant tree species and understory management on post-fire vegetation recovery, assessed in terms of structural development, establishment of natural regeneration, and species richness. In the holm oak stands burned in 1982 (Plot 1 vs Plot 2), the vegetation response of the unmanaged plot was positively influenced by the lack of active post-fire forest management. In spite of the time elapsed since the passage of the fire, notable differences were found according to different post-fire management practices. In the unmanaged plot, the dominance of holm oak was strongly reduced (from 90% to 64% in terms of density and from more than 80% to 41% in terms of basal area), and the relative contribution of other tree species has increased correspondingly. Holm oak individuals were more abundant, higher and larger in the managed plot, probably being favored in the interspecific competition with shrubs due to understory management. The abundance of holm oak regeneration also depended on its great resprouting ability from damaged stumps and/or roots after a disturbance event. The relative contribution of downy oak was by far higher in the unmanaged plot, despite the overall basal area was virtually the same in the two plots. Despite having the same number of stumps, the coppice shoots were much more numerous in the unmanaged plot (5'700 vs 4'000 per hectare). Similar levels of tree regeneration were found but tree individuals were much larger in the unmanaged plot, reaching an average height of 1 m, thus being definitely established. The unmanaged plot showed a higher structural complexity, arising from the higher cover of all the vegetation layers, from herb to tree. The upper layer is formed by thermophilous evergreen and deciduous trees, dominating a structurally complex shrub layer. This multilayered structure, providing a large micro-site variability, has favoured the natural regeneration by different woody species. The shrub layer was richer, with higher and well established individuals, in the unmanaged plot. The overall plant species richness was also higher, as well as the number of species related to pre-forest and forest habitats. As a consequence of the ongoing succession process, many species typical to pre-forest communities (*garrigue* and open *maquis*), such as *Emerus major* subsp. *major* and *Spartium junceum* L., but also several plant species linked with mature, closed and shady forest communities, have been found only in the unmanaged plot (Pasta 1993). The absence of post-fire interventions in the holm oak forest burned in 1982 has favored the compositional and structural evolution of the forest stand, which showed a high degree of resilience, resembling to a natural-like forest ecosystem. About 23 years after wildfire were sufficient for forest ecosystems to recover and to reach a near-steady state condition, thus allowing the entry of ornithochorous woody colonizers. Based on the published data on the vegetation of Mt. Petroso (Troìa 1994) and concerning local vegetation dynamics (La Mantia et al. 2002, Cullotta and Pasta 2004), the steady state is represented by mixed forest stands dominated by holm oak (*Quercus ilex*), where several deciduous broadleaved trees, such as *Acer campestre* L., *Quercus pubescens* and *Fraxinus ornus*, may co-occur. Such plant communities are framed into the phytosociological subassociation *Aceri campestris-Quercetum ilicis helleboretosum bocconei* (Brullo et al. 2009). After a few decades, in fact, forest stands have reached a condition similar to the pre-disturbance state, i.e. showing a great resilience and proving the high resprouting ability of local woody species. Such a fast recovery implies the presence of the necessary ecological interactions with local seed dispersers (da Silveira Bueno 2018). The role played by birds, and in particular by the Eurasian Jay (*Garrulus glandarius* L.), in the dispersal of *Quercus* acorns is well known (e.g.: Gomez 2003). Also, recent field surveys carried out in Sicily have confirmed the importance of such biotic interactions for the process of recovery of Mediterranean vegetation (La Mantia et al. 2015, La Mantia and da Silveira Bueno 2016, da Silveira Bueno 2018). However, the fire took place just once; recurrent wildfires may seriously hamper the regenerating capacity by native woody species, also causing soil erosion and degradation (Diaz-Delgado et al. 2002).

Comparing stone pine forest stands burned in 1982 (Plot 5 and Plot 6), the managed plot was composed by larger and higher individuals (mean DBH = 37 cm and Hm = 17 m), with a lower stand density (531 stems per hectare). The second tree layer was characterized by more developed and abundant holm oak individuals (> 90% in terms of frequency and almost 13% of the overall basal area). The managed plot showed a much higher basal area (66 m^2 ha^{-1} vs 37 m^2 ha^{-1}). In the unmanaged plot, pine individuals were smaller, determining a higher stand density (619 stems per hectare). Holm oak is well established in both plots due to its shade tolerance during seedling stage and the lack of competition after the fire destroyed the shrub cover. However, the higher light availability in the understory of the managed plot has favored its regeneration, as well as that of downy oak and *Ampelodesmos*. It is well recognized that an excessive pine cover may exert inhibitory effects for the growth and survival of thermophilous oaks in the understory of Mediterranean pine plantations (Manor et al. 2008, Pasta et al. 2012). Furthermore, such density values are far beyond the most recommended threshold for similar forest stands, that is 350-400 pines per hectare (Del Favero 2008). After fire passage, occasional regeneration by *Cupressus sempervirens* and *Pinus pinea* were also observed, but their saplings have died probably due to the excessive dominant tree cover, suggesting a very low regeneration performance by planted conifers. This was also due to the inability of stone pine to regenerate immediately after fire, as this species does not bear serotinous cones (Pausas et al. 2008). The regeneration of shrub species was not particularly represented in the managed plot, consisting the regeneration layer almost exclusively of *Ampelodesmos* (> 85% of total individuals). A massive regeneration by *Erica multiflora* is found in the unmanaged plot, where it probably benefited from the lack of understory management practices. The two plots share a similarly low plant species richness, despite a slight prevalence of forest species in the managed plot. Such low richness probably depended not only on a possible allelopathic effect by pines, but also on the lack of the necessary silvicultural treatments, namely thinning. The importance of thinning to enhance the overall biodiversity and to accelerate the renaturalization of Mediterranean plantations is largely acknowledged, for their strong influence on light availability, as well as on overstory-understory competitive interactions (e.g.: Gómez-Aparicio et al. 2009). Most of taxonomic groups appear to be negatively influenced by a too dense dominant canopy (Gil-Tena et al. 2007, Manor et al. 2008), so that exceeding some threshold value the interventions are absolutely required (Pasta et al. 2012). For instance, shade-intolerant plants and their associated biological communities may survive only in presence of a sufficient number of suitable-sized gaps (Heiri et al. 2009). Thinning after fire passage may also have a positive effect to favor the recovery of woody species, but keeping in mind that a too low tree density may significantly reduce the regeneration potential after a new fire event (García-Jiménez et al. 2017).

Comparison plots burned in 1954

The comparison between holm oak stands burned in 1954 (Plot 3 vs Plot 4), characterized by a longer post-fire dynamic, allowed us to appreciate the contrasting role of understory management on post-fire vegetation recovery, assessed in terms of forest structural development, establishment of natural regeneration, and species richness. In these plots, woody vegetation has had more time to develop and be diversified so that tree cover reached 80% and 70%, respectively. Differently from what observed in the holm oak stands burned in 1982, here the managed plot was more diversified both in terms of species composition and structure. This would prove that the management of Mediterranean afforestations subject to wildfires is a quite difficult task and no generalized approach can be established (Vallejo and Alloza 2012). In the managed plot, the contribution of holm oak was greatly reduced (from 87% to 28% in terms of basal area). Conversely, the contribution of other tree species, which attained large sizes (Mean DBH ranges from 21.7 cm of cork oak to 28 cm of downy oak), was significant. Accordingly, total basal area was far higher in the managed plot (almost 47 m^2 ha^{-1} vs 26 m^2 ha^{-1}). The regeneration of tree species was also much more abundant in the regularly managed plot (22'500 vs 6'500 individuals per hectare). As far as regeneration is concerned, in the managed plot there was a far higher number of *Q. ilex* coppice shoots (almost 6'000 per hectare). More similar was the understory condition, consisting of the same number of species with a similar development stage. However, the regeneration was more abundant in the unmanaged plot (5'800 vs 4'500 individuals per hectare), mostly due to the considerable presence of *Teucrium* spp.. The managed plot showed a higher species richness, especially due to the contribution of mantle species (order *Prunetalia spinosae*), as well as of Mediterranean perennial grasses (order *Hyparrhenietalia hirtae*). The same number of species of forest and pre-forest formations was found. The large gap in terms of species-richness between the two unmanaged plots dominated by holm oak probably depended on the different time since last wildfire. However, despite not directly measured, the fire intensity may have played a prominent role. The oldest cork oak individuals have probably survived the fire in 1954 thanks to the bark protection. Such passive strategy makes cork oak one of the most fire resistant Mediterranean oaks (Catry et al. 2010). Their limited diameter increase and the lack of natural regeneration are probably related to local sub-optimal edapho-climatic conditions as cork oak is known to be particularly demanding for such abiotic factors.

Comparison by dominant tree species

The comparison between the plots dominated by holm oak and stone pine (Plots 1-4 vs Plots 5-6), respectively, allowed us to assess the role played by the dominant tree species in the post-fire response. On average, holm oak canopy seemed to have a more positive effect on understory regeneration dynamics and richness than stone pine cover. The tree regeneration in the understory of holm oak forests was particularly favored in terms of abundance, but also a species-specific effect was found, being downy oak the species that benefited the most from holm oak as dominant species. The total regeneration of shrubs was not found to be strongly affected by dominant tree canopy. Conversely, species of forest and pre-forest communities tended to be more common in holm oak stands, while species linked to herbaceous communities tended to prefer stone pine stands. Importantly, species such as *Calicotome*, *Emerus*, *Spartium*, *Cistus salviifolius* and *Teucrium fruticans* were found to be exclusive of the understory of holm oak stands, indicating a crucial role for the conservation of biodiversity.

Comparison managed vs unmanaged

Also interesting was the comparison between managed and unmanaged plots, regardless of dominant tree species (Plots 1, 3 and 5 vs Plots 2, 4 and 6). Despite hosting a similar total number of plant species, these plots showed quite large differences in species identity. Species tied to forest communities were equally represented in both groups of plots. However, the other groups of species were found to be sensitive to forest management. Shrub species belonging to *Cisto-Ericetalia multiflorae* phytosociological order were negatively affected by understory management practices which, by contrast, seemed to favor the occurrence of mantle shrubs (Order *Prunetalia spinosae*) and species of Mediterranean perennial grasslands (Order *Hyparrhenietalia hirtae*).

All the investigated areas should be provided with a management plan including forest interventions aimed at favouring the gradual conversion from stone pine plantations to holm oak mixed forest stands. Such renaturalization process is widely occurring in the Mediterranean Basin (Pausas et al. 2004). For instance, a similar conversion has been observed within the afforestation carried out in 1928 at Casaboli wood (Palermo Mts.), where holm oak have reached a noteworthy size and definitely prevailed over pines, clearly revealing the potential vegetation (climax) (Troìa 1994, Maggiore et al. 2005).

Conclusions

One relevant factor constraining the success of Mediterranean plantations is the occurrence of wildfires, which may negatively affect seedling survival and establishment, seriously hindering the natural regeneration of planted species. Resilience to fire has a crucial importance in driving vegetation dynamics, as well as for the failure or success of forest plantations. Local forest stands appeared to be quite complex and diversified, so that the surveyed plots may represent different patches of a landscape mosaic resulting not only from differences in fire events and dominant tree species but also from post-fire silvicultural practices. Despite the few replicas of our field surveys do not allow making broad generalizations of the investigated natural processes, our research has confirmed the high resilience to wildfire of Mediterranean woody vegetation. In particular, we found that native *Quercus* spp. recovered very well after fire, confirming the possibility to use them for afforestation purposes without prior using *Pinus* spp. Also, species identity, autoecology, and provenance have to be carefully considered as the origin of plant material is recognized to significantly affect the establishment chance by plant species (González-Alonso et al. 2004).

We found that the understory management (understory cleaning and removal of dead biomass) had a role in the forest recovery after fire, influencing the structural and floristic characteristics of the forest stands, as well as the regeneration pattern of woody species. However, also the unmanaged plots were characterized by effective regeneration processes in the understory. This suggests that the option of no management after fire should be considered, in presence of sufficient propagules of native woody species.

A forest ecosystem subject to recurrent wildfires needs to be managed avoiding forest interventions which could hinder the natural dynamics, following the principles of systemic silviculture (e.g.: Ciancio and Nocentini 2011). The knowledge of post-fire vegetation response is crucial if we are to adopt management practices which take into account the ecological characteristics of the plant species, their ability to respond to and their resilience to fire. Future management interventions should be planned and addressed to favor the intrinsic ability of the ecosystem to develop, to mature and to recover after a disturbance event, without looking at any predefined structure and specific composition, following the natural dynamics and promoting the natural regeneration (Ciancio and Nocentini 2011).

The results of our field surveys, including long-term vegetation response, may be used in order to better manage Mediterranean afforested areas prone to frequent wildfires. Such Mediterranean forest ecosystems should be subjected to "*assisted natural regeneration, (re)introduction of tree or shrubby species, and fire-prevention measures*", following an integrated approach (Vallejo and Alloza 2012).

Acknowledgements

We thank two anonymous reviewers for their constructive comments which significantly improved the quality of the manuscript.

References

AA. VV. 1943 - *Enciclopedia Italiana*. Rizzoli Milano Vol. XXIX: 335-336.

Abate B., Catalano R., Renda P. 1982 - *Schema geologico dei Monti di Palermo*. Bollettino della Società Geologica Italiana 97: 807-819.

Álvarez R., Muñoz A., Pesqueira X.M., García-Duro J., Reyes O., Casal M., 2009 - *Spatial and temporal patterns in structure and diversity of Mediterranean forest of* Quercus pyrenaica *in relation to fire*. Forest Ecology and Management 257: 1596-1602.

Badalamenti E., La Mantia T., La Mantia G., Cairone A., La Mela Veca D.S. 2017 - *Living and dead aboveground biomass in Mediterranean forests: Evidence of old-growth traits in a* Quercus pubescens *Willd. s.l. stand*. Forests 8 (6): 187. doi:10.3390/f8060187

Badalamenti E., Pasta S., La Mantia T., La Mela Veca D.S. 2018 - *Criteria to identify old-growth forests in the Mediterranean: A case study from Sicily based on literature review and some management proposals*. Feddes Repertorium 129, 25-37. DOI:10.1002/fedr.201700009

Bellingham P.J., Sparrow C.D. 2000 - *Resprouting as a life history strategy in woody plant communities*. Oikos 89: 409-416.

Bengtsson J., Nilsson S.G., Franc A., Menozzi P. 2000. *Biodiversity, disturbances, ecosystem function and management of European forests*. Forest Ecology and Management 132: 39-50.

Braun-Blanquet J. 1964 - *Pflanzensoziologie*. Grundzüge der Vegetationskunde. Springer-Verlag, Wien and New York. 865 pp.

Bros V., Moreno-Rueda G., Santos X. 2011 - *Does post fire management affect the recovery of Mediterranean communities? The case study of terrestrial gastropods*. Forest Ecology and Management 261: 611-619.

Brullo S., Gianguzzi L., La Mantia A., Siracusa G. 2009 - *La classe* Quercetea ilicis *in Sicilia*. Bollettino dell'Accademia Gioenia di Scienze Naturali 41 (369) (2008): 1-124.

Buhk C., Götzenberger L., Wesche K., Sánchez Gómez P., Hensen I. 2006 - *Post-fire regeneration in a Mediterranean pine forest with historically low fire frequency*. Acta Oecologica 30: 288-298.

Calvo L., Santalla S., Marcos E., Valbuena L., Tárrega R., Luís E. 2003 - *Regeneration after wildfire in communities dominated by* Pinus pinaster, *an obligate seeder, and others dominated by* Quercus pyrenaica, *a typical resprouters*. Forest Ecology and Management 184: 209-223.

Catry F.X., Rego F., Moreira F., Fernandes P.M., Pausas J.G. 2010 - *Post-fire tree mortality in mixed forests of central Portugal*. Forest Ecology and Management 260: 1184-1192.

Ciancio O., Nocentini S. 2011 - *Biodiversity conservation and systemic silviculture: Concepts and applications*. Plant Biosystems 145 (2): 411-418.

Conti F., Abbate G., Alessandrini A., Blasi C. 2005 - *An Annotated Checklist of the Italian Vascular Flora*. Ministero dell'Ambiente e della Tutela del Territorio, Direzione per la Protezione della Natura, Dipartimento Biologia Vegetale "La Sapienza", Università degli Studi di Roma, Palombi Ed. Roma. 420 pp.

Corona P., Ascoli D., Barbati A., Bovio G., Colangelo G., Elia M., Garfì V., Iovino F., Lafortezza R., Leone V., Lovreglio R., Marchetti M., Marchi E., Menguzzato G., Nocentini S., Picchio R., Portoghesi L., Puletti N., Sanesi G., Chianucci F. 2015 - *Integrated forest management to prevent wildfires under Mediterranean environments*. Annals of Silvicultural Research 39 (1): 1-22.

Corona P., Ferrari B., Iovino, F., La Mantia T., Barbati A., 2009 - *Rimboschimenti e lotta alla desertificazione in Italia*. Aracne Editrice Roma. 228 pp.

Crosti R., Ladd P.G., Dixon K.W., Piotto B. 2006 - *Post-fire germination: The effect of smoke on seeds of selected species from the central Mediterranean basin*. Forest Ecology and Management 221: 306-312.

Cullotta S., Pasta S., 2004 - *Vegetazione mediterranea: Sicilia, Sardegna, Calabria. In: "Incendi e complessità ecosistemica. Dalla pianificazione forestale al recupero ambientale"*, Blasi C., Bovio G., Corona P., Marchetti M., Maturani A. (a cura di). Ministero dell'Ambiente e della Tutela del Territorio, Direzione per la Protezione della Natura - Società Botanica Italiana - Commissione per la Promozione della ricerca botanica, Palombi Ed., Roma, pp. 291-307.

Curt T., Adra W., Borgniet L. 2009 - *Fire-driven oak regeneration in French Mediterranean ecosystems*. Forest Ecology and Management 258: 2127-2135.

Da Silveira Bueno R. 2018 - *The role of plant-animal and plant-plant interactions on vegetation dynamics and ecological restoration in a Mediterranean landscape*. PhD Thesis, University of Palermo, Italy.

Del Favero R., 2008 - *I boschi delle regioni meridionali e insulari d'Italia*. Cleup, Padova, 469 pp.

Díaz-Delgado R., Lloret F., Pons X., Terradas J. 2002 - *Satellite evidence of decreasing resilience in Mediterranean plant communities after recurrent wildfires*. Ecology 83 (8): 2293-2303.

Escudero A., Sanz M.V., Pita J.M., Pérez-García F. 1999 - *Probability of germination after heat treatment of native Spanish pines*. Annals of Forest Science 56: 511-520. doi:10.1051/forest:2001137

Espelta J.M., Verkaik I., Eugenio M., Lloret F. 2008 - *Recurrent wildfires constrain long-term reproduction ability in* Pinus halepensis *Mill*. International Journal of Wildland Fire 17: 579-585.

Espelta J.M., Barbati A., Quevedo L., Tárrega R., Navascués P., Bonfil C., Peguero G., Fernández-Martínez M., Rodrigo A. 2012 - *Post-Fire Management of Mediterranean Broadleaved Forests*. In: "Post-Fire Management and Restoration of Southern European Forests", Moreira F., Arianoutsou M., Corona P., De las Heras J. (eds.), Springer, The Netherlands. pp. 171-194.

Fierotti G., 1988 - *Carta dei suoli della Sicilia (scala 1:250.000)*. Regione Siciliana, Ass. TT. AA.-Univ. Palermo, Fac. Agraria, Ist. Agron. Generale, Cattedra di Pedologia.

García-Jiménez R., Palmero-Iniesta M., Espelta J.M. 2017 - *Contrasting effects of fire severity on the regeneration of* Pinus halepensis *Mill. and resprouter species in recently thinned thickets*. Forests 8 (3): 55.

Gil-Tena A., Saura S., Brotons L. 2007 - *Effects of forest composition and structure on bird species richness in a Mediterranean context: Implications for forest ecosystem management*. Forest Ecology and Management 242: 470-476.

Gómez J.M. 2003 - *Spatial patterns in long-distance dispersal of* Quercus ilex *acorns by jays in a heterogeneous landscape*. Ecography 26: 573-584.

Gómez-Aparicio L., Zavala M.A., Bonet F.J., Zamora R. 2009 - *Are pine plantations valid tools for restoring Mediterranean forests? An assessment along abiotic and biotic gradients*. Ecological Applications 19 (8): 2124-2141.

González-Alonso F., Merino-de-Miguel S., García-Gigorro S., Roldán-Zamarrón A., Cuevas J.M., Arino O. 2004 - *Mapping forest-fire damage using ENVISAT*. ESA Bulletin 120: 22-26.

Granados M.E., Vilagrosa A., Chirino E., Vallejo V.R. 2016 - *Reforestation with resprouter species to increase diversity and resilience in Mediterranean pine forests*. Forest Ecology and Management 362: 231-240.

Guiomar N., Godinho S., Fernandes P.M., Machado R., Neves N., Fernandes J.P. 2015 - *Wildfire patterns and landscape changes in Mediterranean oak woodlands*. Science of the Total Environment 536: 338-352.

Hanes T.L. 1971 - *Succession after fire in the chaparral of southern California*. Ecological Monographs 41: 27-52.

Heiri C., Wolf A., Rohrer L., Bugmann H. 2009 - *Forty years of natural dynamics in Swiss beech forests: Structure, composition, and the influence of former management*. Ecological Applications 19: 1920-1934.

Incerti G., Giordano D., Stinca A., Senatore M., Termolino P., Mazzoleni S. 2013 - *Fire occurrence and tussock size modulate facilitation by Ampelodesmos mauritanicus*. Acta Oecologica 49: 116-124.

Kazanis D., Arianoutsou M. 2004 - *Long-term post-fire vegetation dynamics in* Pinus halepensis *forests of central Greece: a functional group approach*. Plant Ecology 171: 101-121.

Keeley J.E., Zedler P.H. 1978 - *Reproduction of chaparral shrubs after fire: comparison of sprouting and seeding strategies*. American Midland Naturalist 99: 142-161.

Keeley J.E., Bond W.J., Bradstock R.A., Pausas J.G., Rundel P.W. 2011 - *Fire as an evolutionary pressure shaping plant traits*. Trends in Plant Science 16 (8): 406-411.

La Mantia T., Cullotta S., Cutino I., Maggiore C.V., Marchetti M., Pasta S., 2002 - *A survey on post-fire behaviour of* Pinus halepensis *and* P. pinea *plantations in Sicily (Italy), with special reference to the re-colonisation processes*. "MedPine 2002" Conference (MAICh Chaniá, Crete, 8-13.09.2002).

La Mantia T., da Silveira Bueno R. 2016 - *Colonization of eurasian jay* Garrulus glandarius *and holm oak* Quercus ilex*: the establishment of ecological interactions in urban areas*. Avocetta 40: 85-87.

La Mantia T., Massa B., Pipitone S., Rühl J. 2015 - *Il ruolo degli uccelli come vettori di dispersione durante le successioni secondarie*. In: "XVII Convegno Italiano di ornitologia. Atti del Convegno di Trento", Pedrini P., Rossi F., Bogliani G., Serra L., Sustersic A. (a cura di). Ed. MUSE, Museo delle Scienze, Trento, pp. 123-124.

Lloret F., Calvo E., Pons X., Díaz-Delgado R. 2002 - *Wildfires and landscape patterns in the Eastern Iberian Peninsula*. Landscape Ecology 17: 745-759.

Maetzke F.G., La Mela Veca D.S., La Mantia T., Badalamenti E., Sferlazza S. 2017 - *The use of species in plantations: renaturalisation and reforestation in Sicily*. In: Chiatante D., Domina G., Montagnoli A., Raimondo F.M. (eds.). Flora Mediterranea, 27: 19-20 ISSN: 1120-4052 printed, 2240-4538 online. International Congress: "Sustainable restoration of Mediterranean forests: analysis and perspective within the context of bio-based economy development under global changes". Palermo, 19-21 April, 2017.

Maggiore C., Cutino I., Marchetti M., Pasta S., La Mantia T. 2005. *La dinamica degli incendi e l'effetto degli interventi selvicolturali sui soprassuoli a pino d'Aleppo e domestico percorsi da incendio in un comprensorio boscato mediterraneo (Sicilia Nord-occidentale)*. Società Italiana di Selvicoltura ed Ecologia Forestale, Atti del IV Congresso Meridiani Foreste. Rifreddo (PZ) 7-10 Ottobre 2003. pp. 237-244.

Manor R., Cohen O., Saltz D., 2008 - *Community homogenization and the invasiveness of commensal species in Mediterranean afforested landscapes*. Biological Invasions 10: 507-515.

McGaughey R.J. 1997. *Visualizing forest stand dynamics using the stand visualization system*. In: Proceedings of the 1997 ACSMASPRS Annual Convention and Exposition, 7-10 April 1997, Seattle, WA, Bethesda, MD, American Society of Photogrammetry and Remote Sensing 4: 248-257.

Moreira F., Viedma O., Arianoutsou M., Curt T., Koutsias N., Rigolot E., Barbati A., Corona P., Vaz P., Xanthopoulos G., Mouillot F., Bilgili E. 2011 - *Landscape – wildfire interactions in southern Europe: Implications for landscape management*. Journal of Environmental Management 92 (10): 2389-2402.

Moreira F., Arianoutsou M., Corona P., De las Heras J. (eds.). 2012 - *Post-Fire Management and Restoration of Southern European Forests*. Managing Forest Ecosystems 24. Springer, The Netherlands. 330 pp.

Moreira F., Ferreira A., Abrantes N., Catry F., Fernandes P., Roxo L., Keizer J.J., Silva J. 2013 - *Occurrence of native and exotic invasive trees in burned pine and eucalypt plantations: Implications for post-fire forest conversion*. Ecological Engineering 58: 296-302.

Moya D., De las Heras J., Ferrandis P., Herranz J.M., Martínez-Sánchez J.J. 2011 - *Fire resilience and forest restoration in Mediterranean fire-prone areas*. Technology and Knowledge Transfer e-Bulletin, Universidad Politécnica de Cartagena 2 (3): 1-5.

Pasta S. 1993 - *Considerazioni teoriche sulle prospettive di ricerca aperte dalla geobotanica ed esempi di applicazioni di questa disciplina al caso specifico dello studio della flora e della vegetazione di un settore dei Monti di Palermo*. Diplome Thesis, UNIPA.

Pasta S., La Mantia T., Rühl J. 2012 - *The impact of* Pinus halepensis *Mill. afforestation on Mediterranean spontaneous vegetation: do soil treatment and canopy cover matter?* Journal of Forestry Research 23 (4): 517-528.

Pausas J.G. 2006 - *Simulating Mediterranean landscape pattern and vegetation dynamics under different fire regimes.* Plant Ecology 187: 249-259.

Pausas J.G., Bladé C., Valdecantos A., Seva J.P., Fuentes D., Alloza J.A., Vilagrosa A., Bautista S., Cortina J., Vallejo R. 2004 - *Pine and oaks in the restoration of Mediterranean landscape: New perspective for an old practive - a review.* Plant Ecology 171: 209-220.

Pausas J.G., Llovet J., Rodrigo A., Vallejo R. 2008 - *Are wildfire a disaster in the Mediterranean basin? - A review.* International Journal of Wildland Fire 17: 713-723.

Pignatti S. 1982. *Flora d'Italia.* Bologna, Edagricole, 3 Voll. 2324 pp.

Raimondo F.M., Domina G., Spadaro V. 2010 - *Checklist of the vascular flora of Sicily.* Quaderni di Botanica ambientale e applicata 21: 189-252.

Retana J., Arnan X., Arianoutsou M., Barbati A., Kazanis D., Rodrigo A. 2012 - *Post-Fire Management of Non-Serotinous Pine Forests.* In: "Post-Fire Management and Restoration of Southern European Forests", Moreira F., Arianoutsou M., Corona P., De las Heras J. (eds.), Springer, The Netherlands. pp. 151-170.

Rivas-Martínez S. 2008 - *Global Bioclimatics (Clasificación Bioclimática de la Tierra).* Available online: http://www.globalbioclimatics.org/book/bioc/global_bioclimatics-2008_00.htm (accessed on 20 February 2017).

Rollan A., Real J. 2011 - *Effect of wildfires and post-fire forest treatments on rabbit abundance.* European Journal of Wildlife Research, Volume 57 (2): 201-209.

Rühl J., Gristina L., La Mantia T., Novara A., Pasta S., 2015. *Afforestation and reforestation: the Sicilian case study.* In: Valentini R., Miglietta F. (Eds.), *The Greenhouse Gas Balance of Italy - An Insight on Managed and Natural Terrestrial Ecosystems.* Springer. 173-184 pp.

Rühl J., Pasta S., La Mantia T. 2005 - *Metodologia per lo studio delle successioni secondarie in ex-coltivi terrazzati: il caso studio di Pantelleria (Canale di Sicilia).* Forest@ 2 (4): 388-398.

San-Miguel-Ayanz J., Moreno J.M., Camia A. 2013 - *Analysis of large fires in European Mediterranean landscapes: Lessons learned and perspectives.* Forest Ecology and Management 294: 11-22.

Sheffer E. 2012 - *A review of the development of Mediterranean pine–oak ecosystems after land abandonment and afforestation: are they novel ecosystems?* Annals of Forest Science 69: 429-443.

Tessler N., Wittenberg L., Provizor E., Greenbaum N. 2016 - *The influence of short-interval recurrent forest fires on the abundance of Aleppo pine (*Pinus halepensis *Mill.) on Mount. Carmel, Israel.* Forest Ecology and Management 324: 109-116.

Trabaud L., Galtié J.-F. 1996 - *Effects of fire frequency on plant communites and landscape pattern in the Massif des Aspres (southern France).* Landscape Ecology 11 (4): 215-224.

Troìa A., 1994 - *La querceta di Monte Petroso (Palermo): proposte di interventi su base naturalistica.* Sviluppo Agricolo 28 (10-11): 75-79.

Troìa A., Bazan G., Schicchi R. 2011 - *Nuove aree di rilevante interesse naturalistico nella Sicilia centro-occidentale: proposta di tutela.* Il Naturalista siciliano 4, 35 (2):257-293.

Vallejo V.R., Alloza J. A. 2012 - *Post-Fire Management in the Mediterranean Basin.* Israel Journal of Ecology & Evolution, 58 (2-3): 251-264.

Yu F.-H., Schütz M., Page-Dumroese D.S., Krüsi B.O., Schneller J., Wildi O., Risch A.C. 2011 - Carex sempervirens *tussocks induce spatial heterogeneity in litter decomposition, but not in soil properties, in a subalpine grassland in the Central Alps.* Flora 206: 373-379.

Genetic resources and forestry in the Mediterranean region in relation to global change

Fulvio Ducci[1*]

Abstract - The purpose of this review is to examine a few aspects of global change effects on forest genetic resources and their interaction. Genetic resources can provide many opportunities for the development of adaptive forest management in the Mediterranean region. At the same time, forestry and its various disciplines can offer manifold chances to develop methods and techniques for the *in situ* and *ex situ* protection, as well as for the correct management of species and populations at risk because of climate change. Among these aspects, the studies on the Marker Assisted Selection are particularly taken into consideration, as well as the phenotypic plasticity and the different types of assisted migration. A special emphasis is given to genetic resources growing at marginal peripheral populations, which need to be safeguarded as possible containers of adaptive diversity. They are subjected, in fact, to an extreme climatic stress more than others.

Keywords - Forest genetic resources, forest reproductive materials, genetics, adaptation, assisted migration, marginal peripheral populations, Mediterranean area, global change, silviculture.

Introduction

This paper aims to examine some key interactions between environmental factors and the management of forest genetic resources (FGR) in a context of rapid global changes.

The Mediterranean region is surely one of the major genetic hotspots for natural resources, closely interacting with human populations and cultures (Cavalli-Sforza 1991, Cavalli-Sforza et al. 1994) and it needs special care from both the scientific community and the policy makers and the public opinion.

Combining genetic, breeding and FGR management experiences, the Author provides some examples and considerations in order to draw attention to the complexity of managing Forest Genetic Resources in the context of the already evident changes occurring in the Mediterranean region.

Climate factors and Forest Trees

Over million years of evolution, trees have developed different behaviors in order to optimize their adaptive traits. Adaptation occurs at different levels and following ways (Larcher 2003).

- *Modulative adaptation*, it is important in sites where wide abrupt fluctuations of ecological factors occur (i.e. the circadian rhythms, as the opening and closing of stomata and chloroplast movements in relation to the minor or major light intensity fall here);

- *Modificative adaptation*, it takes place during morphogenesis in response to the environment and it is generally irreversible. The formation of light-leaves and shade-leaves in many species (e.g. silver fir) is one example. It could be regarded as a synonym of phenotypic plasticity;

- *Evolutionary adaptation*, it is fixed in the genotype according to habitat characteristics, preference of the species or its temperament. The distinction between shade-tolerant and light-demanding trees is an example, as well as between xero-tollerant or mesophyte trees or those adapted or not to acid soils.

Light, temperature and water availability are ultimately the main climatic factors interacting with forest trees, those that can significantly affect tree life and ecosystems where they live, with regard to the changes taking place.

Photosynthesis is the first adaptation process that biological forms invented to store energy

---

[1] Consiglio per la ricerca in agricoltura e l'analisi dell'economia agraria, Forestry Research Centre (CREA-SEL), Arezzo, Italy
* fulvio.ducci@entecra.it

from the sun. This biological process has radically changed Earth's environment, triggering ecological processes and enabling the evolution, which is nothing but the search for species of the right position to take advantage of a trickle flow of energy flowing through the ecosystem (Larcher 2003).

Among the different ecological factors, temperature is one of the most important and can significantly address species distribution. In Mediterranean contexts, temperature, on the ground, up to 70°C. at Aleppo pinewoods in southern Italy were usually measured (Magini 1955).

Clear evidence exists of the adaptive meaning of temperatures influencing the photosynthetic function in trees, in relation to habitats (Taylor et al. 2012). Seedlings collected along an elevation gradient, from 730 to 1460 m above the sea level, showed a photosynthetic optimum temperature reduced to about 4.3° C per 500 m in altitude. These values are closely related to the negative elevation gradient of the measured air temperature that equals to ± 3.9 C per 500 m on average.

Under these conditions, the pressure of natural selection towards a greater photosynthetic rate is very heavy.

A plant is under water stress when the water requirement necessary for transpiration exceeds the amount provided by the roots. The trees react by increasing their own stomatal resistance, i.e. by closing stomata and reducing the CO^2 and water flow.

Climate modifications expected in the future

Since 1980s, global change has started to show its evidence and has been acknowledged as seriously affecting all of Earth ecosystems. Under this framework, forests, biodiversity and genetic resources are under threat, as well as people depending on goods and services (WMO 2007).

Diversity is the main tool for fragmenting risks in the Biosphere and also forest diversity contains adaptive traits which can be useful for the adaptation of species to the changing environment and for supplying the variation to be used in silvicultural activities to develop adaption and mitigation strategies of management.

The effects of global changes, with special regard to climate parameters can endanger forest tree populations because of the persistent and intense modification of patterns within main climatic factors (i.e. intensity, seasonal and geographical distribution).

Water availability, together with air temperature, will change seasonal patterns too rapidly and trees may not be able to adapt to the changes. The repeated drought stress over several years and its cumulative effects can at times be quantified only when the damage is already nearly definitive.

The Mediterranean region, already subjected to highly variable climate regimes, is more and more characterized by unpredictability and by an increasing frequency of extreme events. In this area, the effects of global change and especially of climate are likely to be even more sensitive. It therefore raises the need to manage and preserve the Mediterranean forest tree resources because differentiated and characterized by the presence of adaptive traits possibly useful to develop strategies for adaptation (Ducci et al. 2011 in Besacier et al. 2011, Ducci 2014).

The historical occurrence of many civilizations that significantly influenced the life of ecosystems and their genetic resources is a typical attribute of the Mediterranean region. People have been using and taking advantage of this biodiversity for millennia. Although it resulted in the rise of original habitats and human-designed landscape structures, in fact, the interaction between humans and biodiversity has sometimes led to the decline and disappearance of species and populations (Blondel and Aronson 1999).

Nowadays, climate change effects pose other formidable challenges to biodiversity in this region. Most valuable forest ecosystems and species in the Mediterranean region experienced and survived the most recent post-glaciation climate warming thank to the presence of mountain ranges where suitable ecological conditions were found at different altitudes. But the changes since last glacial periods were not so rapid as the changes predicted today. At least five cold periods have been recorded in Europe over the last 8000 years; their duration was between 100 and 300 years on average and they influenced both the life standard of civilizations and the diffusion-extension and type of forest ecosystems (Pinna 1977).

More recently, the IPCC (Intergovernmental Panel on the Climate Change 2001, 2012) predicted an average temperature increment over the next 50 years in Europe of $2 \div 4$°C, according to the different regions. This change means the upward shifting of climatic and ecological areas in the mountains. For a number of reasons however, this "escape" will not always be sufficient to solve the problem. In fact, the on-going climate modification and its effects will be more complex than we can imagine and many factors will change the patterns and the related feedbacks and interactions.

Changes in the main patterns

The actual changes in climate seem to be very rapid as compared to past alternate warming, gla-

Table 1 - Scenarios for minimum temperature changes estimated according to different scenarios and in different regions of Italy (courtesy of Perini, Salvati et al. 2007). *Legenda*: IS: Initial Situation; S0: lower intensity scenario; S1: averaged intensity scenario; S2: higher intensity scenario.

Minimum temperatures (°C.)

Geografic Region (Italy)	Winter				Spring				Summer				Autumn			
	IS	S0	S1	S2	IS	S0	S1	S2	IS	S0	S1	S2	IS	S0	S1	S2
Nord	2.4	1.1	1.7	1.1	3.8	3.7	3.7	4.6	12.1	12.1	12.4	13.1	5.8	5.7	5.9	6.1
Central	2.3	3.5	3.4	3.8	6.8	6.7	6.9	7.5	15.3	15.3	15.7	16.3	9.8	9.7	10.0	10.4
South	5.7	7.0	7.0	7.2	9.4	9.4	9.5	9.9	18.2	18.2	18.5	19.2	13.0	12.9	13.2	13.5
Sicily	7.8	9.2	9.3	9.5	10.6	10.6	10.5	11.1	19.3	19.2	19.7	20.2	15.2	15.0	15.5	15.8
Sardinia	6.4	7.6	7.7	7.6	9.2	9.2	9.3	9.7	17.7	17.6	18.1	18.5	13.3	13.1	13.5	13.7

cial and micro-glacial periods. But the occurrence of climate change is not new and climate warming started earlier than admitted by media and popular press. Such change was recorded by climatologists since the second half of the XIX century.

At the end of 1800, many glaciers were lost or widely reduced in area on the Alps and other mountain ranges. In 1935, the sea mean temperature, at higher latitudes, increased of 1 to 3.5°C (Pinna 1977). Many similar forecasts and scenarios are recorded in the literature on this topic. Concerning Italy, for instance, Perini and Salvati (2007) produced some scenarios with major details. In particular, three different cases (different intensities of change) were considered by the authors, named respectively *S0* (ΔT (°C): 0.1 Tm 13.0° C.), *S1* (ΔT (°C): 0.2 Tm 13.4° C.) and *S2* (ΔT (°C): 0.8 Tm 14.0° C.) over the reference period 1960 - 1990). Climatic patterns change mainly with latitude, but also they vary according to orographic variation, local morphology and distance from the sea.

The *mean annual temperature* change is actually forecasted to increase from 13°C to 13.4 and 14.8°C respectively, with a maximum increment of 0.8° C.

This perspective of the authors is relatively moderate as compared with current ICCP scenarios (2012).

Minimum temperatures will probably change with different patterns and intensities in different geographical areas of Italy; the South of Italy and the larger Mediterranean islands will be especially affected (Table 1).

The changed patterns are even more pronounced when *maximum temperatures* are considered (Table 2).

Predicted change in *rainfall amount* would be also significantly heavy. About 60% of the annual precipitation could be lost in some areas, especially in winter and spring. The autumn rainfall would predictably be more stable, but both intensity and concentration of precipitation could become a problem due to the increased occurrence of floods. Their *geographical distribution* would also predictably change and, according to the most optimistic scenario (S0), Sicily and Sardinia would reduce total rainfall amounts between 17 and 21%, whilst reductions would amount to about 5% - 9% in the continental regions of Italy.

If both of the two worst scenarios (S1 and S2) are taken into account, the decrease of precipitation would predictably be 20% in northern, 17% in central Italy, 25-26% in the South and Sardinia, and 12% in Sicily (Table 3). An increasing drought towards a 200 kilometres higher latitude would affect Italian central regions.

Main traits of forest genetic resources under climate driving forces

According to Thompson (2005), the climate changes in the Mediterranean region, the most recent ones being occurred during the Quaternary (Huntley and Birks 1983), together with the salinity crisis in late the Messinian Tertiary, have driven the shaping of diversity among and within Mediterranean species.

Together with Balkans and the Iberian Peninsula, the Mediterranean region is indeed considered among the major and complex refuge zones in the Pleistocene. As a result, plant diversity is relatively high in this area (Hampe and Petit 2005, Fady and Conord 2010).

Table 2 - Scenarios for maximum temperature changes in different regions of Italy and in different seasons (courtesy of Perini, Salvati et al. 2007). *Legenda*: IS: Initial Situation; S0: lower intensity scenario; S1: averaged intensity scenario; S2: higher intensity scenario.

Maximum temperatures (°C.)

Geografic Region (Italy)	Winter				Spring				Summer				Autumn			
	IS	S0	S1	S2	IS	S0	S1	S2	IS	S0	S1	S2	IS	S0	S1	S2
Nord	4.6	5.3	5.3	5.8	13.2	13.0	13.5	14.8	22.8	228.8	23.1	24.0	14.5	14.2	14.5	14.6
Central	10.0	11.1	11.2	11.6	16.8	16.6	17.1	18.0	27.1	27.1	27.4	28.4	19.3	19.0	19.4	19.7
South	12.7	14.0	14.2	14.5	18.2	18.1	18.4	19.2	28.3	28.1	28.7	29.7	21.4	21.1	21.7	22.1
Sicily	14.6	15.9	16.0	16.3	18.9	18.8	19.2	20.0	28.8	28.5	29.3	30.0	23.0	22.7	23.2	23.7
Sardinia	13.8	15.0	15.1	15.3	18.5	18.2	18.9	19.8	28.7	28.5	29.2	30.2	22.1	21.8	22.4	22.8

Table 3 - Scenarios for rainfall changes in different regions of Italy in different seasons (courtesy of Perini, Salvati et al. 2007). *Legenda*: IS: Initial Situation; S0: lower intensity scenario; S1: averaged intensity scenario; S2: higher intensity scenario.

	Precipitation (mm)															
Geografic Region (Italy)	Winter				Spring				Summer				Autumn			
	IS	S0	S1	S2	IS	S0	S1	S2	IS	S0	S1	S2	IS	S0	S1	S2
Nord	220	221	152	179	264	250	220	184	263	242	238	204	282	262	212	267
Central	278	270	236	239	199	178	161	163	140	122	120	96	282	277	238	269
South	286	255	241	237	166	148	152	119	78	73	65	47	248	231	226	203
Sicily	223	203	226	247	132	109	113	88	34	21	19	26	186	177	202	191
Sardinia	241	180	198	190	126	124	90	92	34	24	21	19	201	146	177	169

As above mentioned, the Mediterranean region is one of the world's biodiversity hotspots for its outstandingly high plant species richness and endemism (Médail and Diadema 2009). In this area, the presence of genetically valuable populations of forest species belonging to central and northern ranges will be seriously affected and, in a few cases, endangered. These populations are in most cases characterized by high levels of differentiation and are frequently typically marginal.

It is thus likely that forest genetic resources originated in the southern latitudes of Europe have often developed their major traits under difficult environments and survived well under unsuitable ecological conditions because of unusual adaptive traits. They are valuable sources of variation for expanding as well as retreating forests under the climate pressure and for developing adaptive management, silviculture and mitigation initiatives.

The changing climatic patterns may have significant influences on the adaptive traits of species, and any rapid change can cause disequilibrium in growth, physiology and reproductive systems.

In Mediterranean climate areas, forest species adapted their living cycles to the rainfall time; autumn and winter represent for most of them the growing period and provide the best conditions for seed germination.

Besides, most of trees, especially those growing in temperate areas in mountain sites and belonging to mature ecosystems, such as many hardwoods, can activate their annual cycles only strictly following the seasonal climatic change. Thus, growth and reproductive steps can occur only in suitable conditions and/or when adverse conditions can be mitigated or avoided.

As above mentioned, the isotherms will probably shift very rapidly because of the increasing average temperatures. With regard to the ability of tree populations to migrate under the influence of isotherm shifts, Màtyàs (2007) proposed a scenario where mean temperature would increase by about 2°C in 35 years, an estimate more cautious than the Italian evaluations. According to his standpoint, in Europe the south thermal gradient would shift towards the north at the speed of 3 km/year and of 11.5 m in elevation (Màtyàs 2007).

In northern America, the shift is estimated 6-15 km/year in the plains and, in any case, the estimated speed would exceed the potential migration rate of most of forest trees (Aitken et al. 2008).

This change is very rapid, as most of the species have migration rates which can vary between 100 and 400 m/year in the plains (Màtyàs, personal communication), according to seed weight, reproductive system, habitat, ground morphology and other ecological and biological factors at local level. Iverson et al. (2004) estimated this rate at 100 - 200 m/year for northern America.

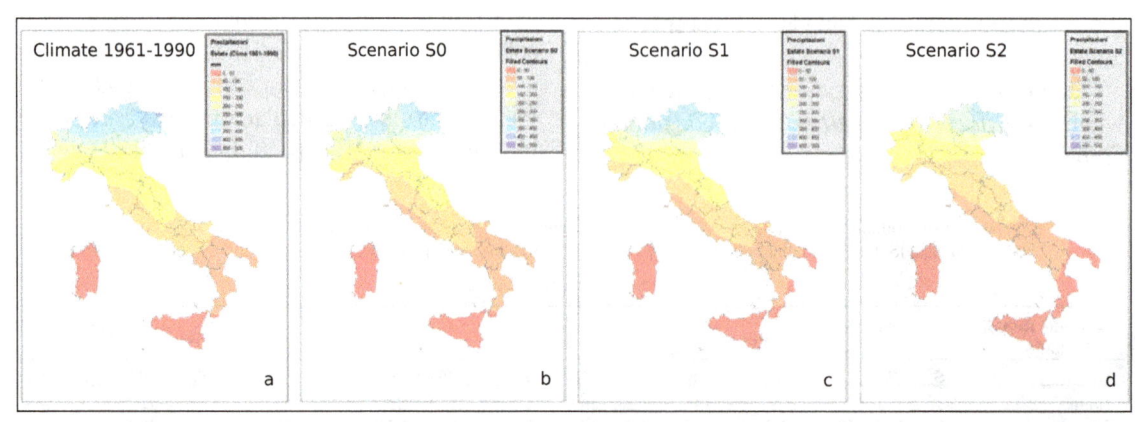

Figure 1 - Maps showing the peculiar situation of decreased summer mean rainfalls in most of central and southern Italy which will be only partially offset by the presence of the Apennines. a) 1961-1990; b) IPCC lower intensity scenario S0; c) IPCC lower average scenario S1; d) IPCC higher intensiry scenario S2.

With special regard to the southern European and the Mediterranean area, the mountain area is widespread and could probably contribute to mitigate changes with the elevation and aspect effects. There, tree species have to migrate upward following the shift of their environmental envelope and niche to survive. Migration towards mountain tops is made more difficult because of gravity. Moreover, mountains are not always so wide and high enough to allow the migration of an adequate gene pool to sufficiently high elevations.

Possible FGR responses to climate change

Tree species and their ecosystems in the Mediterranean region are adapted to several climatic constraints, such as summer drought, late spring frost and severe winter frost, depending on site.

Even the lower range of climate change scenarios predicts a worsening of these conditions, with increases in mean temperature and lengthening of summer drought (Màtyàs 2007, Màtyàs et al. 2009) and frequency/intensity of extreme events (IPCC 2007).

Already in 1937, De Philippis proposed an effective and realistic scheme of relationships between the main climatic and edaphic characteristics of the Mediterranean region and the corresponding change of main forest ecosystems (Fig. 2). In this work, the Author defined the limits of the Mediterranean area and the vegetation type replacing the holm oak forest as the mature forest ecosystem, highlighting, moreover, the possible main trends according to changing patterns.

According to the State of Mediterranean Forests 2013 (Besacier et al. 2013, COFO-FAO 2014), Forest Genetic Resources (FGRs) may respond in various ways to environmental change, including migration to track the geographical shift of areas providing suitable environmental conditions (Parmesan and Yohe 2003):

1. acclimation through phenotypic plasticity (i.e. the change in functional traits expressed by an individual in response to environment change across its lifetime; Chevin et al. 2010);
2. evolutionary adaptation (i.e. a change in gene frequency from one generation to the next resulting in a change in fitness; Kawecki and Ebert 2004);
3. changes in the forest community (van der Putten et al. 2010).

It is often accounted that migration was the dominant factor in shaping genetic diversity during the Pleistocene (Petit et al. 2003). However, despite past glacial and post-glacial migrations of many taxa (inferred from fossil pollen records and genetic data), which suggest a robust capacity for range shifts, today the migration potential of several species is considered insufficient to keep pace with the projected rapid future climatic change (Loarie et al. 2009).

Evolutionary adaptation can also be very rapid: major shifts have been demonstrated over only a few generations, such as increased drought resistance and growth in *Cedrus atlantica* (Lefèvre 2004) and epigenetic-based shifts in bud break phenology in *Picea abies* (Yakovlev et al. 2012). There are examples in the Mediterranean region of local adaptation in trees (although they are mostly phenological), especially from 'common gardens' and ex situ experiments (Savolainen et al. 2007, Vitasse et al. 2009). Phenotypic plasticity has been demonstrated to be an efficient response mechanism to change (e.g. *Cedrus atlantica* in Fallour-Rubio et al. 2009).

In nature, migration is an important way to re-establish populations in suitable conditions (Pitelka et al. 1997). The possibility to migrate or find local refuges exists principally in the northern and eastern sides of the Mediterranean region; migration would be a problem, however, for forest tree species in the southern part of the Mediterranean area, where the only chance of natural migration is towards the mountain top.

Another important limiting factor is drought. In general, it would affect the southern border of natural tree species ranges and its effects would probably be more tangible in the Mediterranean region than elsewhere in central and northern Europe. This region is characterized by several traits which make it an important hotspot of adaptive variation:

1. it is the xeric southern limit of many northern and continental European tree species (mostly in the northern side of the Mediterranean sea);
2. there are typical Mediterranean species of conifers and hardwoods, adapted to the Mediterranean climate;
3. there are also Mediterranean temperate forests, which are very sensitive to change;
4. there are remnants of former climatic changes (colder periods), found in refuges on the mountains (mostly on the southern side of the Mediterranean).

Concerning the above categories, the following short predictions can be formulated according to Pitelka (1997):

> case 1) *the southern populations of northern species*, in many cases fragmented and isolated would be reduced progressively in size

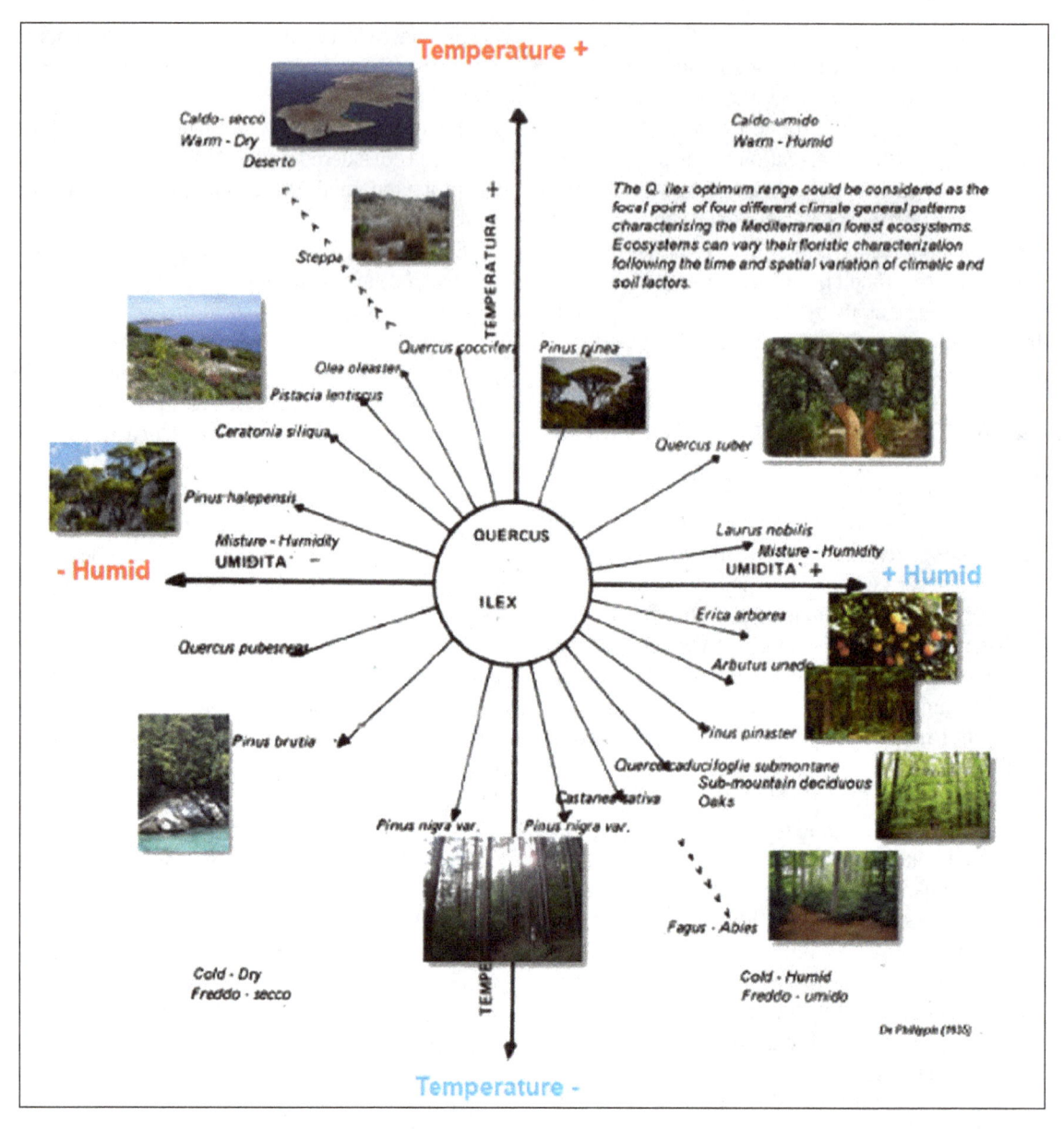

Figure 2 - Studying holm oak and cork oak (*Quercus ilex* and *Quercus suber*) in Mediterranean arboreal vegetation, Pavari highlighted the climatic and edaphic parameters influencing the distribution of these two species, then defined their needs and drew attention to the consequences of heavy and prolonged human action on their distribution (De Philippis 1937 in Pavari 1959, modified).

and eroded and, somewhere, populations would disappear. Indeed, the isotherm shift towards north would probably leaves smaller populations growing at higher elevations. These would remain isolated and genetically eroded. In the case, these populations were differentiated for adaptive traits and their disappearing will determine the loss of important amounts of genetic information. In any case, this information would be really endangered.

case 2) *the Mediterranean forests and ecosystems* (i.e. Mediterranean Pines, evergreen oaks and shrubs communities) could be initially the less endangered during the first steps of climate change. Most of them are

The marginal population of Pinus nigra v. Villetta Barrea in central-southern Italy (M. Marchi and A. Teani - CREA SEL).

characterized by extended and/or continuous populations and, in principle, this species should be more adaptable to drought. Despite this, even the toughest forest trees in the Mediterranean climate, need a minimum amount of water availability to survive. Their resources could be probably endangered in the same way of point (1) due to the progressive drying of climate, to weather extremes and also to related causes (i.e. forest fires and migration of human activities as agriculture and grazing).

case 3) *the deciduous trees forests of hardwoods or pines and other conifers* covering the mountain sites at average elevations. Their water and temperature requirements make them really sensitive to the changes expected in the Mediterranean region. Adaptation and migration towards higher elevations should be consequently implemented using natural (silviculture) and artificial methods. The search of local suitable condition spots for survival would be a good strategy.

case 4) *several relict populations as Cedars and Mediterranean firs and others*, grow in isolated and endangered populations at the higher elevations of all the Mediterranean contour. Most of them are small and/or under genetic erosion since long time. Climate change would probably put them in the first category of priorities for common action for their rescue.

Throughout the Mediterranean region, the selection pressure caused by drought has been always significant over the past millennia and its effects will probably increase with the ongoing climate change. In some areas, however, tree species have reached an equilibrium and/or found microclimates able to supply a refuge for them. The progressive further effects of drought will be probably initially seen in the decline of tree species in many areas and by mortality related to weather extremes. A recent evidence is the outcome of prolonged summer droughts of 2003, 2007, 2010 and 2011.

The prolonged drought decade since the 1980s, resulted in recurrent and frequent pullulations of pests and diseases for many species. Symptoms were initially mistaken as primary causes of mortality, whilst the former stress was probably the predisposing factor.

Temperature, drought and day length, are the most pronounced driving factors for phenological processes (eco-dormancy) in northern areas, and temperature is, among them, the most important and variable factor (Richardson et al. 1974, Linvill 1990). Temperatures can thus be seen as the limiting factor for the northern borders of the natural range of forest species in the area. Dormancy is an important function for tree phenology and for seed germination, related to temperature effect.

Dormancy. which is also related to phenology and seed germination (Walck et al. 2011) can be seriously influenced in many areas. Actually, changes in the "*Chilling Units*" (CU) amount can lead to disequilibrium in phenology for growth, flowering and seed germination, which are crucially important for a successful natural regeneration (Fig. 3). As an example, damages due to late spring frost occurrence are increasing in frequency (Hodkinson et al. 2011).

Each species, or *taxon*, requires a specific average number of CU to activate physiological processes and to break endo-dormancy.

The absence of early lower temperature signals in autumn concluding hardening and too early warmish weeks at the end of winter and in early spring can induce breaking of dormancy too early in many species. In such cases, these species will be not able to avoid damage by late frosts.

Genetic adaptation

Species were so far generally successful to adapt to climate changes (Eriksson et al. 1993).

The mechanisms used to confront such changes have proven effective in a long series of earlier climatic events and changes, even in relatively recent times. Earlier evolution did not account however the most dangerous environmental factor (i.e. humankind and its explosive (and polluting) demographic and industrial development).

Until now, trees have reacted relatively well to the change, and it is commonly thought that genetic variation is still enough to supply materials for the

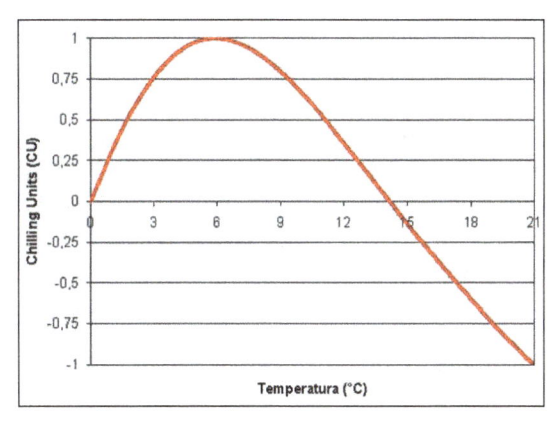

Figure 3 - Average effect of temperature (°C) on the Chilling Units (CU) requirements (Richardson et al. 1974). A CU is the time unit (1 hour) at moderately low temperature (5°C or less according to methods).

adaptation of the populations and to sustain their dynamics (Eriksson 1996 and 1998, Hamrick 2004).

Moreover, even phenotypic plasticity (Schlichting 1986, Pigliucci 2001) and long life cycles of forest trees can support natural regeneration over the decades, in spite of environmental fluctuations. Micro-environmental variability can play an important role in the process of conserving variability and in some cases microclimate refuges can be maintained. Adaptation is estimated to be relatively rapid: 2 o 3 generations can be sufficient for a forest tree population to modify its adaptive patterns.

However, as above mentioned, the speed of environmental change is increasing and there is a general fear that tree species and populations may not be able to adapt rapidly enough.

In the Mediterranean Region, a genetic hotspot region (Fady 2005), forest tree populations are in general small and genetically eroded; they are also generally isolated, and gene flows and gene exchanges between them are often limited.

Most species are distributed by meta-populations, which probably exchanged genes in past times, when the climate allowed their expansion. Nowadays, human activities, at least in Europe, have often interrupted the chance of gene flows definitively, and probably for longer times than similar interruptions in natural conditions.

Thus, in addition to natural factors which may influence the response of trees to environmental change, *social* and *civilization aspects* should be considered as factors which play a role in the spontaneous migration of forest species in places where agricultural areas, industrial trees, roads and human settlements establish insuperable barriers.

In the past, a lot of genetic information was used by forest populations and tree species to adapt to environmental change; the present, rapid climatic changes may exceed the tolerance limits of many species. We have to expect a dramatic loss of variation for populations in the Mediterranean area and, in some cases, we should expect an extinction of the species or part of them, especially in the most extreme conditions of their natural range.

Together with a genetic response, a wide part of the interest of the research community is focused on the assessment of the *quantitative response to change*. In fact, knowing rapidly and experimentally those responses is fundamental to support decision making and the actions aimed at conserving variation.

Conditions deserving special attention are both frequency and intensity of extreme events (frost, drought, high temperatures etc). These extreme events are important factors for the selection within populations and species. It is well known that species response to natural selection is not monolithic, as variation in responses to change can be clearly observed for many species (as many comparative provenance and progeny or clonal tests have widely shown).

Two main complex characteristics are used by forest tree species to be adapted.

Resistance/Tolerance and *Phenotypic plasticity* are the two faces of adaptation and are the basic traits which can be used to study the effects of changes on forest tree populations and their response:

- *Tolerance:* it is the ability of a genotype to preserve its fitness under the pressure of a damage factor. It is genetically settled and fixed by the evolutionary force; it allows each species to occupy a given ecological niche in a given habitat.
- *Phenotypic plasticity:* it is the asymmetric response of genotypes to extreme events. It can be defined as the property of a given genotype to produce different phenotypes in response to distinct environmental conditions.

Both of these properties are presently considered as a key for adaptation to climatic changes.

Among them the role played by evolution in shaping phenotypic plasticity remains still poorly understood (Pigliucci 2005). This property of many phenotypic traits can be used by organisms to start adaptation, which can become the first step of an evolutive process, producing fixation and then divergence and differentiation among populations.

Chambel et al. (2005) discussed this concept as a tool to understand the adaptive processes in forest species.

In a recent review Healy and Schulte (2015) supported the idea that the variety of patterns observed in the gene expression evolution in some species could evidence that a set of traits can occur in organisms, and that each of them can respond to environmental variation through phenotypic plasticity and genetic divergence in different ways. This aspect can influence genetic divergence between species and populations and finally their adaptive potential.

In this field, Santos-del-Blanco et al. (2013) heightened this concept by studying 52 Mediterranean pine populations. They found that pines can represent a model to learn about the adaptive value of allocation to reproduction vs. vegetative growth. That happens thanks to their higher differentiation among-population, from the adaptive point of view, and their ability to cope with environmentally dry and harsh contexts.

Their results followed theoretical predictions and support the idea that phenotypic plasticity for reproduction can be considered as adaptive under stressful environments. Phenotypic plasticity, together with natural selection on reproductive traits, can therefore play a relevant role in the future adaptation of forest tree species in an increasing selective context as in the Mediterranean area. Aranda et al. (2010) and Climent et al. (2009a, 2011, 2013) findings contributed to confirm these concepts as for Mediterranean and Asian pines. This is also important in introduced species adaptation, even in cases where genetic variability seems to be low after their introduction. For instance, Fady et al. (2003) reported about *Juglans regia*, a tree introduced in western Europe by different human migration waves (Pollegioni et al. 2014) and characterized by low genetic variation. They found that, for adaptation, the selection pressure operated by human uses is still high and bud break ranking is significantly correlated with its European *provenances*. As a practical consequence, southern European early bud break plant materials should not be planted under most middle European conditions. Moreover, as in these areas late spring frost can be expected, damages on the apical buds caused by frost are closely correlated to architectural traits (and thus wood quality).

A debate is still ongoing about the real ability of species to adapt in view of so rapid changes, but, pragmatically, strategies and possibly common actions plans have to be established in case they will be in danger of genetic impoverishment or extinction.

Whilst related information on adaptive characteristics of boreal species has gradually been collected over the past years, the lack of knowledge about the Mediterranean species in this regard is still wide. As above mentioned, other influences and pressures due to global change effects as pests and diseases, can affect tree and plant species.

On the sidelines of this work, primarily focused on adaptation to the physical factors of the environment, it is necessary to refer shortly to the complexity of ecosystems hit by the changes. Global change is actually increasing the biodiversity patterns in Europe and in other regions because of the rapid migration of new pathogenic or invasive organisms.

Over their long life-span, forest trees have to develop not only a demographic interaction with individuals of their own and other species, but also with biotic factors migrated or introduced in their natural range, as pathogens, parasites etc. Several case can be recorded where trees have to cope with many new problems.

Many palm species, for instance, are heavily endangered by the attacks of *Rhyncophorus ferrugineus* (*Coleoptera*) in the Region (Gomez and Ferry 1998).

Recently, the economy of Stone Pine (*Pinus pinea* L.) coastal pinewoods has been seriously damaged by fungal pathogens, among them *Diplodia pinea* (Waterman 1943) and the western conifer seed bug *Leptoglossus occidentalis* (Tescari 2001), which have contributed to important losses of stone pine nuts.

Again, most of the chestnut (*Castanea* spp.) production in Europe is severely affected by *Dryocosmus kuriphilus* (Yasumatsu 1951) gall wasp known as chestnut gall wasp, Oriental chestnut gall wasp or Asian chestnut gall wasp. This pest is native to China and it is rapidly spreading to other world regions where competitors are lacking (Graziolin and Santi 2008).

Challenges for management and policy making

Climate change could have implications for the forest ecosystem services needed for human wellbeing, such as water cycling, carbon sequestration and the production of numerous wood and non-wood products (Millennium Ecosystem Assessment 2005).

FGRs also face other challenges: Sala et al. (2000), for example, showed that land-use change and biological invasions remain key drivers of biodiversity change in Mediterranean biomes.

Nevertheless, all forest management decisions should now take climate change into consideration, but how to take the uncertainty associated with climate change into account in management plans is a formidable challenge.

In a region with the attributes of the Mediterranean, where fragmentation is high because of geomorphology and the history of human activities is long, it is unlikely that the migration of plant species and forest types will be fully able to reduce the impacts of climate change on forests. Even where migration is possible, societies may be unwilling to accept massive forest dieback in some areas and the subsequent natural selection of more suitable genotypes, and may demand intervention.

Societies may also be unwilling to accept a substantial reduction in the productivity of high-yield forests as a consequence of phenotypic plasticity.

Some of the challenges that forest managers will face in developing strategies under the uncertainties of climate change are listed below. For each challenge, research can already provide management options.

Possible actions to be undertaken

Assuming that climatic change is ongoing and that its gravity is increasing, the possibility to undertake initiatives for the mitigation of climate change effects and the conservation of forest genetic resources (FGRs) in the Mediterranean region, should be considered.

In agreement with Sala et al. (2000), integrated efforts by climatologists, ecologists, social scientists, and policy makers will be required to get realistic scenarios of future changes in the Earth system, especially the Mediterranean.

These scenarios should be based on quantitative analyses and have to consider studies of the interactions among factors to which local biodiversity is most sensitive in each biome.

The knowledge of the expected effects on biodiversity will help to develop management practices according to the biological, social, and economic characteristics of the area where you are working on.

Choice of priorities

Research initiatives should be prioritized in order to save time and resources.

A first useful approach to apply priorities might be based on a list of *tree species which are presently considered to be seriously endangered.*

A method for prioritizing actions could be based on choosing *model species able to represent different climatic situations or altitude ranges.*

A third way could be based on choosing *model species which represent a number of genera.*

Màtyàs (2007) reported on a valuable comparison carried out among forest tree populations, according to their main characteristics and the structure of populations, and the corresponding urgency levels for action to be undertaken; see Table 4 below.

Mediterranean pines, evergreen oaks and other tree species belonging to Mediterranean habitats may be initially less endangered than central - northern temperate habitat species, actually they are generally better adapted to drought.

However, with the progressive drying of climate, the increase in weather extremes and related events such as forest fires, and shifts in human activities such as agriculture and grazing (particularly in the southern Mediterranean region), also the habitats in this area will come under increasing pressure.

Isolated populations of many mountain Mediterranean species (e.g. *Alnus cordata, Pinus nigra* v. *laricio* and *P. heldreichii*, Cedars and Mediterranean firs) growing at their ecological or geographical margins will be endangered. Most of these populations are small and scattered and have been affected by genetic erosion in the past.

Other isolated populations of widely distributed central and northern European species should receive special attention. This may be the case in southern areas of natural ranges in the Mediterranean area. Here, isolated populations growing on southern aspects of mountain sites could not have space to migrate in altitude to follow future climatic isotherm shifts. All these populations are likely to be already seriously suffering the effects of the present climate change, and they cannot migrate as a response to such pressure.

In some case, these populations are accounted and classified as Marginal Peripheral populations, also known as MaPs (Yeh and Layton 1979, Hampe and Petit 2005, Eckert et al. 2008).

Furthermore, it is to be considered that particular attention and therefore priority should be given to species or populations which are isolated and growing under extreme conditions. Very often, these species and populations, important for specific uses or endemic, are really endangered.

In these populations, the gene flow is generally reduced, as well as the ecosystem below a critical mass, and there are problems related to the low dispersal capacity and the genetic erosion.

This is the typical situation where the habitat has been severely reduced and even small additional disturbances can compromise the survival of the population. Consistent examples are *Cedrus libani*, *Abies nebrodensis*, *Cupressus dupreziana* and a number of other spp.

In situ *conservation*

Foresters should be aware that conserving and managing the genetic variability of forest trees *in situ*, following a dynamic approach, is the basis for selection and for developing adaptive management. It is also important to perpetuate the ecosystem functions and services as well as to ensure more resilience to the ecosystem.

In situ, conservationists and improvers can interact in order to select adaptable basic materials within Marginal populations, allowing the production of adapted offspring. Thus, the reaction of basic materials can be tested *in situ*, being exposed to the temporal and spatial variation of micro-environmental conditions. It is evident that *in situ* selection should be carried out mostly on highly heritable adaptive traits, mainly eco-physiological (drought and frost resistance/tolerance) and phenological.

Table 4 - Comparison between species/population structures and their urgency requirements (Màtyàs 2007, modified).

A. Species/population structures requiring **low urgency** for initiatives	B. Species/population structures requiring **high urgency** for initiatives
1. Continuous distribution	1. Fragmented or isolated distribution
2. High density	2. Small or scattered
3. Naturally regenerated	3. Artificially regenerated
4. Effective gene flows	4. Limited replenishment of the gene pool
5. Spontaneously spreading	5. Low dispersal capacity
6. Extensive, zonal habitats	6. Extreme sites as habitat or small size
7. Reproduction unaffected	7. Disturbances in flowering and seeding

McKown et al. (2014) studied this topic on *Populus trichocarpa* where they found that environmental gradients can shape both the phenotypic adaptive trait variation and the genetic structure.

Gene flow, genetic drift, selection, recombination, and the reproductive system characteristics are factors that act in determining the genetic structure of a population, but the value of the genetic resources of a stand is also strictly determined by the way foresters manage them.

In general, silviculturists do not seem to be aware of these concepts and they also do not seem to be used to account the genetic value of each single tree in the forest and their possible contribution to the structure of the species population they belong to.

Individual trees, through the variation of their fitness and by means of sexual interactions, determine the genetic structure of a population and, together with mutations, they produce variation and are the basis of the evolutive processes, adaptation included.

In silviculture and forest management, as for the *in situ* management of valuable populations, the knowledge of the spatial genetic structure of forests should be a common and shared approach.

A first simple model of structure in a stand is the subdivision of a population in sub-populations or *demes*. According to this model, if a plant population is dense enough, it is likely that pollen and seeds are dispersed not far from the trees that produce them (Sagnard et al. 2011), regardless of the type of dispersion. In this way, the exchange of genes in a population is very slow between groups of trees which are more distant and relatively quicker and easier within these and the entire population can be accounted for a number of sub-populations.

Because of their small size and isolation, isolated demes can differentiate themselves from others due to genetic drift. If their structure is stable for several generations, within demes you may have a certain degree of consanguinity between neighboring trees.

Gene flow between demes counteracts the effect of differentiation. The structuring pattern increases with number of generations and gradually, during that population is consolidating, until a balance between drift and gene flow is reached.

Managers and silviculturists should take in account the genetic structure of populations, as above described, in order to preserve genetic variation and they should therefore have in mind the need to have a good knowledge of the genetic structure and the spatial distribution of the genetic information of the forest they are managing (Lindner et al. 2011, Bolte et al. 2015). This would be necessary to preserve biodiversity and sustainability, as well as to maintain the availability of the selected materials to be used as a source for planting programmes.

As an example, an explorative study (Ducci et al. 2004) showed that in *Quercus cerris* coppices with standards, most of the genetic variation can be found in the left untouched or cut, depending on the practice in progress purpose, dominated stumps. Therefore, new forms of coppice management should be tested in order to preserve an amount of dominated or apparently not useful stumps as a stock of the genetic information belonging to the local gene pool.

Two cases regarding pedunculate oak (*Quercus robur*) and silver fir marginal populations (*Abies alba*) in the Po Valley and in central Italy, respectively, are reported in boxes A and B.

Silviculturists should account the right way to preserve this variation which was in the quoted case study neutral but, at the same time, an indicator of the genetic structure of the populations of concern.

A - A case-study.
The in situ ***management of marginal pedunculate oak populations in the Po Valley.***

Most Italian populations of pedunculate oak, fractionated and usually not extended, are made up of a mature layer, consisting of a few trees per unit area and a nearly total absence of regeneration established.

The consistency of the mature layer decreases progressively, due to the natural ageing of individuals or to heavy meteo events, these leading to the collapse of mature trees. Not being available a continuous monitoring over time of the genetic structure, one can only assume the occurrence of an ongoing process of progressive genetic erosion. This aspect may help to explain the very low observed heterozygosity (Ho), despite the good potential expressed by high values of expected heterozygosity (He) and the high rate of inbreeding.

To address the population to a higher dynamic equilibrium is therefore necessary to carry out silvicultural operations which promote natural regeneration and, more generally, the dynamics of populations. It is necessary to try to encourage the mixing within the population as much as possible, both through improved pollen exchanges between trees and with the active management promoting the occurrence of dissemination and regeneration establishment (internal transfer of seeds and seedlings).

This pro-active practice will influence the following generations, increasing the levels of diversity and interchanges between related groups (demes made by families of siblings or half-siblings), especially within and among peripheral strips of the population, affected by intense genetic erosion.

A well-suited example is given by the small and isolated oak population of Capriano del Colle in the province of Brescia, Po Valley, already described by Ducci (2007).

In relatively recent times gene flows in both pedunculate and sessile oak populations growing in the Po Valley (northern Italy) were interrupted. The main changes occurred during the Renaissance age, when wood exploitation and the use of soil for agriculture had a dramatic rise following the population increase and the diffusion of settlements. Later, the industrial and urban development (XIX and XX century) made the original widespread oak forest covering the plains which were progressively reduced to small-sized, isolated remnants.

The forest of Capriano del Colle is small but relatively large as compared to the forest cover in this area of the Po valley. It has quite a favorable genetic potential useful to restore a dynamic population, now subjected to a high risk of genetic erosion, because of its state of maturity and for the absence of a natural regeneration already established.

Its genetic structure is articulated and clearly identifiable on the ground by clusters (Fig. 4) and characterized by a variation core surrounded by peripheral areas of lower variability. The low mobility of seeds, the ground morphology and the microenvironmental conditions drive the distribution of genetic clusters, following the shape of a small river valley. The site is also characterized by a relatively dry climate as compared to other populations and this may have influenced the genetic parameters, favoring, for instance, certain genotypes rather than others, and negatively affecting the viability of seed and the regeneration establishment in some years.

This attribute should be taken into account in planning any collection activities of reproductive materials.

Seed collections should be performed from trees located in different patches of the population, together with the silvicultural management needed to start population dynamics again and to provide, by targeted practices, sufficient brightness for the development and success of seedlings cohorts under the mature canopy trees. That should be done mainly within the variation core, in order to collect seedlings to be planted in clear areas and in the peripheral population, with lower genetic richness.

Seedlings originated from different areas of the forest, will improve the within population gene flows in the future generations and increase levels of diversity in the most peripheral and genetically eroded population. That will help to preserve the gene pool dynamism, as well as to reduce the likelihood of random rare alleles loss.

As for the *in situ* conservation of forest tree species aimed at **mitigating the effects of climate change**, there is no reliable scientific information available for the Mediterranean area. Many studies have been carried out on the genetics of forest tree populations and on quantitative productive traits in multi-site comparative trials. Such quantitative traits can be considered as indirect indicators for adaptability, but no specific research has been systematically carried out on the effects of climate change.

It is time to start studies on the role of the *in situ* conservation, including analyses on silviculture and tending effects on the genetic structure and therefore on management sustainability.

Konnert and Hosious (2010) discussed on these topics showing the importance of genetic aspects for developing a sustainable forest practice both for silviculture targeted to natural regeneration and for implementing artificial regeneration.

Special attention should be given to mixtures of tree species, and to the control of microclimate, through canopy cover and crown density regulation, to reduce susceptibility to drought.

The central area containing the variation core of the forest of Capriano del Colle (BS).

A peripheral strip of the oak population in Capriano del Colle (BS).

Namkoong et al. (1996, 2002) and Eriksson et al. (1993, 1996, 1998) proposed different methods for monitoring the impact of forest management on genetic diversity by using genetic and demographic indicators to evaluate the efficiency of management actions which drive genetic processes, such as genetic drift, migration and selection, useful to maintain the existing levels of genetic diversity.

The concept of monitoring can be applied at several stages and in different fields of forestry, from the management of forest reproductive materials to plantations and natural populations.

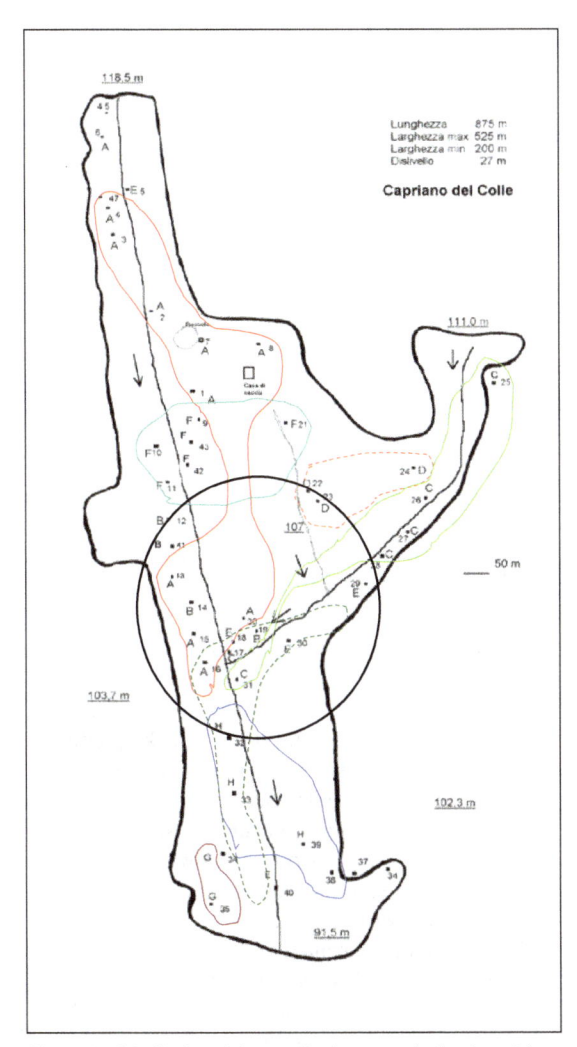

Figure 4 - Distribution of the genetic clusters at the Capriano del Colle (Bs) forest. The central core of genetic variation is circled (Ducci 2007).

Tree spacing and tree density can be managed to control light and then susceptibility to drought. Managers can also change the composition of species diversity to reduce the vulnerability of forests to disturbances as fire, drought, wind, insects or pathogens and find out and choose tree specific compositions better suited to a changed climatic regime.

Changes in tree density will also affect genetic diversity within species (Sagnard et al. 2011). Applying a not extensive silviculture, based on small management areas, will produce patchy alternatives where natural selection occurs. This can also allow managers to maintain the same tree species composition.

Monitoring *in situ*: genetic diversity and adaptability

Diversity is essential for the adaptation of tree populations and monitoring the genetic dynamics of trees is fundamental for developing long-term strategies.

> ### B - *The case of Abies alba MaP populations in central Italy (after the LIFE project RESILFOR)*
>
> In the framework of the Life Nature project entitled "Resilfor - Restoring Silver-fir Forest" (Miozzo et al. 2014) a few permanent plots were established to monitor changes in stand structures and also adaptive and genetic characteristics. These plots were established in northern Apennines, forest of La Verna (Arezzo) and in the forests of Mount Amiata, an ancient volcano (Southwestern Tuscany).
>
> 1) **Forest of La Verna** (community of Chiusi della Verna, Arezzo, Italy)
> - Plot A: Lat. 43°42'30,22N, Long. 11°55'54.34"E, 1 ha, aspect NO, alt. 1166 m;
> - Plot B: Lat. 43°42'31.80"N, Long. 11°56'01.42"E, 1 ha, aspect NE, alt. 1188 m.
>
> 2) **Natural reserve of Pigelleto** (community of Piancastagnaio, Siena, Italy):
> - Plot A: Lat. 42°48'17.30"N, Long. 11°38'46.69"E, 1 ha, aspect NE, alt. 780 m [very low elevation as for the Mediterranean area];
>
> *Forest structure*
> The silver fir in the Apennines is undergoing a slow and steady contraction of its range, in favor of broadleaf trees (mainly beech).
>
> Going into detail about each area, the forest of La Verna is substantially made up of fir - beech even-aged forest, with the sporadic presence of other species.
>
> At Pigelleto, vice versa, chestnut first, along with elm, ash and hornbeam are commonly associated in the chestnut phytoclimatic belt.
>
> The structure of fir populations, included in the permanent monitoring plots, revealed a tendency to form an even-aged groups, even if very irregular structure, probably resulting from not well-planned cuts of the past (Pigelleto), or influenced by natural disasters (storms) such as those occurred in La Verna.
>
> *Genetic structure*
> Genetic data, and especially data of the regeneration, fully confirm these features.

The populations still contain relatively high levels of genetic variability useful to support the evolutionary dynamics for future and new challenges. The analysis, both within and among adult/mature layers and on natural regeneration, showed good levels of panmittic balance in the different demographic classes. A trend to homozygosity is relatively usual within forest species especially those where self-pollination is frequent.

The clustered spatial distribution of diversity on the plot, at the ground level, showed to be lightly structured. This feature confirms what has been observed in small populations of other species such as *Abies nebrodensis* in Sicily and pedunculate oak (*Quercus robur*) in the Po Valley (Ducci et al. 1999, Ducci 2007, see box A): diversity is highest at the center of the population, where genetic clusters tend to overlap. This distribution can be driven by forces such as gravity or direction of local winds, which disperse the seeds in the direction of the slope or in the more sheltered inner valleys.

At La Verna and at Pigelleto, the number of genotypic clusters decreases towards the margins of the populations, while outward the number of individual genotypes increases and clusters are fewer.

In the case of small populations, this phenomenon can be generated by genetic erosion at the edges, as well as determined by local ecological factors.

In the case of larger populations, the same occurrence may depend on trees spreading pollen and seeds in the tested plot area from more distant parts or unknown *demes* of the forest, not covered by the analysis and topographically more distant.

Concerning the natural regeneration pattern, the mixture determined by wind dispersion of seeds and pollen is high. This is confirmed by the spatial overlapping of clusters and by the high variability.

Adaptation

To better characterize the populations from the adaptive and plasticity potentials viewpoint, the phenological-cambial activity has been monitored. This technology is very precise and data reliability is high.

High intra-population variability were highlighted, with respect to the earliness and tardiness (dates of the beginning and end of the vegetative phase) of firs. Some fir trees were able to grow until December under warmer conditions, while others have growth periods much shorter. There are a few trees which are highly sensitive to the availability of summer rains, constantly producing false rings, while others are very regular.

A high plasticity of the species was also observed. The phenotypic plasticity is defined as the ability of a genotype. In this case, the species (i.e. the relationship between the Tyrrhenian fir populations) change the phenotypic behavior, in relation to different environmental factors. In fact, the firs of La Verna and Pigelleto revealed to be enough "plastic" to cope with many different environmental conditions. The phenological monitoring allowed to observe both the plasticity at the spatial scale (the two sites), and at the time scale (the years of monitoring) and to observe the behavior of silver in two environments as diverse as La Verna (cold, relatively dry and internal) and Pigelleto (much more temperate and with a longer growing season).

At Pigelleto, the warmer weather until late December in 2011, resulted in phenological cambial stages, typical of spring and summer time.

The results clearly showed how fir copes with different environmental conditions. Its variation in adaptation and plasticity at an individual and population level will be a source of diversity, useful to preserve the species in these areas and to develop adaptive silvicultural strategies.

Guiding principles for managing these populations

It is fundamental that genetic monitoring has been repeated for several years to pick up different reactions in relation to different climatic and biological seasons and trends.

While plots should be managed with the highest standards of 'close to nature' silviculture, or at least continuing to proceed as made so far [given the function of monitoring areas], on the other hand, it is basic to manage in reducing the influence of beech where this species tends to be invasive. Anyway, this has to be carried out in a controlled way, by maintaining edge micro climate 'ocean like' conditions, because this marginal/cline conditions provide protection to fir offspring in the early stages of development.

Silviculturists have to manage the overripe layers of tree population, isolating adult trees vigorous enough to produce and bear fruiting.

The quantitative assessments of molecular genetic variation at either the neutral or the adaptive level can be therefore used.

Advances are being made in genomics and with bio-statistical tools to improve the efficiency and cost-effectiveness of genetic monitoring and the inference on demographic relationships within populations obtained good results (Schwartz et al. 2006, Chybicki and Burczyk 2010).

Recent advances in forest trees were made about association studies between single nucleotide polymorphism (SNP) and adaptive traits (Holliday et al. 2010, Eckert et al. 2009).

The importance of preserving marginal and/or peripheral populations (MaP)

As above introduced, *Marginal Populations*

(MaP) and the genetic information they contain are especially vulnerable.

With this definition, the populations growing at the edges of the natural range of forest species are determined. Margins can be identified at both lower and higher latitudes and similar situations can be generated by other factors, in relation to the elevation effects on temperature and water availability.

Even those populations having a margin context determined by an environmental factor may be marginal, regardless of location, it includes those produced by human influences at the local level.

It is meaningful that in southern areas of many tree species range, most of the present genetic variation is concentrated at the rear edge populations, concentrated in three main hot spots, respectively located in the Iberian peninsula, in Italy and in the Balkans. That is a clear heritage of the Pleistocene glacial refuges.

Marginal populations, which generally survive in less favorable environmental contexts, are typically isolated (geographically separated) from the central continuous range and can be small and at risk of extinction.

Among them, we can identify:

1) populations of the geographic edge (*leading edge* and *rear edge*) that may also coincide with the marginal ecological requirements of the species,

2) marginal populations, from an ecological viewpoint, living in the central range of species but in ecologically marginal conditions as compared with the species requirements,

3) populations growing at the altitudinal margins.

The pioneer populations in the *leading edge*, migrating along the shifting of climate zones, leave behind the central population (*core population*) in its optimum. They are followed, in turn, by progressively more and more rarefied populations which remain isolated from the central core (*rear edge*, or marginal populations).

The MaP populations, growing at the geographical edges, may be divergent from the genetic viewpoint due to the joint effect of genetic drift and natural selection, but they can also fill the role of containers of genetic diversity among species due to the outcomes of the spatial-temporal environmental variation at the edges and to the consequent dynamics being induced (Hampe and Petit 2005).

The *leading edge* populations can be characterized by dispersion over long distances and probably the founder effect occurrence is frequent. The *rear edge* populations can be subject to extinction whilst the natural range can undergo latitudinal displacement (*trailing edge*) and sometimes leave small populations surviving in relatively suitable environmental conditions, but with reduced extension (*stable edge*).

The same occurrence takes place when the upland forest populations are being accounted. Here, leading or rear fronts can be detected within the limits of high and lower altitude (Ettinger et al. 2011, Batllori et al. 2012).

According to climatic oscillations, the fronts were reversed several times in past ages, and probably, at each time, populations have shuffled or swapped their genetic information.

The importance of these marginal populations resides in the fact that they may contain adaptive traits of special interest for the species because of

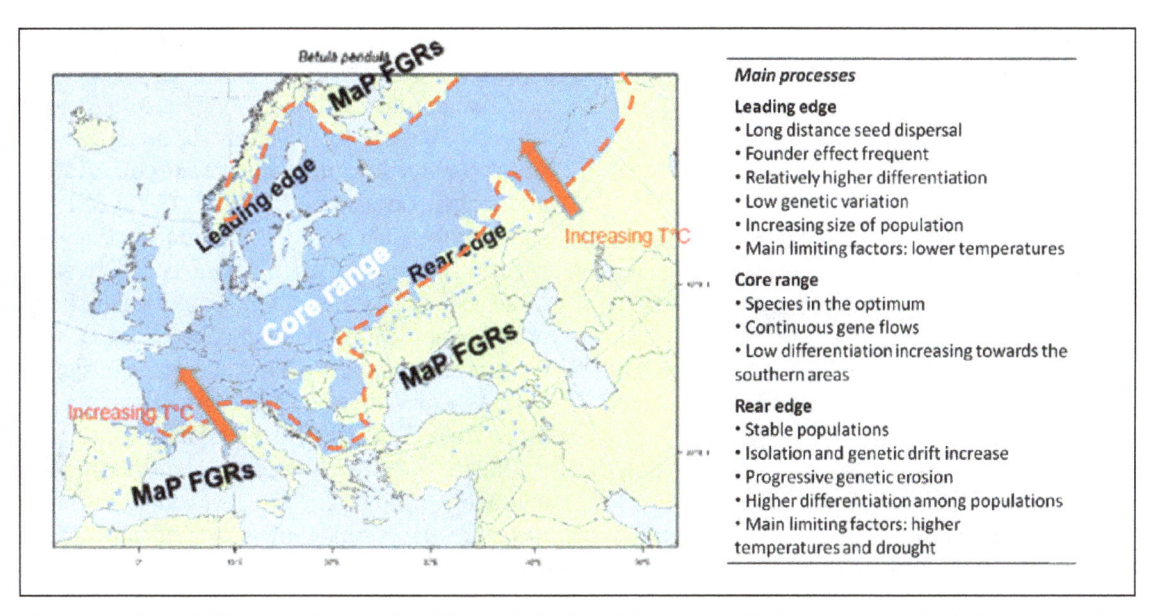

Figure 5 - Variation of different genetic parameters of the species in relation to the stucture of their range and according to the type of edge in advancement or withdrawal (Ducci 2015, after Hampe and Petit 2005).

Figure 6 - Representation of the optimum and of upper and lower margins in the sense of the altitude (Ducci 2015, elaborated after Ettinger et al. 2011).

more marked and variable evolutionary factors at margin areas.

Limiting factors or otherwise factors able of exerting selective pressures, typical of southern marginal areas or low altitude, are related to the effects of high temperatures and aridity.

In northern and higher-altitude areas, the main driving factors are connected to lower temperatures.

While, for boreal species, especially in the northern districts of their distribution areas, the information on the genetic structure of populations is wide, for those having southern and Mediterranean range it is still inadequate.

Ex situ *conservation*

Without scientific experimental information on adaptive indicators, it is difficult to manage genetic resources and strategies for their conservation. For this purpose, it is extremely important to profit from the existence of several *ex situ* experimental networks and collections established in the Mediterranean area.

IUFRO, FAO *Silva Mediterranea* and other important international networks established, in the past, wide international multi-site tests; these concerned mainly conifers (4 genera and some 20 species) and *Quercus suber*. This genetic stock, just recently inventoried in the framework of *Silva Mediterranea* (Pichot 2011, in Besacier et al. 2011) and European projects as TreeBreedex and Trees-4Future, will be a sound base on which future actions may be developed.

Traits to be accounted for the future activities will reflect the present trends in forest and tree breeding research, as such research has proved to be efficient (formerly *Silva Mediterranea* itself, IUFRO, EUFORGEN networks, EU TREEBREEDEX, Tree4Future etc.).

Adaptive characters and *phenotypic plasticity* have to be initially investigated as simply phenotypic quantitative traits; in the following steps, the genetic aspects of these traits should also be investigated as a basis for breeding, improvement and conservation, and for genetic mapping (i.e. QTLs, QTNs, SNPs association techniques etc.). The relationship between variation in complex traits and molecular diversity of genes can be studied following a genomic approach, but the identification of genes responsible for variation remains a slow and time-consuming process, especially in long-lived organisms such as forest trees (Vendramin and Morgante 2005).

New strategies have to be identified, tested and adopted according to a common agreement, to reduce the negative impacts of climate change on tree species and populations and the loss of the genetic information they represent. Moreover, the basic concept to be taken into consideration is that *ex situ* conservation should target, where and when possible, dynamic traits.

Reproductive materials and legislation

Ex situ conservation can be carried out all levels: regional, national and international. The possibility for entering international agreements for safeguarding national genetic resources in another country should be reviewed and better developed. This concept is not new. International field trials and common field experiments can be, partially, considered as forms of conservation abroad of genetic resources.

In the case of the Mediterranean region, seriously endangered and valuable forest tree populations or species should be identified and conserved with the support of the international community.

Efforts should be shared among countries and framed within the networks created to host genetic materials in suitable environments and managing them according to common plans. This point is very important and it should stimulate cooperation.

Conceivably, laws and regulations on management *in situ*, nursery systems and trade in forest reproductive materials, should be modified and integrated into the overall forest management, thus contributing to the mitigation of negative climate change effects.

Furthermore, in relation to the above statements, the Revision of *seed zones* and *provenance regions delineation* should be considered, and the present concepts about relatively static Provenance regions systems should be modified following a more dynamic vision, while also the rapid modifications of climate and phytoecological parameters should be taken into account (Ying and Yanchuk 2006).

Almost universally, forest reproductive materials (FRMs) are used in forest plantation projects according to guidelines written under the assumption that local soil and climatic conditions will remain stable. The organization for Economic Co-operation and Development (OECD) is the main reference for FRM certification and standardization in the Mediterranean region, but the European Directive 1999/105/CE also provides general criteria and guidelines for FRM trade within the EU. According to most climate models, climatic conditions in the region will not remain stable in the next decades and there is need to revise the rules on the delineation of the provenance of the species and the transfer of seeds and other reproductive materials (Konnert et al. 2015).

Under this framework, the recent Survey of World Forest Genetic Resources (2014) with the contribution of FAO *Silva Mediterranea* for Forest Genetic Resources in the Mediterranean region should be accounted. Already in the 90's, Topak (1997) inventoried the FRMs used for reforestation in 17 FAO *Silva Mediterranea* countries adopting the OECD standards.

Moreover, the FAO *Silva Mediterranea* database lists national and international forest tree common gardens[1] in the Mediterranean region. Such basic information is essential for rethinking seed zone delineation and provenance selection in the face of climate change.

The recent establishment of the Nagoya Protocol (2014), focused on the Access to Genetic Resources and Equitable Sharing of Benefits arising from their use. This international instrument adopted by the Conference of Parties to the CBD (Convention on Biological Biodiversity) at its X Meeting held in October 2010 in Nagoya, Japan, was opened for signature on Feb. 2, 2011.

The aim of the Protocol is the fair and equitable sharing of benefits arising from the utilization of genetic resources, including the appropriate access to genetic resources and appropriate transfer of relevant technologies, taking into account all rights to those resources and those technologies and appropriate funding thereby contributing to the conservation of biological diversity and to the sustainable use of its components.

The Protocol contains provisions ruling both the access to genetic resources and the equitable sharing of benefits arising from the use of them. A user who wants to access a genetic resource of another country (e.g. a medicinal plant for studying the active ingredient, or to produce a medication) must follow the provided procedure in the country access, providing that resource.

In addition, a contract must be drawn up providing for the equitable sharing of benefits, between user and provider, arising from the use of the resource in question (e.g. profits, technology, knowledge, and so on). Genetic resources are often associated with the traditional knowledge of indigenous and local communities. Therefore some provisions of the Protocol rule the access to such knowledge and the equitable sharing of benefits procured by their application.

---

[1] 'Common garden': field test in which many individuals (clones, families, populations) of a given plant species sampled from an identified geographic area are grown in a common environment, making it possible to infer genetic information from the observation of phenotypic differences. These networks were early known as "multisite comparative trials".

Assisted migration

The above mentioned legislative tools can help to develop actions aimed at preserving really endangered genetic resources, when no other possibilities can be considered for their rescue. Assisted Migration is an option still to be studied more in depth.

Examples of artificial migrations or translocation in forest trees are spread in Europe since long time. We can mention agriculture crop species as well as fruit crop trees following mankind in his migrations.

Among forest trees, several historical case studies can be found in areas where ancient civilizations passed and left signs of successful translocations. Some of them are millennial. Genetic evidences of the translocation of walnut along the Silk Road were found by Pollegioni et al. (2014), but we can also record widespread examples as chestnut in western Europe mountain ranges, cypress introduced in Tuscany by Etruscans and Greeks, *Pinus pinea* spread by Phoenician and Greeks in all the Mediterranean, and more recently Greek fir, Black pine and finally Douglas fir as one of the most important and successful recent intercontinental translocations.

Despite trees are used to migrate in response to changing climate eras, the present climate projections show the real impossibility of forest tree species to migrate so rapidly. In most cases, probably species will be able to find out adaptation within their gene pools.

Aitken et al. (2011) discussed this topic concerning boreal populations trees. These populations are characterized by adaptive variation patterns and - as above said - they show moderate to heavy clines in phenology and growth along temperature gradients. These adaptive traits appear to be the result of small effects of many genes, and may facilitate rapid local adaptation despite the high gene flow in the northwestern American area. On this way, the gene flow with alleles linked to adaptive traits to warmer climate conditions may promote adaptation and migration at the leading edge, while populations at the rear will likely face extirpation.

In the opinion of the Authors *'widespread species with large populations and high fecundity are likely to persist and adapt, but will likely suffer adaptation lag for a few generations. As all tree species will be suffering lags, interspecific competition may weaken, facilitating persistence under suboptimal conditions'.*

Species with small populations, fragmented ranges, low fecundity, or suffering declines due to introduced insects or diseases should be candidates for facilitated migration.

So, the idea that most of the widespread species own resources and variation are sufficient conditions for adaptation to climate changes is widely diffused. But, in some cases, the natural tools will not be enough to preserve species and populations and human actions would be probably oriented towards the adoption of Assisted migration methods. In this case, transfer guidelines have to be developed rapidly, possibly based on experimental results or already existing cases.

Assisted migration, implies some human interventions to help a species to migrate. This basic concept can be also extended to single endangered populations. Several variants/synonyms can be found in literature of this definition *as assisted population migration, translocation, reinforcement, assisted population/range expansion, assisted long-distance migration* (Ste-Marie et al. 2011). As a very complex development of these variants, Seddon (2010) introduced the concept of migration of groups of species defined as *community construction.*

In the present experiences on assisted migration, a prudential "mimic" approach is generally used. Indeed, Authors highlight the increasing risks of unforeseen influences and consequences when very wide migration distances are being considered (Vitt et al. 20110). For this reason, the within range relocation is mainly diffused.

Anyway, in some extreme cases, a more determined approach should be considered. In Europe and especially in southern areas, an early human activity has left examples of forest tree species relocated and, in a few cases, this presence is reported since several hundred years (i.e. Cypress, Stone pine, Aleppo pine, Chestnut, Firs, Cedars). This artificial old population could be taken into account to test the possible consequences of the long geographic and ecological distances of the past translocations, even from the genetic structure and adaptive viewpoint.

Another basic topic concerns the trade of forest reproductive materials. The present rules concerning forest reproductive materials do not take care of the procedures related to future climate conditions and the associated risks (Williams and Dumrose 2013).

Several techniques, as Assisted gene flow (AGF) between populations and Assisted migration, can help to mitigate any maladaptation due to the climate change effects (Aitken and Whitlock 2013). Several initiatives were undertaken in Canada, where species have access to wide free lands where they can carry out migration strategies.

On the contrary, in Europe situations related to a fragmented landscape are more frequently diffused. In this landscape also the range of many species is fragmented and a huge number of species and populations may be unable to migrate to suitable habitats and get locally extinct.

McLachlan et al. (2007) proposed several examples, comments and considerations about this topic, on how relevant policy analysis and decisions about the opportunity to adopt this approach and at what intensity have to be undertaken.

The authors posed some important and basic questions about the developing of informed policies on Assisted Migration strategies. The first is aimed to identify the demographic trait that should trigger the implementation of assisted migration; a second one is aimed to know how many or which species or other taxa should be prioritized as candidates for translocation, and eventually to know how to manage populations in order to minimize adverse ecological effects.

Human land use may also create an impediment to gene flow among populations. In situations where trees are prevented from migration, human intervention may be necessary to prevent extinction. This action, can take the form of "assisted migration" or "managed relocation", a set of frequently considered controversial techniques useful to ensure the maintenance of (forest tree) populations in a changing global environment through the intentional creation of populations beyond the boundaries of their current presence (Ducci 2011).

The tendency to consider the opportunity to give rise to assisted migration in these cases already exists. As we have seen in the context of the risks for the species and forest populations induced by climate change, migration is an important strategy. Species will be successful in their perpetuation purpose only if able to move and adapt in environments and places where historically they did not previously exist.

In addition to the geographical barriers and the limits imposed by the areal distribution of ecological factors, the effect of human activities on the landscape fragmentation has been the crucial additional factor over the last centuries. It took place with the clearing of the large forest cover and of natural areas.

Nowadays, we have to face an almost permanent discontinuation of possible migration routes which could be useful to establish and/or restore the gene flow by tracking the motion of local climate. This interruption is mainly due to human settlements and agriculture.

In these cases, therefore, there is the need to provide forms of assisted migration, which aims to implement the physical transfer of populations, which are especially recognized for their adaptive or biological value, in areas outside of their natural range.

The main purpose of assisted migration is to preserve the genetic information contained in the original gene pool and restart the evolutionary dynamics along with these demographic - structural patterns as well as to recreate an *ex situ* secondary ecosystem, where they can start dynamics again.

Several definitions of Assisted migration can be found in the literature regarding this topic. Vitt et al. (2009) and S.te Marie et al. (2011) used different key words and definitions to define the concepts of Assisted migration, Assisted colonization, Managed relocation.

A few examples can be cited about species or population translocations: concerning *Pinus albicaulis*, McLane and Aitken (2012) reported their ongoing experiments in Northwestern America to test a model for establishing translocations. Furthermore, in Canada, Gray et al. (2011), tested assisted migration methods on *Populus tremuloides* populations. According to their experimental findings, model projections for this area seem restricted to a short 20-years planning horizon for prescribing seed movement in reforestation programs.

They also found that a safe and realistic climate change adaptation strategy has to be based on a holistic approach to obtain information. Some uncertainty is associated with recommendations for assisted migration, due to the rapid evolution of scenarios.

About Europe, we can cite the experimental case study of *Abies nebrodensis*, the Sicilian fir.

Various inventories and numerous research trials (Morandini 1969, Morandini 1986, Morandini et al. 1994, Virgilio et al. 2000) have shown that this species, although reduced to a small-sized population almost totally unable of originating a viable and dynamic regeneration *in situ*, still contains sufficient variability to enable a rescue attempt (Vendramin 1997, Ducci et al. 1999). This fir was chosen as a model to develop and then to implement an experimental program of assisted migration. Within this program, the residual individuals were transferred in the northern Apennines, in the form of grafted trees, at the beginning of the '90s (Ducci 2011). Two seed orchards were created with these grafts and the obtained seeds made possible to produce siblings that, year after year, are being transplanted into two areas away from any source of pollen contamination. Thus, they will form the first generation in the new environment.

Each of the two areas (Ducci 2014), respecting the climatic requirements of the species, has similar physical environmental characteristics. They are differentiated only with regard to the former vegetation cover, it being in the first case an ancient mixed chestnut and sporadic beech forest with hornbeam, yew and holly. In the second area, an abandoned field ground surrounded by the forest, former conditions ranged from semi-forest cover to full light conditions.

The purpose is to trigger more driving forces, give rise to different dynamics and allow more genetic characteristics to be preserved and to become clear.

In Sicily, Raimondo and Schicchi (2005) carried out a similar programme, both at local and regional level, in the framework of a LIFE Natura project, dedicated to *Abies nebrodensis in situ* and *ex situ* conservation, between 2001 and 2005.

Assisted migration has far-reaching consequences, far beyond the technical problems of physical translocation, all the manifold dimensions of community ecology, conservation and socio-economy being concerned (Richardson et al. 2009). These would need to be addressed also by forest managers and policymakers.

Garzon and Fernandez (2015) tested tools and scenarios for evaluating the sensitivity of target sites and identifying potential sources not yet developed. They used the Spanish and French National Forest Inventories to design scenarios for AM on *Pinus halepensis* Miller and *Pinus pinaster* Aiton, following a projection to 2050. Results suggest that volume and mortality trends are not always correlated with seed sources and targets, that projected provenances mortality do not always follow a southern to northern pattern and that refugia may be useful for compensating for the effects of climate change only in a subset of provenances.

Conclusions

Several indicators confirm that the Mediterranean region is being strongly affected by the climate change.

Over the last decades, temperatures have increased along with the frequency of deep and prolonged drought episodes, while rainfalls reduced by up to 20 % in the Italian peninsula as well as in other regions of the Mediterranean.

By 2100, temperatures are expected to increase from 2°C to 4°C on average, while precipitations could decrease from 4% to 30% and might interact with the already typical vulnerability of Mediterranean countries related to the general environmental degradation which is due to the relatively diffused high human pressure.

These direct and indirect effects of the global change may lead to huge human, social and economic losses.

Concerning forest resources, especially the genetic ones, the observed and predicted effects of climate change, specifically the increased extreme events, bring new threats forth and risk exacerbating the existing pressures, the environmental degradation and the vulnerability of Mediterranean ecosystems, particularly the valuable genetic resources contained within.

Together with climate change effects, several cohorts of natural hazards such as pests, diseases and also the increased frequency and intensity of forest fires, are endangering our resources.

Major, sometimes irreversible, changes are affecting the most vulnerable forest ecosystems where marginal populations of mesic species are growing and have probably developed some valuable adaptive traits (e.g. the rear edges).

The main challenge is thus connected to the opposition to the climate change acceleration, which makes difficult also any adaptation strategy. Only a scientific approach conveying a more in-depth knowledge would be useful to rethink and prepare any kind of strategy. This means that part of the already existing possibly useful data must be re-organized and, again, inferred under a new vision and updated approaches. This requires time and the tools must be adapted to the new needs, and there must be the production of new climatic and phytoclimatic models. Details in mapping have to be improved using smaller scales as well as new software is required for the inference of models and traits related to genomics in adaptation.

The present concept tools (including, modeling, silviculture, nursery legislation, policies etc.) and research networks are still based on a '*static*' or too long-term vision of the environment, whilst a major and rapid dynamism is nowadays perceived (Kerr and Dobrowski 2013).

The generally diffused current low level of silvicultural management of our forests constitutes a real danger for the maintenance of these specific populations, which need to be carefully managed *in situ* with proper and well-focused adaptive cultivation techniques, aimed at preserving their diversity and demographic structure, as well.

Strategies for the management of Mediterranean mountain forests should carefully take into account the issues to strengthen their natural resilience and to equally distribute resources. In this context, urgent initiatives are strongly required to mitigate the impact of climate change on our Mediterranean forest ecosystems and other woodlands.

Acknowledgements

The author wishes to acknowledge the valuable inputs received from the National project funded by the Ministry for Agriculture, Food and Forestry

Policies (MiPAAF) "RGV FAO - International Treaty", the EU projects TreeBreedex and Trees4Future, the Cost Action FP1202 *'Strengthening conservation: a key issue for adaptation of marginal/ peripheral populations of forest tree to climate change in Europe (MaP-FGR)'* (http://www.cost.eu/domains_actions/fps/Actions/FP1202, http://map-fgr.entecra.it/) along with the preparation of the proposal documents and from Colleagues of FAO Silva Mediterranea WG4 on *'Forest Genetic Resources in the Mediterranean Region'* (http://www.fao.org/forestry/24287/en/).

The author wants to thank the anonymous reviewers for their helpful comments.

References

Aitken S.N., Yeaman S., Holliday J.A., TongLi W., Curtis-McLane S. 2008 - *Adaptation, migration or extirpation: Climate change outcomes for tree populations.* Evolutionary Application 1: 95 – 111.

Aitken S.N., Whitlock M.C. 2013 - *Assisted Gene Flow to Facilitate Local Adaptation to Climate Change.* Annual Review of Ecology, Evolution, and Systematics 44: 367-388. DOI: 10.1146/annurev-ecolsys-110512-135747

Aranda I., Alia R., Ortega U., Dantas A., Majada J. 2010 - *Intraspecific variability in biomass partitioning and carbon isotopic discrimination under moderate drought stress in seedlings from four Pinus pinaster populations.* Tree Genet Genomes 6: 169–178.

Bolte B., Ammer C., Löf M., Nabuurs G.-J., Schall P., Spathelf P. 2015 - *Adaptive Forest Management: A Prerequisite for Sustainable Forestry in the Face of Climate Change.* In : "*Sustainable Forest Management in a Changing World - a European Perspective*" Managing Forest Ecosystems 19, Chapter: 8. Publisher: Springer, Editors: Peter Spathelf: 115-139. [Online]. Available: http://www.researchgate.net/profile/Andreas_Bolte [2015] DOI:10.1007/978-90-481-3301-7_8.

COFO-FAO 2014 - *The state of the world's forest genetic resources.* FAO, Rome, 291 p. ISBN 978-92-5-108402-1

Aravanopoulos F. A. 2009 - *Genetic Monitoring for Gene conservation Units.* Document submitted for discussion in the EUFORGEN Steering Committee Meeting 2009. [Online]. Available: http://www.euforgen.org/fileadmin/www.euforgen.org/Documents/SteeringCommittee/SC07_BackgroundDocs/GeneticMonitoringGCUs.pdf [2015].

Benito-Garzòn M. Fernandez-Manjarrés J.F. 2015 - *Testing scenarios for assisted migration of forest trees in Europe.* New Forests (in press) DOI: 10.1007/s11056-015-9481-9.

Demesure B., Eriksson G., Kleinshmit J., Rusanen M.. Stephan R. 1996 - *Noble Hardwood Network.* Report of the first meeting, 24 – 27 March 1996, Escherode, Germany. IPGRI, Rome: 61-75.

Besacier C., Briens M., Duclercq M., Garavaglia V. editors 2013 - *State of Forests 2013.* FAO, Rome, 177 p. ISBN 978-92-5-107984-3 (print), E-ISBN 978-92-5-107538-8 (PDF).

Blondel J., Aronson J. 1999 - *Biology and Wildlife of the Mediterranean Region.* Oxford University Press, 328 p.

Batllori E., Camarero J.J., Gutiérrez E. 2012 - *Climatic Drivers of Trees, Growth and Recent Recruitment at the Pyrenean Alpine Tree Line Ecotone.* In: "*Ecotones Between Forest and Grassland*", R.W. Myster (ed.), Springer Science+Business Media New York, 247 p. DOI: 10.1007/978-1-4614-3797-0_11.

Cavalli-Sforza L.L. 1991 - *Genes, peoples and languages.* Scientific American 265 (5): 104-110.

Cavalli-Sforza L.L., Menozzi P., Piazza A. 1994 - *The history and geography of human genes.* Princeton, N.J., Princeton University Press, 522 p.

Chambel M.R., Climent J., Alía R., Valladares F. 2005 - *Phenotypic plasticity: a useful framework for understanding adaptation in forest species.* Investigatión Agraria. Sistemas y Recursos Forestales 14 (3): 334-344.

Chevin L.-M., Lande R., Mace G.M. 2010 - *Adaptation, Plasticity and Extinction in a Changing Environment: Towards a Predictive Theory.* PLoS Biology 8 (4): DOI: e1000357. doi:10.1371/journal.pbio.1000357

Chybicki I.J., Burczyk J. 2013 - *Seeing the forest through the trees: comprehensive inference on individual mating patterns in a mixed stand of Quercus robur and Q. petraea.* Annals of Botany 112 (3): 561-574. DOI: 10.1093/aob/mct131 .

Climent J., Costa e Silva F., Chambel M.R., Pardos M., Almeida H. 2009a - *Freezing injury in primary and secondary needles of Mediterranean pine species of contrasting ecological niches.* Annals of Forest Science 66: 407-415.

Climent J., San-Martı́n R., Chambel M.R., Mutke S. 2011 - *Ontogenetic differentiation between Mediterranean and Eurasian pines (sect. Pinus) at the seedling stage.* Trees – Structure and Function 25: 175-186.

Climent J., Kidelman Dantas A., Alia R., Majada J. 2013 - *Clonal variation for shoot ontogenetic heteroblasty in maritime pine (Pinus pinaster Ait.).* Trees DOI: 10.1007/s00468-013-0901-1.

De Philippis A. 1937 - *Classificazioni e indici del clima in rapporto alla vegetazione forestale.* Nuovo Giornale Botanico Italiano N.G.B.I. XLIV, Firenze, Italia.

Ducci F.(ed.) 2007 - *Le risorse genetiche della farnia in Val Padana [Genetic resources of peduncolate oak in the Po valley].* CRA, Istituto Sperimentale per la Selvicoltura, Arezzo: 59 – 64.

Ducci F. 2011 - *Abies nebrodensis (Lojac.) Mattei, a model for forest genetic resource conservation.* In: "*Status of the Experimental Network of Mediterranean Forest genetic resources.*" Besacier C., Ducci F., Malagnoux M., Souvannavong O. (Eds.), 2011. CRA SEL, Arezzo, FAO Rome, Italy: 40 - 46.

Ducci F. 2014 - *Species restoration through dynamic ex situ conservation: Abies nebrodensis as a model.* In: "*Genetic considerations in ecosystem restoration using native tree species. State of the World's Forest Genetic Resources – Thematic Study.*" Bozzano M., Jalonen R., Thomas E., Boshier D., Gallo L., Cavers S., Bordács S., Smith P., Loo J. eds., 2014. Rome, FAO and Bioversity International: 225 – 233.

Ducci F., Proietti R., Cantiani P. 2006 - *Struttura genetica e sociale in un ceduo di cerro in conversione [Genetic and social structure within a Turkey oak coppice with standards].*In: "*Selvicoltura sostenibile nei boschi cedui*". Annali CRA – Centro di Ricerca per la Selvicoltura 33 (2002 – 2004): 143 – 158.

Ducci F., Proietti R., Favre J. M. 1999 – *Allozyme assessment of genetic diversity within the relic Sicilian fir Abies nebrodensis (Lojac.) Mattei*. Annals of Forest Science 56: 345 –355.

Eckert C. G., Samis K.E., Lougheed S.C. 2008 - *Genetic variation across species' geographical ranges: the central–marginal hypothesis and beyond*. Molecular Ecology 17: 1170–1188. DOI: 10.1111/j.1365-294X.2007.03659.x.

Eckert A.J., Andrew D., Bower A.D., Wegrzyn J.L., Barnaly Pande B., Kathleen D., Jermstad K. D., Konstantin V., Krutovsky K.V., Bradley St. Clair J., Neale D.B. 2010 - *Association Genetics of Coastal Douglas Fir Pseudotsuga menziesii var. menziesii, Pinaceae)*. Cold-Hardiness Related Traits Genetics 182: 1289–1302. DOI: 10.1534/genetics.109.102350.

Eriksson G. 1996 - *Evolutionary genetics and conservation of forest tree genetic resources*. In: "Noble Hardwood Network. Report of the first meeting", 24-27 March 1996, Escherode, Germany. Turok J., Eriksson G., Kleinschmit J. and Canger S. compilers. IPGRI, Rome: 159 – 167.

Eriksson G. 1998 - *Sampling for genetic resources populations in the absence of genetic knowledge*. Journal of Forest Research 39: 1259–1269.

Eriksson G., Namkoong G., Roberds J.H. 1993 - *Dynamic gene conservation for uncertain futures*. Forest Ecology and Management 62: 15-37.

Ettinger A.K., Ford K.R., Hille Ris Lambers J. 2011 - *Climate determines upper, but not lower, altitudinal range limits of Pacific Northwest conifers*. Ecology 92: 1323-1331.

Fady B. 2005 - *Is there really more biodiversity in Mediterranean forest ecosystems?* Taxon 54 (4): 905-910.

Fady B., Conord C. 2010 - *Macroecological patterns of species and genetic diversity in vascular trees of the Mediterranean basin*. Diversity and Distributions 16 (1): 53-64. DOI: 10.1111/j.1472-4642.2009.00621.x.

Fady B., Ducci F., Aleta N., Becquey J., Diaz Vazquez R., Fernandez Lopez F., Jay-Allemand C., Lefèvre F., Ninot A., Panetsos K., Paris P., Pisanelli P., Rumpf H. 2003 - *Walnut demonstrates strong genetic variability for adaptive and wood quality traits in a network of juvenile field tests across Europe*. New Forests 25: 211–225 .

Fallour-Rubio D., Guibal F., Klein E.K., Bariteau M., Lefèvre F. 2009 - *Rapid changes in plasticity across generations within an expanding cedar forest*. Journal of Evolutionary Biology 22 (3): 553–563.

FAO 2012 - *State of the World's Forests*. FAO, Rome, 60 p. ISBN 978-92-5-107292-9.

FAO 2013 - *State of the Mediterranean Forests 2013*. FAO 2013, 191 p. E-ISBN 978-92-5-107538-8 (PDF).

Gomez M., Ferry S. 1998 - *The Red Palm Weevil in the Mediterranean Area*. Palms "Death and Destruction" 46 (4) [Online]. Available: http://www.palms.org/palmsjournal/2002/redweevil.htm [2003].

Grazioli I., Santi F. 2008 - *Chestnut gall wasp (Dryocosmus kuriphilus): spreading in Italy and new records in Bologna province*. Bulletin of Insectology 61 (2): 343-348.

Gray L. K., Gylander T., Mbogga M.S., Chen P.-Y., Hamann A. 2011 - *Assisted migration to address climate change: recommendations for aspen reforestation in western Canada*. Ecological Applications 21 (5): 1591–1603.

Healy T.M., Schulte P.M. 2015 - *Phenotypic plasticity and divergence in gene expression*. Molecular Ecology 24 (13): 3220–3222. DOI: 10.1111/mec.13246.

Hampe A., Petit R.J. 2005 - *Conserving biodiversity under climate change: the rear edge matters*. Ecology Letters 8 (5): 461 – 467. DOI:10.1111/j.1461-0248.2005.00739.x.

Hamrick J.L. 2004 - *Response of forest trees to global environmental changes*. Forest Ecology and Management 62: 323 - 336.

Hodkinson T.R., Jones M.B., Waldren S., Parnell J.A.N. 2011 - *Climate Change, Ecology and Systematics*. Cambridge University Press, Cambridge: 182 – 184.

Holliday J.A., Ritland K., Aitken S. N. 2010 - *Widespread, ecologically relevant genetic markers developed from association mapping of climate-related traits in Sitka spruce (Picea sitchensis)*. New Phytologist 188: 501– 514.

Huntley B., Birks H.J.B. 1983 - *An Atlas of past and present pollen maps for Europe 0 – 13000 years ago*. Cambridge University Press, Cambridge, UK, 667 p.

IPCC 2001 - *Climate change 2001: Synthesis Report. A Contribution of Working Groups I, II and III to the Third Assessment Report of the Intergovernmental Panel on Climate Change*. Watson R.T. and the Core Writing Team eds. Cambridge Un. Press, Cambridge, UK and NY, USA, 3+98 p.

IPCC 2007 - *Summary for Policymakers*. In: "Climate Change 2007: Impacts, Adaptation and Vulnerability." Contribution of Working Group II to the Fourth Assessment Report of the Intergovernmental Panel on Climate Change, Parry M.L., Canziani O.F., Palutikof J.P., van der Linden P.J. and Hanson C.E., Eds., Cambridge University Press, Cambridge, UK: 7-22.

IPCC 2012 - *Managing the Risks of Extreme Events and Disasters to Advance Climate Change Adaptation*. A Special Report of Working Groups I and II of the Intergovernmental Panel on Climate Change.

Iverson L.R., Schwartz M.W., Prasad A.M. 2004 - *Potential colonization of newly available tree-species habitat under climate change: an analysis for five eastern US species*. Landscape Ecology 19: 787-799.

Kawecki T.J., Ebert D. 2004 - *Conceptual issues in local adaptation*. Ecology Letters 7: 1225–1241.

Kerr J.T., Dobrowski S.Z. 2013 - *Predicting the impacts of global change on species, communities and ecosystems:it takes time*. Global Ecology and Biogeography 22: 261–263.

Konnert M., Hosius B. 2010 - *Contribution of forest genetics for a sustainable forest management*. Forstarchiv 81: 170 – 174. DOI: 10.2376/0300-4112-81-170.

Konnert M., Fady B., Gömöry D., A'Hara S., Wolter F., Ducci F., Koskela J., Bozzano M., Maaten T., Kowalczyk J. *European Forest Genetic Resources Programme (EUFORGEN). 2015. Use and transfer of forest reproductive material in Europe in the context of climate change*. Bioversity International, Rome, Italy, 77 p.

Koskela J., Buck A., Teissier du Cros E. editors 2007 - *Climate change and forest genetic diversity. Implications for soustainable forest management in Europe*. Bioversity International, Rome, Italy, 111 p.

Gowik U., Westhoff P. 2011 - *The Path from C3 to C4 Photosynthesis*. Plant Physiology 155: 56–63.

Larcher W. 2003 - *Physiological Plant Ecology: Ecophysiology and Stress Physiology of Functional Groups*. Springer-Verlag, Berlin, Heidelberg, New York: 52 - 53.

Lefèvre F. 2004 - *Human impacts on forest genetic resources in the temperate zone: an updated review*. Forest Ecology and Management 197: 257-271.

Lindner M., Maroschek M., Netherer S., Kremer A., Barbati A., Garcia-Gonzalo J., Seidl R., Delzon S., Corona P., Kolstroem M., Lexer M. J., Marchetti M. 2010 - *Climate change impacts, adaptive capacity, and vulnerability of European forest ecosystems*. Forest Ecology and Management 259: 698–709.

Linvill D.E. 1990 - *Calculating chilling hours and chill units from daily maximum and minimum temperature observations*. HortScience 25 (1): 14-16.

Loarie S.R., Duffy P.B., Hamilton H., Asner G.P., Field C.B., Ackerly D.D. 2009 - *The velocity of climate change*. Nature 462: 1052–1055.

McLane S.C., Aitken S.N. 2012 - *Whitebark pine (Pinus albicaulis) assisted migration potential: testing establishment north of the species range*. Ecological Applications 22: 142–153. DOI: http://dx.doi.org/10.1890/11-0329.1

Màtyàs C. 2007 - *What do field trials tell about the future use of Forest Reproductive Material?*. In: "Climate change and forest genetic diversity. Implications for soustainable forest management in Europe" Koskela J., Buck A., Teissier du Cros E. editors, 2007. Bioversity International, Rome, Italy: 53 – 68.

Koskela J., Buck A., Teissier du Cros E. 2007. *Climate change and forest genetic diversity. Implications for sustainable forest management in Europe*. Bioversity International, Rome, Italy: 53 – 68.

Magini E. 1955 - *Pinete di pino d'Aleppo [Aleppo pine Pinewoods]*. In: Proceedings National Italian Congress of Silviculture, Florence 1954. Edizioni Accademia di Scienze Forestali, Florence: 50 – 60.

Matyas C., Vendramin G.G., Fady B. 2009 - *Forests at the limit: evolutionary - genetic consequences of environmental changes at the receding (xeric) edge of distribution. Report from a research workshop*. Annals of Forest Science 66 (8). Article number: 800.

McKown A.D., Guy R.D., Klápště J., Geraldes A., Friedmann M., Cronk Q.C.B., El-Kassaby Y.A., Mansfield S.D., Douglas C.J. 2014 - *Geographical and environmental gradients shape phenotypic trait variation and genetic structure in Populus trichocarpa*. New Phytologist 201: 1263-1276.

McLachlan J.S., Hellmann J.J., Schwartz M.W. 2007 - *A Framework for Debate of Assisted Migration in an Era of Climate Change*. Conservation Biology 21 (2): 297–302. DOI: 10.1111/j.1523-1739.2007.00676.x.

Médail F., Diadema K. 2009 - *Glacial refugia influence plant diversity patterns in the Mediterranean Basin*. Journal of Biogeography 36 (7): 1333-1345. DOI: 10.1111/j.1365-2699.2008.02051.x.

Millennium Ecosystem Assessment (MA) 2005 - [Online]. Available: http://www.millenniumassessment.org/en/index.aspx [2014].

Morandini R. 1969 - *Abies nebrodensis (Lojac) Mattei, Inventario 1968*. Pubblicazioni dell' Istituto Sperimentale di Selvicoltura di Arezzo 18, 93 p.

Morandini R. 1986. *Abies nebrodensis (Lojac.) Mattei*. In : "Databook of endangered tree and shrub species and provenances", FAO Forestry Paper 77: 11 – 20.

Morandini R., Ducci F., Menguzzato G. 1994 - *Abies nebrodensis (Lojac.) Mattei - Inventario 1992. [Abies nebrodensis (Lojac.) Mattei - Survey 1992]*. Annali dell' Istituto Sperimentale per la Selvicoltura di Arezzo 22: 5-51.

Namkoong G., Boyle T., El-Kassaby Y.A., Palmberg-Lerche C., Eriksson G., Gregorius H.R., Joly H., Kremer A., Savolainen O., Wickneswari R., Young A, Zeh-Nlo M., Prabhu R. 2002 - *Criteria and indicators for sustainable forest management: assessment and monitoring of genetic variation FGR/37*. FAO, Rome. [Online]. Available: http://www.fao.org/docrep/016/i3010e/i3010e.pdf [2014].

Namkoong G., Boyle T., Gregorious H.R., Joly H., Savolainen O., Ratman W., Young A. 1996 - *Testing criteria and indicators for assessing the sustainability of forest management: Genetic criteria and indicators*. Centre for International forestry research (CIFOR)10, Bogor, Indonesia, 12 p.

Parmesan C., Yohe G. 2003 - *A globally coherent fingerprint of climate change impacts across natural systems*. Nature 421 (6918): 37–42.

Pavari A. 1959 - *Scritti di ecologia, selvicoltura e Botanica forestale*. Pubblicazioni dell' Accademia Italiana di Scienze Forestali, Firenze, Italia: 95- 116.

Perini L., Salvati L., Ceccarelli T., Motisi A., Marra F.P., Caruso T. 2007 - *Atlante Agroclimatico, scenari di cambiamento agroclimatico*. CRA, Ufficio Centrale Ecologia Agraria, Roma: 81 p.

Petit R.J., Aguinagalde I., de Beaulieu J.L., Bittkau C., Brewer S., Cheddadi Ennos R., Grivet D., Lascoux M., Mohanty A., Müller-Starck G., Demesure-Musch B., Palmé A., Martin J. P., Rendell S., Vendramin G.G. 2003 - *Glacial refugia: hotspots but not melting pots of genetic diversity*. Science 300: 1563-1565.

Pigliucci M. 2001 - *Phenotypic Plasticity: beyond nature and nurture*. The Johns Hopkins University Press, Baltimore and London, 328 p.

Pigliucci M. 2005 - *Evolution of phenotypic plasticity: where are we going now?* Trends Ecology and Evolution 20: 481-486.

Pinna M. 1977 - *Climatologia*. UTET, Torino: 401-429.

Pitelka L.F., Gardner R.H., Ash J., Berry S., Gitay H., Noble I. R., Saunders A., Bradshaw R.H.W., Brubaker L., Clark J. S., Davis M.B., Sugita S., Dyer J. M., Hengeveld R., Hope G., Huntley B., King G.A., Lavorel S., Mack R.N., Malanson G.P., Mc Glone M., Prentice I.C., Rejmanek M. 1997 - *Plant migration and climate change*. American Scientist 85, 501 p.

Pollegioni P., Woeste K.E., Chiocchini F., Olimpieri I., Tortolano V., Clark J., Hemery G.E., Mapelli S., Malvolti M.E. 2014 - *Landscape genetics of Persian walnut (Juglans regia L.) across its Asian range*. Tree Genetics & Genomes 10: 1027-1043. DOI 10.1007/s11295-014-0740-2.

Richardson E.A., Seeley S.D., Walker D.R. 1974 - *A model for estimating the completion of rest for 'Redhaven' and 'Elberta' peach trees*. HortScience 9 (4): 331-332.

Sagnard F., Oddou-Muratorio S., Pichot C., Vendramin G.G., Fady B. 2011 - *Effect of seed dispersal, adult tree and seedling density on the spatial genetic structure of regeneration at fine temporal and spatial scales*. Tree Genetics and Genomes 7: 37-48.

Sala O.E., Chapin F.S. III, Armesto J.J., Berlow R., Bloomfield J., Dirzo R., Huber-Sanwald E., Huenneke L.F., Jackson R.B., Kinzig A., Leemans Lodge R.D., Mooney H.A., Oesterheld M., Poff N.L., Sykes M.T., Walker B.H., Walker M., Wall D.H. 2000 - *Global biodiversity scenarios for the year 2100*. Science 287: 1770-1774. DOI: dx.doi.org/10.1126/science.287.5459.1770.

Santos-Del-Blanco L., Bonser S.P., Valladares F., Chambel M.R., Climent J. 2013 - *Plasticity in reproduction and growth among 52 range-wide populations of a Mediterranean conifer: adaptive responses to environmental stress.* Journal of Evolutionary Biology 26 (9): 1912-1924. DOI: 10.1111/jeb.12187

Savolainen O., Pyhäjärvi T., Knürr T. 2007 - *Gene flow and local adaptation in forest trees.* Annual Review of Ecology, Evolution and Systematics 38: 595-619.

Schlichting C.D. 1986 - *The evolution of Phenotypic Plasticity in plants.* Annual Review of Ecology and Systematics 17: 667-693.

Schwartz M.K., Luikart G., Waples R. S. 2006 - *Genetic monitoring as a promising tool for conservation and management.* Trends in Ecology and Evolution 22: 25-33.

Seddon P.J. 2010 - *From reintroduction to assisted colonization: moving along the conservation translocation spectrum.* Restoration Ecology 18: 796-802.

Ste-Marie C., Nelson E.A., Dabros A., Bonneau M.-E. 2011 - *Assisted Migration: introduction to a multifaced concept.* The Forestry Chronicle 87 (6): 724-730.

Taylor S.H., Franks P.J., Hulme S.P., Spriggs E., Christin P.A., Edwards E.J., Woodward F.I., Osborne C.P. 2012. *Photosynthetic pathway and ecological adaptation explain stomatal trait diversity amongst grasses.* New Phytologist 193: 387–396. DOI: 10.1111/j.1469-8137.2011.03935.x.

Tescari G. 2001 - *Leptoglossus occidentalis, coreide neartico rinvenuto in Italia (Heteroptera, Coreidae). [Leptoglossus occidentalis, a neartic Coreyde recorded in Italy (Heteroptera, Coreidae)].* Lavori della Società Veneziana di Scienze Naturali 26: 3-5.

Thompson J.D. 2005 - *Plant evolution in the Mediterranean.* Oxford University Press, Oxford, 293 p. ISBN 0198515332; 0198515340 (PDF).

Topak M. 1997 - *Directory of seed sources of the Mediterranean conifers.* FAO, Rome, 118 p. (http://www.fao.org/docrep/006/AD112E/AD112E00.HTM)

Van der Putten W.H., Macel M., Visser M.E. 2010 - *Predicting species distribution and abundance responses to climate change: why it is essential to include biotic interactions across trophic levels.* Philosophical. Transaction of the Royal Society B 365: 2025-2034. DOI:10.1098/rstb2010.0037.

Vendramin G.G., Morgante M. 2005 - *Genetic diversity in forest trees populations and conservation: analysis of neutral and adaptive variation.* Abstracts of the Meeting "The role of Biotechnology" Villa Gualino, Turin, Italy – 5-7 March, 2005: 129-130. http://unfccc.int/2860.php

Virgilio F., Schicchi R., La Mela Veca D. 2000 - *Aggiornamento dell'inventario della popolazione relitta di Abies nebrodensis (Lojac.).* Naturalista Siciliano 24 (1-2): 13-54.

Vitt P., Havens K., Kramer A.T., Sollenberger D., Yates E. 2011 - *Assisted migration of plants: Changes in latitudes, changes in attitudes.* Biological Conservation 143 (1): 18-27. DOI: 10.1016/j.biocon.2009.08.015

Vitasse Y., Delzon S., Bresson C.C., Michalet R. 2009 - *Altitudinal differentiation in growth and phenology among populations of temperate-zone tree species growing in a common garden.* Canadian Journal of Forest Research 39: 1259–1269.

Walck J., Hidayati S.N., Dixon K.W., Thompson K., Poschold P. 2011 - *Climate change and plant regeneration from seed.* Global Change Biology 17 (6): 2145–2161. DOI: 10.1111/j.1365-2486.2010.02368.

Waterman A.M. 1943 - *Diplodia pinea the cause of a disease of hard Pines.* Journal of Phytopathology 33: 1018-1031.

Williams M.I., Dumroese R.K. 2013 - *Preparing for Climate Change: Forestry and Assisted Migration.* The Journal of Science and Technology for Forest Products and Processes 111 (4): 287–297 http://dx.doi.org/10.5849/jof.13-016

WMO (World Meteorological Organisation) 2007 - *WMO's role in global climate change issues with a focus on development and science based decision making.* World Climate Programme (WCP) and Climate Coordination Activities (CCA), Position Paper, 13 p.

Yakovlev I., Fossdal C.G., Skroppa T., Olsen J.E., Hope Jahren A., Johnsen Ø. 2012 - *An adaptive epigenetic memory in conifers with important implications for seed production.* Seed Science Research 22 (2): 63–76.

Yeh F.C., Layton C. 1979 - *The organization of genetic variability in central and marginal populations of lodgepole pine (Pinus contorta ssp. latifolia).* Canadian Journal of Genetics and Cytology 21: 487–503.

Ying C.C., Yanchuk A.D. 2006 - *The development of British Columbia's tree seed transfer guidelines: Purpose, concept, methodology, and implementation.* Forest Ecology and Management 227: 1–13. DOI: 10.1016/j.foreco.2006.02.028.

Multifunctionality assessment in forest planning at landscape level: the study case of Matese Mountain Community (Italy)

Umberto Di Salvatore[1*], Fabrizio Ferretti[1], Paolo Cantiani[2], Alessandro Paletto[3], Isabella De Meo[4], Ugo Chiavetta[2]

Abstract - The main objective is to improve a method that aims at evaluating forest multifunctionality from a technical and practical point of view. A methodological approach - based on the index of forest multifunctionality level - is proposed to assess the "fulfilment capability" of a function providing an estimate of performance level of each function in a given forest. This method is aimed at supporting technicians requested to define most suitable management guidelines and silvicultural practices in the framework of a Forest Landscape Management Plan (FLMP). The study area is the Matese district in southern Apennines (Italy), where a landscape planning experimentation was implemented. The approach includes the qualitative and quantitative characterization of selected populations, stratified by forest category by a sampling set of forest inventory plots. A 0.5 ha area around the sample plot was described by filling a form including the following information: site condition, tree species composition, stand origin and structure, silvicultural system, health condition, microhabitats presence. In each sample plot, both the multifunctionality assessment and the estimate of the effect of alternative management options on ecosystem goods and services, were carried out. The introduction of the term "fulfilment capability" and the modification of the concept of priority level - by which the ranking of functions within a plot is evaluated - is an improvement of current analysis method. This enhanced approach allows to detect the current status of forest plot and its potential framed within the whole forest. Assessing functional features of forests with this approach reduces the inherent subjectivity and allows to get useful information on forest multifunctionality to support forest planners in defining management guidelines consistent with current status and potential evolutive pattern.

Keywords - forest multifunctionality, Forest Landscape Management Planning, function fulfilment index, silvicultural system, Matese district (Italy)

Introduction

The Sustainable Forest Management (SFM) paradigm - defined at the Montreal Process (1987) - aims to balance social, economic, ecological, and cultural needs of present and future generations (Wyder 2001, Tabbush 2004) and to maintain resources based on the multiple use of forests (Garcıa-Fernandez et al. 2008).

The theoretical and practical development of multiple use forest management (MFM) started in North America and was re-conceptualized in Europe, giving greater emphasis to the concept of forest functions instead that to the concept of forest use. Nix (2012) referred to MFM as "the management of land or forest for more than one purpose, such as wood production, water quality, wildlife, recreation, aesthetics, or clean air". According to this definition, MFM is an approach that combines two or more uses of forests (i.e. wood production, maintenance of proper conditions for wildlife, landscape effects, recreation, protection against floods and erosion, and protection of water supplies).

In Europe the concept of forest multifunctionality was born in 1953 in Germany with the elaboration of the "Theory of Forestry Function" by Viktor Dieterich of the University of Munich. In this theory, the concept of multiple-use was developed and widened through a less anthropocentric vision where the functions have an intrinsic importance (vitality and health of ecosystem).

Over the last years, MFM has been envisioned as a promising and more balanced alternative to sustained yield strategies. Some authors emphasize that the inclusion of multiple values and multiple stakeholders might give SFM a much needed social and financial boost (Campos et al. 2001, Hiremath 2004, Kant 2004, Wang and Wilson 2007). The incorporation of multiple forest values in forest management decisions is one of the important dimensions of SFM (Kant 2007). Nowadays a modern forestry vision requires forests to satisfy demands of many

---

[1] Consiglio per la Ricerca e la sperimentazione in Agricoltura, Apennines Forest Research Unit (CRA-SFA), Isernia, Italy
[2] Consiglio per la Ricerca e la sperimentazione in Agricoltura, Forestry Research Centre (CRA-SEL), Arezzo, Italy
[3] Consiglio per la Ricerca e la sperimentazione in Agricoltura, Forest Monitoring and Planning Research Unit (CRA-MPF), Villazzano di Trento, Italy
[4] Consiglio per la Ricerca e la sperimentazione in Agricoltura, Agrobiology and Pedology Centre (CRA-ABP), Firenze, Italy
* corresponding author: umberto.disalvatore@entecra.it

stakeholder for multiple products and services (Kant 2004, Cantiani 2012).

SFM is a concept in continuous evolution both in time and space (Angelstam et al. 2005, Straka 2009). The multifunctional forest management planning aims to integrate in decision making the non-productive issues of the forest, just as well as the socio-cultural and environmental issues (Vincent and Binkley 1993, Kangas and Store 2002). In such planning approach the logical process that leads to the final management choice becomes considerably complicated (Pukkala 2002). For this reason, the most unambiguous, reproducible and economically sound definition and experimentation of a methodology regarding the planning process is necessary (Paletto et al. 2012).

During the last years, in Italy, forest management planning is not only realized through traditional plans at stand or regional level, but new Forest Landscape Management Plans (FLMP) are gaining importance as well. FLMPs provide alternative scenarios of forest landscape management rather than defining where and when a specific forest practice must be applied (Agnoloni et al. 2009).

Many forest planners have recognized the development of planning systems on a landscape scale as the proper tool to analyse the forest complexity and to define the management guidelines (Kant 2003, Kennedy and Koch 2004, Farcy and Devillez 2005, Cubbage et al. 2007, Schmithüsen 2007).

FLMP addresses long-term forest management issues, with special attention to environmental issues that cannot be properly considered by referring to a single forest management unit (i.e. single forest ownership).

In addition, FLMP provides management recommendations and silvicultural guidelines, according to forest category and silvicultural system (coppice or high forest). These are then divided and adapted for every function (Paletto et al. 2012).

Referring to the method developed by Paletto et al. (2012), devoted to define the forest multifunctionality from a practical point of view to support the forest practitioners, the main objective of this study is to improve this method in several aspects.

Specifically, we implemented the following three issues:

i) introduction of the priority level of every function. Zero priority function no longer exists; instead a priority ranking will be established among all functions;

ii) introduction of an index of the forest multifunctionality level. This index is defined through the capability of function fulfilment which provides an estimates of how much every function is performed in a given forest plot compared to the average performance of the same forest category. This feature introduces the novel concept of the relative performance in a 0 to 10 range.

iii) identification of which forest functions we have to take into consideration is carried out through a participatory process involving local stakeholders and experts.

Furthermore we aimed to test the method proposed by Paletto et al. (2012) in a different forest

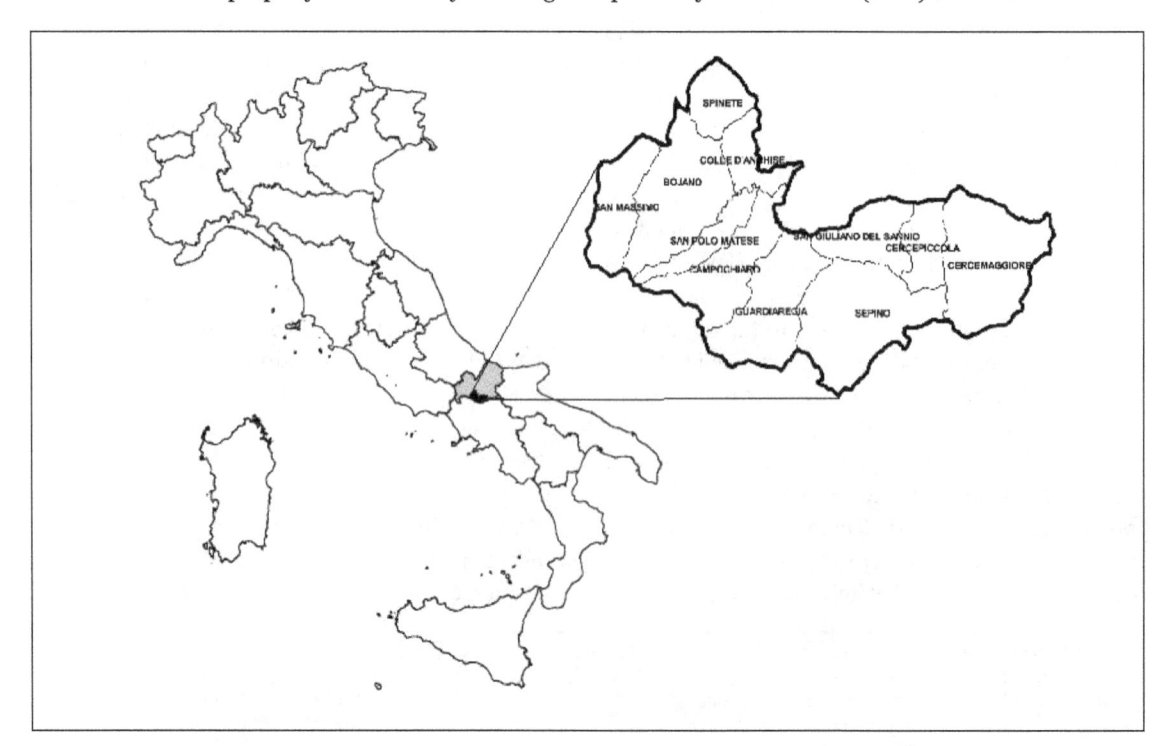

Figure 1 - The study area and its municipalities.

Table 1 - Forest categories distribution in the study area.

Forest categories	Area (ha)	%
Beech forests	4785	29.7
Turkey oak forests	6644	41.2
Downy oak forests	290	1.8
Hop-hornbeam forests	1556	9.7
Chestnut forests	320	2.0
Riparian forests	842	5.2
Holm oak forests	18	0.1
Other broadleaved forests	1020	6.3
Shrublands	406	2.5
Coniferous plantations	212	1.3
TOTAL	**16094**	**100**

environment. Indeed, the application of the method in a different social and ecological context is a further element useful to improve the method and it can provide important suggestions from a practical point of view.

Materials and methods

Study area

The study area is included by the "Comunità Montana" of Matese, located in the Molise Region in Central Italy (Fig. 1). It has a total area of 36,500 ha and includes 11 municipalities.

The altitude ranges from 422 m a.s.l. of Spinete lowland to the 2,050 m a.s.l. of Monte Miletto.

The study area has 15,687 ha of forest lands and 407 ha of other wooded lands (Chirici et al. 2011). Forest area covers 43% of Matese district; the percentage of forest area varies from a maximum of 75% in Guardiaregia municipality to a minimum of 19% in Cercepiccola municipality.

The most forested area is represented by the South-western part of study area; in the North-eastern part forests are more fragmented and juxtaposed with urban and agricultural lands.

In terms of surface (Fig. 1), Turkey oak (*Quercus cerris* L.) forests are the most extended forest category (41.2% of forest area), they are often pure and fertile stands with well-shaped trees. Turkey oak forests are divided into the following forest types: i) mesophilous Turkey oak forests, closed and mainly pure stands growing in very fertile sites; ii) meso-xerophilous Turkey oak forests, with the significant presence of meso-xerophilous species or more rarely mesophilous species such as common hornbeam (*Carpinus betulus* L.), sycamore maples (*Acer pseudoplatanus* L.) and downy oak (*Quercus pubescens* Willd.).

The second forest category is represented by European beech (*Fagus sylvatica* L.) forests which occupy an area of 4,785 ha (29.7% of forest area) and are localized at the highest elevations and northern expositions. European beech forests are divided into the following three forest types (Chirici et al. 2011): i) high-mountainous beech forests, localized just below the timberline, in high slopes or in peak summits often characterized by rocky soils, strong winds, soil aridity and low fertility; ii) mountainous beech forests, which are the beech main forest type characterized by pure and fertile stands, where the understory vegetation is very sparse or absent; iii) sub-mountainous beech forests, localized in the transition zone between beech and Turkey oak forests or more rarely hop-hornbeam forests.

Other significant forest categories are represented by hop-hornbeam forests (9.7% of forest area) and by other broadleaved forests (6.3% of forest area). Finally riparian forests occupy the 5.2% of forest area and are localized along main creeks and rivers at the lowest altitudes.

Considering the economic importance of European beech and Turkey oak forests which occupy the 70.9% of forest area, for the multifunctionality analysis we focused only on these two forest categories which represent our reference population.

Method

We characterized the selected forest categories surveying 117 inventory plots and collecting qualitative and quantitative data.

We carried out an unaligned systematic sample design consistent to the Italian National Forest and Carbon sinks Inventory (INFC, 2004).

We generated a geo-referenced squared grid with 1 km step and random origin. A point with random coordinates was positioned in every square. Finally all points (more than 10,000) were overlapped to Molise forest types map (Chirici et al. 2011) in order to select the reference sample plots (117) in European beech or Turkey oaks forests (Fig. 2).

We described a 0.5 ha area around the sample plot by filling a form including the following information: site condition, tree species composition, stand origin and structure, silvicultural system, health condition, microhabitats presence.

In every sample plots we carried out the multi-functionality assessment and the effect estimation

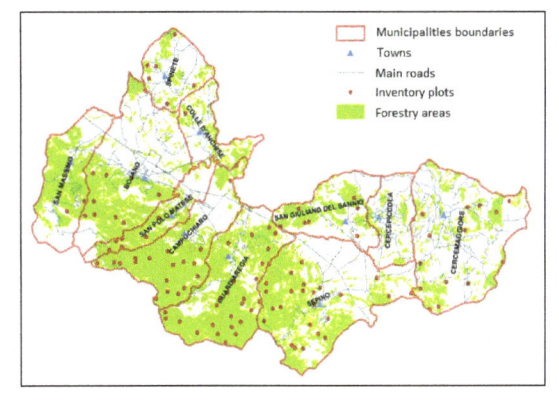

Figure 2 - Distribution of sample plots over the map of forest land use in the study area.

Table 2 - Stakeholders interviewed during the first phase of the participatory process.

Stakeholder	Num.
Majors	10
Forest enterprises	8
Associations	7
Agri-touristic farms	3
Freelance foresters	4
Local Action Groups	3
Forest nurseries	1
Sawmills	1
Mushrooms/truffles canning industry	1

of alternative management options on ecosystem goods and services by the method described below.

From 117 plots, 63 plots (53.8%) were classified as Turkey oak forests and 54 plots (46.2%) as European beech forests. In terms of silvicultural system, instead, 65 plots (55.6%) fell into coppice system and 52 (44.4%) into high forest system. In high forests, the most represented structure were the one-layered (32.5%), this suggesting shelterwood as the most common system. Nonetheless, also more complex high forest structures were found: two-layered (10.3%) and multi-layered (8.5%).

Multifunctionality: silvicultural system and forest category

At the purpose of this study, we considered the ability of forest ecosystem to supply goods and services. As a consequence, multifunctionality was assessed by a forest experts' team in each plot by assigning a value for two parameters: i) function priority level and ii) capability of function fulfilment.

The function priority level is a score aiming to relatively rank all functions considered essential for each plot in the specific context where the forest is located. The score consisted in an integer positive value ranging from 1 to n, where n is the number of all functions considered essential for that specific plot. The most important function takes the value 1 and the less important function takes the value n. An even score is possible if two functions are considered equally important.

The capability of function fulfilment is an estimation of how much that forest can perform every considered function compared to the average performance of the same forest type. The score ranges from 0 (no performance) to 10 (best performance for that forest type).

Forest functions considered in the study area were selected taking into account four aspects at once: i) ecological, social and economic context of the study area, ii) internationally recognized forest functions resulting from a literary review, iii) a participatory process involving local stakeholders, and iv) existing and up to date forest planning at unit level in the study area.

Concerning the participatory process, 39 stakeholders were contacted and interviewed to highlight the most relevant forest functions in the study area (Table 2).

The seven forest functions identified are described below.

- Landscape conservation. Considering the landscape as the result of interaction between human and natural environment (Brady 2003), landscape management is based on multiple values including ecological, economic, cultural and perception aspects (Sepp et al. 1999). Evaluation criteria were: the relative importance of the landscape in the local cultural context and the visibility from road and trail networks.
- Firewood/biomass production. All products (primary and secondary) provided by the forest for domestic heating.
- Timber production: all wood assortments not used for heating.
- Non-wood forest production. The total of non-wood forest products such as truffles, mushrooms, berries, etc.
- Soil and water protection. Direct and indirect protection against natural hazards such as floods, landslides, rock falls, soil erosion, etc. (Führer 2000).
- Touristic/recreational function. Forests provide many recreational opportunities such as trekking, bird-watching, biking, orienteering, plant and animal observing etc. (Krieger 2001).
- Environment conservation. It considers the positive effect that forests have on biodiversity and microhabitat conservation. We evaluated the possibility/opportunity of increasing the number of microhabitats and diversifying forest structure (horizontal and vertical) to promote wildlife biodiversity (FAO 2006).

In a first step, we stratified plots by silvicultural system (coppice or high forest) and by forest category. During a second step, we compared the strata by multifunctionality level indicators using two indicators described below.

Mean priority level and mean fulfilment capability were calculated for each function as:
Where:

$$\bar{v}_{f\,fi} = \frac{\sum_{i=1}^{n} v_{f\,i}}{n}$$

n = total number of plots per stratum;
v_{fi} = priority level or fulfilment capability for the f function in the i-th plot.

This indicator assesses the priority level or the fulfilment capability for every forest category (or silvicultural system) and for each function. Thus it

Table 3 - Silvicultural options.

coppices	1)	traditional coppicing: total harvesting of trees except the release of a variable number of standards with a main dissemination function (Perrin, 1954);
	2)	conversion into high forest: set of techniques aiming at the preparation for the conversion into high forest. The application of these treatments lead to a transitory stand alike to a high forest structure (Bernetti, 2005).
	3)	natural evolution of the stand.
high forests	1)	even-aged high forest regeneration practices: shelterwood, large-medium strips or large-medium groups felling (Kimmins, 2004);
	2)	coppice/high forest integration;
	3)	high forest in continuous regeneration: to get an uneven-aged structure per single tree by selection felling or per small groups by small strips or group shelterwood (Helms, 1998);
	4)	natural evolution of the stand

gives useful indications for operational purposes at a stratum level.

Total mean priority level and fulfilment capability as:

Where:

$$\overline{V}_{F\,FT} = \frac{\sum_{j=1}^{m} \overline{v}_{f\,fi\,j}}{m}$$

m = total number of selected functions;
$\overline{v}_{f\,fi\,j}$ = mean priority level or mean fulfilment capability of the stratum for the j-th function.

This indicator assesses the total multifunctionality value of the stratum giving a synthetic value. It is useful to compare different forest categories and silvicultural systems.

The joint analysis of these indicators provide a synthetic evaluation of the current multifunctionality of the stratum (forest category or silvicultural system) which is the base to analyse future silvicultural options (Paletto et al. 2012)

Performance capability of silvicultural options

In this study we evaluated the capability of each silvicultural option to perform the requested function, that means how much each treatment application can affect the function fulfilment both in the short- and mid-term (Agnoloni et al. 2009).

We considered for each plot the silvicultural options described in Table 3.

In each plot, a team composed by two forest experts evaluated each silvicultural option by giving a synthetic score for the capability of the treatment to perform each function.

In Table 4 we reported the correspondence between the evaluation and the score in 7 classes.

N.P. represents a null fulfilment capability, it is used when a specific silvicultural option is not able to allow the stand to perform a specific function (e.g. natural evolution is evaluated N.P. for firewood production function).

N.A. is used when a specific silvicultural option is technically or legally not applicable in that particular forest context (e.g. coppicing option is evaluated N.A. in the case of a coppice abandoned for more than the legally allowed period to be coppiced, specifically two times the rotation period).

The evaluations were carried out considering the effects of each treatment both in short-term (validity of a management plan, equal to 10 years), and in mid-term (20-30 years).

A degree of function fulfilment of each silvicultural option was calculated for every forest category by the Capability of Function Fulfilment Index. It was calculated as the mean of the product between the index of importance of function and the capability of the silvicultural option to fulfil the function of all sampling points related to the forest category:

Where:

$$C_{s\,fi\,f} = \frac{\sum_{i=1}^{n} I_{f\,i} \cdot c_{s\,i}}{n}$$

n = total number of plots per stratum;
$I_{f\,i}$ = priority level of f function for the i-th plot;
$c_{s\,i}$ = capability of s silvicultural option to fulfil the f function in the i-th plot.

Expert evaluation acquires a relevant importance for forest planning, because experts assess directly in field the possible effects of a silvicultural option which can affect positively or negatively each forest function (Paletto et al. 2012).

Our dataset do not respect all assumptions for parametric analysis and almost all the variables are ordinal and non-normally distributed. Thus, we carried out a non parametric analysis. Specifically the Mann-Whitney (U) test (Mann and Whitney 1947) was utilized to investigate the differences between forest categories (European beech and Turkey oak forests) and between silvicultural systems (coppices and high forests); we set a p-level = 0.01 to separate significant from non-significant differences.

Table 4 - Score to evaluate the fulfilment of each silvicultural option.

Evaluation	Score
Good	5
Average good	4
Average	3
Average poor	2
N.P. = Not performing	1
N.A. = Not applicable	0

Table 5 - Mean values of functions' priority level and fulfilment capability by silvicultural systems.

Function/Silvicultural system	Priority level		Fulfilment capability	
	Coppices	High forests	Coppices	High forests
Landscape conservation	4.98	4.52	7.19	7.13
Firewood/biomass production	6.03	4.31	6.86	6.28
Timber production	2.70	4.31	4.29	6.06
Non-wood forest production	4.00	3.61	6.00	5.85
Soil and water regulation	4.24	4.50	6.90	6.98
Touristic/recreational function	3.52	4.07	5.94	6.44
Nature conservation.	3.81	4.07	6.37	6.57
Mean value	**4.18**	**4.20**	**6.22**	**6.47**
Standard deviation (σ)	1.07	0.32	0.97	0.46

Results

Priority level

Regarding the silvicultural system and considering the full set of functions, we obtained the following results of multifunctionality (V): for coppices the mean priority level was 4.18 ($\sigma = 1.10$), and the mean fulfilment capability was 6.18 ($\sigma = 0.82$); for high forests the priority level resulted 4.22 ($\sigma = 0.82$) and the mean fulfilment capability resulted 6.54 ($\sigma = 0.63$) (Table 5).

Considering the value of priority level for single functions (v), we can note that firewood production is the main function for coppices and the third for high forests. The difference between coppices and high forests is statistically significant (U = 2,466.5, Expected value = 1,701, p-value = 0.0001)

Both landscape conservation and soil and water protection have high priority for both silvicultural systems.

Concerning the fulfilment capability of single functions, high forests fulfil more non-productive functions such as (in order of importance) landscape conservation, soil and water protection, environment conservation and touristic/recreational function.

Another result is the high mean priority level of coppices for (in order of importance) the landscape conservation and the soil and water protection.

Moreover, timber production resulted as one of the less important function for both silvicultural systems. This is probably due to the main use of wood coming from Matese forests i.e. firewood, also when it could be useful for alternative uses.

Nonetheless, timber production resulted more important in high forests than in coppices and this difference is statistically significant (U = 1,037.5, Expected value = 1,701, p-value = 0.0001). Also the fulfilment capability of this function resulted significantly higher for high forests than for coppices (U = 839, Expected value = 1,701, p-value = 0.0001)

Regarding the forest category and considering the full set of functions, we obtained the following results of multifunctionality (V): for Turkey oak the mean priority level was 4.18 ($\sigma = 1.07$), and the mean fulfilment capability was 6.22 ($\sigma = 0.97$); for European beech forests priority level was 4.20 ($\sigma = 0.32$) and the mean fulfilment capability was 6.47 ($\sigma = 0.46$) (Table 5).

Considering the value of priority level for single functions (v) we can note that in Turkey oak forests firewood production is the most important function and significantly more important than in European beech forests (U = 2,447.5, Expected value = 1,690, p-value = 0.0001). Furthermore, Turkey oak forests have a priority level of the landscape conservation function higher than European beech forests (U = 2,311, Expected value = 1,690, p-value = 0.001).

On the other hand, European beech forests showed two prior functions: the most important was the soil and water protection (U = 1,098, Expected value = 1,690, p-value = 0.001), the second was the environmental conservation.

These results reflect very clearly the different geo-morphological position of the two forest categories. Indeed, Turkey oak forests are mainly located at a lower altitude and in sites with lower slopes than European beech forests.

Concerning the productive aspects, European beech forests have a significantly higher priority level for the timber production function (U = 1,203, Expected value = 1,690, p-value =0.006). Instead, for Turkey oak forests the timber production show the lowest priority level among functions considered.

On the other hand, European beech forests show to fulfil better non-productive functions: soil and water protection, landscape conservation and environmental conservation.

Particularly, European beech forests show a significantly higher fulfilment capability than Turkey oak forests for soil and water protection (U = 910.5, Expected value = 1,690, p-value = 0.0001), and for environmental conservation (U = 1,113, Expected value = 1,664, p-value = 0.002).

Fulfilment capability of timber production is for both silvicultural systems at the last rank. None-

Table 6 - Mean values of functions' priority level and fulfilment capability by forest categories.

Function/Silvicultural system	Priority level		Fulfilment capability	
	Turkey oak forests	Beech forests	Turkey oak forests	Beech forests
Landscape conservation	5.37	4.02	7.20	7.12
Firewood/biomass production	5.98	4.31	6.74	6.40
Timber production	2.89	4.13	4.60	5.73
Non-wood forest production	4.05	3.54	5.97	5.88
Soil and water regulation	3.82	5.04	6.54	7.44
Touristic/recreational function	3.54	4.08	6.03	6.35
Nature conservation.	3.58	4.42	6.16	6.87
Mean value	**4.18**	**4.22**	**6.18**	**6.54**
Standard deviation (σ)	1.10	0.46	0.82	0.63

Table 7 - Results of objectives-options matrix for Turkey oak forests.

Option / Function	Coppicing (only coppice)		Conversion into high forest (only coppice)		Even-aged high forest (only high forest)		Coppice/high forest integration (only high forest)		Uneven-aged high forest (only high forest)		Natural evolution (both coppices/high forest)	
	Short-term	Long-term	Short-term	Long-term	Short-term	Long-term	Short-term	Long-term	Short-term	Long-term	Short-term	Long-term
Landscape conservation	23.5	23.9	24.6	25.7	22.3	22.4	9.7	10.0	18.9	19.0	26.7/26.7	25.3/25.6
Firewood/biomass production	29.8	30.5	19.9	23.5	20.3	21.2	6.2	6.8	10.4	10.9	0/0	0/0
Timber production	4.5	5.2	4.8	8.3	13.0	16.1	2.9	3.6	8.7	11.1	0/0	0/0
Non-wood forest production	13.3	14.7	14.2	15.8	14.3	15.0	5.7	5.0	11.4	11.8	13.3/13.5	13.6/13.8
Soil and water regulation	15.5	15.8	16.9	17.6	15.7	16.2	7.1	7.2	14.5	14.7	17.5/17.6	17.4/18.0
Touristic/recreational function	10.6	10.9	12.8	15.0	12.8	13.8	4.0	4.8	10.5	11.4	11.3/11.6	12.8/12.9
Environmental conservation	12.5	12.6	14.0	15.3	13.4	13.2	6.2	6.4	12.9	13.1	15.9/16.0	17.4/17.6
Mean value	**15.7**	**16.2**	**15.3**	**17.3**	**16.0**	**16.9**	**5.8**	**6.3**	**12.5**	**13.2**	**12.1/12.2**	**12.4/12.6**
Standard deviation (σ)	**8.4**	**8.4**	**6.2**	**5.8**	**3.8**	**3.6**	**2.2**	**2.1**	**3.4**	**2.9**	**9.6/9.6**	**9.4/9.5**

theless, the non-parametric test of Mann-Whitney shows statistical significantly differences for this function in the European beech forests (U = 1,153.5, Expected value = 1,690, p-value = 0.003)

From a productive viewpoint, we can confirm that firewood/biomass production is the only product requested by the market for both forest categories. This is particularly relevant for Turkey oak forests.

Silvicultural options and multifunctionality

Concerning Turkey oak fulfilment capability calculated for every silvicultural option (Table 7), results show very high values for firewood/biomass production function by coppicing. This capability increase from short-term (29.8) to long-term (30.5). Coppice system allows to maintain good capability to fulfil soil and water protection (15.5-15.8) and landscape conservation (23.5-23.9).

These results highlight that coppicing and even-aged high forest options are more able to fulfil every function than the integration of both options.

Concerning timber production, results show that longer is the term of application of every option, higher is the capability to fulfil a specific function. This aspect is especially evident for even-aged high forest option which has a capability to fulfil timber production of 13.0 in the short-term and 16.1 in the long-term.

Besides, the experts' team evaluated natural evolution as the optimal option to foster together environmental conservation, soil and water protection, and landscape conservation.

Concerning European beech forests (Table 8), firewood/biomass production by coppicing has a fulfilment capability halved compared to Turkey oak forests. This aspect is due to the position of European beech coppices mainly located at high elevation on sloping and medium-low site-index terrains often going to be naturally converted to high forests.

Coppicing, conversion into high forests and even-aged high forest options are fulfilling firewood production with similar performance.

As already reported for Turkey oak, timber production by even-aged high forests options is fulfilled better in the long-term (20.7) than in the short-term (17.7).

Furthermore, this option allows to maintain good fulfilment capability for non-monetary forest functions such as: soil and water protection, touristic/recreational function, environment conservation and landscape conservation.

Also for European beech forests, natural evolution fulfils forest services such as: landscape conservation, soil protection and water regulation, environmental conservation. The same option fulfils better than others non-wood forest production, too.

Discussion and conclusions

Forest planning in Molise has been and still is very active. Economic planning of regional forests started to be active since the 20th century and also contributed - thanks to its methodological consistency - to create a still lasting standardization of forest planning methods (Cantiani et al. 2010).

This consideration is valid also for the Matese area, where economic targets conditioned forest planners and managers choices, influencing both structure and developmental stages of forest stands.

Concerning Turkey oak high forests, their old customary management has been linked to the railway sleepers production. This context produced the spreading of even-aged stands, initially generated from shelterwood. Nonetheless, because of the

Table 8 - Results of objectives-options matrix for beech forests.

Option / Function	Coppicing (only coppice)		Conversion into high forest (only coppice)		Even-aged high forest (only high forest)		Coppice/high forest integration (only high forest)		Uneven-aged high forest (only high forest)		Natural evolution (both coppices/high forest)	
	Short-term	Long-term	Short-term	Long-term	Short-term	Long-term	Short-term	Long-term	Short-term	Long-term	Short-term	Long-term
Landscape conservation	11.6	12.4	14.8	15.8	15.0	15.4	8.9	9.3	13.1	13.5	18.7/19.7	16.8/17.2
Firewood/biomass production	16.3	16.1	16.2	16.6	15.9	16.2	10.8	12.8	12.7	13.8	0/0	0/0
Timber production	4.3	4.3	6.3	9.8	17.7	20.7	5.1	6.2	14.2	16.8	0/0	0/0
Non-wood forest production	10.5	10.5	12.8	13.8	12.8	12.8	10.2	10.1	11.5	11.4	13.9/14.5	13.4/14.1
Soil and water regulation	14.7	15.9	22.7	22.9	19.2	20.3	13.3	13.7	19.1	19.2	24.0/22.9	22.6/21.7
Touristic/recreational function	7.6	7.6	9.3	11.8	17.2	18.4	7.1	7.9	15.2	15.4	11.1/11.9	18.7/17.9
Environmental conservation	9.2	9.4	16.5	20.1	16.9	17.6	10.9	11.7	16.5	17.0	20.8/22.4	19.9/20.1
Mean value	10.6	10.9	14.1	15.8	16.4	17.3	9.4	10.2	14.6	15.3	12.7/13.1	13.0/13.0
Standard deviation (σ)	4.1	4.3	5.3	4.6	2.1	2.8	2.7	2.7	2.6	2.6	9.6/9.8	9.4/9.2

unsuitable application of thinning and regeneration felling, these stands have been acquiring an irregular structure favouring the invasion of secondary species, these contributing to threaten Turkey oak regeneration (Cantiani et al. 2010).

Only at 36% of Turkey oak high forests plots the presence of Turkey oak regeneration was reported. Turkey oak's natural regeneration can be considered absent at the remaining 64%.

Based on the traditional presence of Turkey oak, the experts considered landscape conservation one of the most important functions performed by this forest type. The more appropriate options to fulfil this function were assumed to be natural evolution and shelterwood system (uniform or by groups); this last option, indeed, allows both natural regeneration, and environmental conservation as well as soil protection and water regulation.

From the perspective of production, Turkey oak high forests are linked exclusively to the increase of local firewood demand. Nonetheless, the fulfilment of timber production was considered to increase from short to long-term management under an even-aged regime. This aspect highlights that an appropriate silvicultural treatment (e.g. selective thinning) and in-depth studies on technological features of Turkey oak wood fibre, can improve its market value.

Quite similar considerations can be performed about European beech high forests. Also this type underwent the shelterwood system for many decades as confirmed by all economic plans in the area. Even though many stands reached their technical maturity, the regeneration process has not been activated yet by consistent silvicultural practices.

Multifunctionality analysis of these forest categories highlighted that non-productive functions in Matese are more important for European beech forests than for Turkey oak ones. Especially, natural evolution has been selected by experts as the most appropriate options to fulfil functions such as: soil and water protection, touristic/recreational function, environmental conservation and landscape conservation.

Nonetheless, wood production function of beech forests must be considered, especially their high capability to fulfil timber and firewood production functions.

Shelterwood has been considered the silvicultural options maximising productivity and maintaining optimal values of fulfilment capability also for non-monetary forest functions.

Concerning coppice system, few are similarities between the two forest categories. In the study area, Turkey oak coppices are more actively managed than European beech coppices. This is confirmed by the low percentage overcoming the maximum legal age to be coppiced (20%) as compared with European beech (40%). This trend is due to the localization of coppices of the two forest categories and to the economic importance of firewood for the area.

These considerations are confirmed also by the multifunctionality indices: firewood production is the most important function for Turkey oak coppices but not for beech coppices. Furthermore, coppicing ensures high values of fulfilment capability also for landscape conservation. This result classifies coppiced Turkey oak patches as very important elements of the Matese landscape.

Firewood production shows on the contrary very low values for European beech coppices: about half of the deciduous oak type. Beech is mainly located at high elevation, in sites with medium-low site-index and their management has been not very active over the last decades leading them towards a natural conversion into high forests. That is why, both the active conversion into high forest and the natural evolution showed to be the most performing option for European beech coppices to fulfil several functions both in the short and in the long-term.

As after Paletto et al. (2012) the synthetic indicators of multifunctionality:
1. provided adequate outputs: the value of overall multifunctionality increased with the altimetric gradient, from Turkey oak forests at low altitudes to European beech forests at higher elevations;
2. the analysis proved the low economic value of the Matese forests and limited to firewood production;
3. the high forest system provided the fulfilment of the highest number of functions;
4. conversion from coppice to high forest may increase the overall value of the Matese forests because of the parallel increase of protective, touristic and productive functions.

The introduction of the fulfilment capability and the modification of the priority level concept - by which we evaluated how much a function is important compared to the others at each time and for the same plot - represented an improvement to the method proposed by Paletto et al. (2012). This enhanced approach allowed to detect the current state of each plot and its potentiality in the framework of the whole forest. The evaluation of forests functional features using the proposed approach reduces the inherent subjectivity.

The proposed method allows to elaborate useful information on forest multifunctionality and to support forest planners in defining management guidelines consistent with current state and the evolutionary potentiality of forest stands.

Acknowledgements

This work is part of the researches implemented within the Agreement Consiglio per la Ricerca e la sperimentazione in Agricoltura (CRA) and Regione Molise (Accordo di programma quadro pluriennale per attività di ricerca e sviluppo nel settore forestale). The "ProgettoBosco" informative system and methodology, elaborated by CRA-MPF and CRA-SEL with the MIPAF "Ri.Selv.Italia" - Subproject 4.2 "Geographic informative system for forest management" was also used. Ri.Selv.Italia was a national, multi-year, "problem solving oriented" project in the forest-wood-environment sector promoted by the Ministry of Agriculture and Forests in concert with Regions.

The authors want to thank the anonymous reviewers for the helpful revision of the paper.

All the authors contributed equally to this work.

References

Agnoloni S., Bianchi M., Bianchetto E., Cantiani P., De Meo I., Dibari C., Ferretti F. 2009 - *I piani forestali territoriali di indirizzo: una proposta metodologica*. Forest@ 6: 140-147.

Angelstam P., Kopylova E., Korn H., Lazdinis M., Sayer J., Teplyakov V., Törnblom J. 2005 - *Changing forest values in Europe. In: Forests in landscapes*. Ecosystem Approaches to Sustainability (Sayer J.A., Maginnis S., eds), Earthscan 59: 59-74.

Bernetti G. 2005 - *Atlante di Selvicoltura*. UTET Torino, 495 p.

Brady E. 2003 - *Aesthetics for the natural environment*. Edinburgh: Edinburgh University Press Ltd., 263 p.

Campos J.J., Finegan B., Villalobos R. 2001 - *Management of Goods and Services from Neotropical Forest Biodiversity: Diversified Forest Management in Mesoamerica*, In: Conservation and Sustainable Use of Forest Biodiversity (CBD Technical Series no. 3), Montreal: Secretariat of the Convention on Biological Diversity (SCBD): 5-16.

Cantiani M.G. 2012 - *Forest planning and public participation: a possible methodological approach*. iForest 5: 72-82. http://dx.doi.org/10.3832/ifor0602-009

Cantiani P., Ferretti F., Pelleri F., Sansone D., Tagliente G. 2010 - *Le fustaie di cerro del Molise. Analisi del trattamento del passato per le attuali scelte selvicolturali*. Annali CRA - Centro di ricerca per la Selvicoltura 36: 25-36.

Chirici G., Di Martino P., Ottaviano M., Santopuoli G., Chiavetta U., Tonti D., Garfi' V. 2011 - *La Carta Forestale Su Basi Tipologiche*. In: Tipi Forestali E Preforestali Della Regione Molise. Alessandria: Edizioni Dell'orso, Isbn: 978-88-6274-280-1: 145-152

Cubbage F., Harou P., Sillsa E., 2007 - *Policy instruments to enhance multi- functional forest management*. Forest Policy and Economics 9 (7): 833-851.

Dieterich V. 1953 - *Forstwirtschaftspolitik: eine Einfuehrung*, 398 p.

FAO 2006 - *Global Forest Resources Assessment 2005*. Rome: FAO Forestry, Paper, 147 p.

Farcy C., Devillez F. 2005 - *New orientations of forest management planning from an historical perspective of the relations between man and nature*. Forest Policy and Economics 7 (1): 85-95.

Führer E. 2000 - *Forest functions, ecosystem stability and management*. Forest Ecology and Management 132 (1): 29-38.

Garcıa-Fernandez C., Ruiz-Perez M., Wunder S. 2008 - *Is multiple-use forest management widely implementable in the tropics?* Forest Ecology and Management 256: 1468–1476.

Helms J. A. 1998 - *The dictionary of Forestry*. The Society of American Foresters, 210 p.

Hiremath A. J. 2004 - *The ecological consequences of managing forests for non-timber products*. Conservation & Society 2 (2).

Kangas J., Store R. 2002 - *Socio-ecological Landscape Planning: An Approach to Multi-Functional Forest Management*. Silva Fennica 36 (4): 867-871.

INFC 2004 - *Il disegno di campionamento. Inventario nazionale delle Foreste e dei Serbatoi Forestali di Carbonio*. MiPAF – Direzione Generale per le Risorse Forestali, Montane ed Idriche. Corpo forestale dello Stato, ISAFA, Trento.

Kant S. 2003 - *Extending the boundaries of forest economics*. Forest Policy and Economics 5: 39-56.

Kant S. 2004 - *Economics of sustainable forest management.* Forest Policy and Economics 6: 197-203.

Kant S. 2007 - *Economic perspectives and analyses of multiple forest values and sustainable forest management.* Forest Policy and Economics 9: 733-740.

Kennedy J.J., Koch N.E. 2004 - *Viewing and managing natural resources as human- ecosystem relationships.* Forest Policy and Economics 6: 497-504.

Kimmins J.P. 2004 - *Forest Ecology. A Foundation for Sustainable Forest Management and Environmental Ethics in Forestry.* Third edition. Prentice Hall, 612 p.

Krieger D.J. 2001 - *Economic Value of Forest Ecosystem Services: A Review.* Washington: The Wilderness Society, 30 p.

Mann H.B. and Whitney D.R. 1947 - *On a Test of Whether one of Two Random Variables is Stochastically Larger than the Other.* Annals of Mathematical Statistics 18 (1): 50–60. http://dx.doi.org/10.1214/aoms/1177730491

Nix S. 2012 - *Multiple use* [Online]. Available at: http://forestry.about.com/cs/glossary/g/multi_use.htm [2013].

Paletto A., Ferretti F., Cantiani P., De Meo I. 2012 - *Multifunctional approach in forest landscape management planning: an application in Southern Italy.* Forest Systems 21 (1): 68-80.

Perrin H. 1954 - *Sylviculture (Tome II). Le Traitement des Forets, Théorie et Pratique des Techniques sylvicoles.* Ecole Nationale des Eaux et Forets. Nancy, 411 p.

Regione Molise Direzione Generale III^ delle Politiche agricole, alimentari e forestali. 2002 - *Piano Forestale Regionale 2002-2006.*

Schmithüsen F. 2007 - *Multifunctional forestry practices as a land use strategy to meet increasing private and public demands in modern societies.* Journal of Forest Science 53 (6): 290-298.

Sepp K., Palang H., Mander U., Kaasik A. 1999 - *Prospects for nature and landscape protection in Estonia.* Landscape and Urban Planning 1-3: 161-167.

Straka T.J. 2009 - *Evolution of Sustainability in American Forest Resource Management Planning in the Context of the American Forest Management Textbook.* Sustainability 1838-854 http://dx.doi.org/10.3390/su1040838

Tabbush P. 2004 - *Public money for public good? Public participation in forest planning.* Forestry 77: 145-156.

Vincent J.R., Binkley C.S. 1993 - *Efficient Multiple Use Forestry may require Land Use specialization.* Land Economics 69 (4): 370-376.

Wang S., Wilson B. 2007 - *Pluralism in the economics of sustainable forest management.* Forest Policy and Economics 9: 743-750

Wyder J. 2001 - *Multifunctionality in the Alps: Challenges and the potential for conflict over development.* Mountain Research and Development 21: 327-330.

Ozone dynamics in a Mediterranean Holm oak forest: comparison among transition periods characterized by different amounts of precipitation

Flavia Savi[1,2*], Silvano Fares[1]

Abstract - Tropospheric ozone (O_3) is one of the most toxic compounds for plants in the atmosphere. The large amount of anthropogenic O_3 precursors in the urban areas promote O_3 formation, thus making Mediterranean forests located in periurban areas particularly vulnerable to this pollutant. O_3 flux measurements have been carried out using the Eddy Covariance technique over a Holm oak forest located 25 Km from Rome downtown, inside the Presidential Estate of Castelporziano (Italy). Two transition periods - early Spring and late Fall - in two consecutive years were examined. The uncommon low precipitation recorded in both transition periods in 2012 allowed to evaluate the influence of water availability on O_3 fluxes during seasons which are not commonly affected by drought stress. Overall, the forest canopy showed to be a net sink of O_3, with peak values of mean daily O_3 fluxes of -8.9 nmol $m^{-2}s^{-1}$ at the beginning of flowering season and -4.6 nmol $m^{-2}s^{-1}$ at the end of Fall. O_3 fluxes were partitioned between stomatal and non stomatal sinks using the evaporative/resistive method based on canopy transpiration in analogy with an Ohm circuit. By comparison of the two years, water availability showed to be an important limiting factor during Spring, since in this season plants are more photosynthetically active and more sensitive to water availability, while in Fall, under conditions of low stomatal conductance, the dependence on water availability was less appreciated.

Keywords - ozone fluxes, O_3, Holm oak, Mediterranean forest, Eddy Covariance, drought stress, pollution

Introduction

Tropospheric ozone (O_3) is a significant environmental problem as it affects human health (Anenberg et al. 2010) and decreases carbon sequestration potentials of forest ecosystems (Fares et al. 2013a). It is also an important greenhouse gas, with a radiative forcing of 0.35–0.37 W m^{-2}, responsible for 5% - 16% of the global temperature increase since preindustrial time (Foster et al. 2007).

O_3 is produced in the atmosphere by photochemical reactions between anthropogenic and biogenic volatile organic compounds (VOC) and nitrogen oxide (NO_x), high irradiance and temperature occurring in the Mediterranean regions promote O_3 formation more than in other area (Paoletti 2006). O_3 removal from forest ecosystems is attributed to both stomatal and non-stomatal sinks, which include deposition on cuticles and soil surface as well as O_3 depleted by gas-phase reactions (Kurpius and Goldstein 2003, Cieslik 2004). The majority of O_3 deposition is often attributed to non-stomatal O_3 sinks (Fowler et al. 2009), especially during the summer season when stomatal conductance is reduced under conditions of drought stress and high vapour pressure deficit (Gerosa et al. 2009).

The objective of this study is to quantify O_3 removal during two transition periods: before the beginning of the driest season, from March 20 to April 14 (early Spring) and before the coldest season, from November 11 to December 6 (late Fall) in a Mediterranean Holm oak forest. O_3 flux dynamics were compared between the uncommonly dry early Spring and late Fall of 2012 with the more wet periods of 2013 in order to highlight whether water availability can have a significant effect on ozone fluxes.

Materials and Methods

Study site description

The study site, named *"Grotta di Piastra"* (41°42' N, 12°21'E), is located within the Castelporziano Estate, 25 km SW from the centre of Rome, Italy. This site is a wild coastal rear dune ecosystem, 1.5 km from the seashore, covered almost prevalently by an even-aged evergreen Holm oak forest (*Quercus ilex* L.). The forest main height was 14 m and the Leaf Area Index (LAI) was 3.69 m^2leaf m^{-2}ground,

---

[1] Consiglio per la Ricerca e la sperimentazione in Agricoltura, Research Center for the Soil-Plant System (CRA-RPS), Rome, Italy
[2] Department for Innovation in Biological, Agro-Food and Forest Systems (DIBAF), University of Tuscia, Italy
* corresponding author: flavia.savi@entecra.it

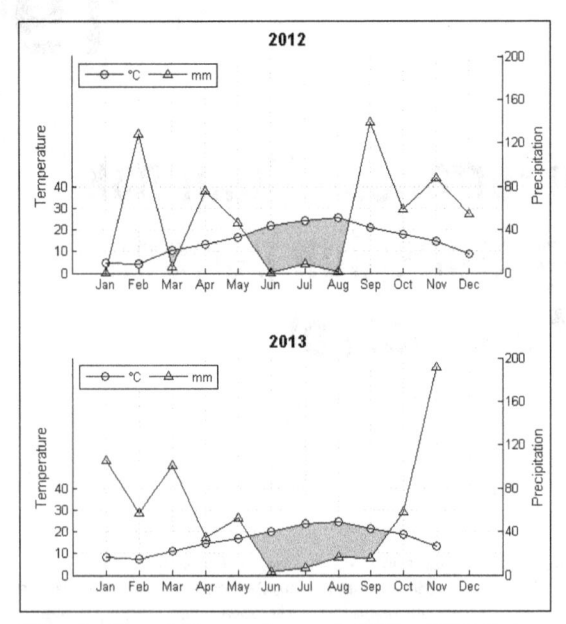

Figure 1 - Bagnouls-Gaussen diagrams for 2012 and 2013. Circles are monthly average temperature (°C) while triangles are monthly cumulated precipitation (mm). Shaded area represent drought period.

measured using a LAI 2000 instrument (Li-Cor, USA). The soil has a sandy texture and low water-holding capacity. The climate is typically Mediterranean: precipitation occurred mainly in Fall and Winter, whereas Summers were hot and dry (Fig. 1). Averaged in the year 1999-2010, annual precipitation was 789.3±230.6 mm and mean monthly temperatures range between 7.3°C and 23.3°C. The wind regime was characterized by winds from the sea (S-SW) blowing during the morning, and winds from the inland (N-NE) in the afternoon.

Meteorology and flux measurement

Measurements were carried out in 2012 and 2013 in early Spring and late Fall, from day of the year 79 to 104 and 315 to 340 of each year.

Air temperature, precipitation, relative humidity, net solar radiation, wind direction, soil humidity and soil temperature were recorded every minute and averaged for 30 min intervals with a Davis vantage pro meteorological station (Davis Instruments Corp. CA, USA).

Flux measurement equipment was installed at 19.7 m height at the top of a scaffold tower. A tridimensional sonic anemometer (Gill Windmaster) was used to measure instantaneous wind speed and temperature fluctuation. H_2O and CO_2 concentrations were measured with a closed-path infrared gas analyzer (LI-7200, Li-Cor, USA). O_3 fast measurements were performed by a chemiluminescence methods which uses coumarin dye reaction with O_3, thanks to a customized instrument developed by the National Oceanic and Atmospheric Administration (NOAA, Silver Spring, MD, Bauer et al. 2000). The chemiluminescence detector was calibrated against 30 min average O_3 concentrations from a UV ozone monitor (Thermo Scientific, mod. 49i). Data were recorded at 10 Hz for all gases using a data logger (CR-3000, Campbell Scientific, Shepshed, UK).

H_2O, CO_2, O_3, latent and sensible heat fluxes were calculated according to the eddy covariance technique:

$$F_c = \overline{w'c'} \qquad (1)$$

where w' and c' are deviations from the 30 minute means of vertical wind velocity and gas concentration, respectively. The method is extensively described in Goldstein et al. (2000) and Fares et al. (2012).

O_3 fluxes were partitioned between stomatal and non-stomatal trough several steps: first stomatal conductance for water (G_{sto}) was calculated from latent heat flux by inverting the Penmann–Monteith equation according to the evaporative/resistive method (Monteith 1981, Fares et al. 2013b, Gerosa et al. 2009). O_3 stomatal conductance (G_{O3}) was calculated from G_{sto}, by correcting for the difference in diffusivity between O_3 and water vapor (Massman 1998). O_3 stomatal fluxes were calculated multiplying G_{O3} by O_3 concentration, assuming a constant vertical flux between the measurement height and the top of canopy and negligible intercellular O_3 concentration (Laisk et al. 1989). The remaining fraction of the O_3 flux is considered as non-stomatal deposition and includes all other deposition pathways.

Fluxes are expressed per unit of ground area per second, positive fluxes indicate upward transfer of mass and energy from the ecosystem to the atmosphere, and negative fluxes indicate transfer from the atmosphere into the ecosystem.

Results and discussion

Meteorology and energy fluxes

Periods examined in this work are both transition phases between a cold and wet season and a dry and warm season, when usually drought stressed does not occur: average in the year 1999-2010, mean precipitation were 61.3±24.6 mm in April and 130.1±54.4 mm in November. Early Spring and late Fall (2012) were dry respect to values collected in 2013 (Fig. 2): 8.0 mm in 2012 versus 58.4 mm of 2013 in the early Spring and 71.4 mm of 2012 versus 160.8 mm of 2013 in late Fall.

Figure 3 shows 2012 and 2013 mean daily course of temperature (°C), net radiation (W m^{-2}), latent heat flux (LE, W m^{-2}), sensible heat flux (H, W m^{-2}) and vapour pressure deficit (VPD, kPa) for the two

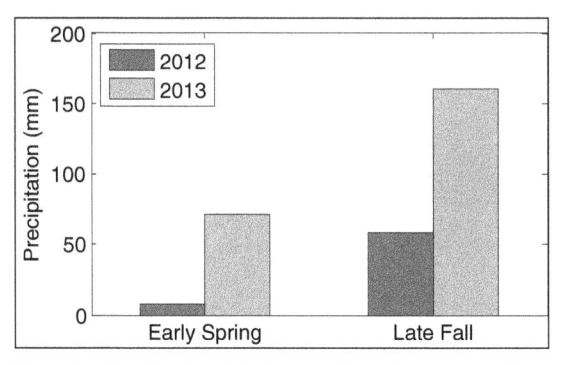

Figure 2 - Cumulated precipitation (mm) from March 20 to April 14 (Early Spring) and from November 11 to December 6 (Late Fall).

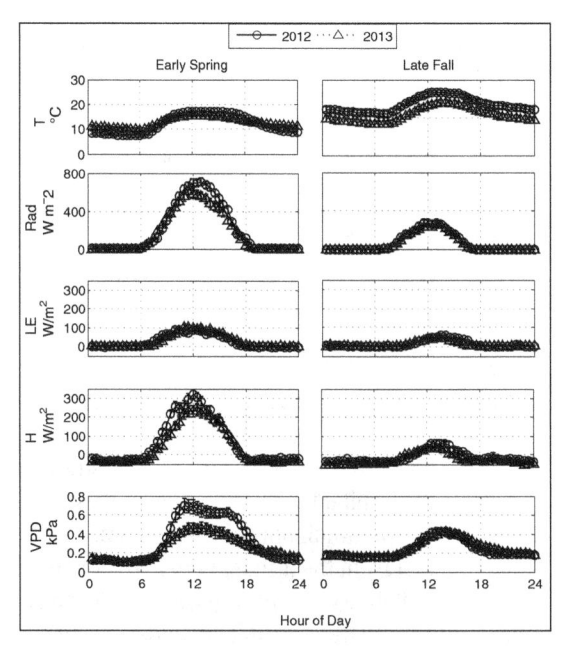

Figure 3 - Averaged daily values (± standard deviation) of temperature (a), net radiation (b), latent heat flux (c), sensible heat flux (d) and vapour pressure deficit (e) for the periods from March 20 to April 14 (Early Spring) and from November 11 to December 6 (Late Fall).

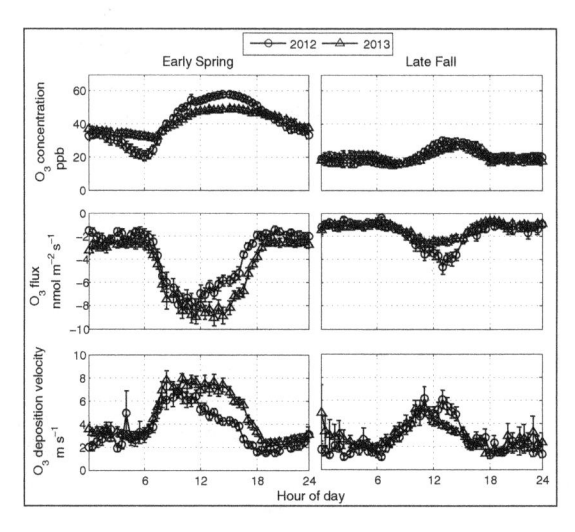

Figure 4 - Averaged daily values (± standard deviation) of ozone concentration (ppb), ozone fluxes (nmol m^{-2} s^{-1}) and deposition velocity (m s^{-1}) at the site in 2012 and 2013 early Spring and late Fall.

periods. Early Spring mean temperature was similar across the two years (12.1±4.2°C and 12.6±2.9°C for 2012 and 2013, respectively) while 2013 late Fall was colder than the 2012 one (13.4±3.9°C and 10.6±4.2°C for 2012 and 2013, respectively). Late Fall night temperatures were higher than those recorded in early Spring 2012. Daytime cloudiness in 2013 early Spring was 19% higher than 2012. H reflected the solar radiation trend. During the early Spring hottest hours LE flux intensities were 16% lower in 2012 than 2013, as expected considering the scarcity of precipitation occurred in 2012. Interestingly, the relation is inverse in late Fall (LE fluxes were 23% minor in 2013 than 2012) suggesting that the water availability did not represent a limiting factor during Fall.

O_3 concentration at the site

No significant differences were observed between mean O_3 concentrations at the top of the canopy for the two years in early Spring (41.4±16.4 ppb and 41.5±10.8 ppb for 2012 and 2013, respectively) neither in late Fall (20.7±13.3 ppb and 20.3±12.8 ppb for 2012 and 2013, respectively). For both periods, O_3 concentration was higher in the warmest hours of the day and decreased during the night (Fig. 4). Overall, O_3 concentration was lower in the late Fall period according to the dependence of O_3 on air temperature (Kurpius and Goldstein 2003, Fares et al. 2010, Finlayson and Pitts 1997). The land-sea wind regime at the site also affected O_3 concentration. Figure 5 shows daytime (a) and nighttime (b) wind direction and O_3 concentration. During the day wind blew prevalently from the sea (S-W), carrying air masses to the forest, while during night wind blew from the city (N-E), transferring polluted air plumes to the forest site, as previously reported by Fares et al. (2009). Air coming from the city was previously characterized by low O_3 concentrations due to the fast reactions with anthropogenic pollutants like nitrogen oxides (NO$_x$, Finlayson and Pitts 1997). This may explain the average low concentrations of O_3 in Castelporziano as compared with periurban area north of Rome (Fares et al. 2009 and 2013b). Moreover, during the few times that wind circulation diverged from its typical pattern (Fig. 5 a, b), O_3 concentration at night was higher, thus confirming that that air masses not directly coming from the urban areas are less depleted in O_3 concentration.

O_3 fluxes and deposition velocity

O_3 fluxes to the forest reached the maximum values during the central hours of the day both in early Spring and in late Fall. The peak values of mean daily O_3 fluxes in early Spring were -8.1±0.7 nmol m^{-2}s^{-1} and -8.9±0.6 nmol/m^{2}s for 2012 and 2013 respectively, while in late Fall they were -4.6±0.7 nmol m^{-2}s^{-1} and

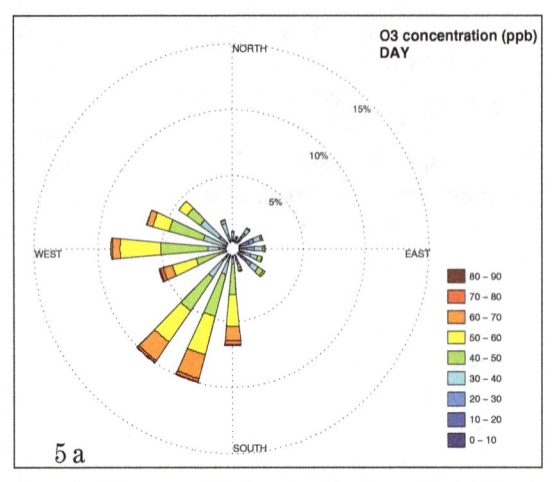

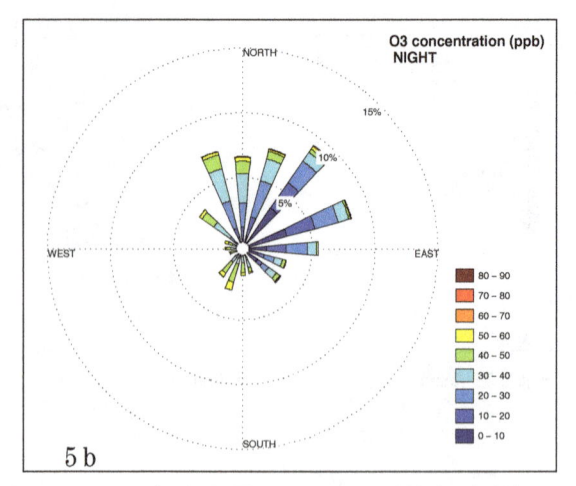

Figure 5 - Wind roses of daytime (a) and nighttime (b) wind directions and ozone concentration (ppb). The frequencies at which the wind blew from each direction is represented by the radial thickness of each slice, while ozone concentration is represented by the color of the filled area. Data are from February 2012 to November 2013.

Table 1 - Summary statistics of O_3 fluxes. For each periods is reported: mean O_3 flux ± standard error, 25th percentile, 75th percentile, skewness, number of observations (n) and percentage of valid observations (N).

Time period year	season	O_3 flux (nmol m^{-2} s^{-1}) mean±se	25th perc.	75th perc.	skewness	n	N (%)
2012	early Spring	-3.92±0.12	-5.88	-1.2	-1.14	991	81
	late Fall	-1.7±0.08	-2.80	-0.01	-1.7	1021	86
2013	early Spring	-4.92±0.10	-7.00	-2.29	-0.92	1123	99
	late Fall	-1.35±0.04	-1.95	-0.09	-1.69	1025	99

-2.5±0.3 nmol m^{-2}s^{-1} for 2012 and 2013, respectively. Late Fall O_3 fluxes for both years were about half of the fluxes measured in early Spring. A summary statistics of O_3 fluxes is reported in Table 1.

A strong correlation between O_3 fluxes and O_3 concentrations was observed in both study periods (Fig. 4), with the exception of the hottest hours of the day in 2012 early Spring. O_3 deposition velocity (flux normalized by concentration) in this period was reduced by 23.6%. We ascribe this behaviour to the reduction in stomatal O_3 fluxes, as previously hypothesized given the dependence of stomatal conductance on moisture content. Stomatal conductance was indeed lower in early Spring 2012 during the central hours of the day (Fig. 6).

O_3 deposition velocities were lower in late Fall than in early Spring for both years (Fig. 4). This indicates that not only O_3 concentration controls the flux magnitude but also plant phenology, which determines low stomatal conductance in Fall, played a leading role in controlling O_3 flux. In order to better understand seasonal effects on O_3 fluxes during Spring and Fall, these were partitioned between stomatal and non-stomatal fluxes (Fig. 7) for measurements performed in 2013 (data for 2012 not available).

In agreement with dynamics of stomatal conductance shown in fig. 6, stomatal contribution to the total O_3 flux was different during the two seasons, with higher stomatal fluxes in Spring, while non-stomatal O_3 fluxes were similar in the two seasons. This result confirmed the predominant role of stomatal control on O_3 removal from the atmosphere in dependence on water availability in a photosynthetically active season (Spring). Therefore in Fall, under conditions of low stomatal conductance, the dependence on water availability was less appreciated.

Individual contribution of stomatal and non-stomatal sinks to total O_3 fluxes also varied during the day. Night values of non stomatal fluxes could have several contributors: gas-phase reaction with VOC and NOx (Finlayson and Pitts 1997), and surface deposition. The latter is probably responsible

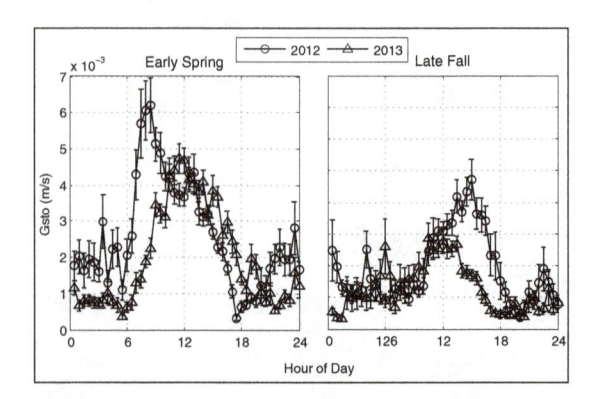

Figure 6 - Averaged daily values (± standard deviation) of stomatal conductance (Gsto, m/s) at the site in 2012 and 2013 early Spring (a) and late Fall (b).

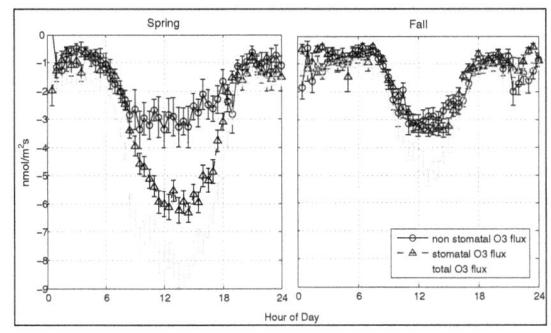

Figure 7 - Averaged daily values (± standard deviation) of total O_3 flux, stomatal O_3 flux and non stomatal O_3 flux measured during 2013 Spring (a) and Fall (b).

for the observed high nocturnal non-stomatal fluxes in Fall, when leaf surface wetness and air humidity have been shown to increase O_3 deposition (Zhang et al. 2002, Altimir et al. 2005).

Conclusion

O_3 fluxes were measured in an periurban Mediterranean evergreen Holm oak forest during transition periods in two different years, characterized by different amount of precipitation.

During the measurement periods, not commonly affected by drought stress, O_3 flux was found to be reduced under conditions of low water availability, when stomatal sink contribution is typically higher. The non-stomatal ozone deposition proved to be an important sink of tropospheric ozone in this Holm oak ecosystem, current studies are aimed at partitioning these non-stomatal sinks between different contributors (e.g. NOx, VOCs, surface deposition).

Overall, our results indicate that the Castelporziano evergreen forest represents a net sink of O_3. This type of ecosystem service must be taken into account while evaluating the complex of the benefits that forest ecosystems can provide to urban and peri-urban areas.

Currently, O_3 fluxes are still measured at the site. A large temporal series will help to elucidate deeply the contribution of the environmental control factors on ozone dynamics.

Acknowledgements

The research leading to these results has received funding from the European Project Marie Curie-CIG-EXPLO3RVOC", the Scientific Commission of Castelporziano "CASTEL2" project, the FO3REST (LIFE10 ENV-FR-208), ECLAIRE (FP7-ENV-2011) and ERA-NET FORESTERRA (PN 291832) projects.

This Publication reflects only the authors views and the European Community is not liable for any use that may be made of the information contained herein.

We thank: the General Secretariat of the Presidency of Italian Republic; the Directorate of Castelporziano Estate; the Scientific Commission of Castelporziano, in particular the President, Prof. Ervedo Giordano; the Multi-disciplinary Center for the Study of Coastal Mediterranean Ecosystems, in particular Ing. Aleandro Tinelli and Dr. Luca Maffei, for technical and logistic support allowing execution of these studies and publication of the data; the team members of the biometeorology laboratory at CRA: Mr. Roberto Moretti, Mr. Valerio Moretti, Mr. Tiziano Sorgi and Mr. Filippo Ilardi, for their help with setting up the experimental station. Castelporziano is one of the Transnational Access site of the FP7 INFRA I3 project ExpeER (contract no. 262060).

The authors want to thank the anonymous reviewers for the helpful suggestions.

References

Altimir N., Kolari P., Tuovinen J.-P., Vesala T., Bäck J., Suni T., Kulmala M., Hari P. 2005 - *Foliage surface ozone deposition: a role for surface moisture?* Biogeosciences Discussions 2: 1739–1793.

Anenberg S.C., Horowitz L.W., Tong D.Q. ,West J.J. 2010 - *An estimate of the global burden of anthropogenic ozone and fine particulate matter on premature human mortality using atmospheric modeling.* Environmental Health Perspective 118: 1189–1195.

Bauer M.R., Hultman N.E., Panek J.A., Goldstein A.H. 2000 - *Ozone deposition to a ponderosa pine plantation in the Sierra Nevada Mountains (CA): a comparison of two different climatic years.* Journal of Geophysical Research 105: 123-136.

Cieslik S. 2004 - *Ozone uptake at various surface types: a comparison between dose and exposure.* Atmospheric Environment 38: 2409–2420.

Fares S., Mereu S., Scarascia Mugnozza G., Vitale M., Manes F., Frattoni M., Ciccioli P., Gerosa G., Loreto F. 2009 - *The ACCENT-VOCBAS field campaign on biosphere-atmosphere interactions in a Mediterranean ecosystem of Castelporziano (Rome): site characteristics, climatic and meteorological condition and eco-physiology of vegetation.* Biogeosciences 6: 1043–1058.

Fares S., McKay M., Holzinger R., Goldstein A.H. 2010 - *Ozone fluxes in a Pinus ponderosa ecosystem are dominated by non-stomatal processes: evidence from long-term continuous measurements.* Agricultural and Forest Meteorology 150: 420-431.

Fares S., Weber R., Park J.H., Gentner D., Karlik J., Goldstein A.H. 2012 - *Ozone deposition to an orange orchard: Partitioning between stomatal and non-stomatal sinks.* Environmental Pollution 169: 258–266.

Fares S., Vargas R., Detto M., Goldstein A.H., Karlik J., Paoletti E., Vitale M. 2013a - *Tropospheric ozone reduces carbon assimilation in trees: estimates from analysis of continuous flux measurements.* Global Change Biology 19: 2427–2443.

Fares S., Schnitzhofer R., Jiang X., Guenther A., Hansel A., Loreto F. 2013b - *Observations of Diurnal to Weekly Variations of Monoterpene-Dominated Fluxes of Volatile Organic Compounds from Mediterranean Forests: Implications for Regional Modeling.* Environmental. Science and Technology 47 (19): 11073–11082.

Finlayson-Pitts B.J., Pitts J.N. 1997 - *Ozone, airborne toxics, polycyclic aromatic hydrocarbons, and particles.* Science 276: 1045–1052.

Fowler D., Pilegaard K., Sutton M.A., Ambus P., Raivone M., Duyzer J., Simpson D., Fagerli H., Fuzzi S., Schjoerring J.K., Grainer C., Neftel A., Isaksen I.S.A., Laj P., Maione M., Monks P.S., Burkhardt J., Daemmgen U., Neirynck J., Personne E., Wichink-Kruit R., Butterbach-Bahl K., Flechard C., Tuovinen J.P., Coyle M., Gerosa G., Loubet B., Altimir N., Gruenhage L., Ammann C., Cieslik S., Paoletti E., Mikkelsen T.N., Ro-Poulsen H., Cellier P., Cape J.N., Horvath L., Loreto F., Niinemets U., Palmer P.I., Rinne J., Misztal P., Nemitz E., Nilsson D., Pryor S., Gallagher M.W., Vesala T., Skiba U., Brueggemann N., Zechmeister-Boltenstern S., Williams J., O'Dowd C., Facchini M.C., de Leeuw G., Flossman A., Chaumerliac N., Erisman J.W. 2009 - *Atmospheric Composition Change: Ecosystems-Atmosphere interactions.* Atmospheric Environment 43: 5193-5267.

Forster P., Ramaswamy V., Artaxo P., Berntsen T., Betts R. et al. 2007 - *Changes in atmospheric constituents and in radiative forcing. In: "Climate Change 2007: The Physical Science Basis. Contribution of Working Group I to the Fourth Assessment Report of the Intergovernmental Panel on Climate Change".* D. Qin, M. Manning, Z. Chen, M. Marquis et al., Cambridge ed. S Solomon, Cambridge Univ. Press 53: 129–234.

Gerosa G., Finco A., Mereu S., Vitale M., Manes F., Denti A.B. 2009 - *Comparison of seasonal variations of ozone exposure and fluxes in a Mediterranean Holm oak forest between the exceptionally dry 2003 and the following year.* Environmental Pollution 157: 1737–1744.

Goldstein A.H., Hultman N.E., Fracheboud J.M., Bauer M.R., Panek J.A., Xu M., Qi Y., Guenther A.B., Baugh W. 2000 - *Effects of climate variability on the carbon dioxide, water, and sensible heat fluxes above a ponderosa pine plantation in the Sierra Nevada (CA).* Agricultural and Forest Meteorology 101: 113–129.

Laisk A., Kull O., Moldau H. 1989 - *Ozone concentration in leaf intercellular air spaces is close to zero.* Plant Physiology 90: 1163-1167.

Manes F., Astorino G., Vitale M., Loreto F. 1997 - *Morpho-functional characteristics of Quercus ilex L. leaves of different age and their ecophysiological behaviour during different seasons.* Plant Biosystem 131 (2): 149-158.

Massman W.J. 1998 - *A review of the molecular diffusivities of H_2O, CO_2, CH_4, CO, O_3, SO_2, NH_3, N_2O, NO, and NO_2 in air, O2 and N2 near STP.* Atmospheric Environment 32: 1111-1127.

Monteith J.L. 1981 - *Evaporation and surface temperature.* Quarterly Journal of the Royal Meteorological Society 107: 1–27.

Kurpius M.R., Goldstein A.H. 2003 - *Gas-phase chemistry dominates O_3 loss to a forest, implying a source of aerosols and hydroxyl radicals to the atmosphere.* Geophysical Research Letters 30 (7): 1371.

Paoletti E. 2006 - *Impact of ozone on Mediterranean forests: a review.* Environmental Pollution 144: 463–474.

Zhang L., Brook J.R., Vet R. 2002 - *On ozone dry deposition, with emphasis on non- stomatal uptake and wet canopies.* Atmospheric Environment 36: 4787–4799.

Comparing growth rate in a mixed plantation (walnut, poplar and nurse trees) with different planting designs: results from an experimental plantation in northern Italy

Francesco Pelleri[1*], Serena Ravagni[2], Elisa Bianchetto[1], Claudio Bidini[1]

Abstract - Results of a mixed plantation with poplar, walnut and nurse trees established in winter 2003 in Northern Italy, are reported. Main tree species (poplar and walnut) were planted according to a rectangular design (10 x 11m), with different spacings and alternate lines. The experimental trial was carried out to verify the following working hypotheses: (i) possibility to combine main trees with different growth levels (common walnut, hybrid walnut, and different poplar clones) and test two different poplar and walnut spacings (5.0 and 7.4 m) in the same plantation; (ii) opportunity to reduce cultivation's workload, in comparison with poplar monoculture, using mixtures with different poplar clones and N-fixing nurse trees; (iii) verifying the growth pattern of two new poplar clones in comparison with the traditional clones cultivated for different purposes in Italy. The use of different valuable crop trees' mixtures intercropped with nurse trees and shrubs (including N-fixing trees) allows to decrease the cultivation's workload. In fact, a heavy reduction of cultural practices - fertilizers, weed control, irrigation and pesticides applications (-61%) are the main concurrent, supplementary benefits. The best growth performances (DBH and tree height), associated with the higher competition towards walnuts, were recorded with the new clones Lena and Neva in comparison with the I214 and Villafranca. The closer spacing (5 m between poplar and walnut trees) was found to be unsuited to get merchantable poplars sized 30 cm without developing a heavy competition towards walnut trees. The wider spacing (7.4 m) resulted vice versa suitable to get poplar trees sized as requested by veneer factories and to maintain an acceptable competitive level with walnut. Within this plantation design, a shorter rotation (8 yrs) is needed for Lena and Neva clones in comparison with I214 and Villafranca (10 yrs). Walnut intercropped with poplar showed cone-shaped crowns, light branching and a good stem quality in comparison with walnut grown in pure plantations. This model of mixed plantation can become an interesting optional choice to walnut's and poplar's monoculture with notable advantages both for farm economics, landscape quality and environment preservation.

Keywords - common walnut, hybrid walnut, mixed plantation, poplar clones, sustainable tree farming

Introduction

Poplar cultivation is the main internal source for timber industry, producing about 50% of the whole roundwood volume (Facciotto et al. 2003) in Italy. Since the early nineties, poplar cultivation is suffering a heavy reduction from 90,000 to 61,381 hectares (Gasparini e Tabacchi 2011, Coaloa et al. 2012) due to several factors: (i) the weak farmers' contractual position when selling a poplar plantation, (ii) the lack of any industry planning useful for programming planting investments, (iii) the wider use by poplar industry of semi-finished timber products imported from other European and extra-European countries, (iv) the increasing production costs, i.e. petrol, pesticides, fertilizers, etc. (Nervo et al. 2007). This condition enhanced research efforts for the innovation of poplar cultivation systems towards less energy-consuming plantations (Coaloa and Vietto 2005) or plantations able to produce, within the same rotation, larger-sized trees, along with reducing production's costs (Buresti and Mori 2006).

Another tree species, traditionally cultivated in Italy for valuable timber production, is common walnut (*Juglans regia* L.). In the past, this species was generally cultivated as a single tree or in linear plantations, intercropped to agricultural crops, both for nuts and timber production (Minotta 1990, 1992). More recently, this twofold production has been replaced by specialized pure plantations for timber production, using a square spacing of 5 to 6 m, or walnut seed orchards (Giannini and Mercurio 1997).

Over the last decades in Europe, under the financial support of EU rules 2080/92 and of Rural Development Plan, many mixed plantations were established with walnut and other valuable broad-leaved and nurse trees (Becquey 1997, Becquey and Vidal 2006, Buresti and Mori 2006, Kelty 2006, Tani et al. 2006, Clark et al. 2008, Mohni et al. 2009). Recently in Italy, France, and North America, mixed

---

[1] Consiglio per la Ricerca e la sperimentazione in Agricoltura, Forestry Research Centre, Arezzo (CRA-SEL), Italy
[2] Associazione Arboricoltura da Legno Sostenibile per l'Economia e l'Ambiente (A.A.L.S.E.A.)
* corresponding author: francesco.pelleri@entecra.it

plantations with walnut and poplar clones have been tested both in plantation forestry (Zsuffa et al. 1977, Buresti et al. 2008a, Vidal e Becquey 2008a, Paquette et al. 2008) and in agro-forestry systems (Balandier and Dupraz 1999, Rivest et al. 2010). These experiences pointed out the opportunity of testing the simultaneous cultivation of both species, characterized by different rotations (short poplar, medium-long walnut).

In this type of plantation, named in Italy *polycyclical plantation* (after Buresti et al. 2008b and Buresti and Mori 2012), the main crop trees with the same rotation are spaced at a definitive distance, to reach the merchantable size requested by industry within a shorter cycle. The reciprocal distance between main walnut and poplar trees must allow each tree to complete its rotation without the establishment of any heavy interspecific competition. Furthermore, when interplanting distance is correct, a positive competition can arise and the fast growing tree species can influence in a beneficial way the shape of lower growth and shade tolerant trees because of their low-covering crown (Schütz 2001, Pommerening and Murphy 2004, Kelty 2006, Buresti et al. 2008a, Buresti et al. 2008b, Vidal and Becquey 2008a).

A further benefit arising from this type of plantation is economical, by allowing a better distribution of income, poplar being harvested within 8-10 years and walnut within 20-30 years (Vidal and Becquey 2008b).

Results obtained in a poplar and walnut *polycyclical plantation*, aged 9 years, are here reported. The plantation has been carried out to verify the following hypotheses:

1) opportunity to get poplars sized 30 cm DBH testing two different distances from walnut (5 and 7.4 m) without heavy interspecific competition;
2) possibility of poplar cultivation at lower costs as compared with the traditional technique, reducing tending operations, using more than one poplar clone and mixtures with N-fixing trees;
3) test poplar clones different from the I214 traditionally planted in Italy, to verify their own productivity and ability to be competitive on the wood market;
4) opportunity of harvesting additional valuable trees established as nurse trees.

Materials and Methods

Site description and plantation management

Planting operations started in February 2003 in a flood plain of the Oglio river in the Mantua Province (San Matteo alle Chiaviche). The area is characterized by a good site-index, deep, silty-sandy and moderately alkaline soils, subjected to periodical flooding. A sub-continental climate with cold winters and hot wet summers characterizes the area (mean annual temperature 13.6° C., mean annual precipitation 790 mm with maximum in autumn and minimum in winter; mean rainfall reaches 145 mm in summer with a dry period of one month only).

This mixed plantation, extended over 14 hectares, is characterized by the following main crop trees: four poplar clones (Lena, Neva, I214 e Villafranca), common walnut (*Juglans regia* L.) and hybrid walnut (*Juglans x intermedia* MJ209) (Tab. 1). The common walnut's trees have been planted initially in pairs, each tree being located at 1 m, in order to select shortly the best tree of each couple. In this way, the chance to get a good quality tree in the expected position increases (Buresti et al. 2001, 2003). Walnut and poplar main trees are planted according to a rectangular design (10 x 11 m) using alternate lines. In all, 90 poplars, 45 hybrid walnuts, and 45 common walnut couples were planted per hectare. The plantation is divided into two areas, depending on the planting design adopted (A and B). Each area is divided in monoclonal plots of about 1.5 hectare for each clone. In design A, walnut trees (both hybrid and common walnuts) are set in a row with poplar at 5 m. In design B, walnut trees are staggered to poplars by 7.4 m. Nurse trees were black alder (*Alnus glutinosa* (L.) Gaertn.) and hazel (*Corylus avellana* L.); along with these other four valuable nurse trees were introduced, i.e. narrow-leaved ash (*Fraxinus angustifolia* L.), wild service tree (*Sorbus torminalis* (L.) Cranz.), English oak (*Quercus robur* L.), and pear (*Pyrus*

Table 1 - Main characteristics of poplar clones. Modified from Facciotto et al. (2003).

Clone	origin	main characteristics
Lena	Populus deltoides Bartr.	higher productivity in comparison with I214; resistance to leaf diseases (rusts and aphids); sensitive to wind; high shrinkage and nervousness in comparison with I214.
Neva	Hybrid Populus X canadensis Monch	sensitive to spring defoliation and rusts; resistance to bronzing; bark necrosis and brown spots; higher productivity in comparison with I214; sensitive to wind
I 214	Hybrid Populus X canadensis Monch	resistance to spring defoliation; virus and brown spots; green wood density lower to Lena and Neva; resistance to wind; easy to prune
Villafranca	Populus alba L. clone	resistance to main diseases (rusts. spring defoliation. bronzing. viruses and aphids); hard to prune; slower growth in comparison with I214

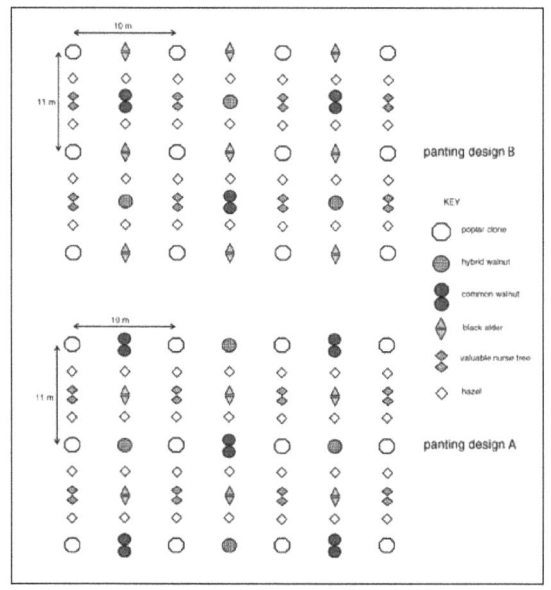

Figure 1 - The planting designs (A and B).

pyraster Burgsd.) (Fig.1). Valuable nurse trees, in addition to their own tending role, act as sort of insurance for the final outcome of the plantation, because they are able to replace any main crop tree in case of death, damage or if they do not achieve the expected development (Buresti and Mori 2004). Consequently, the valuable nurse trees must be subjected, at least for 4 to 6 years (pruning phase), to the same tending operations as the main crop trees (especially pruning).

In this plantation, the operations have been considerably reduced when compared with traditional monoclonal poplar plantations by:

1. Reduction of both fertilizers using N-fixing nurse trees, and pesticides using multi-clonal mixed plantations. These plantations allowed the reduction of spreading disease risk by poplar leaf rust (*Melampsora* spp.) or poplar leaf and shoot blight (*Venturia populina* (Vuill.) Fabric.), as already verified in other areas intensively cultivated with poplars (Coaloa and Vietto 2005).
2. Stimulation of poplar radial growth, with consequent reduction of the rotation, by planting more spaced trees that develop longer and well-lighted crowns over the whole cycle.

Over the first two years, only one treatment against wood insects' poplar-and-willow borer (*Cryptorrhynchus lapathi*) was performed; in the following two years a localized treatment to control longhorn beetle (*Saperda carcharis* L.) was carried out in the outside strips (about 20 m) to prevent the diffusion of this insect from neighbouring traditional poplar plantations. Weed control was carried out by polyethylene mulching and using both chemical and mechanical weeding during the first four years. On average two mechanical weeding and one chemical weeding per year were carried out.

No use of fertilizers or irrigations was made, whereas neighbouring traditional I214 plantations are managed intensively with one initial fertilization, 3-5 pesticide treatments per year and periodical emergency irrigations.

Selection of best trees between walnut's couples was accomplished at age 5 when trees differences in vigour and shape were already evident. Pruning of main trees and of valuable nurse trees was carried out up to the age of 6, i.e. when the farmer decided to promote walnut trees because of the good stem quality and clear boles at least up to 3 meters. Poplars were pruned up to 5.5 meters.

Field survey and data analysis

DBH surveys were performed yearly between the age of 4 and 9 (2006 to 2011). In the pruning phase (2006-2007) a sample of walnut and poplar clones only (at least 30 trees per species, clone and spacing) were considered. Since 2008, when the pruning ended, a complete survey of walnut and poplar DBH was made. Total tree and clear bole height of walnut and poplar clones were measured at ages 6 and 9 (2008 and 2011); in 2011 only 50% of poplar heights were measured. In 2011 an individual stem quality evaluation was carried out for walnut trees, ranking stems per quality classes, from A (veneer) to D (firewood) using the method set up for tree farming plantations proposed by Nosenzo et al. (2008) that assigns a quality class, valuing length, trunk axiality, knots presence and other defects. Only DBHs were measured on valuable nurse trees in 2008 and 2009.

First, the normality of distributions was evaluated by mean of Kolmogorof-Smirnov's test, then data were processed with the analysis of variance (ANOVA) comparing separately poplar clones, common walnut and hybrid walnut, but considering both planting designs (A and B). The comparison of means was performed by the Tukey's Test (HSD), with a significance of 0.05 (Statistica 2005). The walnut's distribution per stem quality classes was compared by χ^2 Pearson's test. The performances of valuable nurse trees and walnut trees were evaluated by comparing mean DBH at 2009 (age 7).

A comparison between traditional poplar plantation (Colaoa and Vietto 2005) and this new type of poplar and walnut plantation was carried out comparing the presence (1)/ absence (0) of cultural practices undertaken during the poplar rotation (10 years).

Results

Poplar

Stem diameter - Poplar clones showed different growth rates at the age of 7-8 (years 2009-10) Lena and Neva clones reached the commercial size (DBH=30 cm), whilst I214 and Villafranca clones reached the same size 1 or 2 years later (Fig. 2). In 2011, the ANOVA of DBH showed significant differences among the clones and blocks (designs) ranging from 38.0, 36.1, 31.5 and 30.1 cm respectively for Lena, Neva, I214 and Villafranca (Tab. 2).

DBH increments keep high, i.e. 3-4 cm yr^{-1}. In 2009, Lena pointed out significant higher increments as compared with all clones in both designs, differences that decreased in the following years. Less productive clones showed more steady increments while a heavy DBH increment reduction was highlighted in Lena and, to a lesser extent, in Neva. In the last two years the I214 clone showed higher DBH increments (Fig. 3).

Tree height - The ANOVA of height is significant for both years (2008 and 2011) pointing out, in 2011, a superior height growth of Lena and Neva clones that reached on average 24.4 m and 24.2 m respectively, in comparison with 20.8 m and 18.2 m of the I214 and Villafranca clones (Tab. 3). At the age of 9 all poplar clones had built up green and efficient, up to 12-18 m long, crowns in comparison with poplars cultivated as usual, where crowns are narrow and shorter because of their progressive elevation due to the heavy competition (poplar stem density = 277 tree ha^{-1}).

Table 2 - ANOVA of poplar clones DBH 2011 (9 yrs).

Poplar clone	DBH (cm)	±SD (cm)	HSD	
Lena	38.0	2.2	a	
Neva	36.1	2.7	b	
I214	31.5	3.0	c	
Villafranca	30.1	2.1	d	
ANOVA	df	MS	F	P value
Clone	3	2559.1	516.4	0.0000
design (block)	1	896.2	180.8	0.0000
clone x design	3	65.9	13.3	0.0000
error	716	5.0		

Walnut

Stem diameter - Data analysis pointed out significant differences for walnut DBH grown with different spacings, it being lower in design A compared to B. Significant variations in DBH were also found for both walnuts originated by the different intercropped poplar clones (Fig. 4 and 5). Hybrid walnut reached on average lower DBH (13.2 cm and 13.7 cm) when intercropped with Lena and Neva in comparison with I214 and Villafranca intercropping (14.9 cm and 15.4 cm) in 2011 (age of 9). Common walnut reached 12.6 and 12.9 cm when intercropped with Lena and Neva and 13.1 and 13.6 cm when intercropped with I214 and Villafranca (Tab. 4).

A reduction of DBH increment was observed for both walnuts and designs within all the intercropped poplar clones in the last years. Since 2010, significant differences among intercropping and designs were found for both walnuts. A mean reduction of DBH increment from 1.6 to 0.6 cm in Lena and from 1.5 to 0.7 cm in Neva intercropping was observed for hybrid walnut over 2009-2011, whilst a lower DBH variation of annual increment was detected with the less productive clones: from 1.9 to 1.2 cm in I214 and from 2.0 to 1.1 cm in Villafranca intercropping (Fig. 6). In common walnut DBH increment variations were less significant and only walnut intercropped with I214 in design B was significantly different. A mean reduction of common walnut DBH increment from 1.7 to 0.9 cm in Lena and from 1.8 to 1.1 cm in Neva intercropping was observed along the full observation period; whilst a DBH increment variation from 2.0 to 1.3 cm and from 1.9 to 1.2 cm was recorded, respectively in I214 and Villafranca intercropping (Fig. 7). The trend of DBH increment highlights the different specific growth pattern of both walnuts. Hybrid walnut has an early more sustained growth followed by a progressive reduction due to poplar competition, up to a radial (DBH) growth lower than common walnut (Fig. 8).

Tree height - The ANOVA showed significant differences in height with common walnut for both intercropping types in 2008 and 2011, and for planting design in 2011 only. In hybrid walnut significant differences in height were noticed for the different

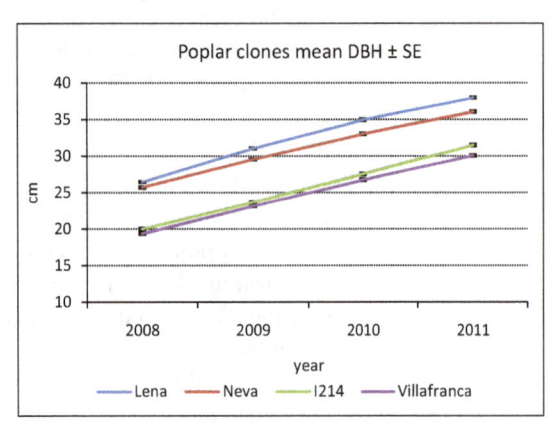

Figure 2 - Poplar clones DBH growth trend ± SE.

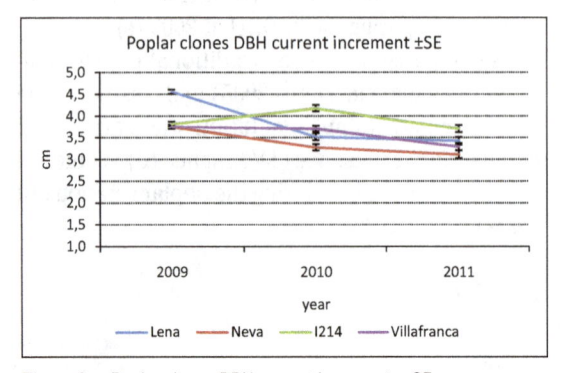

Figure 3 - Poplar clones DBH current increment ± SE.

Table 3 - ANOVA of poplar clones tree height 2008 - 2011 (6-9 yrs).

Poplar clone	h2008 (m)	±SD (m)	HSD		h2011 (m)	±SD (m)	HSD	
Lena	17.43	1.43	a		24.38	2.06	a	
Neva	16.80	1.21	b		23.16	1.35	a	
I214	14.09	1.49	c		20.75	1.30	b	
Villafranca	12.13	0.99	d		18.23	2.00	c	
ANOVA	df	MS	F	P value	df	MS	F	P value
clone	3	1111.5	780.50	0.0000	3	798.6	285.58	0.0000
design (block)	1	114.5	80.40	0.0000	1	4.1	1.48	0.2242
clone x design	3	26.3	18.50	0.0000	3	25.2	9.02	0.0000
	726	1.4			357	2.8		

intercroppings in 2008 and 2011, and for planting design in 2011 only.

The best performances of common walnut at the end of 2011 were measured in the design B with Lena and Neva intercropping (13.7 and 14.4 m) while the worst results were found in design A with I214 (11.2 m) and Villafranca (11.6 m) (Tab 5). In 2011, the best performances of hybrid walnut were recorded in design B with Lena and Neva intercropping (14.9 and 14.5 m) and the worst results were achieved in design A with I214 (13.2 m) and Villafranca (12.8 m) (Tab. 6).

Stem quality - The quality of hybrid walnut showed to be significant superior ($\chi^2 = 31.2$ p<0.001) as compared with common walnut. The 72% of hybrid trees belong to classes A and B, suitable for veneer and first quality saw-timber production, while only 53% of common walnut trees reached the same category (Tab.7). Stem pruning was carried out up to the height of 3 m. Over this height, crowns were left to grow free. The percentage of trees suitable to produce valuable assortments could increase for both walnut species at higher plantation ages, because defects due to knots and small stem curves could be overestimated in trees aged 9 years.

Valuable nurse trees

Close to the end of the pruning phase (2009), the owner decided to concentrate operations only on walnut trees showing faster growth and/or fitting the expected stem quality, neglecting valuable nurse trees. Among these, the best DBH growth performance was recorded by narrow-leaved ash followed by pear, oak, and wild service trees. Mean DBH values measured in 2009 in walnuts and valuable nurse trees are reported in Fig. 9.

Cultivation practices

The comparison between cultivation practices undertaken in the two plantation types (pure and polycyclical) during the poplar rotation are showed in Tab. 8 and 9. In the mixed walnut-poplar plantation, the absence of any irrigation and fertilization, a heavy reduction of the use of pesticides (-90%) and of mechanical weed control (-60%) was recorded. On the other hand, the chemical weed control along walnut lines, the selection between walnut couples, and the increase of pruning (+25%), were necessary in this new type of plantation.

Discussion

At the age of 9 (2011), Lena and Neva clones reached notable diameters (34.7 to 38.4 cm) but showed already evidence of an incremental reduction, whilst the I214 and Villafranca maintained a more regular growth course with lower DBH (29.1 to 33.3 cm). These growth patterns highlight on one side the early establishment of intraspecific competition within the fastest growing clones, especially Lena, and on the other side, the delayed occurrence of the same condition within the less productive clones (I214 and Villafranca).

The year 2009 showed to be the right time for

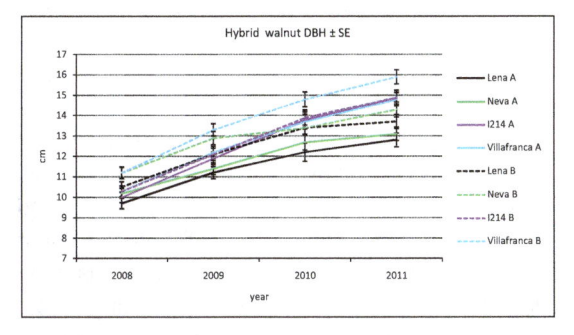

Figure 4 - Hybrid walnut DBH growth trend ± SE per intercropping and design.

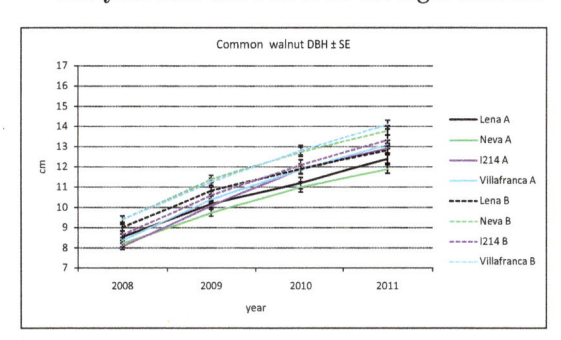

Figure 5 - Common walnut DBH growth trend ± SE per intercropping and design.

Table 4 - DBH 2011 (9 yrs) common and hybrid walnut ANOVA per planting design and intercropped poplar clones.

intercropping	design	Hybrid walnut DBH (cm)	±SD (cm)	HSD		DBH	±SD (cm)	HSD (cm)
Lena	A	12.8	1.9	c		12.39	1.35	de
Neva	A	13.1	2.4	c		11.90	1.03	e
I214	A	14.9	2	ab		12.92	0.86	bcd
Villafranca	A	14.8	2.3	ab		13.11	1.68	bcd
Lena	B	13.7	2.2	bc		12.82	1.25	cd
Neva	B	14.3	2.1	bc		13.80	1.43	ab
I214	B	14.9	2.3	ab		13.34	1.59	abc
Villafranca	B	15.9	2.2	a		14.09	1.65	a
ANOVA	df	MS	F	P value	df	MS	F	P value
design	1	50.93	10.58	0.0013	1	70.06	36.31	0.0000
intercropping	3	79.69	16.55	0.0000	3	14.91	7.73	0.0001
intercropping x design	3	5.70	1.18	0.3160	3	9.42	4.88	0.0025
error	314	4.81			314	1.93		

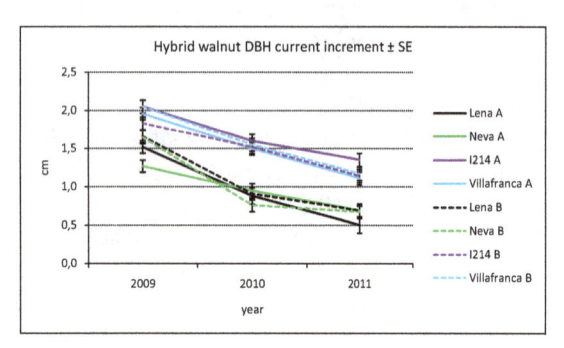

Figure 6 - Hybrid walnut DBH current increment per intercropping and design.

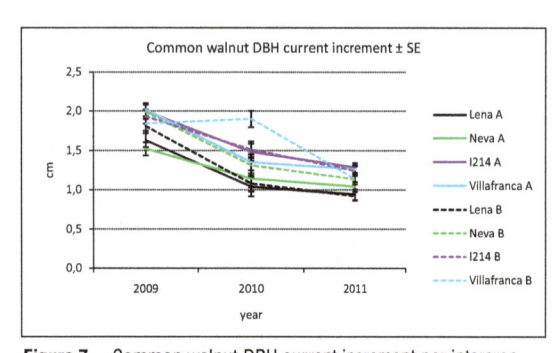

Figure 7 - Common walnut DBH current increment per intercropping and design.

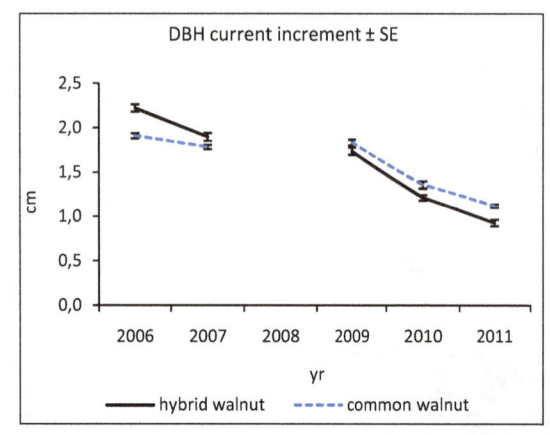

Figure 8 - Comparison between common and hybrid walnut DBH increment.

poplars harvesting, especially for the more productive clones - Lena and Neva - (Facciotto et al. 2003) already sized enough for marketing (DBH ≥30 cm), but adverse concomitant local market conditions didn't make profitable their felling to the farmer. Their maintenance for a longer time affected negatively walnut DBH increment, that is -63% compared with 2009 (hybrid walnut intercropped with Lena), -46% (common walnut) and -36% with the I214 intercropping in both walnuts spp. (Tab. 10). At the end of 2011, walnuts have still long, green functional crowns (12-18 m) and looked able to react to poplars' harvesting as compared with pure walnut plantations undergoing a late thinning, these generally showing a lower reaction (De Meo et al. 1999, Marchino and Ravagni 2007). Despite that, no data exist at now about the incremental DBH reaction of walnut following a period of heavy competition in this type of plantation. This points out the need of using larger poplar-walnut interplanting distances, especially if we are going to use both slower growing clones needing rotations of about 10 years to produce poplars sized more than 40 cm (Buresti and Mori 2013). The basic importance of defining suited planting designs and correct walnut-poplar distances is in this way underlined.

Both walnut trees (common and hybrid) showed a fast DBH growth. Hybrid walnut presented a superior stem quality in accordance with Paris et al. (2003), 72% of stems being suitable for good quality veneer and saw timber production. On the contrary, common walnut had only 53% of stems in the best quality classes (Tab. 7). Despite the worse performance in terms of growth and stem quality, the use of common walnut is more frequent since its timber is more appreciated by the Italian industry and because this species has been traditionally cultivated in Italy since the Roman period. The achievement of a good standard in the trunk quality has been determined by poplar's presence (i.e. fast growing species with low-covering crown) able to nurse, under his light canopy, walnuts characterized by

Table 5 - Tree height 2008 - 2011 (6-9 yrs) common walnut ANOVA per planting design and intercropped poplar clones.

intercropping	design	Common walnut							
		h2008 (cm)	±SD (cm)	HSD		h2011 (cm)	±SD (cm)	HSD	
Lena	A	8.26	1.06	a		13.13	1.60	a	
Neva	A	7.89	0.98	a		12.53	1.22	a	
I214	A	7.08	0.64	b		11.20	1.08	b	
Villafranca	A	7.80	1.15	a		11.63	1.50	b	
Lena	B	8.38	1.33	ab		13.71	1.63	a	
Neva	B	8.55	1.55	a		14.42	1.99	a	
I214	B	7.39	1.19	c		12.38	1.91	b	
Villafranca	B	7.74	0.89	bc		12.05	1.30	b	
ANOVA		gdl	MS	F	P value	gdl	MS	F	P value
design		1	4.86	3.86	0.0503	1	80.51	33.62	0.000
intercropping		3	16.97	13.48	0.0000	3	68.94	28.79	0.000
intercropping x design		3	1.70	1.35	0.2571	3	8.97	3.75	0.011
error		281	1.26			306	2.39		

Table 6 - Tree height 2008 - 2011 (6-9 yrs) hybrid walnut ANOVA per planting design and intercropped poplar clones.

intercropping	design	Hybrid walnut							
		h2008 (cm)	±SD (cm)	HSD		h2011 (cm)	±SD (cm)	HSD	
Lena	A	8.36	0.97	a		13.57	1.65	a	
Neva	A	8.14	1.40	a		13.69	2.03	a	
I214	A	8.11	0.75	a		13.22	1.44	a	
Villafranca	A	7.88	1.05	a		12.81	1.54	a	
Lena	B	8.57	1.40	ab		14.92	2.18	a	
Neva	B	9.08	1.38	a		14.51	2.03	ab	
I214	B	7.94	1.06	b		14.09	1.67	ab	
Villafranca	B	8.06	0.98	b		13.41	1.28	b	
ANOVA		df	MS	F	P value	df	MS	F	P value
design		1	4.87	2.97	0.0862	1	65.07	21.1	0,0000
intercropping		3	6.01	3.66	0.0131	3	20.45	6.63	0,0002
intercropping .x design	3	3.15	1.92	0.1274	3	2.06	0.67	0.5722	
error		238	1.64			307	3.08		

Table 7 - Distribution of common and hybrid walnut per stem quality classes and relative χ^2 test.

Stem quality classes	common walnut		hybrid walnut	
	n	%	n	%
A	35	10.8	63	19.4
B	137	42.4	170	52.5
C	97	30.0	71	21.9
D	54	16.7	20	6.2
tot	323	100.0	324	100.0

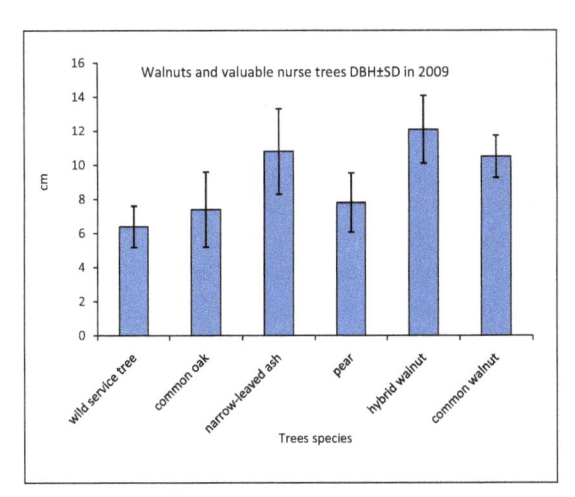

$\chi^2 = 31.19$ p=0.000

Figure 9 - Comparison among walnuts and valuable nurse trees DBH ±SD at the age of 7 (2009).

slender, cone-shaped crowns and small-sized, easily self-pruned branches (Buresti et al. 2008b). Valuable nurse trees were no longer managed (except the best phenotypes) because of the good walnut performance both in terms of growth and stem quality. That is why their cultivation (pruning) has been abandoned in this trial. In any case, according to Kelty (2006) and Buresti and Mori (2004), they can support an improvement of biodiversity and stability of mixed plantations and their intercropping is a sort of farmer's insurance for the final outcome of the plantation.

Results highlight that the concurrent cultivation of walnut and poplar species is possible and advisable because of the manifold positive outcomes: reduced environmental impact of the plantation type as compared with traditional poplar monoculture, since the trial pointed out a clear reduction of external inputs in terms of (i) fertilization (-100%), (ii) irrigation (-100%), (iii) use of pesticides (-90%) as compared with the traditional cultivation.

This was possible by using multi-clonal mixed plantations with N-fixing nurse trees that considerably reduce both the spreading of pests and diseases and also the consequent amount of chemical treatments. The use of fertilizers was cut down, too.

All these achievements are quite positive as

Table 8 - Presence (1) or absence (0) of cultural practices in traditional poplar plantations - modified from Colaoa and Vietto (2005).

Traditional poplar plantation cultural operations	1°	2°	3°	4°	5°	6°	7°	8°	9°	10°	total
mechanical weed control	1	1	1	1	1	1	1	1	1	1	10
irrigation	1	1	1	1	1	1	1	1	1	1	10
pesticide	1	1	1	1	1	1	1	1	1	1	10
chemical weed control	0	0	0	0	0	0	0	0	0	0	0
pruning	1	1	1	0	1	0	0	0	0	0	4
selection double walnut	0	0	0	0	0	0	0	0	0	0	0
fertilization	1	1	1	1	0	0	0	0	0	0	4
total interventions	5	5	5	4	4	3	3	3	3	3	38

Table 9 - Presence (1) or absence (0) of cultural practices in the studied plantation.

Polyciclical plantation cultural operations	1°	2°	3°	4°	5°	6°	7°	8°	9°	10°	total
mechanical weed control	1	1	1	1	0	0	0	0	0	0	4
irrigation	0	0	0	0	0	0	0	0	0	0	0
pesticides	0	1	0	0	0	0	0	0	0	0	1
chemical weed control	1	1	1	1	0	0	0	0	0	0	4
pruning	0	1	1	1	1	0	0	0	0	1	5
selection double walnut	0	0	0	0	1	0	0	0	0	0	1
fertilization	0	0	0	0	0	0	0	0	0	0	0
total interventions	2	4	3	3	2	0	0	0	0	1	15

Table 10 - Reduction of current DBH increments between 2009 and 2011

Intercropped clones	hybrid walnut DBH increment reduction %	common walnut DBH increment reduction %
Lena	63	46
Neva	53	37
I214	36	36
Villafranca	43	38

compared with the number of cultivation practices needed in a traditional poplar's plantations along the whole poplar rotation (Colaoa and Vietto 2005). Their occurrence was limited to less than one half in the studied plantation. The abatement of these cultivation costs has a notable economic advantage both for farmers and towards the environment.

A limitation to the use of the Lena and Neva clones is nowadays the shortage of market for these clones in Italy, in spite of their significant superior growth, the lower cultivation needs, in comparison with I214, and the suitable mechanical timber characteristics. Unfortunately, the higher shrinkage of timber in these clones decreases the value for veneer production, this determining a difficult marketing. At present, veneer industry fully prefers timber from I214 plantations (Facciotto et al. 2003). According to Vidal and Becquey (2008b), the mixture between poplar and walnut appears to be interesting from the economic point of view too. Further research trials are necessary to value the overall productivity of the two main tree species at the end of walnut rotation and the analytical evaluation of costs.

Conclusive remarks

The recent experimental trials on mixed *polycyclical plantations* (walnut, poplar and nurse trees) highlighted the notable potential of this type of plantation in flood plain areas favourable to the cultivation of both main species, providing interesting productive outcomes as well as ecological-environmental considerations. These mixed plantations, more resistant to external disturbances and less demanding in terms of energetic inputs (fertilization, pesticides, irrigation, etc.) proved to be innovative and more sustainable for poplar and walnut cultivation. It can become an advantage for the farmer, especially within the current unfavourable period for poplar timber marketing and for the quality of flood plain environment.

The experiences with this plantation design are at an early stage and more comparative tests, widespread trials and screening of results are necessary. The goals in progress are: development of suitable cultivation models with wide inter-distances walnut-poplar (≥ 7.4 m), enough to allow: (i) the early achievement of poplar's cultivation goals before heavy competition with walnut becomes established; (ii) the use of a fast growing species, i.e. poplar to nurse walnut trees provided with slender, conical-shaped crowns, and light branching.

Acknowledgements

Special thanks to Enrico Buresti, who conceived this type of plantation and to Francesco Mattioli who managed with competence and care the full plantation cycle. The authors want to thank also the anonymous referees for their helpful suggestions.

References

Balandier P. and Dupraz C. 1999 - *Growth of widely spaced trees. A case study from young agroforestry plantations in France.* Agroforestry Systems 43: 151-167.

Becquey J. 1997 - *Les noyers à bois.* Institut pour le Développement Forestier (IDF), Paris, 144 p.

Becquey J. and Vidal C. 2006 - *Quels accompagnements ligneux choisir pour les plantations de noyer?* Forêt Entreprise 170: 35-38.

Buresti E., Mori P., Ravagni S. 2001 - *Arboricoltura da legno con il ciliegio: ridurre i rischi adottando la doppia pianta.* Sherwood - Foreste ed Alberi Oggi 73: 11-16.

Buresti E., Mori P., Ravagni S. 2003 - *Quando diradare la doppia pianta. Un'esperienza con la farnia (Quercus robur L.).* Sherwood - Foreste ed Alberi Oggi 85: 21-24.

Buresti Lattes E. and Mori P. 2004 - *Ruolo delle piante, specie e tipologie d'impianto in arboricoltura.* Sherwood - Foreste ed Alberi Oggi 98: 15-18.

Buresti Lattes E. and Mori P. 2006 - *Legname di pregio e biomassa nella stessa piantagione.* Sherwood - Foreste ed alberi oggi 127: 5-10.

Buresti Lattes E., Mori P., Pelleri F., Ravagni S. 2008a – *Des peupliers et des noyers en mélange, avec des plants accompagnateurs.* Forêt – Enterprise 178: 26-30.

Buresti Lattes E., Cavalli R., Ravagni S., Zuccoli Bergomi L. 2008b - *Impianti policiclici di Arboricoltura da legno: due esempi di progettazione e utilizzazione.* Sherwood - Foreste ed Alberi Oggi 139: 37-39.

Buresti Lattes E. and Mori P. 2012 - *Piantagioni policicliche. Elementi di progettazione e collaudo.* Sherwood - Foreste ed Alberi Oggi 189: 12-16.

Clark J., Hemery G.E., Savill P. 2008 - *Early growth and form of common walnut (Juglans regia L.) in mixture with tree and shrub nurse species in southern England.* Forestry 81: 631-644.

Coaloa D., Tonetti R., Napolitano L. 2012 - *Novità della misura 221B del PSR della Lombardia: ciclo medio lungo anche per il pioppo.* Sherwood – Foreste ed Alberi Oggi 186: 23-27.

Coaloa D. and Vietto L. 2005 - *Pioppicoltura ecologicamente disciplinata.* Sherwood – Foreste e Alberi Oggi 113: 23-28.

De Meo I., Mori P., Pelleri F., Buresti E. 1999 - *Prime indicazioni sugli interventi di diradamento nelle piantagioni di Arboricoltura da Legno.* Sherwood – Foreste ed Alberi Oggi 43: 15-20.

Facciotto G., Minotta G., Zambruno G.P. 2003 - *Analisi della produttività dei cloni di pioppo Dvina, Lena e Neva.* In: Atti IV congresso Nazionale SISEF, Meridiani Forestali, Università degli Studi della Basilicata, Rifreddo (PZ), 07-10 Ottobre 2003: 151-158.

Gasparini P. and Tabacchi G. (a cura di) 2011 - *L'Inventario Nazionale delle Foreste e dei serbatoi forestali di Carbonio INFC 2005. Secondo inventario forestale nazionale italiano. Metodi e risultati.* Ministero delle Politiche Agricole, Alimentari e Forestali; Corpo Forestale dello Stato, Consiglio per la Ricerca e la sperimentazione in Agricoltura, Unità di ricerca per il Monitoraggio e la Pianificazione Forestale. Edagricole - Il Sole 24 ore, Bologna, 653 p.

Giannini R. and Mercurio R. 1997 - *Il Noce comune per la produzione legnosa.* Avenue media, Bologna, 302 p.

Kelty M.J. 2006 - *The role of species mixture in plantation forestry.* Forest Ecology and Management 233: 195-204.

Marchino L. and Ravagni S. 2007 - *Effetti del diradamento in impianti puri di noce.* Sherwood - Foreste ed alberi oggi 139: 40-41.

Minotta G. 1990 - *La cultura del noce da frutto e a duplice attitudine produttiva in collina e montagna.* Monti e boschi 41 (1): 27-33.

Minotta G. 1992 - *La nocicoltura a duplice attitudine.* In: Atti del 12° convegno pomologico "La coltura del noce", Caserta 4 luglio: 140-150.

Mohni C., Pelleri F., Hemery G.E. 2009 - *The modern silviculture of Juglans regia L.: a literature review.* Die Bodenkultur 60 (3): 19-32.

Nervo G., Magni M., Gazza F., Petrucci B., Carovigno R., Berti S., Ducci P. 2007 - *Libro bianco pioppicoltura.* Commissione nazionale per il pioppo, 143 p.

Nosenzo A., Berretti R., Boetto G. 2008 - *Piantagioni da legno. Valutazione degli assortimenti ritraibili.* Sherwood - Foreste e Alberi Oggi 145: 15-19.

Paquette A., Messier C., Périnet P., Cogliastro A. 2008 - *Simulating light availability under different hybrid poplar clones in a mixed intensive plantation system.* Forest Science 54 (5): 481-489.

Paris P., Ducci F. Brugnoli E., De Rogatis A., Fady B., Malvolti M.E., Olimpieri G., Pisanelli A., Proietti R., Scartazza A. Cannata F. 2003 - *Primi risultati di prove comparative di accessioni di noce da legno.* In: Atti del III congresso Nazionale SISEF, Alberi e foreste per il nuovo millennio, Viterbo: 181-188.

Pommerening A. and Murphy ST. 2004 - *A review of the history, definition and methods of continuous cover forestry with special attention to afforestation and restocking.* Forestry 77 (1): 27-44.

Ravagni S. and Buresti E. 2003 - *Piantagioni con pioppo e noce comune: accrescimento e sviluppo dopo i primi anni.* Sherwood - Foreste ed alberi oggi 94: 19-24.

Rivest D., Oliver A., Gordon A.M. 2010 - *Hardwood intercropping system: combining wood and agricultural production while delivering environmental services.* [Online] URL:http://www.agrireseau.qc.ca/Agroforesterie/documents/Hardwood_Intercropping_Systems_(1Mo).pdf [2013]

Schütz J.P. 2001 - *Opportunities and strategies of transforming regular forest to irregular forest.* Forest Ecology and Management 151: 87-94.

Tani A., Maltoni A., Mariotti B., Buresti Lattes E. 2006 - *Gli impianti da legno di Juglans regia realizzati nell'area mineraria di S. Barbara (AR). Valutazione dell'effetto di piante azotofissatrici accessorie.* Forest@ 3(4): 588-597.

Statistica 2005 - *Statistica 7.1 version*, StatSoft Edition, Tulsa, OK, USA

Vidal C. and Becquey J. 2008a - *Enseignements de deux plantations mélangée de peuplier I214 et de noyer hybride.* Institut pour le Développement Forestier – Forêt Entreprise 178: 31-36.

Vidal C. and Becquey J. 2008b - *Le mélange peuplier-noyer est il économiquement intéressant?* Institut pour le Développement Forestier – Forêt Entreprise 178: 37-39.

Zsuffa L., Anderson H. W., Jaciw P. 1977 - *Trends and Prospects in Ontario's Poplar Plantation Management.* The Forestry Chronicle 53(4): 195-200.

Comparison between people perceptions and preferences towards forest stand characteristics in Italy and Ukraine

Oksana Pelyukh[1*], Alessandro Paletto[2], Lyudmyla Zahvoyska[1]

Abstract - Understanding people's perceptions and preferences towards forest stand characteristics can bring many benefits to forest managers in the short term. This study aims to identify and compare people's perception and preferences of forest stand characteristics in Trentino province (Italy) and Rakhiv region (Ukraine). These regions were chosen as study areas for two main reasons: both are in mountain areas and local communities are strictly dependent on the forest resource. Data were collected through a questionnaire administered to a sample of local people. The collected data were statistically analysed to highlight the preferred type of forests related to different stand characteristics. The results of comparative analysis confirmed the importance of socio-demographic characteristics in shaping respondents' preferences. The results show that respondents in both case studies prefer mixed forests with a random distribution of trees with different diameter sizes. However, respondents from Trentino province prefer open forests, while respondents from Rakhiv region prefer closed one. The present study increased the level of knowledge about people's preferences in Italy and Ukraine for different forest stand characteristics. This information can be used by decision makers (forest managers and planners) to improve the recreational attractiveness of forest stands.

Keywords - analysis of perceptions; participatory forest management; questionnaire survey; Trentino province (Italy); Rakhiv region (Ukraine).

Introduction

In the last century, the multifaceted phenomenon of climate change and increasing human pressure on natural resources questioned previous forest management paradigms and now it requires holistic and critical thinking and decision-making in actions (Rockström et al. 2009, Steffen et al. 2015, Waters et al. 2016). In conditions of increasing likelihood and impact of environmental risks (e.g., extreme weather events, failure of climate-change mitigation and adaptation, major biodiversity loss and ecosystem collapse, major natural disasters) (The Global Risks Report, 2018), the adaptive complexity in forest management and silviculture (Fahey et al. 2018) has become an objective to mitigate, adapt and promote a forest ecosystem resilience to perturbations. The adaptive complexity in silviculture has coincided with a recognition among scientists and practitioners of the necessity of applying a multi-functional forest management planning (Paletto et al. 2012a) based on public participation (Cantiani 2012, Paletto et al. 2015, Pelyukh et al. 2018) because it can increase the social acceptance of the decisions and reduce conflicts among forest users.

In this context, it is important to understand and analyze people's perceptions and preferences, and local knowledge to support decision makers in the sustainable forest management and maintenance of forest resources use in an effective way (Lewis and Sheppard 2005, Šišák 2011, Zahvoyska 2014, Nijnik et al. 2017). People's preferences for forest stand characteristics can be defined as the degree to which a person prefers a feature rather than other features (Sheppard and Meitner 2005). In the last two decades, some studies have provided insight into individual values towards main forests stand characteristics, such as tree species composition, horizontal and vertical stand structure, canopy cover and deadwood distribution (Tahvanainen et al. 2001, Tyrväinen et al. 2003, Edwards et al. 2012, Paletto et al. 2013, Jankovska et al. 2014, Pastorella et al. 2014, Pelyukh and Zahvoyska 2018).

Moreover, being aware of people's perceptions and preferences regarding the forest stand characteristics is important for designing and implementing management decisions (Jensen and Koch 1998, Lee 2001, Cantiani et al. 2002, Heer et al. 2003, Edwards et al 2012, Zahvoyska and Bas 2013). This aspect is particularly significant in fragile mountain areas characterized by a strong relationship between society and natural resources such as the Alps and the Carpathians. The Italian Alps and the Ukrainian Carpathians are characterized by a strong link between local communities and forests (Notaro and Paletto 2011, Soloviy and Melnykovych 2014, Melnykovych et al. 2018).

Understanding the people's beliefs and perceptions about forest stand characteristics is a key factor in the success and attractiveness of planned activities (Mill et al. 2007, Zahvoyska et al. 2017). Given these considerations, the aim of the present study is to increase knowledge about people's preferences for different forest stand characteristics to overcome the current knowledge gap and provide key information for decision makers which could help in increasing recreational attractiveness of forest stands.

Materials and methods

Study area

People's perceptions and preferences regarding forest stand characteristics were investigated in two study areas (Fig. 1): Trentino province (Italy) and Rakhiv region (Ukraine). These regions were chosen as study areas for two reasons: both are located in mountain areas and local communities are strictly dependent on the forest goods and ecosystem services. The Trentino province (46° 04' 00"N; 11° 07' 00'E) - located in the Italian Alps (North-East of Italy) - has a population of 539,175 inhabitants and a total land area of 6,207 km^2 (density of 86.9 inh./km^2). The altitude of Trentino is between 65 m and more than 3000 m a.s.l. with around 70.0% of total land area located above 1500 m a.s.l. The main town in this Italian province is the Trento municipality characterized by a population of 114,236 inhabitants (density of 723 inh./km^2) divided into 12 districts. In

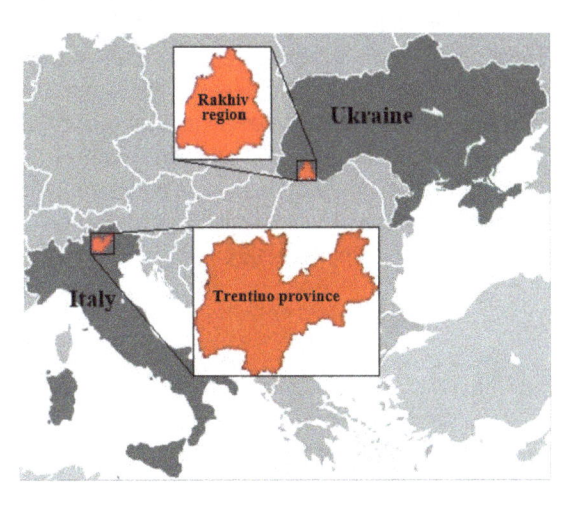

Figure 1 - The geographical location of study areas.

the Trentino province, the forest area covers 63.0% of total land area (390,463 ha) and most forests are public (76.0%), while private forests cover the remaining 24.0%. The main forest types are Norway spruce (*Picea abies* (L.) Karst.) forests with 32.0% of forest area, followed by European beech (*Fagus sylvatica* L.) forests with 14.0% and European larch (*Larix decidua* Mill.) forests with 13.0% (Odasso et al. 2018). The forest management is based on the close-to-nature principles and all public and common forests are managed through a forest unit management plan. The total growing stock is estimated in 60,000,000 m^3 of which 475,392 m^3 yr^{-1} are harvested annually (around 50% of annual volume increment) (Gandolfo and Comin 2017).

Rakhiv region (48° 3' 24.72"; 24° 11' 48.75") is located in the South-East of the Transcarpathian oblast, in the Ukrainian Carpathians. The altitude of Rakhiv region is between 500 m and 2,061 m a.s.l., the climate is temperate-continental in the lower parts, cold and wet in the upper ones. Rakhiv region occupies 1,892 km^2, with a population of 93,053 inhabitants (population density of 49 inh./km^2). Population of Rakhiv region mostly live in the rural area (57.8% of the total) and characterized by a high economic and socio-cultural dependence on forest resources. Forests in the Rakhiv region cover 125,800 ha (66.5%) represented mainly by highly productive stands of Norway spruce (*Picea abies* (L.) Karst.), European beech (*Fagus sylvatica* L.), Silver fir (*Abies alba* Mill.) and in mixture with valuable species such as Sycamore maple (*Acer pseudoplatanus L.*), Elm (*Ulmus glabra* Huds.), European ash (*Fraxinus excelsior* L.) and others (Oliynyk et al. 2015). 78.5% of the total forest area belongs to different categories of protection zones and 21.5% belongs to the commercial forest category. The forest management is based on the close-to-nature principles although clear cutting is also carried out. All forests in the territory of Rakhiv region are managed through a forest unit management plan which is renewed every 10 years. The average growing stock of forests is 370 m^3 ha^{-1}. The current annual increment of Rakhiv region forests is 6.0 m^3 ha^{-1} yr^{-1}.

Survey

In this study, people's perceptions and preferences towards forest stand characteristics were collected through a questionnaire survey. A structured questionnaire was administered to a sample of the population in both study areas.

The questionnaire was structured in 10 questions and divided into two thematic sessions. The questionnaire was divided into thematic sections in order to avoid the fatigue of respondents (Nielsen et al. 2007). The first thematic session focused on the personal information of respondents such as gender, age, level of education, place of residence (location). The second thematic session dealt with people's perceptions regarding forest stand characteristics as well as the recreational attractiveness of

Table 1 - Survey questions about forest stand characteristics.

Question	Type of question	Answer option
1. What kind of tree species do you prefer in a forest?	Single choice question	1. Broadleaf forest with less than 20.0% conifer trees 2. Conifer forest with less than 20.0% broadleaf trees 3. Mixed forest
2. Which kind of forest structure do you prefer?	Single choice question	1. Regular distribution of trees in the space; trees with similar diameters and heights 2. Random distribution of trees in the space; trees with similar diameters and heights 3. Random distribution of trees in the space; trees with a variety of diameters and heights
3. Do you prefer open or closed forest?	Single choice question	1. Open forest (10.0–40.0% canopy cover) 2. Closed forest (more than 40% canopy cover)
4. In your opinion, what kind of recreational resources do you find important in a forest?	Specifying level of importance using 10-point Likert scale (1 = very low importance, 10 = very high importance)	1. Paths 2. Picnic benches and tables and barbecues 3. Fitness trails and sports equipment 4. Panoramic views 5. Food vendors 6. Unspoiled nature 7. Parking areas 8. Places of historical and religious interest
5. What goods and services do you look for in a forest?	Specifying level of importance using (1 = very low importance, 10 = very high importance)	1. Hiking and trekking 2. Hunting activities 3. Sporting activities 4. Cultural heritage 5. Relaxation 6. Landscape contemplation 7. Naturalness 8. Timber and firewood harvesting 9. Harvesting of nonwood forest products (edible nuts, berries, fruits, mushrooms, herbs, spices and condiments, aromatic plants)

a forest.

People's perceptions of forest stand characteristics were tested considering three main macro-characteristics: tree species composition, forest structure and canopy openness (Tab. 1).

The following two questions focused on perceptions of recreational infrastructures in a forest and forest goods and services. To rate the importance, we proposed our respondents to use a 10-point Likert scale format (from 1 = very low to 10 = very high value) (Likert 1932). All questions were short, simple and realistic to minimize the time needed to fill in the questionnaire and thus motivate respondents to do so.

The survey was focused on local people (residents) because its main objective was to investigate the preferences of individuals belonging to the same community and living in the mountain area. Therefore, tourists were not considered in this study.

In the Trentino province, the questionnaire was administered to a sample of residents of the Trento municipality. The respondents were asked to return the completed questionnaire within six weeks and were given three options - return by mail, hand deliver to a prearranged collection center, or have collected (by appointment) by survey staff - to maximize the number of completed questionnaires. The questionnaire was administered in the Trentino province from November 2005 to June 2006 (8 months).

In the Rakhiv region, the questionnaire was administered to a sample of respondents in the period from 16 to 30 April 2018 (two weeks). The sample of respondents was sized considering the main social-demographic characteristics of the Rakhiv region such as the gender, age, residence. The questionnaire was administered face-to-face to respondents by a single interviewer. This administration system was chosen because the face-to-face administration could provide a higher response rate, higher quality of data acquired and a better opportunity to explain the questions unclear to respondents (De Leeuw 1992, Goyder 1985).

Data analysis

The collected data were statistically processed by study areas considering the following variables: gender, age, level of education, location, and study area. To test the differences among the groups the χ^2 test was used. The data collected using Likert scale response format were statistically compared using the Kruskal-Wallis and Mann–Whitney non-parametric tests to highlight the influence of socio-demographic characteristics of respondents on their answers.

The non-parametric Kruskal-Wallis test assesses for statistically significant differences on a contin-

Table 2 - Socio-demographic characteristics of respondents in the two study areas.

Characteristics	Trentino province (Italy)		Rakhiv region (Ukraine)		Total	
	n	%	n	%	n	%
Gender:	346		308		654	58.0
Male	232	67.0	147	47.7	379	42.0
Female	114	33.0	161	52.3	275	
Age:	344		308		652	
18-35 years old	48	14.0	57	18.5	105	16.1
36-55 years old	139	40.4	123	39.9	262	40.2
56-75 years old	120	34.9	99	32.2	219	33.6
>75 years old	37	10.7	29	9.4	66	10.1
Level of education:	341		308		649	
None	4	1.2	2	0.7	6	0.9
Elementary school	109	32.0	22	7.1	131	20.2
High school	158	46.3	123	39.9	281	43.3
University or post-University degree	70	20.5	161	52.3	231	35.6
Residence:	318		308		626	
Urban area	244	76.7	122	39.6	366	58.5
Rural area	74	23.3	186	60.4	260	41.5

uous dependent variable by a grouping by values of the independent variable (with three or more groups). In this research, the non-parametric Kruskal-Wallis test was applied to determine the statistically significant differences based on respondents' age and level of education.

The Mann-Whitney U test is used to compare differences between two independent groups when the dependent variable is either ordinal or continuous, but not normally distributed. In this study, the non-parametric Mann-Whitney U test was used to determine the statistically significant differences by gender and location.

All statistical analysis of collected data was carried out using XLStat 2012.

Results

A total of 654 questionnaires were collected in the two study areas: 346 in the Trentino province and 308 in the Rakhiv region. The response rate was very different in the two study areas due to the administration system adopted: 100% in Rakhiv region and 35% in Trentino province. The main socio-demographic characteristics of the respondents by study areas are reported in Table 2.

Regarding the gender, 379 respondents are men (58.0%) and 275 respondents are women (42.0%). The percentage of women in the Rakhiv region is higher than the one in the Trentino province (52.3% vs. 33.0%). Most respondents from Trentino province live in the urban area of Trento municipality

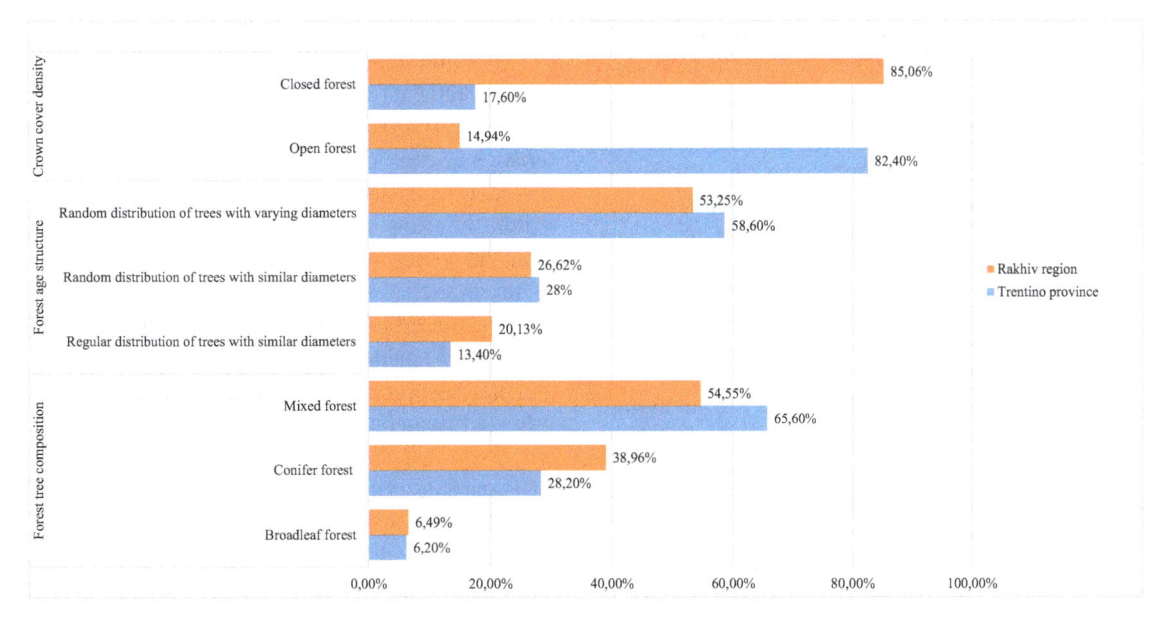

Figure 2 - Perception of forest stand characteristics by respondents from Trentino province and Rakhiv region respondents.

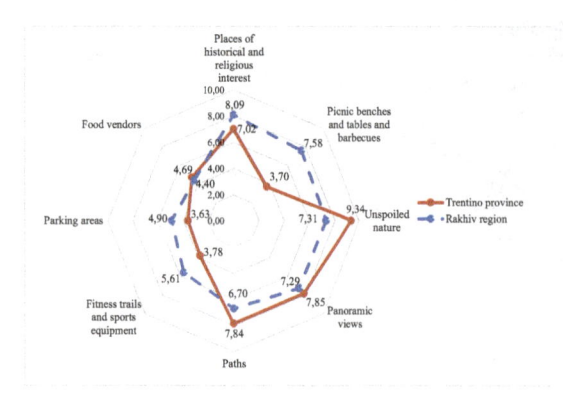

Figure 3 - Stated preferences concerning importance of recreational resources in forests by Trentino province and Rakhiv region respondents
Note. Values present the mean of scores on a 10-point scale.

(76.7%), while the most respondents from Rakhiv region (60.4%) live in the rural area. The results concerning age of respondents show that the mean values are close for the two study areas. In both case studies, the majority of respondents have an age between 36 and 55 years old (40.4% in Trentino province and 39.9% in Rakhiv region).

The results concerning the level of education indicate a quite high degree in both study areas: 46.3% in Trentino province and 39.9% in Rakhiv region have a high school diploma and 20.5% in Trentino province and 52.3% in Rakhiv region have a university or post-university degree.

In both study areas, respondents assigned a higher preference for mixed forests (65.6% in Trentino province and 54.6% in Rakhiv region) (Fig. 2). In addition, women showed a greater preference for mixed forests (71.6% in Trentino province and 57.8% in Rakhiv region) than men (62.7% in Trento province and 51.0% in Rakhiv region). However, the χ^2 test shows no statistical differences between male and female in both case studies.

Investigating people's preferences for tree species composition, the results show that Italian young people (18–35 years old) preferred broadleaf forests (14.6%) more than other age groups (5.9% for ages 36–55, 3.4% for ages 56–75, and 5.4% for ages over 75), while Ukrainian young people showed the smallest preference regarding these forests (3.5%). The highest preference for broadleaf forests in Rakhiv region was expressed by respondents 36-55 years old (13.8%). Interesting that in both regions elder people preferred conifer forests (43.2% in Trentino province and 55.2% in Rakhiv region). A statistically significant difference for tree species composition among the age groups was observed in both Trentino province (χ^2 test: $p=0.003$, $\alpha=0.05$) and Rakhiv region (χ^2 test: $p=0.001$, $\alpha=0.05$).

The majority of the respondents expressed a preference for a random distribution of trees in the space with varying diameters (58.6% in Trentino province and 53.2% in Rakhiv region) (Fig. 2). Again, for both regions a statistically significant difference between men and women was identified, with women showing a stronger preference for uneven-aged forests in Trentino province (χ^2 test: $p=0.045$, $\alpha=0.05$) and Rakhiv region (χ^2 test: $p=0.005$, $\alpha=0.05$).

The results concerning the preference regarding open vs. closed forest (canopy openness) show that Italian respondents (82.4%) prefer open forest (less than 40.0% of canopy cover), while Ukrainian respondents (85.1%) prefer closed forest (more than 40.0% of canopy cover) (Fig. 2). Taking into consideration the gender, the results also show a great difference between the two study areas. Italian women show an even higher preference for open forests than men (χ^2 test: $p=0.048$, $\alpha=0.05$). Ukrainian men prefer closed forest higher than women (χ^2 test: $p=0.344$, $\alpha=0.05$). The χ^2 test showed no statistically significant differences concerning canopy openness preference for age and location.

According to respondents' assessment of recreation infrastructures using a 10-point Likert scale, the most important recreational aspects for residents in the Trentino province are unspoiled nature (mean=9.34), panoramic views (7.85), and paths (7.85), while for residents in the Rakhiv region the most important recreational aspects are places of historical and religious interest (mean=8.09), picnic benches/tables and barbecue areas (7.58), and unspoiled nature (7.31) (Fig.3).

Respondents from both study areas indicate that the less important aspects are: fitness trails and sports equipment (mean value of 3.78 in Trentino province and of 5.61 in Rakhiv region), parking areas (3.63 and 4.90) and food vendors (4.69 and 4.40). Observing the data by socio-demographic characteristics of respondents, the results concerning the recreational infrastructures show small differences within the same study area (Tab. 3). In Rakhiv region, the people living in rural areas assigned a higher importance to unspoiled nature rather than people living in urban areas. In the Trentino province, paths are considered more important by women and older people (56-75 years old, and more than 75 years old) than by men and young people.

Regarding goods and ecosystem services provided by forests to society, respondents in Trentino province ranked naturalness (mean=8.86), hiking and relaxation (8.84 each) as the most important, while respondents in Rakhiv region assigned the highest values to relaxation (mean=8.69), cultural

Table 3 - Top three highly preferred recreational resources in forests for respondents from the two study areas (mean value).

	Characteristics	Trentino province (Italy)	Rakhiv region (Ukraine)
Gender:	Male	Unspoiled nature (9.35) Food vendors (7.87) Paths (7.55)	Historical and religious interest (8.20) Picnic benches, table and barbecues (7.39) Unspoiled nature (7.34)
	Female	Unspoiled nature (9.33) Paths (8.40) Food vendors (7.81)	Historical and religious interest (8.09) Picnic benches, table and barbecues (7.58) Unspoiled nature (7.31)
Age:	18-35 years old	Unspoiled nature (9.50) Food vendors (7.67) Paths (7.00)	Historical and religious interest (8.08) Picnic benches, table and barbecues (7.6) Unspoiled nature (7.31)
	36-55 years old	Unspoiled nature (9.37) Food vendors (8.04) Paths (7.72)	Historical and religious interest (8.09) Picnic benches, table and barbecues (7.58) Unspoiled nature (7.31)
	56-75 years old	Unspoiled nature (9.22) Paths (8.23) Food vendors (7.68)	Historical and religious interest (8.09) Picnic benches, table and barbecues (7.58) Unspoiled nature (7.31)
	>75 years old	Unspoiled nature (9.42) Paths (8.28) Food vendors (8.00)	Historical and religious interest (8.09) Picnic benches, table and barbecues (7.57) Unspoiled nature (7.30)
Level of education:	None	Unspoiled nature (10.00) Food vendors (9.75) Paths (8.50)	Historical and religious interest (7.5) Parking area (6.5) Unspoiled nature and food vendors (5.5)
	Elementary school	Unspoiled nature (9.24) Paths (8.04) Food vendors (7.76)	Historical and religious interest (8.64) Unspoiled nature (7.64) Panoramic view (6.82)
	High school	Unspoiled nature (9.43) Paths (7.82) Food vendors (7.81)	Historical and religious interest (7.3) Picnic benches, table and barbecues (7.8) Unspoiled nature (7.31)
	University or post-University degree	Unspoiled nature (9.26) Food vendors (7.91) Paths (7.44)	Picnic benches, table and barbecues (7.63) Panoramic view (7.47) Unspoiled nature (7.29)
Residence:	Urban area	Unspoiled nature (9.33) Paths (7.87) Food vendors (7.81)	Historical and religious interest (8.09) Picnic benches, table and barbecues (7.58) Unspoiled nature (7.31)
	Rural area	Unspoiled nature (9.27) Food vendors (8.17) Paths (7.99)	Historical and religious interest (8.28) Unspoiled nature (7.47) Picnic benches, table and barbecues (7.40)

heritage (8.20) and harvesting of non-wood forest products (8.11) (Fig. 4).

Respondents from Trentino province estimated the importance of harvesting of timber, firewood, and nonwood forest products, sporting and hunting activities lower than respondents from Rakhiv region. The least important for respondents from both study areas is hunting activities (mean=0.66 and 4.49).

Observing the data by socio-demographic characteristics of respondents, the results concerning the ecosystem services provided by forests show small differences within the same study area (Tab. 4). The non-parametric Kruskal–Wallis test found no statistically significant differences regarding age in both case studies. Concerning the level of education, a statistically significant difference was found for naturalness ($p=0.009$, $\alpha=0.01$) and nonwood forest products ($p=0.001$, $\alpha=0.01$) in the Trentino province. In the Rakhiv region a statistically significant difference was found for the level of education with regard to three forest goods and services: sporting activities ($p<0.0001$, $\alpha=0.01$), cultural heritage ($p=0.002$, $\alpha=0.01$) and landscape contemplation ($p=0.004$, $\alpha=0.01$). Results in both study areas show that unlike those with a tertiary education, people with lower levels of education assigned a higher value to all services.

The non-parametric Mann–Whitney test found a statistically significant difference for gender in answers of Trentino province respondents: women expressed a preference for hiking and trekking ($p=0.002$, $\alpha=0.01$); men expressed one for hunting activities ($p=0.000$, $\alpha=0.01$) and in Rakhiv region: men expressed a preference for hiking and trekking ($p=0.003$, $\alpha=0.01$), hunting activities ($p=<0.0001$, $\alpha=0.01$), relaxation ($p=<0.0001$, $\alpha=0.01$) and landscape contemplation ($p=<0.0001$, $\alpha=0.01$).

With regard to the location, a statistically significant difference was observed in answers of Rakhiv region respondents: rural inhabitants preferred forest goods and services related to the direct use such as hunting activities ($p=0.005$, $\alpha=0.01$), cultural heritage ($p=0.002$, $\alpha=0.01$), timber and firewood harvesting ($p<0.0001$, $\alpha=0.01$) and harvesting of non-wood forest products ($p<0.0001$, $\alpha=0.01$).

Table 4 - Top three highly preferred forest goods and services for respondents from the two study areas (mean value).

	Characteristics	Trentino province (Italy)	Rakhiv region (Ukraine)
Gender:	Male	Naturalness (8.79) Relaxation (8.73) Landscape contemplation (8.66)	Relaxation (8.69) Cultural heritage (8.20) Nonwood forest products (8.13)
	Female	Hiking and trekking (9.29) Naturalness (9.29) Relaxation (9.07)	Relaxation (8.64) Nonwood forest products (8.38) Cultural heritage (8.33)
Age:	18-35 years old	Landscape contemplation (8.96) Naturalness (8.83) Relaxation (8.75)	Relaxation (8.68) Cultural heritage (8.19) Nonwood forest products (8.15)
	36-55 years old	Naturalness (8.81) Landscape contemplation (8.76) Relaxation (8.74)	Relaxation (8.68) Cultural heritage (8.19) Nonwood forest products (8.13)
	56-75 years old	Naturalness (9.19) Relaxation (9.10) Hiking and trekking (9.05)	Relaxation (8.68) Cultural heritage (8.19) Nonwood forest products (8.12)
	>75 years old	Naturalness (8.97) Hiking and trekking (8.73) Landscape contemplation (8.67)	Relaxation (8.69) Cultural heritage (8.20) Nonwood forest products (8.14)
Level of education:	None	Landscape contemplation (9.00) Nonwood forest products (8.25) Naturalness (8.00)	Cultural heritage (8.50) Relaxation (8.00) Timber and firewood harvesting (7.50)
	Elementary school	Naturalness (9.04) Hiking and trekking (8.87) Landscape contemplation (8.82)	Relaxation (9.45) Cultural heritage (9.32) Landscape contemplation (8.86)
	High school	Naturalness (9.09) Relaxation (9.00) Hiking and trekking (8.92)	Relaxation (8.61) Nonwood forest products (8.40) Cultural heritage (8.13)
	University or post-University degree	Hiking and trekking (8.71) Relaxation (8.64) Naturalness (8.56)	Relaxation (8.64) Nonwood forest products (7.88) Cultural heritage (8.1)
Residence:	Urban area	Hiking and trekking (8.94) Naturalness (8.94) Relaxation (8.81)	Relaxation (8.68) Cultural heritage (8.19) Nonwood forest products (8.13)
	Rural area	Landscape contemplation (8.93) Naturalness (8.92) Relaxation (8.85)	Nonwood forest products (8.78) Relaxation (8.76) Cultural heritage (8.45)

Discussion

In the international literature, studies on people's preferences towards tree species composition, conducted in different cultural and environmental contexts, show a high heterogeneity in the preferences. However, a common point for all studies is that European people prefers mixed forests (Gundersen and Frivold 2008, Paletto et al. 2013, Pastorella et al. 2014, Giergiczny et al. 2015, Grilli et al. 2016, Filyushkina et al. 2017) and willingness to pay for visiting mixed forests is higher compared to pure conifer forests or broadleaf forests (Grilli et al. 2014). The results of our study confirm that people from both study areas prefer mixed forests.

Ribe (1989), Gundersen and Frivold (2008), Tahvanainen et al. (2011), Edwards et al. (2012) showed that forest age structure is an important forest characteristic. Gundersen and Frivold (2008), analyzing the results of 53 surveys of the Finnish, Swedish, and Norwegian residents' preferences towards forest landscapes, found that the tree size (diameter and height) is an important forest stand characteristic too. Edwards et al. (2012) investigating public opinions on the forest stand characteristics revealed that the most important characteristic for choosing a resting place was the size of trees, and therefore their age: respondents prefer old-growth forests with few trees.

The results of the present survey also show that respondents from both study areas prefer the random distribution of trees in the space with different tree size. These results are in accord with recent studies indicating that respondents prefer uneven-age forests than even-age ones (Nielsen et al. 2007, Filyushkina et al. 2017).

The canopy openness affects the recreational attractiveness in forests. Closed forests have a low recreational value for respondents due to the low possibility for visual and physical penetration of the forest stand. This is confirmed by Ribe (1989), who believes that the low recreational attractiveness of young forests is due to their high stand density. The semi-open forest provides a better visual penetration and sense of safety than high dense forests (Heyman 2012, Kaplan and Kaplan 1989). Comparison of the results from the two study areas shows that closed forests are preferred by respondents from Rakhiv region more than by respondents from Trentino province. This result may be explained by the fact that in the Ukrainian Carpathians, illegal cutting is frequent (Soloviy et al. 2011). Therefore,

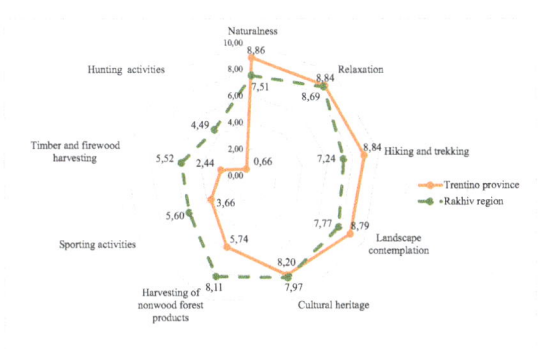

Figure 4 - Stated preferences concerning importance of forest goods and services by Trentino province and Rakhiv region respondents
Note. Values present the mean of scores on a 10-point scale.

local people often associated the low forest stand density with illegal actions or overharvesting.

Many studies have highlighted that people's preferences towards forest stand characteristics depend on many variables, which are partly shaped by the influence of cultural, regional and socio-economic factors (Ribe 1989, Gobster 1999). People's preferences can be influenced by affiliation to certain social groups (Lindhagen 1996, Misgav 2000, Roovers et al. 2002, Tyrväinen et al. 2003), age (Jensen 1999, Kaplan and Kaplan 1989), gender (Tyrväinen et al. 2003), recreational activity (Ribe 1989, Lindhagen 1996, Roovers et al. 2002. Tyrväinen et al. 2003). Ecological knowledge occupies an important position among factors that affect people's preferences. Psychological research (Kaplan and Kaplan 1989, Jensen 1993, Gobster 1999, Daniel 2001, Carlson 2001) confirms that people with a sufficient level of knowledge about forest ecosystems - people with higher education, people who often visit the forest or people who take an active part in forest management - are more likely to give higher preferences to those forest stand characteristics that will characterize it as a natural one. We found that the main factors that influence people's preferences are gender and age, while the level of education, and place of residence have a secondary importance in explaining the different perceptions.

In addition, comparison of the results from the two study areas shows that gender is an important factor that influences people's perception. In both cases, women prefer forests with the highest level of naturalness (mixed forest with uneven-aged structure). These results are in line with those of previous studies (Brown and Reed 2000, Buckingham-Hatfield 2000, Tarrant and Cordell 2002, Kumar and Kant 2007, Paletto et al. 2012b), which investigate relationship between gender and nature (including forest value) and confirm that women prefer environmental and aesthetic values while for men economic and recreational values of forest are more important.

All forest goods and services are highly appreciated by male and female of both regions. This may be due to the strong relationship that exist between local communities and forest resources in Italian Alps (Notaro and Paletto 2011) and Ukrainian Carpathians (Soloviy and Melnykovych 2014).

The majority of Rakhiv region population lives in rural areas; therefore, a special role in their wellbeing has harvesting of firewood and non-wood forest products (Soloviy and Melnykovych 2014, ENPI EAST FLEG II 2015, Melnykovych et al. 2018). Probably, for this reason the respondents from Rakhiv region assessed these groups of forest goods and services much higher than respondents from Trentino province.

Conclusion

Our study shows preliminary results about people's preferences towards forest stand characteristics in two mountain areas in Italy and Ukraine. In the future steps, the sample will be increased in both case studies in order to have a balanced number of respondents for each socio-demographic characteristic (gender, age, level of education and residence). Currently, a weakness of the sample is that most of the Trentino respondents live in the urban areas, while most of Rakhiv region respondents live in the rural areas.

In summary, the results of this survey show that people from the Trentino province prefer open mixed forests with an irregular structure, while people from the Rakhiv region prefer closed mixed forests with an irregular structure. In addition, forests with places of historical and religious value have a high importance for Ukrainian respondents.

People from both study areas like to have facilities in the forests, but at the same time would like these forests to be little frequented by other visitors, to have a greater feeling of forest naturalness. Our study also confirms previous findings and contributes additional evidence that suggests the importance of socio-demographic characteristics in shaping respondents' preferences. A statistically significant difference concerning tree species composition was identified in both regions for different age groups: while younger people prefer mixed forests, the elder people prefer conifer forests.

The results of this survey can support forest managers in at least two major aspects. Firstly, to understand local people's values towards different forest stand characteristics and integrate these values into multi-functional forest management planning. Secondly, to avoid possible conflicts between local community and forest enterprises through detection of their interests in recreational attractiveness of forest stands. These two aspects are fundamental for implementing policy and management strategies aimed at sustainable forest management in the Italian Alps and Ukrainian Carpathians.

Acknowledgements

This article is based upon a research conducted under the COST Action "Climate-Smart Forestry in Mountain Regions" CLIMO CA15226, coordinated by Prof. Roberto Tognetti. The results of present study were presented at the IV Congresso Nazionale di Selvicoltura, Torino (Italy) 5-9 November 2018. The authors thank Isabella De Meo, Maria Giulia Cantiani and Federica Maino for their support in the Italian case study. Authors would like to thank the reviewers for their thoughtful comments and efforts towards improving our manuscript.

References

Brown G., Reed P. 2000 - *Validation of a forest values typology for use in national forest planning.* Forest Science 46 (2): 240-247.

Buckingham-Hatfield S. 2000 - *Gender and environment.* Routledge, London. 137 p.

Cantiani M.G. 2012 - *Forest planning and public participation: A possible methodological approach.* iForest-Biogeoscienes and Forestry 5 (2): 72-82. doi: 10.3832/ifor0602-009

Cantiani M.G., Bettelini D., Mariotta S. 2002 - *Participatory forest planning: A chance of communication between forest service and local communities.* In: Büchel M., Nipkow F., Güntensperger M., (Eds.). Forestry Meets the Public: Seminar and Workshop Proceedings. Bern, Switzerland: Swiss Agency for the Environment, Forests and Landscape: 249-263.

Carlson A. 2001 - *Aesthetic preferences for sustainable landscapes: seeing and knowing.* In: Sheppard S.R.J., Harshaw H.W. (Eds.), Forest and Landscapes. Linking Ecology, Sustainability and Aesthetics. IUFRO Research Series, No. 6. CAB International, Oxon UK: 31-41.

Daniel T.D. 2001 - *Aesthetic preference and ecological sustainability.* In: Sheppard, S.R.J., Harshaw, H.W. (Eds.), Forest and Landscapes. Linking Ecology, Sustainability and Aesthetics. IUFRO Research Series, No. 6. CAB International, Oxon UK: 15-29.

de Leeuw E.D. 1992 - *Data Quality in Mail, Telephone and Face to Face Surveys.* Netherlands Organization for Scientific Research, Amsterdam. 125 p.

Edwards D.M., Jay M., Jensen F.S., Lucas B., Marzano M., Montagne´ C., Peace A., Weiss G. 2012 - *Public preferences for structural attributes of forests: Towards a pan-European perspective.* Forest Policy and Economics 19: 12-19. doi: 10.1016/j.forpol.2011.07.006

ENPI EAST FLEG II. 2015 - *National report on Forest products dependence of rural communities in Ukraine.* [Online] Available: http://www.fleg.org.ua/docs/781. [2018, October 31]

Fahey R.T., Alveshere B.C., Burton J.I., D'Amato A.W., Dickinson Y.L., Keeton W.S. & Saunders M. R. 2018 - *Shifting conceptions of complexity in forest management and silviculture.* Forest Ecology and Management 421: 59-71. doi: 10.1016/j.foreco.2018.01.011

Filyushkina A., Agimass F., Lundhede T., Strange N., Jacobsen J.B. 2017 - *Preferences for variation in forest characteristics: Does diversity between stands matter?* Ecological Economics 140: 22-29. doi: 10.1016/j.ecolecon.2017.04.010

Gandolfo C., Comin P. 2017 – *Servizio Foreste e Fauna. Relazione sull'attività svolta nel 2016. Provincia autonoma di Trento.* Servizio foreste e fauna, Trento.

Giergiczny M., Czajkowski M., Zylicz T., Angelstam P. 2015 - *Choice experiment assessment of public preferences for forest structural attributes.* Ecological Economics 119: 8-23. doi: 10.1016/j.ecolecon.2015.07.032

Gobster P.H. 1999 - *An ecological aesthetic for forest landscape management.* Landscape Journal 18: 54-64. doi: 10.3368/lj.18.1.54

Goyder J. 1985 - *Face-to-face interviews and mailed questionnaires: the net difference in response rate.* The Public Opinion Quartely 49 (2): 243-252. doi: 10.1086/268917

Grilli G., Paletto A., De Meo I. 2014 - *Economic valuation of forest recreation in an Alpine valley.* Baltic Forestry 20 (1): 167-175.

Grilli G., Jonkisz J., Ciolli M., Lesinski J. 2016 - *Mixed forests and ecosystem services: investigating stakeholders' perceptions in a case study in the Polish Carpathians.* Forest Policy and Economics 66: 11-17. doi: 10.1016/j.forpol.2016.02.003

Gundersen V.S., Frivold L.H. 2008 - *Public preferences for forest structures: a review of quantitative surveys from Finland, Norway and Sweden.* Urban Forestry & Urban Greening 7 (4): 241-258. doi: 10.1016/j.ufug.2008.05.001

Heer C., Rusterholz H.P., Baur B. 2003 - *Forest perception and knowledge of hikers and mountain bikers in two different areas in northwestern Switzerland.* Environmental Management 31: 709-723. doi: 10.1007/s00267-003-3002-x

Heyman E. 2012 - *Analysing recreational values and management effects in an urban forest with the visitor-employed photography method.* Urban Forestry & Urban Greening 11: 267-277. doi: 10.1016/j.ufug.2012.02.003.

Jankovska I., Straupe I., Brumelis G., Donis J., Kupfere L. 2014 - *Urban forests of Riga, Latvia –pressures, naturalness, attitudes and management.* Baltic Forestry 20 (2): 342-351.

Jensen F.S. 1999 - *Forest recreation in Denmark from the 1970s to the 1990s.* The Research Series: 26. Danish Forest and Landscape Research Institute, Hoersholm, Denmark. 166 p.

Jensen F.S. 1993 - *Landscape managers' and politicians' perception of the forest and landscape preferences of the population.* Forest and Landscape Research 1 (1): 79-93.

Jensen F.S., Koch N.E. 1998 - *Measuring forest preferences of the population: A Danish approach.* In: Terrasson D., editor. Public Perception and Attitudes of Forest Owners Towards Forest in Europe [in French]. Commentaires et syntheses du groupe de travail COST E3-WG1, 1994/1998. Antony, France: Cemagref e´ditions: 39–82. doi.org/10.3188/szf.2000.0011

Kaplan R., Kaplan S. 1989 - *The Experience of Nature.* Cambridge University Press, Cambridge. 370 p.

Kumar S., Kant S. 2007 - *Exploded logit modeling of stakeholders' preferences for multiple forest values.* Forestry Policy and Economics 9 (5): 516-526. doi: 10.1016/j.forpol.2006.03.001

Lee T.R. 2001 - *Perceptions, Attitudes and Preferences in Forests and Woodlands.* Technical Paper 18. Edinburgh, United Kingdom: Forestry Commission. 166 p.

Lewis J.L., Sheppard S.R.J. 2005 - *Ancient values, new challenges: Indigenous spiritual perceptions of landscapes and forest management.* Society & Natural Resources 18: 907-920. doi: 10.1080/08941920500205533

Likert R. 1932 - *A technique for the measurement of attitudes.* Archives of Psychology 22 (140): 1-55.

Lindhagen A. 1996 - *Forest recreation in Sweden. Four case studies using quantitative and qualitative methods.* Swedish University of Agricultural Sciences, Department of Environmental Forestry, Uppsala. 109 p.

Melnykovych M., Nijnik M., Soloviy I., Nijnik A., Sarkki S., Bihun Y. 2018. *Social-ecological innovation in remote mountain areas: Adaptive responses of forest-dependent communities to the challenges of a changing world.* Science of The Total Environment 613: 894-906. doi: 10.1016/j.scitotenv.2017.07.065

Mill G.A., Van Rensburg T.M., Hynes S., Dooley C. 2007 - *Preferences for multiple use forest management in Ireland: Citizen and consumer perpectives.* Ecological Economics 60 (3): 642-653. doi: 10.1016/j.ecolecon.2006.02.005

Misgav A. 2000 - *Visual preference of the public for vegetation groups in Israel.* Landscape and Urban Planning 48: 143-159. doi: 10.1016/S0169-2046(00)00038-4

Nielsen A.B., Olsenb S.B., Lundhede T. 2007 - *An economic valuation of the recreational benefits associated with nature-based forest management practices.* Landscape and Urban Planning 80: 63-71. doi: 10.1016/j.landurbplan.2006.06.003

Nijnik A., Nijnik M., Kopiy S., Zahvoyska L., Sarkki S., Kopiy L., Miller D. 2017 - *Identifying and understanding attitudinal diversity on multi-functional changes in woodlands of the Ukrainian Carpathians.* Climate Research 73 (1-2): 45-56. doi: 10.3354/cr01448

Notaro S., Paletto A. 2011 - *Links between mountain communities and environmental services in the Italian Alps.* Sociologia Ruralis 5: 137-157. doi: 10.1111/j.1467-9523.2011.00532.x

Odasso M., Miori M., Gandolfo C. 2018 - *I tipi forestali del Trentino: descrizione e aspetti dinamici.* Provincia autonoma di Trento. Servizio foreste e fauna, Trento.

Oliynyk Ya., Zapototsky S., Braichevsky Yu., Galagan O. 2015 - *Rakhiv district: nature, population, economy.* Kiev, Kiev Polytechnic University. 254 p. [in Ukrainian with English summary].

Paletto A., Cantiani M. G., De Meo I. 2015 - *Public Participation in Forest Landscape Management Planning (FLMP) in Italy.* Journal of Sustainable Forestry 34 (5): 465-482. doi: 10.1080/10549811.2015.1026447

Paletto A., De Meo I., Cantiani M.G., Maino F. 2013 - *Social perceptions and forest management strategies in an Italian alpine community.* Mountain Research and Development 33 (2): 152-160. doi: 10.1659/MRD-JOURNAL-D-12-00115.1

Paletto A., Ferretti F., Cantiani P., De Meo I. 2012a - *Multifunctional approach in forest landscape management planning: an application in Southern Italy.* Forest systems 21(1): 68-80. doi: 10.5424/fs/2112211-11066

Paletto A., Maino F., De Meo I., Ferretti F. 2012b - *Perception of Forest Values in the Alpine Community of Trentino Region (Italy).* Environmental Management 8: 414-422. doi: 10.1007/s00267-012-9974-7

Pastorella F., Avdagić A., Cabaravdić A., Osmanović M., Paletto A. 2014 - *Does mountain forest characteristics influence visual appeal? A study case in an Alpine Valley in Italy.* In: Proceedings International Conference Natural Resources, Green Technology & Sustainable Development. 26th-28th November 2014, Faculty of Food Technology and Biotechnology, University of Zagreb, Croatia.

Pelyukh O., Zahvoyska L. 2018 - *Investigation of Lviv region population's preferences regarding recreational forest using choice experiment method.* Scientific Bulletin of UNFU 28 (9): 73-80. doi.org/10.15421/40280915 [in Ukrainian with English summary].

Pelyukh O., Zahvoyska L., Maksymiv L. 2018 - *Analysis of stakeholders' interaction in the context of secondary Norway spruce stands conversion in the Ukrainian Carpathians.* In: Proceedings of the IUFRO unit 4.05.00 International symposium. Zagreb, 10-12 May: 22-24.

Ribe R.G. 1989 - *The aesthetics of forestry: what has empirical preference research taught us?* Journal of Environmental Management 13: 55-74. doi: 10.1007/BF01867587

Rockström J., Steffen W. et al. 2009 - *Planetary boundaries: exploring the safe operating space for humanity.* Ecology and Society 14 (2): 32-64. doi: 10.5751/ES-03180-140232

Roovers P., Merny M., Gulinck H. 2002 - *Visitor profile, perceptions and expectations in forests from a gradient of increasing urbanisation in central Belgium.* Landscape and Urban Planning 59 (3): 129-145. doi: 10.1016/S0169-2046(02)00011-7.

Sheppard S.R.J., Meitner M. 2005 - *Using multi-criteria analysis and visualisation for sustainable forest management planning with stakeholder groups.* Forest Ecology and Management 207: 171-187. doi: 10.1016/j.foreco.2004.10.032.

Šišák L. 2011 - *Forest visitors' opinions on the importance of forest operations, forest functions and sources of their financing.* Journal of Forest Science 57: 266-270.

Soloviy I., Chernyavskyy M., Genyk Ya. 2011 - *Environmental, economic and social impact of inefficient and unsustainable forest practices and illegal logging in Ukraine.* Liga-Press, Lviv. 396 p. [in Ukrainian with English summary].

Soloviy I., Melnykovych M. 2014 - *Contribution of forestry to wellbeing of mountain forest dependent communities' in the Ukrainian Carpathians.* In: Proceedings of the Forestry Academy of Sciences of Ukraine. Collection of Research Papers 12: 233-241.

Steffen W., Richardson K., Rockstrom J., Cornell S.E., Fetzer I., Bennett E.M., Biggs R., Carpenter S.R., de Vries W., de Witt C.A., Folke C., Gerten D., Heincke J., Mace G.M., Persson L.M., Ramanathan V., Reyers B., Sorlin S. 2015 - *Planetary boundaries: Guiding human development on a changing planet.* Science 347 (6223): 736-746. doi: 10.1126/science.1259855

Tahvanainen L., Tyrväinen L., Ihalainen M., Vuorela N., Kolehmainen O. 2011 - *Forest management and public perceptions – visual versus verbal information.* Landscape and Urban Planning 53: 53-70. doi: 10.1016/S0169-2046(00)00137-7

Tarrant M.A., Cordell H.K. 2002 - *Amenity values of public and private forests: examining the value-attitude relationship.* Environmental Management 30 (5): 692-703. doi: 10.1007/s00267-002-2722-7

The Global Risks Report 2018. 13th Edition. World Economic Forum, Geneva. 80 p.

Tyrväinen L., Silvennoinen H., Kolehmainen O. 2003 - *Ecological and aesthetic values in urban forest management.* Urban Forestry & Urban Greening 1 (3): 135-149. doi: 10.1078/1618-8667-00014

Waters C.N., Zalasiewicz J. et al. 2016 - *The Anthropocene is functionally and stratigraphically distinct from the Holocene.* Science 351 (6269): p. aad2622. doi: 10.1126/science.aad2622

Zahvoyska L., Bas T. 2013 - *Stakeholders' perceptions of mountain forest ecosystem services: the Ukrainian Carpathians case study.* The Carpathians: Integrating Nature and

Society Towards Sustainability Springer, Berlin, Heidelberg: 353-367.

Zahvoyska L., Pelyukh O., Maksymiv L. 2017 - *Methodological considerations and their application for evaluation of benefits from the conversion of even-age secondary Norway spruce stands into mixed uneven-aged woodlands with a focus on the Ukrainian Carpathians.* Austrian Journal of Forest Science 134: 251-281.

Zahvoyska L.D. 2014 - *Theoretical approaches to determining economic value of forest ecosystems services: benefits of pure stands transformation into mixed stands.* In: Proceedings of the Forestry Academy of Sciences of Ukraine. Collection of Research Papers 12: 201-209. [in Ukrainian with English summary]

Natural capital and bioeconomy: challenges and opportunities for forestry

Marco Marchetti[1*], Matteo Vizzarri[1], Bruno Lasserre[1], Lorenzo Sallustio[1], Angela Tavone[1]

Abstract - Over the last decades, the stock of natural capital has been globally reduced by human-induced effects such as climate change, and land use and cover modifications. In particular, the continuous flow of goods and services from ecosystems to people is currently under threat if the current human activities still remain unsustainable. The recent bioeconomy strategy is an important opportunity to halt the loss of biodiversity and the reduction of services provision, from global to local scale. In this framework, forest sector plays a fundamental role in further enhancing the sustainable development and the green growth in degraded environments, such as marginal and rural areas. This paper provides an overview of the bioeconomy-based natural resources management (with a focus on forest ecosystems), by analyzing the related challenges and opportunities, from international to national perspective, as in Italy. At first, the role of forest sector in addressing the purposes of green growth is analyzed. Secondly, the most suitable tools to monitor and assess natural capital changes are described. Finally, the most important research contributions within the bioeconomy context are reported. To create the suitable conditions for bioeconomy and green growth, the following insights have to be denoted: (i) a deeper understanding of natural capital and related changes; (ii) the improvement of public participation in decision-making processes, especially at landscape scale; (iii) the effective integration of ecological, socio-cultural, and economic dimensions while managing natural resources.

Keywords - Natural capital, bioeconomy, forest ecosystems, ecosystem services, land use and cover change

The need for bioeconomy-based natural resources management

The concepts of "green-growth" and "bioeconomy" have been developed on the consciousness that population is expected to rapidly raise in the next 40 years (Rosegrant et al. 2012). This trend most probably will cause an increase of pressures on natural resources use and a growing inequality for their distribution among people, especially with regards to wild and seminatural ecosystems, soil, water resources, and croplands, and, as a consequence, an erosion of the largest part of the Ecosystem Services (ES) strictly related to Land Use and Cover Change (LUCC), the main driver of global change.

Overcoming these situations specifically requires responsibility in subsidiarity and innovation in order to achieve concerted changes in lifestyles and resource use, across all levels of society and economy (EU 2012). There are a number of key-drivers for the development of a green economy, as follows (Rosegrant et al. 2012): (i) the demand for renewable biological resources and bioprocesses; (ii) the need for improving the management and the sustainable use of renewable resources; (iii) facing substantial challenges, such as e.g. energy and food security, in the context of increasing unpleasant social phenomena like the neocolonialism (i.e. "land grabbing") or the prevalence of export-driven cropping systems, and several constraints on water, productive lands and carbon emissions (e.g. Sheppard et al. 2011); (iv) the rapid uptake of biotechnologies in agricultural productions; and (v) the opportunity to reduce environmental degradation through more sustainable production procedures. Other important challenges derive by the fact that the bioeconomy proposal is not about protecting the environment, but instead it is about promoting the economy – in spite of clear indications of the harmful impacts that are already resulting from massive new demand for biomass, including soil loss (a long-term renewable resource), biodiversity at gene, species, stand and landscape level, as well as escalating hunger and conflict (Hall et al. 2012).

Taking under consideration the past human-induced changes and their consequences on the increasing depletion of nature, the current stock of natural capital is almost compromised and is passing through several safety thresholds of planetary boundaries (Hughes et al. 2013), such as the CO_2 atmospheric composition, i.e. gaining 395 ppm in 2013, despite a tipping point of 350 ppm (Hansen et al. 2013). But also soil and forests and water are strongly threatened. The key necessary condition for

---

[1] Department of Biosciences and Territory (DiBT), University of Molise, Italy
* corresponding author: marchettimarco@unimol.it

achieving sustainability lies at least on the constancy of the natural capital stock over the time (Pearce et al. 1990). In this way, natural capital properly refers to "a stock that yields a flow of valuable goods and services into the future" and can be differentiated into "renewable natural capital (active and self-maintaining using solar energy, such as forest growing as known since the XVIII century) and non-renewable natural capital (passive)" (Costanza and Daly 1992). For instance, to sufficiently unravel the past anthropogenic effects on natural resources and the more recent shifting from Holocene to Anthropocene era, Ellis and Ramankutty (2008) globally identified and mapped the "Anthromes", namely Anthropogenic Biomes. In this way, the evaluation of ecosystem functioning (including biodiversity as main supporting element; see e.g. Cardinale 2013) is extremely important to globally reduce the impacts of the main drivers of change. For this purpose, monitoring the land use changes (one of the most accelerators of human-induced environmental modifications; Foley et al. 2005) is useful to orient the current overexploitation of natural resources towards a more "resilience-based" trajectory (e.g. Ellis et al. 2013).

Green economy and natural resources: the role of forest sector

Beside these general considerations, in forestry the green economy benefit starts when and occurs through management tools and investments that could limit trade-off effects of traditional multi-functionality and expand the ES availability for the society with a scope of fairness within and among generations (see also Atkisson 2012). Indeed, green economy improves human well-being and social equity, and significantly reduces environmental risks and ecological scarcities (UNEP 2011a). Sustainably managed forests play an essential role in the carbon cycle and provide essential environmental and social values, and ES, beyond their contribution as a source of wood, such as biodiversity conservation, protection against erosion, watershed protection and employment in often fragile rural areas. In this perspective, in order to promote the effectiveness of green economy in managed forests, the UNECE Committee on Forests and the Forest Industry (COFFI) and the FAO European Forestry Commission (EFC) decided to take action and prepared the Rovaniemi Action Plan for the Forest Sector in a Green Economy (ECE/TIM/SP/35). This Action Plan consists of 5 pillars with their respective goals, which are: (i) sustainable production and consumption of forest products (patterns of production, consumption and trade of forest products are truly sustainable); (ii) a low carbon forest sector (the forest sector makes the best possible contribution to mitigation of, and adaptation to, climate change); (iii) decent green jobs in the forest sector (the workforce is able to implement sustainable forest management, and the forest sector contributes to achieving the social goals of the green economy by providing decent jobs); (iv). long-term provision of forest ES (forest functions are identified and valued and payments for ES - PES (Payment for Ecosystem Services)– are established, thus encouraging sustainable production and consumption patterns); (v) policy development and monitoring of the forest sector in relation to a green economy (policy-makers and institutions in the forest sector promote sustainable forest management, in a way that is adequate to mainstream the green economy in forest sector policies).

To operationalize these broad guidelines, it is recommended to follow the Ecosystem Approach (EA). EA is a method for sustaining or restoring natural systems and their functions and values. It is goal-driven approach, and is based on a collaboratively-developed vision of desired future conditions that integrates ecological, economic and social factors (Inter-Agency Ecosystem Management Force 1995). Furthermore, EA is not a static model but is a holistic process for integrating and delivering in a balanced way the three objectives of the Convention on Biological Diversity (CBD): conservation and sustainable use of biodiversity, and equitable sharing of the benefits (Maltby 2000). Therefore, only an ecosystem-based management of natural resources can halt the loss of biodiversity and the degrade of resources quality. This is exactly one of the purposes of the Bioeconomy Strategy, properly aimed at improving the knowledge base and fostering innovation to increase productivity, while ensuring sustainable resource use and alleviating stress on the environment (COM 2012).

According to the evolution of classical economic theories, the need to consider forests both as factors of production and ecological infrastructures is always stronger. In particular, the contribution of forest management and land use planning (especially in fragile forest areas, as mountain environments) in the context of green economy growth has to consider also the biodiversity of forest ecosystems and the related ES as results of complex ecological processes and interactions amongst different ecosystems in a holistic view (Ciancio and Nocentini 2004, Mace et al. 2012).

At European level, Bengtsson et al. (2000) argued that the next generation of forestry practices would need to: (i) deeper understand natural forest dynamics; (ii) analyze the role of biodiversity (i.e.

key species and functional groups) in supporting the ecosystem functionality; (iii) implement and adapt management prescriptions in accordance with natural dynamics; (iv) consider ecology, forestry, economy, and social fields in order to establish a value of the important ES from forest ecosystems. Furthermore, in line with these good practices, forest management needs to avoid the impact of disturbances (such as e.g. anthropogenic eutrophication, toxic pollution, habitat loss, disconnection from adjacent ecosystems, species invasion, climate change, etc.), which can induce long-term ecosystem changes (see e.g. Ellis et al. 2013).

Although natural resources have an intrinsic value for improving sustainability, the vision of the natural capital has become the subject of ethical and conceptual discussion and debate, especially in conservation topics. This led to divisions between those who intend the conservation of nature as such, by virtue of its intrinsic or existence value with an assessment meaning (Soulé 2013), and those, instead, who intend it as an element of supporting for human well-being (e.g. Reid et al. 2006, Kareiva and Marvier 2012, Toledo and Barrera-Bassols 2014), translatable, therefore, in an instrumental value. Nevertheless, in recent years, the concept that the integration of different views and philosophies underlies the conservation, protection and restoration of natural resources has been clarified (Tallis and Lubchenko 2014). Therefore, it is important to remind that the value of a stock of natural resources, such as in particular a forest, is more than the sum of various functions that are assigned to that forest from time to time, which means recognizing that forest has intrinsic value (Ciancio and Nocentini 2004).

In order to further improve the contribution of the forest sector and its intrinsic awareness for a responsible green economy, it is essential to assess (EFI 2014): (i) the forest products market changes and, in particular, the C substitution rate stored in forest products (in general throughout the whole production chain, including the entire Life Cycle Assessment - LCA), and its trade-offs with other ES; (ii) the changes in cultural and non-marketed ES, which are difficult to price, such as tourism and recreation, and aesthetic, historical and cultural values, etc.; (iii) the current and future investments in the business sector related to forests and timber production, taking into account the enhancement of multi-functionality and a responsible and sustainable management; (iv) the changes in the ownership of the forest and the enterprise sector, considering the participation as a strong element of identity, belonging, proximity and protection of the territory; (v) the global demand for expertise services in forest governance, forest administration, inventory and information systems, as well as in forest education. Therefore, the major challenges for the forest sector in the context of the green economy partly refer to land use change and market failures, or to forest policy and planning. The socio-economic processes play a key role in ecosystem modifications, thus directly influencing human welfare (Ellis et al. 2013). For example, all the forestry activities are increasingly knowledge-intensive and address challenges, such as those related to natural resources assessment and monitoring in a context of global change (EFI 2014). In a context of change, the preservation of intrinsic and utilitarian values of natural capital has to be encouraged, as it is a key element for the reconciliation and the building of a sustainable, responsible and resilient human-nature relationships.

Linking natural and cultural capital

The need of a strong interconnection between the natural and cultural capital assets is well expressed in the "Chart of Rome" (CoR, Presidenza Italiana del Consiglio dell'Unione Europea 2014), whose aim is to broaden the scope of nature and biodiversity policy without changing it, but rather mainstreaming it into other policies related to the territory and the economy. Although the main target groups of CoR are scientists, stakeholders and policy-makers, its message is also for citizens. It is a European initiative and develops on the EU cornerstones of Natura 2000 and the EU Biodiversity Strategy to 2020. The primary role is the promotion of a better conservation and valorization of the natural and cultural diversity. Moreover, the CoR acts as a platform for further collaborations on biodiversity in general, and in particular on ES, as well as on their societal implications (i.e. climate mitigation, clean water, clean air, protection against floods and erosion).

Furthermore, it finds its roots in the CBD, specifically with regards to protecting and encouraging the customary use of biological resources in accordance with the traditional cultural practices that are compatible with conservation or sustainable use requirements (UNEP 1992). CoR is strongly connected also with the Convention for the Safeguarding of Intangible Cultural Heritage, because communities and groups are able to constantly recreate their intangible cultural heritage, since it is the product of the interaction between nature and history, and it is transmitted from throughout generations, according to the environment they live in. In this way, people enhance their own sense of identity and continuity, and, as a consequence, promote the respect for cultural diversity and human creativity (UNESCO 2003). Another bridge built by the CoR with the EU biodiversity-related policies is the Green Employment Initiative (COM/2014/446). This initiative aims at indicating the way for job creation potential in the green economy sector with reference to skills, education and training, green public procurement, promotion of entrepreneurship, increasing of data quality (including statistical definition of employment in the environmental sector) and promotion of social dialogue.

CoR is strongly related to the adaptive capacity of human populations to deal with and modify the natural environment (Berkes and Folke 1992), the natural capital, which is composed by the ecosystems. Therefore, healthy and resilient ecosystems can provide society with a full range of economically valuable goods and services. To maintain healthy ecosystems, the following responsible actions are needed (Presidenza Italiana del Consiglio dell'Unione Europea 2014): (i) making use of good knowledge and data on biodiversity, ecosystems, their structures and functions, and on links with ES and associated benefits; (ii) maintaining, restoring and enhancing capacities to provide a range of goods and services and associated benefits; (iii) exploring natural capital as a solution to major challenges such as those related to urban areas, climate change and adaptation, agriculture and soil, forestry, hydrological risks, tourism and recreation. In this sense, good knowledge, research and data gathering on biodiversity and ecosystems are essential, because they make the knowledge base accessible to citizens and decision-makers, thus ensuring that policy-makers continue to understand and consider complex environmental state and dynamics.

In addition, cultural and economic scientists (e.g. Throsby 1999) contributed to identify cultural capital as a set of three main features, such as (Sukhdev et al. 2014): (i) knowledge, including traditional and scientific dimensions; (ii) capacities, as the way knowledge is retained, increased, elaborated and developed; and (iii) practices and human activities producing tangible and intangible flows of goods and services.

In order to maintain a positive link between cultural and natural capital, the following goals have to be reached (Presidenza Italiana del Consiglio dell'Unione Europea 2014): (i) taking into account social and cultural dimension of ecosystem management; (ii) promoting locally adapted knowledge, capacities and activities with positive impacts on natural capital; and (iii) connecting benefits, goods and services from ecosystems (supply) with patterns of culture, society and economy (demand). Moreover, green infrastructures can contribute to these goals, since they connect natural and semi-natural areas with urban and rural areas. They are also drivers of a transition towards a green economy and are able to guarantee many natural, cultural, social and economic linkages. In Italy, the recent report concerning the socio-economic assessment and monitoring of natural capital and Protected Areas (PA) is the first attempt to contribute to the pillars of green economy at national level (MATTM and Unioncamere 2014). The report results mainly reveal what is the current condition about biodiversity conservation, what ES are correlated to cultural capital and local communities, and how sustainable practices effectively contribute to the green economy concerns.

Even green economy-related contributions are increasing, the concepts of natural capital, ES, and cultural capital require further operational definition and understanding. A knowledge-based improvement of the concept and its operationalization are in line with the EU nature and biodiversity strategies, directives and overall policies, which are expected to enhance and promote biodiversity conservation, the sustainable use of natural resources, while improving communication, mainstreaming and policy consideration in a wide societal and political context.

Monitoring changes of natural capital: land use and ecosystem services relationship

An important issue in many debates concerning the policies and the governance of the landscape is the ES assessment. Public interest in ES assessment has been starting since the milestone work on the economic assessment of natural resources made by Costanza et al. (1997). Mostly after the CBD (UNEP 1992), biodiversity and ES in general were placed at the base of the most important global, European and national processes focusing on the enhancement and preservation of natural resources and ecosystems as sources for multiple services and benefits for the society (see European Biodiversity Strategy to 2020 (EP 2011/2307(INI) and the Italian Biodiversity Strategy (MATTM, Decree 6 June 2011)).

Although the ES concept is already central in conservation policies and environmental impact assessments (Burkhard et al. 2010), useful methodologies for its practical application are still needed, in order to support the sustainability in natural resources management. Following the needful for quantifying the natural capital and ES, both biophysical and economic aspects have to be considered. If the goal is to measure the efficiency of natural resources management as a whole, then the quantification of benefits from ecosystems is necessary, especially to preserve the stocks of natural capital useful to generate ES. Indeed, the approach of the Millennium Ecosystem Assessment (2005) is based on the notion that the resource management involves the study of the relations between the ES and their quantitative estimation. As a consequence, there is nowadays a considerable interest to establish innovative approaches to calculate ES at different spatial and temporal scales.

Among terrestrial ecosystems, forests (including other wooded lands) are one of the most important sources of services and benefits for the entire humanity. Forests (Vizzarri et al. 2013): (i) protect biodiversity, providing habitats to more than half of the plant and animal known species; (ii) play a significant role in regulating biogeochemical cycles and, consequently, in the mitigation of climate change at different spatial scales; (iii) generate a large set of goods and products (timber and non-timber); (iv) host and protect sources and catchment areas accessible to man, often characterized by high quality water; (v) protect the traditional, cultural and spiritual values of many societies in the world.

In particular, considering the provisioning services, forests can assure the availability of wood for building, firewood and other non-timber forest products (e.g. cork, tannin, mushrooms, truffles, berries, etc.), which represent important economic components for the economies of many Countries. In addition, forest soil and topsoil have an enormous capacity to filter out most of the chemical components of pollutants and to reduce the surface runoff, thus preventing and reducing the risk of erosion and slope instability. In many cases, the presence of forest areas reduces the need of treatment (and, therefore, of the related costs) for the production of drinking water available to the local population, as shown in several case study around the world (Dudley and Stolton 2003).

Amongst the regulation services, forests are integrated in climate mitigation processes. In particular, forest stands have a threefold relationship in the face of climate change, as follows: they are adapting themselves to the effects of climate change, but at the same time, are subject of the general causes (emission source, from deforestation) and of the solution (major terrestrial sink). Indeed, among the different contributions of forests to climate change mitigation, there is the absorption of carbon from the atmosphere. Especially in "fragile" landscapes (such as mountain areas), forests are of primary importance to protect infrastructures and buildings from disasters, like avalanches, landslides, debris flows, rolling stones, and erosion processes in general. The vegetation strongly affects the water supply to the ground directly intercepting rainfall, attenuating the incident solar radiation and by controlling the evapotranspiration rate.

Supporting services are considered intermediate services as predisposing conditions so that a final service can be provided. In this case, forest biodiversity is the key element to support the provision of all other services, as it directly affects the properties of self-regulation and adaptation of forest ecosystems, and the capacity of a forest to produce timber or to be resilient and resistant against natural or anthropogenic disturbances. In this context, the role of biodiversity is essential for enabling to the availability of other services, because it (Vizzarri et al. 2013): (i) supports ecosystems in the structural, compositional, and functional diversification; (ii) influences the productivity, stability and resilience of ecosystems; (iii) increases the cultural and aesthetic value due to the presence of particular organisms and habitats; (iv) indirectly provides diversified products for rural populations (food, fiber, etc.).

Around the forest ES provision, forest landscapes have also intrinsic traditional, cultural and spiritual values, because they result from a profound historical interaction between man, its activities, and the surrounding nature. In addition, forest landscapes offer unique experiences, such as combinations of suggestive images (e.g. the colors of the vegetation, the behavior of wildlife, remote and unspoiled landscapes, etc.), echoing sounds (e.g. the birds chirping, the hum of insects, superior animal sounds, etc.) and strong scents (e.g. the smell of flowers or berries, etc.).

Considering forests as natural integrated systems, inside and outside ecological processes play a key role in governing the energy and material flows between ecosystems and man. Therefore, the potential of "supply" of services by a forest ecosystem is closely linked to its "health", namely the balance of its resilience characteristics, durability, low vulnerability and stability over time. The analysis and quantification of forest ES may be in conflict with an economic approach, because the intrinsic values that people attribute to ecosystem structures and processes are often not corresponded by economic "market" value (Farber et al. 2002). Consequently, the quantification and economic evaluation of forest ES must take into account the following critical issues: (i) how to separate "stocks" from "flows"; (ii) counting for potential beneficiaries of a given service, as well as its durability and availability in time; (iii) distinguishing the production of the service that may potentially be used with the one that is currently being consumed. The use of indicators can be an effective strategy to "quantitatively" measure and monitor complex phenomena such as ecological ones. In the ES assessment, indicators should be as inclusive as possible and properly selected on the basis of ecosystem properties and structures. They should also be easy to understand, allowing easy communication between institutions, technicians, professionals, and stakeholders at the local scale (Vizzarri et al. 2013).

While analyzing and evaluating forest ES, the anthropogenic impact on ecosystem functioning and, therefore, its ability to provide a set of services (and, consequently, benefits) must be considered. During the evolutionary history, humans excelled due to their ability to model ecosystems throughout the use of tools and techniques, which are beyond the capabilities of other living organisms (Smith 2007). Therefore, the importance of the "human factor" is essential: currently more than 75% of the land in the world shows disturbance caused by human action, with less than a quarter remained as wild land, able to support only 11% of the net terrestrial primary productivity (Ellis and Ramankutty 2008). Consistently, some scientific theories define Anthropocene as the current time that the Earth is living (Zalasiewicz et al. 2008). Lambin et al. (2001) stated that LUCC: (i) has an heavy impact on biodiversity at a global scale; (ii) contributes to climate change at the local and regional level; (iii) represents the main source of soil degradation and water depletion; (iv) alters ES and affects the capacity of natural systems to support human needs. There is indisputable evidence linking changes in the use / land cover to the loss of ES, especially in cases of services as carbon sink, hydrological processes and climate change. A complete ES assessment must be considered as spatially explicit, because it serves as a basis to implement LUCC (and therefore the human impact), as well as to provide a complete overview of offered services, including their current availability and future-oriented simulation (modeled according to various hypothetical scenarios). Furthermore, mapping ES can facilitate the economic evaluation, and provide the balance (trade-off) amongst multiple ES, which is necessary to support decision-making and landscape planning processes (Chirici et al. 2014).

The use of monitoring tools, such as Land use / Land Cover Inventories (Inventario dell'Uso delle Terre in Italy – IUTI, Corona et al. 2012) allows to identify and quantify in a quick way and at low cost the key dynamics characterizing the landscape changes, as well as the monitoring of their impact in ecological and functional terms (Sallustio et al. 2013, Marchetti et al. 2012b, Corona et al. 2012). As an example, for the period 1990-2008 in Italy the following important changes have been identified: (i) the forest area has increased of about 500,000 ha. At that time, the urban areas have expanded of the same amount, especially to the detriment of agricultural land, which recorded a loss of about 800,000 ha; and (ii) the registered urban sprawl can be mainly referred to the downhill and plain territories, and correlated to the increasing pressure on already fragmented and degraded ecosystems. The recovery of human-modified landscapes is necessary to create a socio-economic cohesion between urban and forest area. Furthermore, re-creating the lost agricultural fabric offers enormous ecological potential, including e.g. the reduction of fragmentation and degradation (especially of soil), a significant increase of biodiversity (creation of corridors and ecological niches) and the recovery of an important band transition having the function of mitigation systems between natural and manmade assets (vacant land or derelict land; Marchetti and Sallustio 2012). Delivering and keeping the identity to the rural landscape increases the awareness about the primary sources location of power and energy in urban areas, thus enhancing processes of historical and cultural identity, and improving health and social welfare.

It is important to note that the trends observed at the national level in Italy are not very different from those observed within the National Parks, both for land cover modifications and services provided (Marchetti et al. 2012a, Marchetti et al. 2013a). This trend directly reflects on the landscape planning development, especially taking into account the problem of maintaining grasslands, pastures and agricultural activities of extensive type, which are important for the historical, economic and cultural landscapes heritage, and are essential elements for the conservation of the environmental mosaic, which is typical of the Italian peninsula and of its biodiversity (Marchetti et al. 2013b). Taking apart how the urban sprawl develops over the time, it is important to deeper understand in which way policy instruments and regulations are currently used and implemented in these areas (also within PA- Protected Areas). For instance, the abandonment of silvicultural practices within National Parks and High Conservation Value Forests (HCVFs; Maesano et al. 2011) can reduce the forests growth and productivity, making them less resilient while facing natural disturbances (pest outbreaks, forest fires, etc.).

While contrasting the urban sprawl phenomena, agriculture represents a key activity, because it is able of recreating a balanced landscape by preserving areas which are not built-up and, where possible, by restoring ecological integrity of degraded and fragmented environments (i.e. mountain areas). Farming is the essential and long-lasting territorialization factor, as well as the energy basis of the life cycle in the country. However, it can become central to a regenerative vision of the landscape only if integrated with the ecological characteristics. The productive function of the countryside must be flanked by the importance of the concept of its capacity to be a producer of social cohesion, of a good and healthy environment where people can live a quality lifestyle, feeling a sense of belonging. By the contrary, from the urban point of view, there is mainly the problem of defining, perceiving and recognizing the countryside as an area where food and energy come from, according to conceptual models which focus on the ecological footprint (Wackernagel and Rees 2004, Iacoponi 2001).

Moreover, the participatory aspect is necessary in order to carry out one of the founding principles of the European Landscape Convention (Council of Europe 2000), as well as that of the Italian Constitution, which underlines the fundamental need of enabling local participation in decision-making processes at landscape level (articles 3 and 9). Participation has not to be considered as a simple accessory to democracy, but as a real possibility that local communities have, on different levels, to influence and orient the decision-making processes within a given area, irrespective of their individual, specific interests (Settis 2010). Indeed, the engagement of stakeholders may increase the likelihood that environmental decisions are perceived as holistic and fair, accounting for a diversity of values and needs and recognising the complexity of human-environmental interactions (Richards et al. 2004). Furthermore, in a shared management strategy of the landscape, which takes local interests and concerns into account primarily at an early stage, it may be possible to inform the project design with a variety of ideas and perspectives. In this way, public participation increases the likelihood that local needs and priorities are successfully met (Reed 2008). By establishing common ground and trust between stakeholders, participatory processes have the capacity to transform adversarial relationships and find new ways for participants to work together (Stringer et al. 2006). This may lead to a sense of ownership over the process and outcomes, thus enhancing long-term support and active implementation of decisions (Richards et al. 2004).

Considering the above-mentioned issues, it is important to remark that managing the landscape is another of the many duties carried out by the agricultural establishments, with economic and labour-related repercussions, which factors that cannot be ignored in transitional periods such as that of today. The main goal is to create a new culture, which, while starting with the enterprises, can stimulate interaction amongst businesspeople, public authorities and professionals in order to shape new ways for organizing the land. This takes into account the close connections between urban areas, nature and the world of farmers to guarantee that the principles of sustainable development are fulfilled. This action way can be possible if local and scientific knowledge are integrated to provide a more comprehensive understanding of complex and dynamic natural systems and processes (Reed 2008).

Perspectives for the future implementation of bioeconomy

In this composite changing world, the availability of data and easily upgradeable models that can describe these processes are important, since they allows the creation of future scenarios supporting public and private decision makers, in planning and designing the responsible growing of green economy and its activities. The possibility to calculate uncertainty and accuracy of models being used, the substantial reduction of errors of commission and omission are common issues in the field of land use inventories and maps, especially while focusing on practical forest management (Corona 2010). The evaluation of LUCC effects on biodiversity and ES should be the main element in supporting planning processes. Even if it could appear as a choice linked to particular sensitivity or marketing issues for administrators or ordinary citizens, it is now clear that this must be the *modus operandi*, as already established at international level.

Indeed, many efforts have been made to include the evaluation of the ES within decision-making contexts. For example, in 2012 the IPBES (Intergovernmental Science-Policy Platform on Biodiversity and Ecosystem Services) was established, as a tool for linking the scientific community to policy makers, putting the first track on what are the needs and requirements in applied contexts (http://www.ipbes.net/). Similarly, at the European level, the Action 5 of the EU Biodiversity Strategy to 2020 requires that the Member States start to map and assess the state of ecosystems and their services within their own boundaries in order to support natural capital conservation. For the development of a knowledge framework to support the contexts and needs of different States, the Working Group "Mapping and Assessment on Ecosystems and their Services" (http://ies.jrc.ec.europa.eu/news/468/155/Mapping-and-Assessment-of-Ecosystems-and-their-Services.html) was established. At national level, the first results obtained in research projects such as the "MIMOSE" (Development of innovative models for multi scale monitoring of ES indicators in Mediterranean forests) are promising. MIMOSE specifically aims to develop an innovative monitoring approach to estimate the capacity of a given forest area to provide ES under different management scenarios (Chirici et al. 2014). Key elements of this approach are connected to an integrated set of ES indicators and methods oriented to their spatial estimation. In this perspective, the primary project purpose is to bridge the gap between the concept of ES and their operational implementation in the management of forest ecosystems and environmental planning. The results of the project are expected to provide a real contribution for the incorporation of ES in decision-making processes and the forest landscape management and planning, thus providing an opportunity to understand the trade-offs between the different forest ES. This is expected to be useful to inform local stakeholders, sensitizing them towards a certain management that maximizes net benefits from ecosystems for the society.

For the forest sector, the most important challenges are to find innovative approaches for managing forest resources, in a way that simultaneously increases wood and non-wood production, improves the food security and energy supply against poverty, and safeguards other environmental services and biodiversity (Alexandratos and Bruinsma 2012). Under the current (unsustainable) conditions, forest resources cannot continue to contribute to the natural capital flows in the future, thus reducing the transferring of important services to people, especially in degraded environments, and reducing the ecosystem capacity to sustain the green growth. As a consequence, monitoring changes in forest cover (e.g. Hansen et al. 2010) and relative attributes (e.g. Butchart et al. 2010) is extremely important to make the future-oriented management guidelines coherent with the bioeconomy bases. More recently, several authors pointed out the urgent need to put the bases for a persistent monitoring of forests and their services (Maes et al. 2012). However, further research is required to bridge the gap between ecologic and economic fields (Cardinale et al. 2012), especially considering the emerging international commitments, both at European (EP 2012) and global scale (UNEP 2011b, UNEP 2014).

In this perspective, the nodal points lie in the efficiency evaluation of conservation strategies, in the assessment and monitoring of ES, and in the ability to translate these measures in estimating the cost implications. Similarly, the analysis of ES shall provide an integrated and holistic approach, which has to be able to grasp the complexity of functional processes For this purpose, there are several tools available for orienting conservation policies, such as e.g. the use of biophysical indicators (e.g. Noss 1999), the mapping of natural resources and habitats (e.g. Weiers et al. 2004), and the implementation of economic instruments for the market of "natural products" (e.g. Engel et al. 2008). Time and spatial scales (at which conservation strategies are planned and the effects assessed) are also key issues in mapping ES and related changes. It should be always kept in mind how the resilience of natural systems and their adaptability and susceptibility to change go far beyond the administrative limits or times of programming and planning. Indeed, there is also a "resilience thinking", which describes the collective use of a group of concepts to address the dynamics and development of complex socio-ecological systems (Folke et al. 2010). This implies a profound reflection on how, where and who has to deal with conservation, preferring detailed, solid and shared strategies to "niche" policies (Pressey et al. 2007).

Furthermore, the economic evaluation, despite much closer to a utilitarian view of natural resources, is currently the most effective tool to persuade and influence the people choices, especially waiting for the consolidation of a collective consciousness, more sensitive to the issue of conservation and use of natural resources in general. In this perspective, it is therefore also necessary to review the strategic role of PA. It is no longer enough to establish new PA or expand the existing ones, but it is necessary to strengthen and make more efficient and effective the management in existing ones (Watson et al. 2014). PA must be not only "Shrines of Nature", but real laboratories in which testing the best practices to enhance the natural and cultural capital can be to be exported and implemented in heavily populated surrounding matrix.

The forest sector can offer many opportunities in the context of bioeconomy, such as: (i) the proper and effective implementation of Criteria and Indicators for Sustainable Forest Management, (C&I-SFM; see also EFI 2013); (ii) the expansion of PA network; (iii) the development of initiatives related to projects for reducing global emissions (e.g., Reducing Emissions from Deforestation and forest Degradation, REDD+; http://www.un-redd.org/); (iv) the acceptance of PES in the current economic and productive systems; (v) the implementation of policies aiming to more active management and sustainable conservation of natural capital. Within this context, the research is essential to (Vizzarri et al. 2013): (i) analyze the degree of complexity, the value and quality of forest ES through innovative tools that can simulate the complexity of ecosystems themselves (process-based modelling and mapping); (ii) collect the most complete set of available information relating to the health and resilience of forest ecosystems (new techniques for monitoring and detection); (iii) consider the active involvement of stakeholders in planning decisions and forest management through statistical analysis multi-criteria techniques (agent-based techniques); (iv) reduce the uncertainty associated with estimating the value of ES, as well as reducing the gap between ecological and socio-economic research.

By the other hand, among the critical issues currently found in scientific research in the context of the bioeconomy applied to forest resources, worthy of mention are: (i) the limited availability of spatialized data on a national scale; (ii) the deficient multidisciplinarity in analyzing forest ES; (iii) the absence of widespread and consistent use of models, quantitative analysis and evaluation of ecological, economic, and socio-cultural indicators related to the provision of services delivered by forest ecosystems; (iv) the lack of implementation of EU policies at the local level.

In order to determine, and subsequently improve the competitiveness and the role of the forest sector in relation to other productive sectors as part of the bioeconomy, governments, public administrations, and sector managers need a complete picture of the stock, streams, and balance of costs and benefits of services provided by forest ecosystems.

Therefore, investments have to be oriented towards the improvement of management practices in existing forests and agroforestry systems, in order to ensure the continuous supply of the widest range of services provided. In this context, the development of new methods for supporting planning processes and especially to improve the ability to transfer the skills and knowledge to policymakers are essential elements for implementing the pillars of bioeconomy and green growth, also in the forest sector.

At conclusion, the future-oriented research is expected to be interdisciplinary and multi-purpose, and able to translate theories and concepts in models and methods particularly suitable for analyzing the *status quo* and the potential impact of different policy scenarios and management on ecosystem resilience. In the frame of bioeconomy, research is called to provide scientific bases, models and decision support tools for implementing sustainable growth and local development, which have their roots on paradigms less anthropocentric and more focused on coupling human and natural systems.

Aknowledgements

The authors want to thank the anonymous reviewers for their useful comments and suggestions.

References

Alexandratos N., Bruinsma J. 2012 - *World agriculture towards 2030/2050: the 2012 revision.* (No. 12-03, p. 4), Rome, FAO: ESA Working paper.

Atkisson A. 2012 - *Life beyond growth. Alternatives and complements to GDP-measured growth as a framing concept for social progress.* 2012 Annual Survey Report of the Institute for Studies in Happiness, Economy, and Society–ISHES (Tokyo, Japan) [Online]. Available: http://alanatkisson.wordpress.com/2012/02/29/life-beyond-growth/ [2012, July 31].

Bengtsson J., Nilsson S.G., Franc A., Menozzi P. 2000 - *Biodiversity, disturbances, ecosystem function and management of European forests.* Forest Ecology and Management 132 (1): 39-50. doi: 10.1016/S0378-1127(00)00378-9

Berkes F., Folke C. 1992 - *A systems perspective on the interrelations between natural, human-made and cultural capital.* Ecological Economics 5: 1–8.

Burkhard B., Petrosillo I., Costanza R. 2010 - *Ecosystem services: bridging ecology, economy and social sciences.* Ecological Complexity 7: 257-259.

Butchart S.H.M., Walpole M., Collen B., Van Strien A., Scharlemann J.P.W., Almond R.E.A., Baillie J.E.M., Bomhard B., Brown C., Bruno J., Carpenter K.E., Carr G.M., Chanson J., Chenery A.M., Csirke J., Davidson N.C., Dentener F., Foster M., Galli A., Galloway J.N., Genovesi P., Gregory R.D., Hockings M., Kapos V., Lamarque J-F., Leverington F., Loh J., McGeoch M.A., McRae L., Minasyan A., Morcillo M.H., Oldfield T.E.E., Pauly D., Quader S., Revenga C., Sauer J.R., Skolnik B., Spear D., Stanwell-Smith D., Stuart S.N., Symes A., Tierney M., Tyrrell T.D., Vié J-C., Watson R. 2010 - *Global Biodiversity: Indicators of Recent Declines.* Science 328 (5982): 1164-1168. doi: 10.1126/science.1187512

Cardinale B.J. 2013 - *Towards a general theory of biodiversity for the Anthropocene.* Elementa: Science of the Anthropocene 1: 14.

Cardinale B.J., Duffy J.E., Gonzalez A., Hooper D.U., Perrings C., Venail P., Narwani A., Mace G.M., Tilman D., Wardle D.A., Kinzig A.P., Daily G.C., Loreau M., Grace J.B., Larigauderie A., Diane S., Srivastava D.S., Naeem S. 2012 - *Biodiversity loss and its impact on humanity.* Nature 486 (7401): 59-67. doi: 10.1038/nature11148

Chirici G., Sallustio L., Vizzarri M., Marchetti M., Barbati A., Corona P., Travaglini D., Cullotta S., Lafortezza R., Lombardi F. 2014 – *Advanced Earth observation approach for multiscale forest ecosystem services modeling and mapping (MIMOSE).* Annali di Botanica 4: 27–34.

Ciancio O., Nocentini S. 2004 - *Biodiversity conservation in Mediterranean forest ecosystem.* EFI Proceedings 51: 163-168.

COM 2012 – *Innovating for Sustainable Growth: A Bioeconomy for Europe*, SWD (2012), Brussels, 13 February 2012.

Corona P., Barbati A., Tomao A., Bertani R., Valentini R., Marchetti M., Fattorini L., Perugini L. 2012 - *Land use inventory as framework for environmental accounting: an application in Italy.* iForest - Biogeosciences and Forestry 5 (4): 204–209. doi: 10.3832/ifor0625-005

Corona P. 2010 - *Integration of forest mapping and inventory to support forest management.* iForest - Biogeosciences and Forestry 3 (1): 59–64. doi: 10.3832/ifor0531-003

Costanza R., D'Arge R., De Groot R., Farber S., Grasso M., Hannon B., Limburg K., Naeem S., O'Neill R.V., Paruelo J., Raskin R.G., Sutton P., Van den Belt M. 1997 - *The value of the World's ecosystem services and natural capital.* Nature 387: 253-260. doi:http://dx.doi.org/10.1016/S0921-8009(98)00020-2

Costanza R., Daly H.E. 1992 - *Natural capital and sustainable development*. Conservation Biology 6 (1): 37-46. doi: 10.1046/j.1523-1739.1992.610037.x

Council of Europe 2000 - *European Landscape Convention*, Florence, 20 October 2000.

Dudley N., Stolton S. 2003 - *Running Pure: The importance of forest protected areas to drinking water*. World Bank/WWF Alliance for Forest Conservation and Sustainable Use, 114 p. [Online]. Available: http://wwf.panda.org/about_our_earth/all_publications/?8443/Running-Pure-The-importance-of-forest-protected-areas-to-drinking-water [2014, December 9].

Ellis E.C., Kaplan J.O., Fuller D.Q., Vavrus S., Klein Goldewijk K., Verburg P.H. 2013 - *Used planet: a global history*. In: Proceedings of the National Academy of Sciences 110 (20): 7978-7985. doi: 10.1073/pnas.1217241110

Ellis E.C., Ramankutty N. 2008 - *Putting people in the map: anthropogenic biomes of the world*. Frontiers in Ecology and the Environment 6 (8): 439-447. doi: http://dx.doi.org/10.1890/070062

Engel S., Pagiola S., Wunder S. 2008 - *Designing payments for environmental services in theory and practice: An overview of the issues*. Ecological economics 65 (4): 663-674. doi: 10.1016/j.ecolecon.2008.03.011

EC (European Commission) 2014 - *Green Employment Initiative: Tapping into the job creation potential of the green economy*, Brussels, 2.7.2014, COM(2014)446. [Online]. Available: http://ec.europa.eu/transparency/regdoc/rep/1/2014/EN/1-2014-446-EN-F1-1.Pdf [2014, December 9].

EC (European Commission) 2012 - *Innovating for sustainable growth. A bioeconomy for Europe. Brussels*. Brussels, 13.2.2012, COM(2012)60. [Online]. Available: http://ec.europa.eu/research/bioeconomy/pdf/201202_innovating_sustainable_growth.pdf [2014, December 9].

EFI (European Forest Institute) 2014 – *Future of the forest-based sector: structural change towards bioeconomy*. Lauri Hetermäki Editor, p. 110. [Online]. Available: http://www.efi.int/files/attachments/publications/efi_wsctu_6_2014.pdf [2014, December 9].

EFI (European Forest Institute) 2013 - *Implementing Criteria and Indicators for Sustainable Forest Management in Europe*. Jouni Halonen Editor, p. 132. [Online]. Available: http://www.ci-sfm.org/uploads/CI-SFM-Final_Report.pdf [2014, December 9].

EP (European Parliament) 2012 - *Our life insurance, our natural capital: an EU biodiversity strategy to 2020*. (2011/2307(INI)). [Online]. Available: http://ec.europa.eu/environment/nature/biodiversity/comm2006/pdf/EP_resolution_april2012.pdf [2014, December 9].

Farber S.C., Costanza R., Wilson M.A. 2002 - *Economic and ecological concepts for valuing ecosystem services*. Ecological Economics 41 (3): 375-392. doi: 10.1016/S0921-8009(02)00088-5

Foley J.A., DeFries R., Asner G.P., Barford C., Bonan G., Carpenter S.R., Chapin F.S., Coe M.T., Daily G.C., Gibbs H.K., Helkowski J.H., Holloway T., Howard E.A., Kucharik C.J., Monfreda C., Patz J.A., Prentice I.C., Ramankutty N., Snyder P.K. 2005 - *Global Consequences of Land Use*. Science 309 (5734): 570-574. doi: 10.1126/science.1111772

Folke C., Carpenter S.R., Walker B., Scheffer M., Chapin T., Rockstrom J. 2010 - *Resilience thinking: integrating resilience, adaptability and transformability*. Ecology and Society 15 (4): 20.

Hall R., Ernsting A., Lovera S., Alvarez I. 2012 - *Bio-economy versus Biodiversity*. Global Forest Coalition, 18 p.

Hansen J., Kharecha P., Sato M., Masson-Delmotte V., Ackerman F., Beerling D.J., Hearty J.P., Hoegh-Guldberg O., Hsu S.L., Parmesan C., Rockstrom J., Rohling E.J., Sachs J., Smith P., Steffen K., Van Susteren L., Von Schuckmann K., Zachos J.C. 2013. *Assessing "Dangerous Climate Change": required reduction of carbon emissions to protect young people, future generations and nature*. PloS one 8 (12): e81648. doi: 10.1371/journal.pone.0081648

Hansen M.C., Stehman S.V., Potapov P.V. 2010 - *Quantification of global gross forest cover loss*. Proceedings of the National Academy of Sciences 107 (19): 8650-8655. doi: 10.1073/pnas.0912668107

Hughes T.P., Carpenter S., Rockström J., Scheffer M., Walker B. 2013 - *Multiscale regime shifts and planetary boundaries*. Trends in ecology & evolution 28 (7): 389-395. doi:10.1016/j.tree.2013.05.019

Iacoponi L. 2001 - *La Bioregione. Verso L'integrazione dei processi socioeconomici ecosistemici nelle comunità locali*. Edizioni ETS, Pisa, 116 p.

Inter-Agency Ecosystem Management Force 1995 - *The ecosystem approach: healthy ecosystems and sustainable economies*. National Technical Information Service, Department of Commerce, Springfield, US [Online]. Available: http://www.denix.osd.mil/nr/upload/ecosystem1.htm [2014, 8 December].

Kareiva P., Marvier M. - 2012. *What Is Conservation Science?*. BioScience 62: 962-969. doi: 10.1525/bio.2012.62.11.5

Lambin E.F., Turner B.L., Geist H.J., Agbola S.B., Angelsen A., Bruce J.W., Coomes O.T., Dirzo R., Fischer G., Folke C., George P.S., Homewood K., Ibernon J., Leemans R., Li X., Moran E.F., Mortimore M., Ramakrishnan P.S., Richards J.F., Skånes H., Steffen W., Stone G.D., Svedin U., Veldkamp T.A., Vogel C., Xu, J., 2001 - *The causes of land-use and land-cover change: moving beyond the myths*. Global environmental change 11 (4): 261-269. doi: 10.1016/S0959-3780(01)00007-3

Mace G.M., Norris K., Fitter A.H., 2012 - *Biodiversity and ecosystem services: a multilayered relationship*. Trends in ecology & evolution 27 (1): 19-26. doi: 10.1016/j.tree.2011.08.006

Maes J., Egoh B., Willemen L., Liquete C., Vihervaara P., Schägner J.P., Grizzetti B., Drakou E.G., La Notte A., Zulian G., Bouraoui F., Paracchini M.L., Braat L., Bidoglio G. 2012 - *Mapping ecosystem services for policy support and decision making in the European Union*. Ecosystem Services 1 (1): 31-39. doi: 10.1016/j.ecoser.2012.06.004

Maesano M., Giongo M.V., Ottaviano M., Marchetti M. 2011 - *Prima analisi a livello nazionale per l'identificazione delle High Conservation Value Forests (HCVFs)*. Forest@-Journal of Silviculture & Forest Ecology 8 (1): 22-34. doi: 10.3832/efor0649-008

Maltby E. 2000 - *Ecosystem Approach: from principle to practice*. Ecosystem Service and Sustainable Watershed Management in North China International Conference, Beijing, P.R. China, August 23–25, 2000, 20 p.

Marchetti M., Ottaviano M., Sallustio L. 2013a - *La contabilità ambientale del servizio di sequestro del carbonio*. In: "Il nostro capitale. Per una contabilità ambientale dei Parchi nazionali italiani", Franco Angeli Editore, Milano: 82-91.

Marchetti M., Ottaviano M., Pazzagli R., Sallustio L. 2013b - *Consumo di suolo e analisi dei cambiamenti del paesaggio nei Parchi Nazionali d'Italia*. Territorio 66: 121-131.

Marchetti M., Sallustio L., Ottaviano M., Barbati A., Corona P., Tognetti R., Zavattero L., Capotorti G. 2012a - *Carbon sequestration by forests in the National Parks of Italy*. Plant Biosystems. An International Journal dealing with all Aspects of Plant Biology: Official Journal of the Società Botanica Italiana 146 (4): 1001-1011. doi: 10.1080/11263504.2012.738715

Marchetti M., Bertani R., Corona P., Valentini R. 2012b - *Cambiamenti di copertura forestale e dell'uso del suolo nell'inventario dell'uso delle terre in Italia*. Forest@ 9 (1): 170-184. doi: 10.3832/efor0696-009

Marchetti M., Sallustio L. 2012 - *Dalla città compatta all'urbano diffuso: ripercussioni ecologiche dei cambiamenti d'uso del suolo*. In: "Il progetto di paesaggio come strumento di ricostruzione dei conflitti". Franco Angeli Editore: 165-173.

Marchetti M., Barbati A. 2005 - *Cambiamenti di uso del suolo*. In: "Stato della biodiversità in Italia", Palombi, Roma: 108-115.

MATTM (Ministero dell'Ambiente e della Tutela del Territorio e del Mare), Unioncamere 2014 - *L'Economia reale nei Parchi Nazionali e nelle Aree Protette. Fatti, cifre e storie della Green Economy*, Rapporto 2014. Roma, 254 p.

MATTM (Ministero dell'Ambiente e della Tutela del Territorio e del Mare) 2011 – *National Biodiversity Strategy*. Decree of Italian Republic, 6 June 2011. [Online]. Available: http://www.cbd.int/doc/world/it/it-nbsap-01-en.pdf [2014, December 9].

MEA (Millennium Ecosystem Assessment) 2005 - *Ecosystems and human well-being: current state and trends*. Island Press, Washington, DC, 155 p.

Noss R.F. 1999 - *Assessing and monitoring forest biodiversity: a suggested framework and indicators*. Forest Ecology and Management 115 (2): 135-146. doi: 10.1016/S0378-1127(98)00394-6

Pearce D, Barbier E, Markandya A. 1990 - *Sustainable development: economics and environment in the Third World*. Earthscan Publications, London, UK.

Presidenza Italiana del Consiglio dell'Unione Europea 2014 – *Carta di Roma sul Capitale Naturale e Culturale*. [Online]. Available: http://www.minambiente.it/sites/default/files/archivio/allegati/biodiversita/conference_ncc_carta_roma_ita.pdf [2014, December 9].

Pressey R.L., Cabeza M., Watts M.E., Cowling R.M., Wilson K.A. 2007 - *Conservation planning in a changing world*. Trends in ecology & evolution 22 (11): 583-592. doi: 10.1016/j.tree.2007.10.001

Reed M.S. 2008 - *Stakeholder participation for environmental management: a literature review*. Biological Conservation 141: 2417–2431. doi: 10.1016/j.biocon.2008.07.014

Reid W., Mooney H., Capistrano D., Carpenter S., Chopra K., Cropper A., Dasgupta P., Hassan R., Leemans R., May R., Pingali P., Samper C., Scholes R., Watson R., Zakri, A., Shidong Z. 2006 - *Nature: The many benefits of ecosystem services*. Nature 443 (7113): 749. doi: 10.1038/443749a

Richards C., Blackstock K.L., Carter C.E. 2004 - *Practical approaches to participation*. SERG Policy Brief 1. Macauley Land Use Research Institute, Aberdeen, 24 p.

Rosegrant M.W., Ringler C., Zhu T., Tokgoz S., Bhandary P. 2012- *Water and food in the bioeconomy: challenges and opportunities for development*. Agricultural Economics 44 (1): 139-150. doi: 10.1111/agec.12058

Sallustio L., Vizzarri M., Marchetti M. 2013 - *Trasformazioni territoriali recenti ed effetti sugli ecosistemi e sul paesaggio italiano*. Territori (18): 46-53.

Settis S. 2010 - *Paesaggio Costituzione cemento. La battaglia per l'ambiente contro il degrado civile*. Einaudi (Paesaggi), Torino, 328 p.

Sheppard A.W., Gillespie I., Hirsch M., Begley C. 2011 - *Biosecurity and sustainability within the growing global bioeconomy*. Current Opinion. Environmental Sustainability 3 (1): 4-10. doi: 10.1016/j.cosust.2010.12.011

Smith B.D. 2007 - *The ultimate ecosystem engineers*. Science 315: 1797-1798. doi: 10.1126/science.1137740

Soulé M. 2013 - *The "New Conservation"*. Conservation Biology 27 (5): 895-897. doi: 10.5822/978-1-61091-559-5_7

Sukhdev P., Wittmer H., Miller D. 2014 - *The Economics of Ecosystems and Biodiversity - TEEB: Challenges and Responses*. In: D. Helm and C. Hepburn (eds), "Nature in the Balance: The Economics of Biodiversity." Oxford University Press, Oxford, 16 p.

Tallis H. Lubchenko J. 2014 – *Working together: a call for inclusive conservation*. Nature 515: 27-28. doi: 10.1038/515027a

Throsby D. 1999 - *Cultural Capital*. Journal of Cultural Economics 23: 3–12. doi: 10.1023/A:1007543313370

Toledo V.M., Barrera-Bassols N. 2008 - *La Memoria Biocultural*, Icaria, 232 p.

UNECE 2014 - *Rovaniemi Action Plan for the Forest Sector in a Green Economy*, Geneva, (ECE/TIM/SP/35). [Online]. Available: http://www.foresteurope.org/sites/default/files/01.The%20Rovaniemi%20Action%20Plan_Arnaud%20Brizay.pdf [2014, December 9].

UNEP 2014 - *Towards a global map of natural capital: key ecosystem assets*. [Online]. Available: http://www.unep-wcmc.org/system/dataset_file_fields/files/000/000/232/original/NCR-LR_Mixed.pdf?1406906252 [2014, December 9].

UNEP 2011a - *Towards a Green Economy: Pathways to Sustainable Development and Poverty Eradication*. [Online]. Available: http://www.unep.org/greeneconomy/Portals/88/documents/ger/ger_final_dec_2011/Green%20EconomyReport_Final_Dec2011.pdf [2014, December 9].

UNEP 2011b - *Year in Review 2010. The Convention on Biological Diversity*. [Online]. Available: http://www.cbd.int/doc/reports/cbd-report-2011-en.pdf [2014, December 2014].

UNEP 1992 - *Convention on Biological Diversity (CBD)*. Rio de Janeiro, 5 June 1992. [Online]. Available: https://www.cbd.int/doc/legal/cbd-en.pdf [2014, December 9].

UNESCO 2003 – *Convention for the safeguarding of the intangible cultural heritage*. Paris, 17 October 2003 (MISC/2003/CLT/CH/14). [Online]. Available: http://portal.unesco.org/en/ev.php-URL_ID=17716&URL_DO=DO_TOPIC&URL_SECTION=201.html [2014, December 9].

Vizzarri M., Lombardi F., Sallustio L., Chirici G., Marchetti M. 2013 - *I servizi degli ecosistemi forestali ed il benessere dell'uomo: quali benefici dalla ricerca?* Gazzetta Ambiente 6: 9-18.

Wackernagel M., Rees W.E. 2004 - *L'impronta ecologica. Come ridurre l'impatto dell'uomo sulla terra*. Edizioni Ambiente, Milano, 200 p.

Watson J.E., Dudley N., Segan D.B., Hockings M. 2014 - *The performance and potential of protected areas.* Nature 515 (7525): 67-73. doi:10.1038/nature13947

Weiers S., Bock M., Wissen M., Rossner G. 2004 - *Mapping and indicator approaches for the assessment of habitats at different scales using remote sensing and GIS methods.* Landscape and Urban Planning 67 (1): 43-65. doi: 10.1016/S0169-2046(03)00028-8

Zalasiewicz J., Williams M., Smith A., Barry T.L., Coe A.L., Bown P.R., Brenchley P., Cantrill D., Gale A., Gibbard P., Gregory F.J., Hounslow M.W., Kerr A.C., Pearson P., Knox R., Powell J., Waters C., Marshall J., Oates M., Rawson P., Stone P. 2008 - *Are we now living in the Anthropocene?* GSA Today 18 (2): 4–8. [Online]. Available: http://www.geosociety.org/gsatoday/archive/18/2/ [2014, December 9].

Marginal/peripheral populations of forest tree species and their conservation status

Fulvio Ducci[1*], Ilaria Cutino[1], Maria Cristina Monteverdi[1], Nicolas Picard[2], Roberta Proietti[1]

Abstract - The Mediterranean region includes 13 countries among Europe, Near Orient, and Africa. This area is a huge "hot spot" of cultures, religions, socio-economical situations, and of habitats and biodiversity. The report illustrates the geographical and ecological features of the region. Forest ecosystems and vegetation traits, with particular focus on forest species growing at the edge of their distribution range, are here compiled. The accuracy of reports, shows the interest and attention that the Mediterranean countries have for the different and complex situations of marginality that characterizes the presence of many forest species in this region. In this area the occurrence of 166 marginal and peripheral (MaP) populations of different species has been detected. Most of populations are characterised by vulnerability and fragility. Many MaP survive in environmental refugia and /or in isolated stands. However, most of the MaP populations identified by FP1202 experts are located in protected areas and also sometimes registered as seed sources, although Mediterranean region appears heterogeneous with respect to protection measures.

Keywords - Forest genetic resources; forest tree marginal populations; MaPs; marginality; COST Action FP 1202 MaP FGR.

Geographical characteristics of the Region

Extension and borders and main characteristics

This compilation refers to the Mediterranean region and it is based on data and information supplied by the following Cost Action FP1202 countries: Croatia, Cyprus, France, Greece, Israel, Italy, Lebanon, Morocco, Portugal, Spain, Tunisia and Turkey.

Including part of Europe, Asia and Africa, with four extended peninsulas, many islands, several mountain ranges and active volcanoes, the Mediterranean area is considered as a huge "hot spot" where different biogeographic regions and many habitats coexist.

The Mediterranean bioclimate covers most of southern Europe and part of Middle Orient in Asia and the Maghreb region in northern Africa (Fig. 1). This area is strongly influenced by the sea's presence and several mountain ranges contribute to characterize the wind circulation as well as precipitation and temperature distribution and regime.

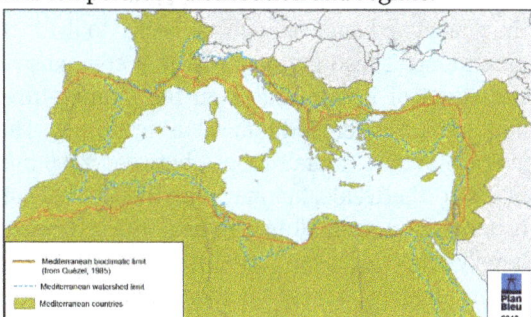

Figure 1 - The Mediterranean contour according to Quezel 1985 (FAO 2013).

Orography

The most important mountain chains in the region are high enough to influence climate at subregional level. They are mainly crossing the region in NW-SE direction as Pyrenees, Apennines, Dynaric and Albanian Alps and Balkan Mountains at the extreme eastern border. The mountain ranges of Atlas, Taurus and Caucasus are instead oriented in the direction of longitude.

Human presence

All the Mediterranean basin is densely populated. FAO (UNDEP 2011 in FAO 2017) estimates the human population in the area in 2020 to reach approx. 550 million inhabitants (Fig. 2). The human presence has been established in the area since very long time. Most of major and ancient civilizations were born in this area 5'000 years ago and pressure on habitats has been always very hard. They interacted with the Mediterranean ecosystem which the

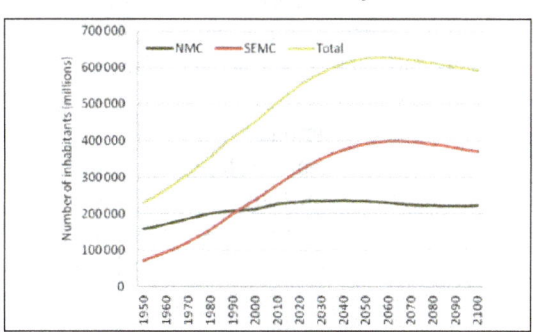

Figure 2 - Population growth in Mediterranean countries, 1950-2010. Source: United Nations, Department of Economic and Social Affairs, Population Division, 2011.

---

[1] CREA Research Centre for Forestry and Wood, Arezzo (Italy) Coordinator of WP4 Forest genetic resources in the Mediterranean region, FAO Silva Mediterranea, Rome, Italy.
[2] FAO Silva Mediterranea Secretariat, Rome, Italy.
*fulvio.ducci@crea.gov.it

resilience is in general very low compared to other temperate systems.

This area has to be considered a hotspot of cultures, religions and socio-economical situations across all its history. Here the forest landscape has been severely reduced by wood utilizations, fires, grazing, agricultural and urban use since the Greek and Roman expansion. Many original forests became gradually converted into coppices in most temperate areas, in other cases native species were partially substituted or integrated by others as cypress, stone pine and chestnut in the ancient times following the human migrations.

Interesting historical examples of active forest management can be found everywhere around the Mediterranean. Beginning from the XI century religious orders started a regulated silvicultural management of forests including plantations. Ancient sea states and kingdoms established permanent and active management of local forest resources.

Most of the ancient Mediterranean forests were fragmented by the human activities and genetic characteristics are nowadays influenced by this situation.

Along with the above-mentioned processes, in some areas of the European side, two main waves of industrial activities occurred respectively in the 15th (Renaissance) and 19th (Industrialization) centuries with a consequent larger deforestation. In some areas intensive forms of silviculture were developed such as clonal poplar cultivation in Italy and exotics. The reforestation activities have expanded in the Mediterranean since the second half of 1800' and early 1900' according to the area (AA. VV. 1924 – 1935). Forests are currently expanding in several areas, in the last 70 years reforestation projects favored the expansion of forest tree farming activities.

Geographic barriers to gene flow

The geographic structure why five important genetic hotspots: the Iberian, the Italian, the Balkan, the Middle Orient and the North African one. These hotspots derive from refugia during the past glacial waves and served as re-colonization gene sources in the interglacial periods.

The region is nearly totally mountainous. In several cases, i.e. in Italy, the peculiar orientation of Apennines and in Alps the glacial valleys determined bottlenecks and northwards migration routes. The case of Oaks is typical, their migration towards northern regions along the eastern part is separate from the western one and Alps represent the main bottleneck. Everywhere, mountains plaid an important role in preserving mesophilic species meta-populations at higher elevation during the interglacial.

Other barriers

Gene flow may become reduced within Mediterranean area and on a local scale by additional factors like topography, local climate condition, altitudinal and edaphic characteristics that can originate isolation, bottleneck or in some cases asynchrony in flower phenology among populations of the same species and genetic incompatibilities. Furthermore, human activities (forest fires, overgrazing, over exploitation, illegal logging, atmospheric pollution, deforestation, habitat fragmentation, expansion of urban areas and other infrastructures) create barriers and hinder the natural gene flow.

Ecological aspects

Climatic characteristics of the Regions and availability of databases and maps at the regional level

An exact climate definition of this large and varied geographical area is difficult. In general, the Mediterranean climate shows a strong seasonal contrast between a hot and dry summer period, and a rainy autumn and spring season with relatively moderate frost episodes. Late frosts are in general frequent. In the region's north-western borders (Italy, France, Spain) climate is temperate with Atlantic or more continental influence.

In this region occurr winter clushes between cold-dry air masses from north-eastern and polar latitudes and hot-humid air of subtropical origin. Atlantic rainy perturbations passes at lower latitudes, involving also the Mediterranean area.

In summer, the Azores and the African subtropical anti-cyclones occurr at higher latitudes influencing the precipitation's distribution and the intensity and lenght of dry periods. Such a climatic complexity together with the geographic characteristics explain the climatic differences between the various parts of the area, from the alpine regions to the semi-desert or desert ones.

Altitude also plays an important role considering the gradients that develop from the sea to the high mountains. These gradients largely compensate the effect of low latitudes on the temperature, determining lower isotherm values in altitude. The average annual rainfall ranges between 2800 and in the northern side, but may drop to 350 - 400 mm (Cyprus) and even 20 mm in nearly desert areas, largely occurring in the Near-East and North Africa.

The dry period ranges from at least two months per year in the western Mediterranean, up to five or six months in the eastern Mediterranean. Even in the driest areas, sudden and intense rainfalls occur, causing considerable runoff phenomena and soil erosion. In July and August, mean temperatures varies between 29°C and 22°C, while in winter range between 10°C and 3°C. Snow is observed every year at higher elevation between December - April on average.

Winter winds from Balkans can be very cold and dry, are humid from the Atlantic regions and very hot and dry from Africa. In summer winds greatly increase evaporation and the drought effect (ANPA, 2001).

Besides these climate features, the high unpredictability of extreme events like late frosts, hot waves, storms and exceptional prolonged drought are typical of the Mediterranean region (Blondel and Aronson 1999, Grove and Rackham 2001 in ANPA 2001).

Table 1 - Datasets and maps on climatic data of Mediterranean countries.

Country	Available information
Croatia	http://prognoza.hr/karte.php?id=ecmwf; http://klima.hr/ocjene_arhiva.html.
Cyprus	Meteorological Service (MS) upon request.
France	https://www.umr-cnrm.fr/spip.php?article788 (SAFRAN).
Greece	Dig - Digital weather data of Greece at weather stations [on-line, real time point weather data and climatic statistics for temperature, precipitation, wind, cloud cover, humidity), source: Hellenic National Meteorological Service/104 georeferenced stations]; Surface wind, Rainfall, Snowfall, Cloudiness, Air temperature, Atmospheric pressure [on-line surfaces, 3-hour step, source: Hellenic Center for Marine Research POSEIDON system] http://poseidon.hcmr.gr/weather_forecast.php?area_id=gr.
Israel	http://www.ims.gov.il/IMSEng/CLIMATE (1981-2000 for temperature and humidity, and 1970/1971 - 1999/2000 for rainfall); http://www.ims.gov.il/IMSEng/CLIMATE/TopClimetIsrael/.
Italy	Observational 2000-2010 (ECAD project) and provisional 2020-2030 (Agroscenari project) data on hourly/daily basis for Italy/Europe (10 km): CNR IBIMET; Observational data 1950-today on hourly/daily/monthly basis for Italy (30 km): SIAN www.cra-cma.it; SCIA www.scia.sinanet.apat.it; Observational data 1975-today on daily basis for Europe (25 km): http://mars.jrc.ec.europa.eu/mars/About-us AGRI4CAST/Data-distribution/AGRI4CAST-Interpolated-Meteorological-Data; Precipitation observational data 1950-2012 on monthly basis for the Globe (0.25°, 0.5°, 1.0°): http://climatedataguide.ucar.edu/guidance/gpcc-global-precipitation-climatology-centre; Evapotranspiration and aridity data 1950-2000 on monthly basis for the Globe (1 km): http://csi.cgiar.org/Aridity/.
Turkey	http://www.ogm.gov.tr/Sayfalar/Ormanlarimiz/T%C3%BCrkiye-Orman-Varl%C4%B1%C4%9F%C4%B1-Haritas%C4%B1.aspx); http://www.ogm.gov.tr/Sayfalar/OrmanHaritasi.aspx.

The following table (Tab. 1) shows the available datasets and maps at regional/national coverage level.

Soil characteristics and availability of databases and maps at the regional level

The geology of the Mediterranean area is originated by the subduction of the African Plate underneath the Eurasian Plate, thus explaining the formation of the present orography.

Lime stones are predominant.

Paleo-soils are abundant, mainly produced by the disintegration of limestone rocks of ancient maritime origin. Red soils named terrarossa are present in all the region and characterize many areas of the Mediterranean landscape.

Karst formations are also frequent in Spain, Italy. Balkans and Anatolia with podzols, vertisoils, red Mediterranean soils, calcic-magnesic soils (dominant soils), brown and isohumic soils, saline and hydromorphic soils and also poorly evolved soils and arid soils are the main types.

Volcanic soils are also frequent in this area: they are pearly dark-coloured, derived from effusive rocks, often giving rise to very high fertility.

The most fertile areas are those of alluvial plains or deriving from siliceous matrix, which are, however, few in the region.

Table 2 shows the available datasets and maps at regional/national coverage level.

Possible future modifications due to climatic change

The global warming is affecting also this area, showing worrying future scenarios for both nature and humankind. Since mid-1980s, an increase in mean temperature was recorded together with a progressive increase in the frequency of extreme events as: heat waves, prolonged drought periods, floods and retreats of alpine glaciers. Italy is particularly at risk under the current climate change, owing to its position within the transition zone between North Africa and continental Europe. Indeed, experts warned about the desertification risk in southern regions and possible climate tropicalization in the rest of the country. Since 2010, an acceleration of the water cycle, the rise of alluvial phenomena and the tropicalization of the Mediterranean were observed. Warming and drought impacts on Mediterranean forests are already ongoing (Giorgi and Lionelli 2008).

Future scenarios assume significant temperature increases of 1°C in winter and more than 2°C in summer, relatively to both maximum and minimum temperatures (Fig. 3). Precipitation is projected to decrease by 2 to 8% and, in extreme cases, by 20%. These results will lead to the isotherm shift. Hot periods are expected to increase by more than 2 weeks/year. The annual number of consecutive dry days are expected to increase by 9 days on average (Giannakopoulos et al. 2010, Hadjinicolaou et al. 2011, AA. VV. 2013). As a result, the Mediterranean environment will probably suffer major changes, causing in some cases irreversible effects and affecting the most vulnerable forest ecosystems as the forest populations at the edge of the species distribution area (FAO 2013). Water scarcity will probably affect several Mediterranean countries in

Table 2 - Datasets and maps on soils data of Mediterranean countries.

Country	Available information
Croatia	http://prognoza.hr/karte.php?id=ecmwf. http://klima.hr/ocjene_arhiva.html.
Cyprus	Geological maps of Cyprus and Geochemical Atlas of Cyprus are being published by the Department of Geological Survey.
France	http://acklins.orleans.inra.fr/.
Greece	European Soil Data Base (ESDB) [polygon vector data of soil attributes, source: JRC_ESDAC, scale 1:1000000, also in raster and Google maps]; Land Use/Cover Area frame Statistical Survey (LUCAS) Soil dataset [vector data of 732 points in Greece/7819 panned, source: JRC_ESDAC]; Soil Threat Maps (Erosion risk/PESERA, Topsoil Organic Carbon Content, Natural susceptibility of soils to compaction, Saline and sodic soils) [raster data, source: JRC_ESDAC, scale 1:1,000,000]; Soil Map of Greece (Edafologikos Xartis Ellados) [image, source: Institute of Geology and Mineral Exploration (I.G.M.E), scale 1:1,000,000, year: 1967]; Physiographic map of Greece ('Nakos' map) [image,source: Forest Research Institute in Athens/Soil Science Laboratory, scale 1:50,000, year: 1982]; Soil Map of East Macedonia - Thrace Region (Soil Taxonomy Classification Map) [Aristotle University of Thessaloniki, School of Agriculture, Lab of Applied Soil Sciences, scale: 1:200,000, year:2010]; http://eusoils.jrc.ec.europa.eu/library/maps/country_maps/metadata.cfm?mycountry=GR; Slope and aspect derivatives from ASTER* satellite Digital Elevation Model (DEM) [grid raster, cell-size: 30 m, source: NASA/Jet Propulsion Laboratory, year: 2008-2009] *Advanced Spaceborne Thermal Emission and Reflection Radiometer: http://asterweb.jpl.nasa.gov/gdem.asp; Slope and aspect from Physiographic map of Greece ('Nakos' map) [image, source: Forest Research Institute in Athens/Soil Science Laboratory, scale 1:50,000,year:1982].
Israel	http://onlinelibrary.wiley.com/doi/10.1111/j.1365-2389.1963.tb00926.x/pdf
Italy	Land use/land cover CORINE 1990-2000-2006 and changes for Europe (rst 100 m or vec1:100.000): http://www.eea.europa.eu/data-and-maps/data#c11=&c17=&c5=all&c0=5&b_start=0&c12=corine+land+cover; Digital elevation model for the Globe: 1km http://www.eea.europa.eu/data-and-maps/data/digital-elevation-model-of-europe/; 90 m http://srtm.csi.cgiar.org/; 30 m http://asterweb.jpl.nasa.gov/gdem.asp; Soils of Europe (different resolutions) in terms of: soil threats data: http://eusoils.jrc.ec.europa.eu/library/themes/ThreatsData.html; soil profile data: http://eusoils.jrc.ec.europa.eu/projects/spade/; European soil database: http://eusoils.jrc.ec.europa.eu/ESDB_Archive/ESDB/index.htm; Soil projects: http://eusoils.jrc.ec.europa.eu/projects/ProjectsData.html; Protected sites Natura 2000 of Europe: http://www.eea.europa.eu/data-and-maps/data/natura-2; Desertification map of Europe (1 km): http://www.eea.europa.eu/data-and-maps/data/sensitivity-to-desertification-and-drought-in-europe; Phytoclimatic map of Italy: prof. C. Blasi – University of Rome "La Sapienza"; Naturalistic maps of Italy GIS NATURA: prof. M. Gatto – Italian Ministry of Environment; Environmentally sensitive area of Italy (Climagri project) (1 km): CRA-CMA www.cra-cma.it; Soil quality, vegetation quality and climate quality maps of Italy (Climagri project) (1km): CRA-CMA www.cra-cma.it; Free satellite data for environmental applications: MODIS 250-500-1000 m https://lpdaac.usgs.gov; SPOT VGT 1 km http://free.vgt.vito.be.

the next years. This will have serious consequences on the social context and will increase pressures on the environment and cause land degradation. Moreover, it should be noted that overcrowded forest stands, due to the lack of forest management, may be more vulnerable to natural hazards such as pests, diseases and forest fires.

Vegetational aspects

Diffusion of forests

More than 25'000 plant species, about half of which are endemic.

According to FAO statistical definitions, wooded lands comprise "forests" and "other wooded lands". In the northern side of the Mediterranean, "other wooded lands" cover about a half of the total forest area; in North Africa, they cover about a third of the wooded lands. Northern Mediterranean countries are covered by both temperate and typically Mediterranean forests. In the southern side, the Mediterranean forest types are rapidly changing into deserts.

The total Mediterranean forest area (Fig. 4) is estimated over 85 million hectares, corresponding to 2% of the world's forest area, about 4'000 million ha (FAO 2015). 55% of the total Mediterranean forest area currently occurs in the northern part of the region.

Prevalent forest types

Most forest areas are usually exploited or over exploited. Many zones (semi-natural) show regressive secondary succession stages.

Concerning the two ecological extremes, natural forests (primary, undisturbed by man) are present only in very rare fragments whilst artificial forests have been widely planted along the 20th century to manage erosion, timber and non wood productions (Scarascia-Mugnozza et al. 2000).

Different types of forests were classified by Quézel (1985):
xerothermic-Mediterranean;
thermo-Mediterranean;
meso-Mediterranean;
supra-Mediterranean;
montane –Mediterranean;
oro-Mediterranean.

The various forest tree species form pure or mixed forests composed by evergreen broadleaved (about 60%), mixed mesic hardwoods and coniferous. However, the majority of Mediterranean forests is composed by oaks and pine stands.

In drier and compromised equilibria conditions, scrubs constitute typical ecosystems characterised by local variants: *maquis, garigue, macchia* and *phrygana*. In the Iberian peninsula, the *dehesa* is a characteristic agroforestry system with scattered trees embedded in pastures. In addition, the forests contain a wide range of aromatic, wild and medicinal plant species.

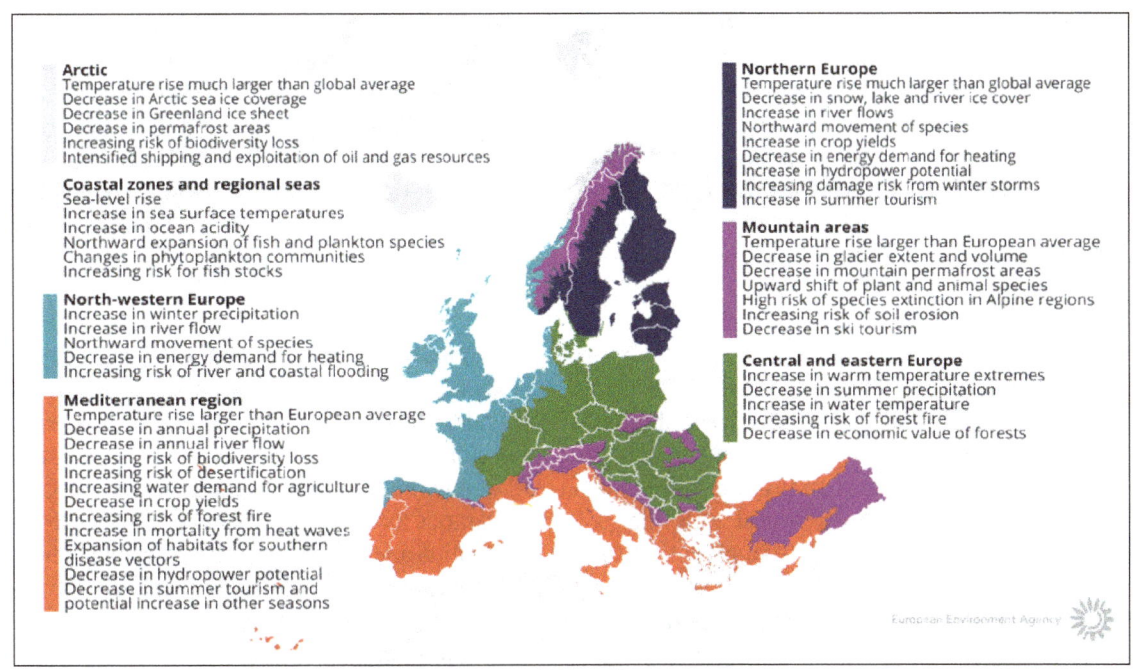

Figure 3 - Scenarios of future modification due to climatic change (EEA 2012).

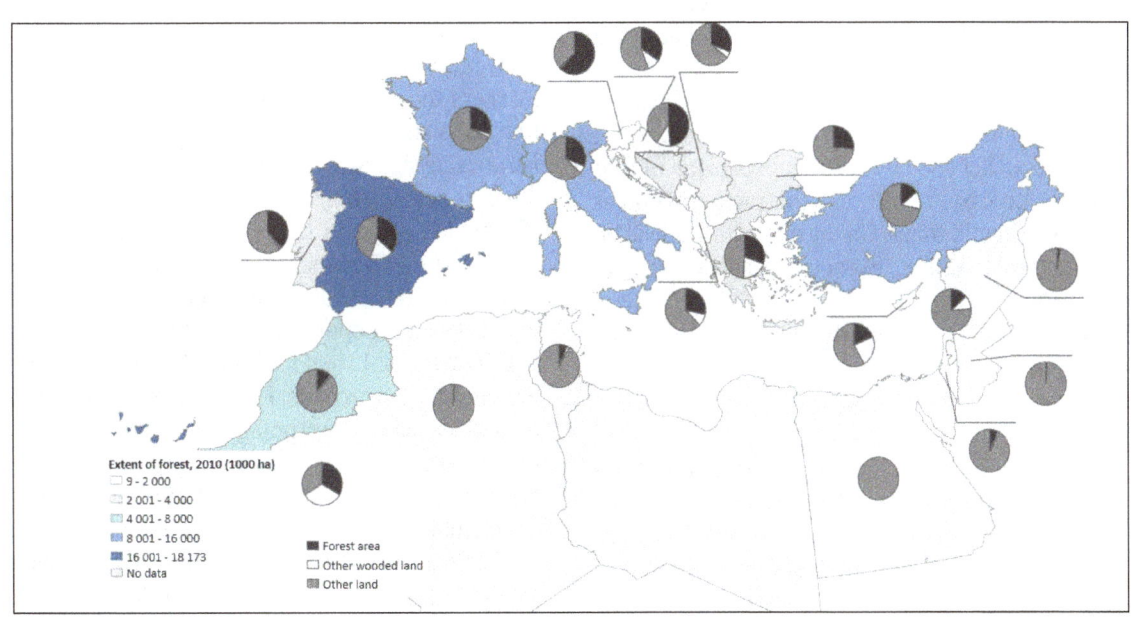

Figure 4 - Extent of forest in Mediterranean region. (FAO 2013).

Common and/or representative species

The distribution of species changes according to altitude and latitude.

Meso-, supra- and montane-Mediterranean forests are characterised by the following trees:: *Fagus sylvatica, Carpinus* spp. and *Ostrya* spp., *Prunus avium, Quercus robur* and *Q. petraea, Q. cerris* and *pubescences* and other oaks, *Acer* spp., *Alnus* spp., *Betula pendula, B. pubescens, Malus sylvestris, Sorbus* spp., *Tilia* spp., *Ulmus* spp., *Picea abies, Larix decidua, Abies* spp., *Pinus sylvestris, P. cembra, P. mugo* and *P. nigra., Cedrus* spp., *Taxus baccata, Juniperus* spp..

In the more xeric Mediterranean zone: *Quercus ilex, Q. pubescens, Q. suber, Q. trojana, Q. coccifera, Q. cerris, Q. calliprinos, Q. infectoria, P. nigra, P. halepensis, P. pinaster, P. brutia, Cupressus sempervirens.*

In lowland area and close to rivers are present: *Populus* spp. and *Salix* spp., *Alnus glutinosa, Fraxinus* spp. and *Q. robur.*

Some species diffusion was strongly influenced by humans management, like for *P. pinea, C. sempervirens, Castanea sativa* and *Juglans regia.*

Expected modifications due to climatic change

The impact of climatic change is likely to vary among the Mediterranean regions. In the Mediterranean region, Euro-Siberian or temperate and the Saharo-Sindic climatic regions overlap (Sánchez et al. 2004). Due to the isotherm shift, species climatic niches will probably be displaced by approximately 180 km northward and 150 m at higher elevations (Perini et al. 2007, Plan Bleu 2009 in FAO 2013). A similar shift is expected for pathogens and their vectors.

Studies have been conducted on the effects of climate change on spatial and temporal distribution of some ecosystems and species, monitoring networks were established in order to cover the various Mediterranean regions (Belgacem et al. 2013, Khouja et al. 2010, Ghandour et al. 2007, Marchi et al. 2016).

The increased frequency of prolonged dry seasons will affect survival even in more drought adapted tree species, as well as reproductive biology and natural regeneration will be affected.

Such changes could increase genetic erosion for a wide range of species, causing in some cases the extinction of local gene pools.

Global change effects, including forest fires, could lead to an accelerated desertification. As an example, in Lebanon by 2080, an expansion of arid zones is forecasted jointly to a contraction of cooler and more humid zones. On the opposite Tunisia, by 2030 natural forests are expected to increase in area.

Forest species at the edge of their distribution range

Species

In Middle East the Lebanese mountains are characterized by the presence of a considerable number of northern species, which may be regarded as relics of past mild and cooler periods and are still growing sporadically in spots such as *O. carpinifolia, A. tauricolum* (*Acer hyrcanum* subsp. *tauricolum* (Boiss. and Balansa) Yalt), *A. hermoneum, Rhododendron ponticum, F. ornus*.

P. halepensis has its easternmost and southernmost limits in Greece and in Israel, where *P. brutia* finds its westernmost limit. Still in Greece, *F. sylvatica, A. alba, P. abies, J. deltoides, J. phoenicea, J. drupacea* and *Q. ithaburensis, B. pendula, C. sativa* and *P. avium* meet their southern limit.

Pinus pinaster is distributed in western Mediterranean with limits between Italy and Portugal.

P. sylvestris, which is growing in high mountains, meets its southern limit in Italy, Spain and Greece.

Q. petraea and *Q. robur* reach their south-western limit in Italian and Iberian peninsula.

Israel is the northernmost limit for some tree species such as *Cyperus papyrus* and *Acacia albida*.

P. heldreichii and *A. cordata* are examples of species in fragmented populations which are marginal and isolated at the same time. Special cases are represented by unique isolated populations of several *Abies* species. They are remnants of ancient ancestral trees nowadays evolved into endemic species such as e.g. *A. pinsapo, A. numidica, A. nebrodensis, A. equitrojani*.

Kind of marginality occurring in the area

Many Mediterranean forest populations growing around the Mediterranean are potentially adapted to stressful environments because of their "marginality" status (Fady et al. 2016). Most of the marginal populations located at the rear edge of the species distribution areas can be considered marginal from both geographical and ecological points of view. Because of their possible adaptive potential, southern populations can be valid candidates as sources of better adapted reproductive materials.

Genetic information available on marginality

Genetic information is available for neutral markers and also for their growth, survival or adaptive traits surveyed in field, indicating populations harbouring significant genetic diversity. Information must be acknowledged to the implementation of EU funded projects like Walnuts, WBrains, Fairoak, Oakflow, Dynabeech, SeedSource, Evoltree, Noveltree, Treebreedex, Trees4future, international networks and cooperation projects coordinated by IUFRO WP 2.02.13 and FAO Silva Mediterranea and national funded projects. Forest genetic diversity in natural and naturalized populations have been studied on oaks, walnut, alder, wild cherry, beech, pines, firs, spruce and poplars. Most of the EU countries are members of the European Technology Platform (ETP) 'Plants for the Future', a forum for the plant sector, including plant genomics and biotechnology. Among EU-funded databases are presently actives FORGER and TreeBreedex, which are focused on the sustainable management of forest genetic resources in Europe, and ProCoGen promotes a functional and comparative understanding of the conifer genome implementing applied aspects for more productive and adapted forests.

Thanks to Evoltree genetic resources are available that are routinely used in tree genomics on *Pinaceae, Fagaceae* and *Salicaceae* (Greece, Italy, Spain, Turkey).

Genetic variation has been widely studied based on molecular markers for the following species: *Abies* spp., *Acer* spp., *A. cordata, A. glutinosa, B. pendula, B. pubescens, C. betulus, C. sativa, C. sempervirens, F. sylvatica, Fraxinus* spp., *J. regia*,

L. decidua, P. avium, P. abies, P. halepensis, P. brutia, P. nigra, P. pinaster, P. pinea, P. sylvestris, Quercus spp.

Adaptive traits were studied for *Abies* spp., *P. halepensis, P. brutia, P. nigra, P. sylvestris, C. sativa, C. libani* and *Populus* spp.

Populations studied for all the above-mentioned studies include both core populations and marginal ones.

Most important marginal populations

The Mediterranean marginal populations represent reservoirs of genetic resources and cultural and landscape heritage. Many of them are characterised by vulnerability and fragility and in most cases are already protected within parks and reserves. A great number of species populations are starting to be affected by climate change effects.

Most of these populations survive in refugia on the mountains, in isolated stands. Very often they should be considered as marginal populations. The adaptive capability of these populations is usually important. The national priorities for protection purposes can change by country. Generally, gene resources conservation needs and nature conservation strategies are the common drivers for the priorities' choices. Table 3 lists the most relevant species recognised for their condition of marginality.

Forest ecosystems and protected areas

Measures of environmental protection

The Mediterranean region appears inhomogeneous as regards protection measures and common initiatives at national and international level. Despite the heavy human pressure on forests, natural forests have locally been conserved by protected area networks: these surfaces vary between 38.17% of the total country area in Croatia and 0.22% in Turkey, (Tab.4).

Increasing the protected area network supports also forest genetic resources (FGR) conservation.

Many forests are currently protected in the framework of European networks such as "Habitat" directive (since 1992), Natura 2000, and at the national level e.g. National Parks and Biogenetic reserves.

Furthermore, many forest species are included in *ex situ* conservation programmes or units, such as seed banks, gene banks, arboreta and provenance trials (Besacier et al. 2011).

Some countries (Cyprus, France, Greece and Italy) have endorsed numerous programmes, conventions and European directives relevant to nature conservation and mitigation actions with reference to climate change:

- National Strategy for Biodiversity
- Partnership REDDplus (Reducing Emissions from Deforestation and Degradation in Developing Countries).
- IPGRI/Bioversity - EUFORGEN Programme and framework of genetic conservation units (GCUs) within the EUFGIS Project (Tab. 5).
- Development of a national strategy for adaptation to climate change adverse impacts in Cyprus.

Measures for protection/exploitation/valorisation of already existing MaPs

The concept of marginality is relatively recent. At the beginning of the COST Action FPS 1202, MaP FGR was still an idea to be defined and developed. Most of the marginal populations belonging to mesophilic species are residuals of ancient glacial refugia and their small extension make them invisible to the present methods of forest inventorying, usually structured by forest types and aimed at timber production and/or carbon stocking evaluation. On the other hand, in the Mediterranean region the national registers of forest basic materials are important sources of information regarding MaPs. In fact, most of the registered seed stands can be considere as marginal at least from thr geographic point of view. Other sources of information can be found for some species in the EUFORGEN/ EUFGIS network of gene conservation units. Information also comes from programmes finalised to conservation and sustainable use of forest genetic resources, like GENFORED. These actions provide maps of species distribution, available to download in shape file format as well as .jpeg.

Protected areas such as national parks, nature protection areas, biosphere reserve areas, natural monuments, nature parks, have very often multiple functions. The gene conservation units represent one of the most relevant among the various functions. The gene conservation units actually provide an efficient coverage of the Mediterranean area and, in many cases, are properly monitored.

Obviously great differences exist by country and species. An international shared list of MaPs is still missing. In 1991 FAO published a list of Mediterranean forest basic materials (Topak 1991) which should be updated and extended to broadleaved tree species. Many interesting populations with marginality traits can be found there.

The French Commission on Forest Genetic Resources (CRGF) has set up a network of *in-situ* gene conservation units for the following species: *A. alba, F. sylvatica, P. abies, P. pinaster, P. sylvestris, Populus nigra, Q. petraea, U. laevis.*

Ex situ conservation activities are ongoing for

Table 3 - Overview of important marginal and peripheral species identified by the FP1202 experts for the Mediterranean area.

Species	CY	ES	FR	GR	HR	IL	IT	LB	MA	PL	TN	TR
Abies alba			X		X		X					
Abies nebrodensis							X					
Abies cephalonica				X								
Abies cilicica								X				X
Abies borisii regis				X								
Abies bornmuelleriana												X
Abies equi-trojani												X
Acacia saligna	X											
Acacia tortilis											X	
Acer tauricolum								X				
Acer hermoneum								X				
Acer obtusifolium								X				
A. syriacum								X				
Acer monospessulanum										X		
Ailanthus altissima	X											
Alnus cordata							X					
Alnus orientalis	X											
Betula pendula			X									
Betula pubescens			X									
Castanea crenata x C. sativa		X										
Castanea sativa		X	X	X	X					X		
Cedrus atlantica									X			
Cedrus brevifolia	X											
Cedrus libani								X				X
Ceratonia siliqua								X			X	
Cercis siliquastrum								X				
Corylus colurna												X
Cupressus sempervirens	X			X								
Cupressus atlantica									X			
Eucalyptus spp.	X											
Fagus orientalis				X								X
Fagus sylvatica		X	X	X	X		X					
Fraxinus excelsior							X					
Fraxinus ornus								X				
Fraxinus spp.			X									
Juglans regia		X					X					
Juglans spp.		X										
Juniperus drupacea								X				
Juniperus excelsa	X							X			X	
Juniperus foetidissima	X											
Juniperus phoenicea	X											
Juniperus thurifera				X								
Juniperus turbinata										X		
Larix decidua				X				X				
Laurus nobilis								X				
Malus sylvestris				X								
Malus trilobata								X				
Ostrya carpinifolia								X				
Picea abies			X	X	X		X					
Pinus halepensis		X	X	X		X	X		X		X	
Pinus brutia				X			X	X				
Pinus canariensis		X										
Pinus cembra			X				X					
Pinus heldreichii,							X					
Pinus mugo							X					
Pinus mugo ssp. uncinata		X	X									
Pinus nigra		X		X			X					
Pinus nigra ssp. pallasiana	X											
Pinus nigra ssp. laricio			X									
Pinus nigra salzmannii			X									
Pinus pinaster		X	X					X		X		
Pinus pinaster ssp. iberia									X			
Pinus pinaster ssp. maghrebiana									X			
Pinus pinaster ssp. renoui											X	
Pinus pinea		X	X					X	X		X	X
Pinus radiata		X										
Pinus sylvestris		X	X	X				X		X		
Pistacia lentiscus								X				
Pistacia terebinthus ssp. palaestina								X				
Pistacia spp.		X										
Platanus orientalis	X						X					
Populus alba		X										
Populus spp.			X	X								
Prunus avium			X	X			X					
Prunus mahaleb								X				
Prunus ursina								X				
Pyrus communis		X										
Quercus suber								X	X	X	X	
Quercus afares											X	
Quercus alnifolia	X											
Quercus brantii												
Quercus calliprinos								X				

Table 3 - Overview of important marginal and peripheral species identified by the FP1202 experts for the Mediterranean area.

Species	CY	ES	FR	GR	HR	IL	IT	LB	MA	PL	TN	TR
Quercus canariensis											X	
Quercus cedrorum								X				
Quercus cerris					X		X	X				
Quercus cerris ssp. pseudocerris								X				
Quercus frainetto					X							
Quercus ilex							X					
Quercus infectoria								X				
Quercus infectoria ssp. veneris	X											
Quercus petraea			X	X	X		X					
Quercus pinnatifida								X				
Quercus pyrenaica										X		
Quercus pubescens					X		X					X
Quercus robur			X	X			X			X		
Quercus trojana							X					
Salix spp.			X									
Sorbus flabellifolia							X					
Sorbus torminalis							X					
Sorbus spp.			X				X					
Taxus baccata			X				X					
Tetraclinis articulata									X			
Tilia spp.			X									
Ulmus laevis			X									
TOTAL	13	16	27	15	8	1	27	28	6	9	9	7

Table 4 - Overview of numbers, surfaces and percentage coverage of protected areas within Mediterranean Region according dataset of IUCN, UNEP-WCMC (2017).

Country	Protected Areas Number (terrestrial and marine)	Land Area Protected (km^2)	Total Land Area (km^2)	Coverage (%)
Croatia	1'194	21'703	56'855	38,17
Cyprus	88	1'690'	9'063	18,65
France	4'611	141'362	548'954	25,75
Greece	1'260	46'509	133'012	34,97
Israel	273	4'180	20'958	19,94
Italy	3'878	64'791	301'335	21,5
Lebanon	34	268	10'329	2,59
Morocco	325	125'477	407'208	30,81
Portugal	440	21'101	92'141	22,9
Spain	4'038	142'140	507'013	28,03
Tunisia	148	12'283	155'230	7,91
Turkey	18	1'709	782'238	0,22

Table 5 - Overview of in situ genetic conservation units in the countries within the Mediterranean Area and the number of species registered in EUFGIS for each country (http://portal'eufgis'org/).

Country	Number GCU (in situ)	Number of tree species in EUFGIS
Croatia	19	8
France	95	8
Greece	15	5
Italy	209	35
Portugal	9	8
Spain	43	5
Turkey	271	36
Total	**661**	**105**

Pinus nigra salzmannii, Populus nigra, P. avium, Sorbus domestica, U. glabra, U. laevis, U. minor.

Some specific tests on assisted migration and active gene flow techniques are ongoing in Italy, on *A. nebrodensis*, an extreme case of marginal species within the *Abies complex*, and on *Q. robur* and *Q. petraea*. In this latter case, the marginality and isolation were generated by anthropic activities in northern Italy. These long-term experimental tests are aimed to monitor and verify the effects of the translocation actions on the gene pools. Aside the scientific purposes, the experimental sites satisfy also the urgent need to preserve a very endangered genetic material.

References

AA.VV. 1924 – 1935 - *Silva Mediterranea*, Buletin de la "Silva mediterranea", years I - X (1924 – 1935). Archive of CREA FL of Arezzo, 814 p.

ANPA 2001 - *La biodiversità nella regione biogeografica mediterranea*. 147 p.

Belgacem A., Ouled M., Tarhouni M., Louhaichi M. 2013 - *Effect of protection on plant community dynamics in the Mediterranean arid zone of southern Tunisia: a case study from Bou Hedma national park*. Land Degradation & Development 24.1: 57-62.

Besacier C., Ducci F., Malagnoux M., Souvannavong O. 2011 - *Status of the experimental network of Mediterranean forest genetic resources*. Arezzo, Italy, CRA SEL and Rome, FAO. 205 p.

EEA 2012 European Environment Agency 2012 - *Climate change, impacts and vulnerability in Europe 2012. An indicator-based report*. Report N. 12/2012, Copenhagen Denmark. 300 p.

EUFGIS. http://portal.eufgis.org/

Fady B., Aravanopoulos FA., Alizoti P., Mátyás C.,von Wühlisch G., Westergren M., Belletti P., Cvjetkovic B., Ducci F., Huber G., Kelleher CT., Khaldi A., Dagher Kharrat MB., Kraigher H., Kramer K., Mühlethaler U., Peric S., Perry A., Rousi M., Sbay H., Stojnic S., Tijardovic M., Tsvetkov I, Varela MC., Vendramin GG., Zlatanov T. 2016 - *Evolution-based approach needed for the conservation and silviculture of peripheral forest tree populations*. Forest Ecology and Management 375: 66–75.

Food and Agriculture Organization of the United Nations (FAO) 2010 - *Global Forest Resources Assessment 2010 Main report*. Rome. 340 p.

Food and Agriculture Organization of the United Nations (FAO) 2013 - *State of Mediterranean Forests 2013*, Rome.173 p.

Food and Agriculture Organization of the United Nations (FAO) 2015 - *Global Forest Resources Assessment 2015*. FAO Forestry Paper N. 1 UN, Rome. 244 p.

Food and Agriculture Organization of the United Nations (FAO) 2017 - *Urban and Peri-urban Forestry Working Group (WG7)* Available: http://www.fao.org/forestry/86889/en/. [2014, August 6]

Ghandour M., Khouja M. L., Toumi L., Triki S. 2007 – *Morphological evaluation of cork oak (Quercus suber): Mediterranean provenance variability in Tunisia*. Annals of Forest Science. 64, 549 – 555.

Giannakopoulos C., Hadjinicolaou P., Kostopoulou E., Varotsos K.V., Zerefos C. 2010 - *Precipitation and temperature regime over Cyprus as a result of global climate change*. Advances in Geosciences, 23: 17–24.

Giorgi F., Lionelli P. 2008 - *Climate change projections for the Mediterranean region*. Global and Planetary Change, Vol. 63, Iusse 2-3: 90 -104.

Hadjinicolaou P., Giannakopoulos C., Zerefos C., Lange M., Pashiardis S., Lelieveld J. 2011 - *Mid-21st century climate and weather extremes in Cyprus as projected by six regional climate models*. Regional Environmental Change 11: 441-457.

IUCN, UNEP-WCMC 2017 - *The World Database on Protected Areas (WDPA)*. [July release] Cambridge (UK): UNEP World Conservation Monitoring Centre. Available: www.protectedplanet.net

Khouja ML., Benjanâa M. L., Franceschini A., Khaldi A., Nouri M., Selmi H., 2010. *Observations sur le dépérissement de différentes provenances de chêne-liège dans le site expérimental de Tebaba au Nord-Ouest de la Tunisie*. IOBC-WPRS Bulletin, 57 : 53-59. ISSN: 1027-3115. URL: http://www.iobcwprs.org/pub/bulletins/bulletin_2010_57_table_of_contents_abstracts.pdf

Marchi, M., Nocentini, S., Ducci, F. 2016. *Future scenarios and conservation strategies for a rear-edge marginal population of* Pinus nigra *Arnold in Italian central Apennines*. Forest Systems, Volume 25, Issue 3, e072. http://dx.doi.org/10.5424/fs/2016253-09476.

Quézel P. 1985 - *Definition of the Mediterranean region and origin of its flora*. In C. Gomez-Campo, ed., Plant conservation in the Mediterranean area. Dordrecht, the Netherlands, W. Junk. 269 p.

Sánchez E., Gallardo C., Gaertner M. A., Arribas A., Castro M. 2004 - *Future climate extreme events in the Mediterranean simulated by a regional climate model: A first approach*. Global and Planetary Change, 44 (1-4), pp. 163-180.

Scarascia-Mugnozza G., Oswald H., Piussi P., Radoglou K. 2000 - *Forests of the Mediterranean region: gaps in knowlwdge and reserch needs*. Forest Ecology and Management 132: 97-109

Topak M. 1991 - Directory of seed sources of the Mediterranean Conifers. FAO – Silva Mediterranea, Rome, July 1997: 218 p. [http://www.fao.org/docrep/006/AD112E/AD112E00.HTM].

Permissions

All chapters in this book were first published in ASR, by CREA Forestry and Wood; hereby published with permission under the Creative Commons Attribution License or equivalent. Every chapter published in this book has been scrutinized by our experts. Their significance has been extensively debated. The topics covered herein carry significant findings which will fuel the growth of the discipline. They may even be implemented as practical applications or may be referred to as a beginning point for another development.

The contributors of this book come from diverse backgrounds, making this book a truly international effort. This book will bring forth new frontiers with its revolutionizing research information and detailed analysis of the nascent developments around the world.

We would like to thank all the contributing authors for lending their expertise to make the book truly unique. They have played a crucial role in the development of this book. Without their invaluable contributions this book wouldn't have been possible. They have made vital efforts to compile up to date information on the varied aspects of this subject to make this book a valuable addition to the collection of many professionals and students.

This book was conceptualized with the vision of imparting up-to-date information and advanced data in this field. To ensure the same, a matchless editorial board was set up. Every individual on the board went through rigorous rounds of assessment to prove their worth. After which they invested a large part of their time researching and compiling the most relevant data for our readers.

The editorial board has been involved in producing this book since its inception. They have spent rigorous hours researching and exploring the diverse topics which have resulted in the successful publishing of this book. They have passed on their knowledge of decades through this book. To expedite this challenging task, the publisher supported the team at every step. A small team of assistant editors was also appointed to further simplify the editing procedure and attain best results for the readers.

Apart from the editorial board, the designing team has also invested a significant amount of their time in understanding the subject and creating the most relevant covers. They scrutinized every image to scout for the most suitable representation of the subject and create an appropriate cover for the book.

The publishing team has been an ardent support to the editorial, designing and production team. Their endless efforts to recruit the best for this project, has resulted in the accomplishment of this book. They are a veteran in the field of academics and their pool of knowledge is as vast as their experience in printing. Their expertise and guidance has proved useful at every step. Their uncompromising quality standards have made this book an exceptional effort. Their encouragement from time to time has been an inspiration for everyone.

The publisher and the editorial board hope that this book will prove to be a valuable piece of knowledge for researchers, students, practitioners and scholars across the globe.

List of Contributors

Alice Angelini, Walter Mattioli and Luigi Portoghesi
University of Tuscia - Department for Innovation in Biological, Agro-food and Forest systems (DIBAF), Viterbo, Italy

Paolo Merlini and Piermaria Corona
Consiglio per la Ricerca e la sperimentazione in Agricoltura, Forestry Research Centre (CRA-SEL), Arezzo, Italy

Saša Orlović and Srdjan Stojnić
University of Novi Sad, Institute of Lowland Forestry and Environment, Novi Sad, Serbia

Mladen Ivanković
Croatian Forest Research Institute, Jastrebarsko, Croatia

Vlatko Andonoski
University Ss. Cyril and Methodius - Faculty of Forestry, Skopje, Macedonia

Vasilije Isajev
University of Belgrade, Faculty of Forestry, Belgrade, Serbia

Alice Angelini and Luigi Portoghesi
University of Tuscia, Department for Innovation in Biological, Agro-food and Forest systems (DIBAF), Viterbo, Italy

Piermaria Corona and Francesco Chianucci
Consiglio per la ricerca in agricoltura e l'analisi dell'economia agraria, Forestry Research Centre (CRA-SEL), Arezzo, Italy

Antoine Harfouche, Sacha Khoury, Francesco Fabbrini and Giuseppe Scarascia Mugnozza
Department for Innovation in Biological, Agro-food and Forest systems (DIBAF), University of Tuscia, Italy

Bill Mason and Thomas Connolly
Forest Research UK, United Kingdom

Serena Ravagni, Claudio Bidini and Francesco Pelleri
Consiglio per la ricerca in agricoltura e l'analisi dell'economia agraria, Forestry Research Centre (CREA-SEL), Arezzo, Italy

Elisa Bianchetto
Consiglio per la Ricerca e la sperimentazione in Agricoltura e l'analisi dell'economia agraria, Centro per l' Agrobiologia e la Pedologia (CREAABP), Firenze, Italy

Angelo Vitone
Dottorando, Dipartimento di Bio-scienze e Territorio (DiBT), Università degli Studi del Molise

Dejan Stojanović and Bratislav Matović
Institute of Lowland Forestry and Environment, University of Novi Sad, Antona Cehova 13d, Novi Sad, Serbia

Tom Levanič
Slovenian Forestry Institute, Vecna pot 2, Ljubljana, Slovenia

Andres Bravo-Oviedo
Forest Research Center - National Institute of Agricultural and Food Research and Technology & Sustainable Forest Management Research Institute University of Valladolid-INIA, Madrid, Spain

Diego Giuliarelli
Università degli Studi della Tuscia, Viterbo

Elena Mingarelli, Piermaria Corona, Francesco Pelleri and Francesco Chianucci
Consiglio per la ricerca in agricoltura e l'analisi dell'economia agraria, Centro di ricerca per le foreste e il legno, Arezzo

Alessandro Alivernini
Consiglio per la ricerca in agricoltura e l'analisi dell'economia agraria, Centro di ricerca per lo studio delle relazioni tra pianta e suolo, Roma

Gianfranco Fabbio
Consiglio per la ricerca in agricoltura e l'analisi dell'economia agraria, Centro di ricerca per le foreste e il legno, Arezzo (Italy)

Giovanni Santopuoli and Marco Marchetti
University of Molise, Dipartimento di Bioscienze e Territorio, Campobasso (Italy)

List of Contributors

Jader Nunes Cachoeira and Marcos Giongo
Universidade Federal do Tocantins, Centro de Monitoramento Ambiental e Manejo do Fogo (Brazil)

Marcelo Ribeiro Viola
Universidade Federal de Lavras, Soil and Water Engineering Group (Brazil)

Luca Salvati
Consiglio per la ricerca in agricoltura e l'analisi dell'economia agraria, Centro di ricerca per le foreste e il legno, Arezzo (Italy)

Lorenza Gasparella and Anna Barbati
University of Tuscia, Dipartimento per l'innovazione nei sistemi Biologici, Agroalimentari e Forestali (UNITUS-DIBAF), Viterbo (Italy)

Michele Munafò
Istituto Superiore per la Protezione e la Ricerca Ambientale, Roma (Italy)

Raoul Romano
Consiglio per la ricerca in agricoltura e l'analisi dell'economia agraria, Research Centre for Agricultural Policies and Bioeconomiy, Roma (Italy)

Isabella De Meo
Consiglio per la ricerca in agricoltura e l'analisi dell'economia agraria Research Centre for Agriculture and Environment, Firenze (Italy)

Fabrizio Ferretti and Alessandro Paletto
Consiglio per la ricerca in agricoltura e l'analisi dell'economia agraria Research Centre of Forestry and Wood, Arezzo (Italy)

Maria Giulia Cantiani
Università degli Studi di Trento, Department of Civil, Environmental, and Mechanical Engineering, Trento (Italy)

Leone Davide Mancini, Anna Barbati and Piermaria Corona
University of Tuscia, Dipartimento per l'innovazione nei sistemi Biologici, Agroalimentari e Forestali (UNITUS-DIBAF), Viterbo (Italy)

Kyriaki Kitikidou, Elias Milios and Kalliopi Radoglou
Democritus University (Greece)

Nicola Puletti and Cristiano Castaldi
CREA Research Centre for Forestry and Wood, Arezzo, Italy

Francesco Chianucci
CREA Research Centre for Agriculture and Environment, Rome, Italy

Ilaria Cutino
CREA Difesa e Certificazione, Firenze (Italy)

Salvatore Pasta
Departement de Biologie, Université de Fribourg, Fribourg (Switzerland)

Concetta Valeria Maggiore, Emilio Badalamenti and Tommaso La Mantia
Department of Agricultural, Food and Forest Sciences, Università degli Studi di Palermo, Palermo (Italy)

Fulvio Ducci
Consiglio per la ricerca in agricoltura e l'analisi dell'economia agraria, Forestry Research Centre (CREA-SEL), Arezzo, Italy

Umberto Di Salvatore and Fabrizio Ferretti
Consiglio per la Ricerca e la sperimentazione in Agricoltura, Apennines Forest Research Unit (CRA-SFA), Isernia, Italy

Paolo Cantiani and Ugo Chiavetta
Consiglio per la Ricerca e la sperimentazione in Agricoltura, Forestry Research Centre (CRA-SEL), Arezzo, Italy

Alessandro Paletto
Consiglio per la Ricerca e la sperimentazione in Agricoltura, Forest Monitoring and Planning Research Unit (CRA-MPF), Villazzano di Trento, Italy

Isabella De Meo
Consiglio per la Ricerca e la sperimentazione in Agricoltura, Agrobiology and Pedology Centre (CRA-ABP), Firenze, Italy

Silvano Fares
Consiglio per la Ricerca e la sperimentazione in Agricoltura, Research Center for the Soil-Plant System (CRA-RPS), Rome, Italy

Flavia Savi
Consiglio per la Ricerca e la sperimentazione in Agricoltura, Research Center for the Soil-Plant System (CRA-RPS), Rome, Italy
Department for Innovation in Biological, Agro-Food and Forest Systems (DIBAF), University of Tuscia, Italy

Francesco Pelleri, Elisa Bianchetto and Claudio Bidini
Consiglio per la Ricerca e la sperimentazione in Agricoltura, Forestry Research Centre, Arezzo (CRA-SEL), Italy

Serena Ravagni
Associazione Arboricoltura da Legno Sostenibile per l'Economia e l'Ambiente (A.A.L.S.E.A.)

Oksana Pelyukh and Lyudmyla Zahvoyska
Department of Ecological Economics, Ukrainian National Forestry University, Lviv (Ukraine)

Alessandro Paletto
CREA Research Centre for Forestry and Wood

Fulvio Ducci, Ilaria Cutino, Maria Cristina Monteverdi and Roberta Proietti
CREA Research Centre for Forestry and Wood, Arezzo (Italy) Coordinator of WP4 Forest genetic resources in the Mediterranean region, FAO Silva Mediterranea, Rome, Italy

Index

A
Ampelodesmos, 143, 146-148, 150, 153
Anthocyanin Reflectance Index 1, 138

B
Basal Area, 20-25, 36-41, 71, 76, 132, 147-150
Betula Pendula, 36-37, 42
Biogenic Volatile Organic Compounds, 189

C
Calicotome, 147, 151
Calicotome Infesta, 147
Castanea Sativa, 1, 6, 66-67
Chestnut Blight, 1
Cistus Creticus L, 147
Cistus Salviifolius, 147, 151
Cistus Spp, 143
Clonal Seed Orchard, 11-12, 15
Conventional Breeding, 27-28, 33
Coppice, 1-5, 30, 57-87, 146-150, 165, 175, 180, 182-183, 185-187
Crataegus Monogyna Jacq, 148
Cryptorrhynchus Lapathi, 197

D
Dendroecology, 51
Dieback, 51-52, 56, 163
Drought Stress, 30, 35, 53, 55, 87, 156, 175, 189, 193
Dryocosmus Kuriphilus, 163, 176

E
Emerus Major, 147, 149
Erica Multiflora L, 147-148
Erysiphe Alphitoides, 12

F
Forest Biomass Feedstock, 27
Forest Canopy Structure, 19
Forest Ecology, 2, 4, 6, 8, 10, 12, 14, 16, 18, 20, 22-26, 28, 30, 32, 34, 38, 40, 42, 46, 48, 50, 52, 54, 56, 58, 60, 62, 66, 68, 70, 72, 74, 76, 78, 80, 82, 84-88, 90, 92, 94, 96, 98, 100, 102, 104, 106, 108-109, 112, 114, 116, 118, 120-122, 124, 126, 128-129, 132, 134-135, 138, 140-142, 144, 146, 148, 150, 152-154, 156, 158, 160, 162, 164, 166, 168, 170, 172, 174, 176-178, 180, 182, 184, 186-188, 190, 192, 194, 196, 198, 200, 202-203, 206, 208, 210, 212-214

Forest Genetic Resources, 7-8, 13-14, 16-17, 86-87, 155, 157-159, 163, 171, 175-176
Forest Landscape Management Planning, 110, 120-121, 179, 188, 213
Forest Multifunctionality, 100, 179-180, 187
Forest Reproductive Materials, 155, 167, 171-172
Forest Stand Dynamics, 43, 86, 144, 153
Fraxinus Ornus, 145-147, 149

G
Generative Seed Orchard, 9-10, 12
Greenhouse Gas Emissions, 128

H
Hybrid Walnut, 195-202

I
Individual Tree-ring Widths, 52
Ink Disease, 1

L
Lignocellulosic Biofuels, 30

M
Marginal Peripheral Populations, 155, 164
Mediterranean Biomes, 163
Metagenome Sequencing, 32
Mid Coppice Rotation, 58-59, 61
Miscanthus, 30, 34

N
Needle Clustering, 20
Next-generation Ecotilling, 28
Normalized Difference Vegetation, 138

P
Pedunculate Oak, 9, 12, 14-17, 49, 51-56, 87, 165, 168
Peri-urban Forest, 109, 130-131, 133-135
Picea Glauca, 28, 34
Pinus Brutia, 130-131, 133-135
Pinus Halepensis Mill, 112, 143, 152-154
Pinus Pinaster Aiton, 143, 174
Pinus Pinea, 131, 143-145, 148, 150, 163, 172
Pinus Sylvestris, 13, 15, 36-37, 39, 41-43, 143
Plant Recruitment, 18

Polycyclical Plantation, 196
Pro-active Silviculture, 64, 68
Pubescens, 37, 42, 66, 111, 143, 145, 147-149, 152, 181

Q
Q. Pubescens, 145, 147-148
Q. Suber, 66, 145, 147
Quercus Ilex, 21, 67, 82, 84, 111, 143-145, 147-149, 153, 160, 189, 194

R
Red-edge Normalized Difference Vegetation, 138
Rhamnus Alaternus L, 143
Rhyncophorus Ferrugineus, 163

S
Saperda Carcharis L, 197
Sentinel-2, 136-139, 141-142
Short Rotation Coppice, 30
Single-tree Selection, 57
Spartium, 147, 149, 151
Sporadic Tree Species, 57-63, 78-79, 86-87
Stand Dynamics Of Outgrown Coppice Forests, 69
Stem Analysis, 70, 73
Sustainable Forest Management, 14, 36, 42-43, 51, 63, 69, 100, 110, 121, 128, 136, 175-177, 179, 188, 204, 212-213
Sustainable Tree Farming, 195

T
Teucrium Fruticans, 147-148, 151
Tree Farming, 44-45, 49-50, 77, 195, 197
Tree Species Composition, 19, 57, 167, 179, 181, 204, 206, 208, 210-211

V
Vegetation Dynamics, 24, 127, 143-144, 146, 149, 151-154

CPSIA information can be obtained
at www.ICGtesting.com
Printed in the USA
BVHW011242270822
645617BV00003B/72